DEVELOPMENTAL AND CELL BIOLOGY SERIES 24
EDITORS
P.W. BARLOW  D. BRAY  P.B. GREEN  J.M.W.SLACK

# THIS SIDE UP

T0243090

# Developmental and cell biology series

SERIES EDITORS

Dr P. W. Barlow, *Long Ashton Research Station, Bristol*
Dr D. Bray, *MRC Cell Biophysics Unit, King's College, London*
Dr P. B. Green, *Dept of Biology, Stanford University, USA*
Dr J. M. W. Slack, *ICRF Developmental Biology Unit, University of Oxford*

The aim of the series is to present relatively short critical accounts of areas of developmental and cell biology where sufficient information has accumulated to allow a considered distillation of the subject. The fine structure of the cells, embryology, morphology, physiology, genetics, biochemistry and biophysics are subjects within the scope of the series. The books are intended to interest and instruct advanced undergraduates and graduate students and to make an important contribution to teaching cell and developmental biology. At the same time, they should be of value to biologists who, while not working directly in the area of a particular volume's subject matter, wish to keep abreast of developments relative to their particular interests.

BOOKS IN THE SERIES

R. Maksymowych *Analysis of leaf development*

L. Roberts *Cytodifferentiation in plants: xylogenesis as a model system*

P. Sengel *Morphogenesis of skin*

A. McLaren *Mammalian chimaeras*

E. Roosen-Runge *The process of spermatogenesis in animals*

F. D'Amato *Nuclear cytology in relation to development*

P. Nieuwkoop & L. Sutasurya *Primordial germ cells in the chordates*

J. Vasiliev & I. Gelfand *Neoplastic and normal cells in culture*

R. Chaleff *Genetics of higher plants*

P. Nieuwkoop & L. Sutasurya *Primordial germ cells in the invertebrates*

K. Sauer *The biology of* Physarum

N. Le Douarin *The neural crest*

J. M. W. Slack *From egg to embryo: determinative events in early development*

M. H. Kaufman *Early mammalian development: parthenogenic studies*

V. Y. Brodsky & I. V. Uryvaeva *Genome multiplication in growth and development*

P. Nieuwkoop, A. G. Johnen & B. Albers *The epigenetic nature of early chordate development*

V. Raghavan *Embryogenesis in angiosperms: a developmental and experimental study*

C. J. Epstein *The consequences of chromosome imbalance: principles, mechanisms, and models*

L. Saxen *Organogenesis of the kidney*

V. Raghavan *Developmental biology of fern gametophytes*

R. Maksymowych *Analysis of growth and development in* Xanthium

B. John *Meiosis*

J. Bard *Morphogenesis*

# THIS SIDE UP

## SPATIAL DETERMINATION IN THE EARLY DEVELOPMENT OF ANIMALS

ROBERT WALL

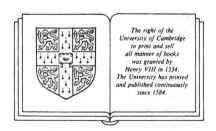

The right of the
University of Cambridge
to print and sell
all manner of books
was granted by
Henry VIII in 1534.
The University has printed
and published continuously
since 1584.

CAMBRIDGE UNIVERSITY PRESS

CAMBRIDGE
NEW YORK PORT CHESTER
MELBOURNE SYDNEY

CAMBRIDGE UNIVERSITY PRESS
Cambridge, New York, Melbourne, Madrid, Cape Town, Singapore,
São Paulo, Delhi, Dubai, Tokyo, Mexico City

Cambridge University Press
The Edinburgh Building, Cambridge CB2 8RU, UK

Published in the United States of America by Cambridge University Press, New York

www.cambridge.org
Information on this title: www.cambridge.org/9780521361156

First published 1990

*A catalogue record for this publication is available from the British Library*

*Library of Congress Cataloguing in Publication data*

Wall, Robert.
    This side up: spatial determination in the early development of
animals/Robert Wall
        p.   cm. - (Development and cell biology series)
    Includes bibliographical references.
    ISBN 0 521 36115 X
    1. Developmental biology.   2. Organizer (Embryology)   I. Title.
II. Series.
    QL971.W35 1990
    591.3 '3-dc20

ISBN 978-0-521-36115-6 Hardback
ISBN 978-0-521-01726-8 Paperback

# Contents

# Preface

My work for this book really began 20 years ago when I worked for the late Professor C.H. Waddington producing summaries of current research on development. I became particularly interested in the origins of spatial patterns, where I felt that I could detect signs of common features in many different developing systems. Since that time I have tried to make notes on all published studies of early development as they appeared, whatever the animal group concerned. I have also gone back in time to look again at many seminal studies published up to 100 years ago.

In selecting from this huge literature, I have been guided mainly by the principle that the work should have a potential relevance for the spatial determination problem. No doubt I have failed to 'pick the winners' in some cases, but the alternative was to risk losing the thread in a book of unwieldy size. My choices will seem to many to be particularly idiosyncratic in the physiological and biochemical sections, where I have presented some classical data (the meaning of which is still unclear) while omitting some modern studies. I justify this on the basis that the former will one day have to be encompassed in theories of spatial determination, while the latter may prove not to be relevant and have in any case been considered in depth in Davidson's *Gene Activity in Early Development*.

At the present time, the data relevant to the spatial determination problem still derive primarily from experimental embryology, and this is reflected in this book. In several cases I have assessed particular approaches to the problem first by considering data exclusively obtained with a particular much-studied group, and then in a comparative survey of other animal groups. For example, concepts of determination based upon intercellular signalling are considered first for sea urchins, and then extended to other embryos including that of *Drosophila* where we seem at last to be uncovering a molecular basis for developmental phenomena. Molecular data are presented quite fully in such cases (although the survey of literature for this book was completed at the end of August 1987), and the final chapter considers the evidence for common mechanisms as well as common patterns in determination.

ix

# Acknowledgements

I have worked as a teacher while writing this book and have been grateful for the flexibility of colleagues in London and Bedfordshire which has allowed me to find the time to complete it. Barry Shephard helped to make it possible for me to study in London. I have used a large number of public libraries in researching this book, and would like in particular to acknowledge the help of the libraries at the Open University and University College London as well as branches of the Science Reference and Information Service. Dr Glyn Williams of University College London also helped me to trace Hofmeister's studies of colloids and later developments of the work. My greatest debt is, however, to my good friend Richard Black who, among other things, has given me innumerable lifts to the Open University, typed and retyped the manuscript on a word-processor and helped me to photograph many of the figures. Other photographs were prepared at Cambridge University Library. Dr Jonathan Slack read and made many helpful comments upon two different drafts of the book, but cannot be held responsible for my final decisions on its content. Much advice and encouragement also came from Dr Adam Wilkins at Cambridge University Press. I apologise to my wife and family for the times when the work has made me uncommunicative or miserable, and thank them for their help and forebearance.

Original prints of published figures were sent to me by Drs M.E. Akam, R.C. Angerer, J.M. Arnold, L.W. Coggins, K. Dan, M.R. Dohmen, T. Ducibella, H. Eyal-Giladi, G. Freeman, W.J. Gehring, H. Grunz, B.E. Hagström, W.R. Jeffery, F.C. Kafatos, K. Kalthoff, C.B. Kimmel, E.B. Lewis, M. Lohs-Schardin, D.L. Luchtel, P.M. Macdonald, A.P. Mahowald, N. Satoh, T. Sawada, L. Saxén, T. Shimizu, L.D. Smith, J.A.M. van den Biggelaar, J.R. Whittaker, M. Wilcox, R.I. Woodruff and K. Yamana.

# 1

## Oogenesis

A study of the determination of spatial pattern in embryos clearly has to start long before fertilisation, in particular to find out how much pattern has been laid down during the development of the egg within the mother. This stage, oogenesis, has been extensively studied in recent years so that we now know quite a lot about the acquisition of organelles and molecules which will be used later in the embryo.

Brief background information about these processes will be given here. Evidence for patterned arrangements of materials within oocytes will be considered next. Finally, we will ask whether such patterns imply that the parts of the oocyte are already determined, i.e. have they specialised so that they are only capable of forming certain parts of the embryo?

### The events of oogenesis

Oogenesis, of course, includes the division of the primordial germ cells to produce successive generations of oogonia, their eventual transformation to oocytes and the meiotic divisions of the oocytes. Most of the obvious preparation of the egg occurs in the prophase of the first meiotic division during which the primary oocyte grows, mainly by an increase in its stores of yolk, and there may be obvious signs of activity in the nucleus and elsewhere. The volumes of animal eggs are usually orders of magnitude greater than those of somatic cells, and the nucleus of the oocyte is also enlarged and commonly called the germinal vesicle. Figure 1.1 gives a good impression of the increasing complexity of the oocyte and its investments during six stages of oocyte development in the frog *Xenopus laevis*. Massive numbers of organelles accumulate in temporal and spatial patterns which show at least some specificity. Some will have obvious roles in embryonic development, e.g. as sources of protein and energy (yolk) or in the release of energy (mitochondria), while the roles of other organelles characteristic of oocytes, such as cortical granules and annulate lamellae, will be discussed later (see especially pages 7 and 31–2). In most cases the accumulation of yolk (vitellogenesis) is a relatively late

1

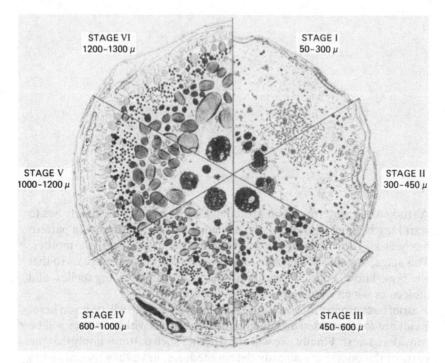

STAGE VI
1200-1300 μ

STAGE I
50-300 μ

STAGE V
1000-1200 μ

STAGE II
300-450 μ

STAGE IV
600-1000 μ

STAGE III
450-600 μ

Fig. 1.1. A diagrammatic representation of six stages in *Xenopus* oocyte development, not drawn to scale. (From Dumont, 1972, where further details may be obtained.)

and rapid event (in *Xenopus* mainly at stage IV) following a longer previtellogenic stage.

The genome in the primary oocyte has already been replicated in preparation for the meiotic divisions and so there is a tetraploid ($4n$) amount of DNA present in the nucleus. At some stage of first meiotic prophase (often at diplotene), the chromosomes commonly take on the lampbrush form, in which loops of chromatin are seen to project from the chromosome axis. RNA is synthesised rapidly there (Gall & Callan, 1962), with RNA polymerase molecules packed far more densely together than is usual on somatic cell DNA (Miller & Bakken, 1972). About 5% of the genome is present in the lampbrush loops of the newt *Triturus* (Callan, 1963), and an enormous amount of genetic information is transcribed. As in most cells, much is lost during processing within the nucleus, but even so a variety of animal eggs retain RNAs with a complexity of some tens of millions of nucleotides (Davidson, 1986). The *Xenopus* egg is supplied with some 20,000 different poly(A)$^+$RNAs, for most of which there are about $10^6$ copies (Perlman & Rosbash, 1978).

Among the messenger RNAs transcribed in the oocyte, several have been identified which will have important roles in development. These include mRNAs for histones (Gross *et al.*, 1973), tubulin (Raff *et al.*, 1972) and ribonucleotide reductase (Noronha, Sheys & Buchanan, 1972; see too Standart *et al.*, 1985). Much of the protein synthesis occurring in early embryos is supported by mRNA derived from the oocyte – until blastula stages in sea urchins (Gross, 1964) and *Xenopus* (Crippa, Davidson & Mirsky, 1967), and even in tadpoles of the rapidly developing tree frog *Engystomops* (Hough *et al.*, 1973) and perhaps ascidians (Lambert, 1971). Many specific mRNAs are present in the maternal pool and in the embryo but are absent from adult organs, and it has been proposed that these code for 'morphogenesis proteins' needed to construct an early embryo (Hough-Evans *et al.*, 1977). The lampbrush form could allow this set to be built up as quickly as possible using the full $4n$ genome.

Perhaps the most seductive evidence for this interpretation of lampbrush chromosomes is the existence of exceptions which seem to prove the rule. These are the meroistic ovaries of many insects where oocytes are syncytially connected to sister cells acting as nurse cells. Figure 1.2 shows two kinds of meroistic development and compares them with the simpler panoistic system. Where nurse cells are present, the oocyte chromosomes appear relatively inactive and certainly no lampbrush stage is seen. The nurse cell nuclei, on the other hand, produce RNA very fast indeed and there is histochemical (Bonhag, 1955) and autoradiographic evidence (Bier, Kunz & Ribbert, 1967) for massive transfers of RNA to the oocyte (see Fig. 1.2). This seems to include at least the bulk of the oocyte mRNAs (Winter, 1974; Paglia, Berry & Kastern, 1976; Capco & Jeffery, 1979). In such cases there are many nuclei transcribing genomic information for the oocyte and each of these nurse cell nuclei may be highly polyploid; and it is in just such cases that oogenesis is particularly rapid. The lampbrush form and the meroistic system therefore seem to be alternative mechanisms to produce the same end – the rapid transcription of the genomic information for early development.

There are, in fact, other cases which indicate the high 'cost', in terms of information, of gametogenesis and early development. In some embryos many chromosomes are actually eliminated early from all somatic lineages, only the germ-line retaining the full chromosome complement (e.g. *Ascaris*: Boveri, 1887; gall-midges: Kahle, 1908). In the oocytes of one gall-midge the 'extra' chromosomes are particularly active in RNA synthesis (Kunz, Trepte & Bier, 1970). Similarly there is evidence that both X chromosomes are active in the oocytes of mammals (Epstein, 1969), while one is soon inactivated in all somatic lineages of female embryos (see Lyon, 1972).

One problem for the views summarised above arises from the dynamics

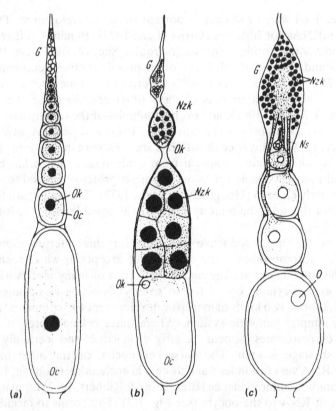

Fig. 1.2. Three types of insect ovary: (*a*) panoistic; (*b*) meroistic polytrophic; (*c*) meroistic telotrophic. Nuclei very active in RNA synthesis are shown black and those relatively inert white; RNA transferred to the cytoplasm is shown as black dots. G, germarium; Oc, oocyte; Nzk, nurse cell nucleus; Ns, trophic cords. (After Bier, 1967.)

of transcript accumulation. Lampbrush loops are present at all six of Dumont's stages in *Xenopus* (see Davidson, 1986) and the mRNAs present at all stages code for the same proteins (Darnbrough & Ford, 1976). Yet, by the start of vitellogenesis (in Stage II), the final quantity of oocyte poly(A)$^+$RNA, and of a large number of specific RNA sequences, has already been accumulated (Golden, Schafer & Rosbash, 1980). Moreover, these early transcripts have half-lives estimated as 2–3 years (Ford, Mathieson & Rosbash, 1977), which should allow them to persist into the embryo. In these circumstances it is difficult to understand why the lampbrush form is retained after Stage II, but both synthesis and degradation rates for poly(A)$^+$RNAs seem to increase continuously through Stage III (when the most extended lampbrush loops are seen) to reach their maximum in full-grown oocytes (Dolecki & Smith, 1979).

There are a few instances of stage-specific transcription in lampbrush loops (Macgregor & Andrews, 1977), but much greater qualitative changes in the mRNA population are seen in sea urchins (Hough-Evans *et al.*, 1979) where true lampbrush matrices do not seem to exist (see Davidson, 1986).

Still more fundamental is the problem that much of the poly(A)$^+$RNA produced for the sea urchin or amphibian egg is untranslatable (Posakony *et al.*, 1981; Thomas *et al.*, 1981). In many ways the maternal RNA molecules resemble nuclear RNAs in other cells, and they may require further processing in the embryo before being translated or may act rather to regulate the expression of other genes in development (Thomas *et al.*, 1981; Shiokawa, 1983). It is also possible, as pointed out by Colman (1983), that the high transcriptional activity of oocytes leads to un-regulated synthesis or incorrect splicing of transcripts. It is relevant here to note that concentrations of the small nuclear RNAs, thought to be involved in mRNA processing, apparently fall once their synthesis stops in the young oocyte (Fritz *et al.*, 1984).

Informational RNAs may seem likely candidates to confer spatial organisation on the oocyte, but development of course also requires relatively massive amounts of other classes of RNA. There are, in fact, efficient and specialised methods of providing oocytes with the RNA species in ribosomes and with transfer RNA. To provide the huge amounts of ribosomes required by large oocytes, the major genes (rDNA) present at the nucleolus organisers are replicated – the first known and probably most dramatic case of specific gene amplification – providing thousands of copies of the nucleolar core ring in amphibian oocytes (Painter & Taylor, 1942; Gall, 1968; Perkowska, Macgregor & Birnstiel, 1968). Even some smaller oocytes, including mammalian ones, show a more limited amplification (Brown & Dawid, 1968; Wolgemuth, Jagiello & Henderson, 1980). All of this rDNA can later be transcribed repeatedly as rRNAs (Miller, 1973), which are assembled with 5S RNA and ribosomal proteins into ribosomes. This can occur with such rapidity that the nucleoli bleb out of the nucleus. The oocytes of fish and amphibians show another highly unusual feature in the system used to synthesise the 5S RNA of ribosomes (Ford, 1971; Wegnez, Monier & Denis, 1972; Mazabraud, Wegnez & Denis, 1975; Denis *et al.*, 1980). This is encoded by a particular 5S gene present in multiple copies in the genome but only activated in oocytes. Most 5S RNA is produced in early oogenesis and is stored in 7S particles or 42S complexes (which also contain transfer RNAs) until rRNAs are produced.

Despite the fact that mRNAs and translational machinery are present in abundance in oocytes, only a very low proportion of the ribosomes (Woodland, 1974; Davis, 1982) or the mRNAs (Rosbash & Ford, 1974; Paglia *et al.*, 1976) is usually found to be engaged in protein synthesis there. Translation appears to be restricted in a variety of ways according to the species and even the stage of oogenesis. Most of the ribosomes in the oocytes of lizards (Taddei *et al.*, 1973) and some mammals (Burkholder, Comings & Okada, 1971) are found in large crystalline aggregates or lattices which are probably an inactive storage form. 'Free' mRNAs are complexed with proteins in mRNPs, and in the eggs of sea urchins (Jenkins *et al.*, 1978) and the chironomid midge *Smittia* (Jäckle, 1980*a*) there is evidence that some of these proteins 'mask' the mRNA preventing its translation. In *Drosophila* there may be both a general limitation of the translational apparatus (Goldstein, 1978) and a selective exclusion of some mRNA species from polysomes (Mermod, Schatz & Crippa, 1980). As *Xenopus* oocytes reach full size, the block seems to be switched from mRNPs (Richter & Smith, 1984; Taylor, Johnson & Smith, 1985) to other components required for translation (Laskey *et al.*, 1977), possibly a specific initiation factor (Audet, Goodchild & Richter, 1987). mRNAs in the tobacco hornworm oocyte apparently lack a $5'$methylated cap structure which would make them untranslatable (Kastern & Berry, 1976).

Among the protein species synthesised in oocytes are many which will be required in large amounts by early embryos. As a preparation for the rapid nuclear multiplication of cleavage at least some oocytes form large stores of DNA polymerases (Tato, Gandini & Tocchini-Valentini, 1974), histones (Adamson & Woodland, 1974; Woodland & Adamson, 1977), other proteins required for chromatin assembly (Laskey *et al.*, 1978; Kleinschmidt *et al.*, 1983) and tubulin needed in the mitotic apparatus and elsewhere (Raff *et al.*, 1971; Miller & Epel, 1973; Bibring & Baxendall, 1977). Another cytoskeletal protein formed is actin (Ruddell & Jacobs-Lorena, 1984). At the time of ribosomal assembly in *Xenopus*, more than 30% of the proteins synthesised are ribosomal proteins (Hallberg & Smith, 1975). Large amounts of RNA polymerases are also present in oocytes (Wassarman, Hollinger & Smith, 1972; Roeder, 1974; Kastern, Underberg & Berry, 1981), despite the fact that transcription rates are, at first, low in embryos. Levels of free and organelle-bound enzymes rise through oogenesis (Miller & Epel, 1973), suggesting again a preparation for extensive metabolic activity in the embryo. Even fibronectin, which has a primarily extracellular role from blastula stages in amphibians, is already formed in oocytes (Darribère *et al.*, 1984).

A major event of oogenesis is the accumulation of the protein yolk. The proteins are normally formed elsewhere in the maternal body and can be taken up from blood serum even against a concentration gradient (Knight & Schechtman, 1954; Telfer, 1954): in vertebrates the liver synthesises these proteins. In a variety of animal oocytes, uptake occurs into coated vesicles (Anderson, 1964; Roth & Porter, 1964: another phenomenon first described in oocytes). The proteins are then packed in a crystalline array in yolk platelets (see Wallace, 1972). In some cases, protein yolk seems formed at least partly in the oocyte (crayfish: Ganion & Kessel, 1972), nurse cells (some polychaetes: Emanuelsson & Anehus, 1985) or follicle cells (*Octopus*: O'Dor & Wells, 1973; *Drosophila*: Brennan *et al.*, 1982). Other storage materials such as glycogen granules and lipid droplets are commonly classified as yolk (see Raven, 1961) but have received less study. Eggs of most placental mammals are often described as lacking yolk, but even in these there is evidence (discussed by Schultz, Letourneau & Wassarman, 1979) for uptake of exogenous proteins by oocytes and breakdown of proteins in early development.

The accumulation of yolk has many effects on development beyond the supply of respiratory substrates and precursor materials. The extent of yolk is often the major factor determining egg size which ranges from less than 100 μm diameter in yolk-poor eggs up to many centimetres in some birds. Where development requires interaction of the areas of the embryo, this has great implications for the distances over which interaction must occur, a topic to which we shall repeatedly have to return. Likewise, we shall many times see effects of yolk upon cleavage pattern. Suggestions of more specific developmental roles for yolk (discussed by Counce, 1973) seem less likely, however, in view of the development of yolkless embryo fragments (Hadorn & Müller, 1974: see also p. 184).

Oocytes also contain a full range of other organelles. There are very large numbers of mitochondria (e.g. see Marinos & Billett, 1981), which will be involved in energy release in the embryo. Other organelles characteristic of oocytes are multivesicular bodies, annulate lamellae and cortical granules. Annulate lamellae are stacked membraneous complexes with an organisation similar to the nuclear membrane. For example, they possess 70 nm diameter pores (see review of Kessel, 1983). Kessel has long suggested that the role of annulate lamellae is to activate stored genetic information very much in the same way as polysomes can assemble at normal nuclear pores. Cortical granules (see review of Guraya, 1982) usually form in late previtellogenesis as Golgi derivatives and move to positions below the plasma membrane: their role is considered in the next chapter (p. 31). Oocytes also have a cytoskeleton which can be quite complex: in amphibian oocytes, for instance, there are large amounts of microtubule protein (Pestell, 1975), actin-containing microfilaments (Franke *et al.*, 1976) and several kinds of intermediate filaments (Franz *et al.*, 1983; Godsave *et al.*, 1984*a,b*). Potentially, a cytoskeleton could act as a framework for the localisation of developmental information, and we shall have repeated cause to consider this possibility.

In meroistic ovaries the nurse cells supply many organelles as well as RNA to the oocyte. Often the nurse cells are almost totally absorbed into the oocyte by the time oogenesis is complete. The materials are trans-

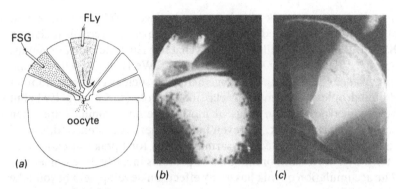

Fig. 1.3. The fate of fluoresceinated compounds injected into the nurse cells of the silkmoth *Hyalophora*. Acidic proteins like serum globulin, FSG (*a*) and methylcarboxylated lysozyme (*b*) are transferred to the oocyte, but lysozyme, FLy (basic), is not (*a*, *c*). Follicle diameters are 500 μm. ((*a*) from Woodruff & Telfer, 1980; (*b*) from Telfer *et al.*, 1981; (*c*) courtesy of Dr R. I. Woodruff.)

ported along quite long trophic tubes in telotrophic ovaries, and in many cases these contain a well-developed system of microtubules (Macgregor & Stebbings, 1970) although their role is not yet clear. In *Drosophila*, contractions by a microfilament system in the nurse cells seem likely to drive their contents into the oocyte (Gutzeit, 1986*b*). In several other species, a factor of importance for transport, and potentially for spatial determination, is the electrical polarity within the ovariole. Nurse cells are electronegative to oocytes, and proteins appear to be electrophoresed across the intercellular bridges: acidic proteins (negatively charged) moving from nurse cells to oocyte, and basic ones (positively charged) from oocyte to nurse cells (Woodruff & Telfer, 1973, 1980; Telfer, Woodruff & Huebner, 1981: see Fig. 1.3). Most soluble proteins and organelles are negatively charged and so should be driven into the oocyte, but positively charged proteins such as histones would require an acidic carrier if they are transported. Such an 'electrophoretic' system is highly unlikely to operate in *Drosophila* (Bohrmann *et al.*, 1986*a,b*), and it may well be that the problem of transport from the nurse cells has been solved in many different ways by different species (Gutzeit, 1986*c*).

There are instances where follicle cells also have open bridges with oocytes and appear to pass materials to them (lizards: see Andreuccetti, Taddei & Filosa, 1978), but usually they have no direct cytoplasmic contact with the oocyte. However, the membranes between the two cell types often show extensive interdigitation and specialised junctions and at least small molecules are able to pass between them (Browne, Wiley & Dumont, 1979). The follicle cells of mammals appear to promote the growth (Eppig, 1977) and block the maturation (Dekel & Beers, 1978; Larsen, Wert & Brunner, 1987) of oocytes in this way. Often the major

(a)                                                                (b)

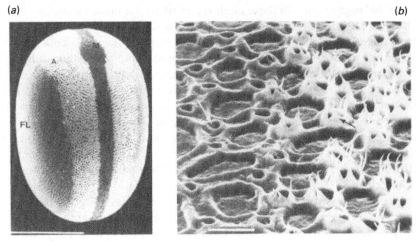

Fig. 1.4. Scanning electron micrographs of the outer surface of a silkmoth (*Antheraea*) eggshell after the removal of the follicular cells. (*a*) at low magnification four surface types are seen: flat (FL), aeropyle crown (A) stripe (S) and micropyle (M). Scale bar = 1 mm. (*b*), detail of the border between the FL and A regions in a tilted view. Scale bar = 10 μm. (From Mazur, Regier & Kafatos, 1982.)

products of follicle cell syntheses are components of the extracellular coats around oocytes (e.g. Paul *et al.*, 1972), and such coats may show spatial pattern. The silkmoth chorion, for instance, shows a characteristic pattern of surface sculpturing (Fig. 1.4) which is the result of a complex programme of follicle cell syntheses controlled both in time and space (Mazur, Regier & Kafatos, 1980; Regier, Mazur & Kafatos, 1980; Bock, Campo & Goldsmith, 1986). The elongated shape of most insect eggs also seems to be imposed by the follicular epithelium using a system of circumferentially oriented microtubules attached to desmosomes to resist circumferential expansion (Tucker & Meats, 1976).

## Visible organisation and its origins

In the well-known case of the *Xenopus* oocyte, illustrated in Figure 1.1, spatial organisation first becomes obvious at stage II when cortical granules and pigment granules accumulate peripherally, producing an inside–outside or concentric organisation. At stage IV, polar organisation is also obvious as a gradient in the amount of yolk and the sizes of the yolk granules, and with the germinal vesicle and pigment granules restricted to the less yolky hemisphere (Dumont, 1972). These same two kinds of organisation are seen in many other types of oocyte; while in others only concentric patterns are ever obvious.

Spatial organisation of these kinds is almost always made visible by the graded distribution of protein yolk. Where yolk concentration increases from one pole to the other, the oocyte is said to be telolecithal. Yolk may be either denser (protein yolk) or less dense (lipid yolk) than other ooplasmic constituents, so a yolk gradient affects how released oocytes and eggs float. This, in turn, has resulted in further spatial differentiation of some oocytes – most obviously in amphibian ones which have pigmented upper halves and unpigmented lower ones so that the egg is cryptically coloured whether viewed from above or below. The origin of the yolk gradient here is apparently directed transport of platelets out of the upper half following uniform yolk uptake over the whole oocyte surface (Danilchik & Gerhart, 1987). In many groups, the germinal vesicle moves to the less yolky pole, and, when meiosis resumes, the polar bodies are given off here. The first cleavages also normally begin at this point, which early embryologists saw as the most active part. They also recognised the nutritional significance of the opposite yolky pole in many cases, and so described the two poles respectively as animal and vegetal. It is easy to understand the choice of such terms if we consider cases such as the chick where the whole embryo arises at the animal pole and grows by using the more vegetal yolk. Usually, the polar differences are far less dramatic than this, but most of our knowledge of spatial differentiation in oocytes still involves the animal–vegetal axis.

There are, however, some animal groups where the animal–vegetal axis is not obvious in oocytes. These include mammalian oocytes with little yolk and some other small oocytes with a central mass of yolk (centrolecithal oocytes). Such oocytes include those of ctenophores, ascidians and *Chaetopterus*. In such cases all other organelles are enriched in an outer ectoplasm. The best-known centrolecithal eggs are the characteristically elongated insect eggs where a central yolky area is surrounded by a relatively yolk-free periplasm. Here the polar bodies are given off close to one end of the egg, known only as the anterior pole, and the opposite end is called the posterior pole. The fact that oocytes and eggs exist which are not telolecithal shows that a yolk gradient cannot be an indispensable factor for the polarity of developmental potential which will be found to exist in most embryos. Polar organisation is detectable in the plasma membrane (Moody, 1985) and cortex (Schroeder, 1985) of the full-sized starfish oocyte, where it may only arise as the germinal vesicle migrates to the future animal pole. Insect oocytes provide the clearest examples of bilateral organisation, seen for instance in the detailed shape of the oocyte and in the exact position of its germinal vesicle (see Gill, 1964). In most other oocytes no dorso-ventral axis is recognisable, though the dorsal side of the unfertilised cephalopod egg is distinguishable by its shape (Watasé, 1891).

The distribution of yolk of course affects the distribution of all other materials because a gradient of yolk in one direction implies a gradient of non-yolk cytoplasm in the opposite direction. Even the animal–vegetal distribution of water and simple ions in *Rana* oocytes is affected in this way (Tluczek, Lau & Horowitz, 1984). Among the organelles affected are the ribosomes, their numbers decreasing from the animal to the vegetal pole of amphibian oocytes (see Brachet, 1965).

If the local distribution of materials is to have more far-reaching consequences for development, it is likely that the materials involved will be the informational nucleic acids or proteins. Davidson & Britten (1971) have proposed that some at least of these molecules act as gene regulatory agents, that they are distributed in specific patterns within the ooplasm, and that, when cleavage distributes nuclei throughout this ooplasm, each type of regulatory agent acts on the local nuclei. The result, either directly or indirectly, would be the local activation of genes appropriate for the development of that part within an embryo. The models are described more fully on p. 317; here we will examine the evidence for local distribution of mRNAs and proteins within oocytes.

When the general distribution of poly(A)$^+$ RNAs is revealed by *in situ* hybridisation with [$^3$H]poly(U), significant spatial differences are seen at various stages of oogenesis in the insect *Oncopeltus* (Capco & Jeffery, 1979) and in *Xenopus* (Capco & Jeffery, 1982). In both cases, however, these differences are lost before the egg is laid. In *Chaetopterus* the ectoplasm shows a great enrichment of poly(A)$^+$RNAs (Fig. 1.5(*a*)) which persists into the early embryo (Jeffery & Wilson, 1983). Ascidian eggs have quite even cytoplasmic concentrations of such RNAs with higher levels in the germinal vesicle (Jeffery & Capco, 1978). Of course, we really need to know the distribution of specific mRNAs, and this is available for only a few immature oocytes. In *Xenopus*, King & Barklis (1985) compared the mRNA populations of animal, equatorial and vegetal regions by translating them *in vitro*, finding that 17 mRNAs were unevenly distributed, most often with their highest concentration vegetally (see also Smith, R.C., 1986; Melton, 1987). One mRNA concentrated at the animal pole encodes a subunit of a mitochondrial ATPase (Weeks & Melton, 1987*a*), and both this and a vegetally localised species (Melton, 1987) retain their distribution in the egg. In *Drosophila*, one transcript accumulates anteriorly in early oocytes but is later evenly distributed (Aït-Ahmed, Thomas-Cavallin & Rosset, 1987). In *Styela*, Jeffery, Tomlinson & Brodeur (1983) have reported that histone mRNAs are evenly distributed throughout the ooplasm and germinal vesicle, but actin mRNAs are concentrated in the outer areas and in the germinal vesicle (Fig. 1.5(*b*)). It would be interesting to know whether the mRNA for alkaline phosphatase is localised in the yolky endoplasm of ascidian oocytes, since there is evidence that this mRNA is formed during oogenesis and the enzyme is later concentrated in endodermal cells

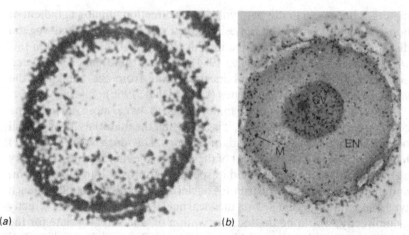

Fig. 1.5. RNA distributions within oocytes demonstrated by *in situ* hybridis-
ation. (*a*) poly (A)⁺RNA in *Chaetopterus*; × 330. (*b*), actin mRNA in *Styela*.
EN, endoplasm; GV, germinal vesicle; M, myoplasm. × 820. ((*a*) from Jeffery
& Wilson, 1983; (*b*) from Jeffery *et al.*, 1983.)

(Whittaker, 1977; but see also p. 104). Some other ooplasms which
appear to determine developmental fates in embryos appear during
oogenesis and show concentrations of RNA of an unknown type. These
include an accumulation of vesicles, the vegetal body, in *Bithynia*
(Dohmen & Verdonk, 1979*a*; and see pp. 35 and 82), the polar granules
in insects (Mahowald, 1971*b*; see p. 27) and similar granules in anurans
(Mahowald & Hennen, 1971; see p. 28). All of these special materials are
at the vegetal poles of the oocytes.

Present indications are that few specific proteins are localised within
oocytes. No differences were detected between the protein species of the
anterior and posterior halves of *Drosophila* oocytes (Gutzeit & Gehring,
1979). Only one cytoplasmic protein was found exclusively in the vegetal
half of axolotl oocytes (Jäckle & Eagleson, 1980), while a small group of
proteins is being formed faster in vegetal than animal halves in *Xenopus*
(Smith, 1986). Radially localised antigens have also been described in
*Xenopus* oocytes (Wylie *et al.*, 1985*a*). Jäckle & Eagleson (1980) also
found 14 proteins restricted to the germinal vesicle, which is in the animal
half. Many of these are probably the proteins formed by oocytes and with
roles in embryonic nuclear activities (see p. 6), as most of these seem at
least preferentially accumulated in the germinal vesicle (see Woodland &
Adamson, 1977; Wassarman & Mrozak, 1981). Materials restricted
within the germinal vesicle will be released near the animal pole at
maturation and so will then show a gradient distribution. These will
include the amplified rDNA of amphibian oocytes but it seems not to be

transcribed then (Brachet, Hanocq & van Gansen, 1970) and probably plays no further part in development.

In many oocytes polarity is first seen in the formation of an aggregate of cytoplasmic organelles referred to as the Balbiani body. Clearly a morphological axis is then present passing through the nucleus and the body, and yolk deposition, etc seem later to follow this axis so that the pole beyond the nucleus becomes the animal pole and the pole beyond the Balbiani body the vegetal pole. Cytoskeletal elements are probably responsible for maintaining the position and integrity of the body (Paleček *et al.*, 1985; Wylie *et al.*, 1985*a*), and the position of the nucleus (Gutzeit, 1986*a*). Wilson (1925, p. 108) pointed out that the Balbiani body is often beside the centriole and that the centriole : nucleus axis had been suggested by van Beneden as the basis of polarity in all cells. In *Xenopus*, a juxtanuclear aggregation mainly of mitochondria is recognisable even in primordial germ cells, and is little changed in oogonia and early oocytes, when the chromosomes attach to the nuclear membrane adjacent to the aggregate (Al-Mukhtar & Webb, 1971; see also Kühn, 1971, pp. 151–2). The nucleoli move to the opposite side of the nucleus according to Coggins (1973), who also found that this polarisation is related to position within a group of syncytially connected early sister oocytes. The nucleoli move to the nuclear pole away from the centre of the cell group, while chromosomes attach on the more 'central' side (Fig. 1.6).

In meroistic insect ovaries the position of the nurse cells (sister cells of the oocyte) is consistently related with oocyte polarity: they are attached at the anterior pole (Hegner, 1917). The polarity of such oocytes is independent of that of the ovariole or of the mother fly (Gill, 1964) and is unchanged by the removal of follicle cells (Junquera, 1983; but see the work of Frey & Gutzeit, 1986, noted on p. 21). In other cases contacts with follicle cells may determine polarity. Only the vegetal area of the oocyte in the mollusc *Ilyanassa* is in contact with follicle cells and lacks a mucopolysaccharide coat (Taylor & Anderson, 1969). Circumstantial evidence suggests an opposite arrangement in another mollusc, *Lymnaea*, with the vegetal pole being free of follicle cell contacts (Ubbels, Bezem & Raven, 1969). In *Drosophila* there is a pair of follicle cells at both the anterior and the posterior poles which have distinctive immunofluorescence properties (Fig. 1.7; Brower, Smith & Wilcox, 1981; see also White, Perrimon & Gehring, 1984). These could have a role in polarity determination, or local functions in the formation of the micropyle and the posterior chorion protuberance. Micropyle formation in fish eggs certainly involves a local association with a specific follicle cell (Mark, 1890): and the funnel-shaped canal through the jelly coat of sea urchin oocytes may also betray a local difference in follicular organisation (see Schroeder, 1980*a,b*). Finally, the general attachment point of an

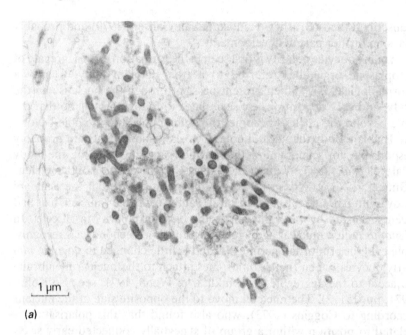

1 µm

(a)

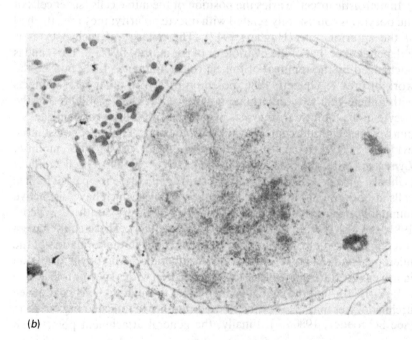

(b)

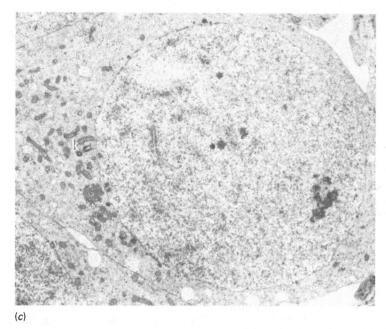

(c)

Fig. 1.6. Polar organisation in *Xenopus* oocytes. (*a*), at leptotene/zygotene the chromosomes attach to the side of the nucleus towards the Balbiani body. (*b*), at zygotene nucleolar fragments are at the opposite side of the nucleus. (*c*), a pachytene oocyte showing the centrioles (k) and with the centre of the cell nest to the left of the photograph. ((*a*, *b*) from Al-Mukhtar & Webb, 1971; (*c*) from Coggins, 1973.)

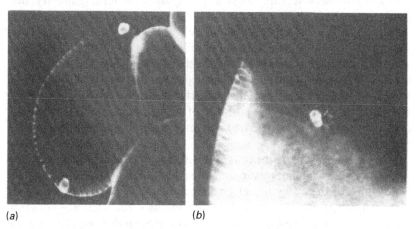

(a)                              (b)

Fig. 1.7. Binding patterns of a fluoresceinated antibody (derived from a mouse cell line) on to *Drosophila* follicles. (*a*) at stage 7 anterior and posterior polar follicular cells show especially bright fluorescence. (*b*), at stage 10, the anterior cells lie next to the oocyte after migrating between the nurse cells. (*a*) × 320; (*b*) × 310. (From Brower *et al.*, 1981.)

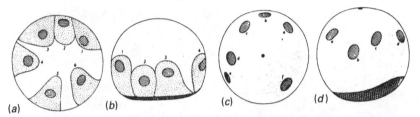

Fig. 1.8. Follicular arrangement and egg organisation in *Lymnaea*. (*a*), (*b*) positions of the six inner follicle cells seen from the oocyte's inner pole (*a*) or the side (*b*). The area of contact with the gonad wall is cross hatched. (*c*), (*d*) the organisation of the oviposited egg seen from the animal pole (*c*) or the side (*d*). a–f are six subcortical 'patches' of different staining, and the vegetal pole plasm is cross hatched. The black dot marks the animal end of the first maturation spindle. (From Raven, 1963*a*.)

oocyte within the ovarian wall seems sometimes to determine polarity. In many cases the attachment point becomes the vegetal pole – in the nemertine *Cerebratulus* (Wilson, 1900), in holothurians (lit. in Lindahl, 1932), sea urchins (Jenkinson, 1911*b*; Lindahl, 1932), *Ascaris* (see Kühn, 1971, p. 150), *Dentalium* (see Wilson 1925, p. 1097), several lamellibranchs (van den Biggelaar & Guerrier, 1983) and probably ascidians (Child, 1951). The attachment point is claimed to become the animal pole in amphioxus (Conklin, 1932) and in the hydrozoan *Amphisbetia* (Teissier, 1931), but the latter author also states that polar bodies are produced at the free pole.

In *Drosophila*, where the oocytes already show bilateral symmetry, the dorso-ventral axis also seems determined by the immediate environment, with the side towards the outside of the ovariole becoming ventral (see Gill, 1964). In the snail *Lymnaea* the follicle cells may determine dorso-ventrality although this only becomes visible as the egg passes through the reproductive tract. At that time six patches of selectively stainable cytoplasm accumulate beneath the surface, in a dorso-ventral pattern which apparently reflects the positions of the six nuclei of the inner follicle cells (Raven, 1966; see Fig. 1.8). Raven's interpretation of this will be considered on p. 39.

Interactions among sister cells and with neighbouring follicle cells are also concerned when meroistic ovaries make the developmental decision as to which of the syncytial cells will become the oocyte (for review see Gruzova, 1982). In *Drosophila* the connections – 'ring canals' – between the cells show the lineage of the group and Brown & King (1964) have shown that it is one of the two cells with four canals which becomes the oocyte (Fig. 1.9). This probably gives it an advantage in nutrient supply, but a posterior position in the egg chamber and contact with follicle cells are other properties of the future oocyte (King, 1972). In the telotrophic

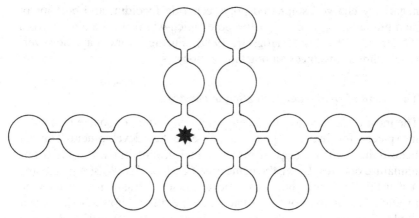

Fig. 1.9. Diagram to illustrate the relationships between the oocyte (*) and the 15 syncytially connected nurse cells in a *Drosophila* egg chamber. The cells are represented by circles lying in a single plane and the ring canals are lengthened for clarity. (After Brown & King, 1964.)

ovarioles of the beetle *Creophilus*, mother cells are seen in which the posterior pole is defined by basophilic inclusions and by contact with prefollicular tissue. Following divisions it is the most posterior cell, retaining these properties, which becomes the oocyte (Kloc & Matuszewski, 1977).

Meroistic ovaries raise a final question for this section: are the currents which electrophorese materials into oocytes (p. 8) sufficiently strong to stratify their charged constituents? Calculations of the voltage gradient (Overall & Jaffe, 1985) suggest that the answer should be 'yes'. However, stratification has not yet been detected for proteins (Gutzeit & Gehring, 1979) or cations (Heinrich, Kaufmann & Gutzeit, 1983). Current may flow through the perivitelline space, rather than the oocyte, in older *Drosophila* follicles (Gutzeit, 1986c), and continues to flow when the nurse cells have collapsed or become detached in *Dysdercus* (Dittmann, Ehni & Engels, 1981). Indeed, currents have also been detected around simple follicles such as the panoistic ones of cockroaches where current enters the ventral side (Kunkel, 1986). Robinson (1979) described current entry at the animal pole of *Xenopus* follicles and preliminarily of medaka fish follicles. He again suggested that the current is of sufficient magnitude to provide a reference axis for charged particles but again no such separation has been seen for proteins (Jäckle & Eagleson, 1980). In any case the current here falls almost to zero at the start of maturation. To date, a strong case for the electrophoretic separation of developmentally important materials within an egg can be made only for the fucoid algae which are organisms really beyond the scope of this book. There, current flow through the egg is probably responsible for the localisation of the negatively charged sulphated polysaccharide fucoidin, and perhaps of membraneous organelles, at the pole which will form the algal rhizoid (Quatrano, Brawley & Hogsett, 1979). Further work may, however, reveal similar instances among animal oocytes.

### The extent of spatial determination in oocytes

The most direct test for local determination within an oocyte is to cut it into pieces, fertilise these pieces and observe their development. This has been done with the eggs of many animal groups but with only a few immature oocytes. Using the nemertine *Cerebratulus*, Wilson (1903) and Yatsu (1904) found that most oocyte nucleate fragments went on to produce apparently normal larvae while the others showed deletions or duplications of a relatively few larval structures. Wilson (1929) also used eggs of the polychaete *Chaetopterus* in which maturation had just started. Here the fragments were produced by centrifugation which also stratified the larger organelles so that the fragments had very different compositions, but still they frequently produced normal larvae. This high degree of regulation is particularly impressive because fragments from both the above species will act like parts of a mosaic by cleavage stages (see Chapter 3).

These data do not, however, mean that oocytes are totally devoid of pattern. For one thing, the possibility of concentric organisation has rarely been tested since fragments usually contain some outer and some inner material. It does in fact seem that some of the cortical cytoskeleton must be present if a *Chaetopterus* fragment is to develop (Swalla, Moon & Jeffery, 1985). With the ooplasmic flows that occur by the time a zygote is formed (see pp. 32 *et seq.*) an inside:outside difference can be (and probably is, e.g. in ascidians) transformed to give considerable spatial information. Secondly, there may still be information defining a polar axis. An analogy would be a magnet: it certainly has polarity, but this does not prevent its halves still acting as a whole magnet. Shirai & Kanatani (1980) have recently provided evidence that the polar axis is indeed fixed in starfish oocytes. Even if the germinal vesicle is moved by centrifugation and polar bodies form in an abnormal position, the original polarity is retained in the embryo.

If oocytes contain extra-nuclear information of significance for developmental patterns, then it should also be detectable by maternal effects on development. The nuclei of sperm and egg make effectively equivalent contributions to the genomes of most animals, so if the embryo shows more similarity with its mother this indicates a cytoplasmic contribution from the oocyte. Hybrids between two sea urchin species were first used to investigate this problem (lit. in Davidson, 1968). Cleavage rate, the external form of the early embryo, and the number

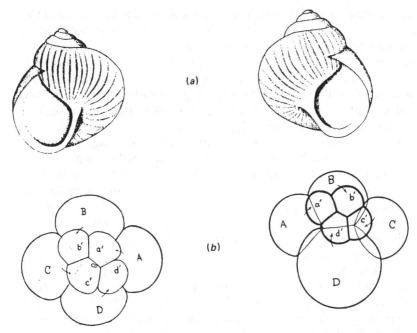

Fig. 1.10. Spiral patterns in gastropods: (*a*) the shells of different species may be left wound (left) or right wound (right); (*b*) the difference is already obvious at third cleavage from the direction of the spiral cleavage forming the micromeres. In species like *Lymnaea peregra*, both kinds of organisation may be found. (After Morgan, 1927.)

(not position) of cells entering certain developmental pathways were shown to be maternally controlled, but Boveri (1903; see also Wilson, 1925, p. 1107), who pioneered this work, considered that maternally formed spatial information was enough only to define animal–vegetal polarity and perhaps dorso-ventrality. The extent of maternal control over development seems similar in most animal groups, but is probably much greater in ascidians. Perhaps in correlation with their rapid development even the form of the adhesive papillae on larvae seems determined by the maternal species, and this is true in androgenetic hybrids where the egg nucleus is removed so that the entire genome comes from the paternal species (Minganti, 1959*a*). Maternal factors affect eye colour in larvae of an annelid (Fischer, 1974) and even in adult *Drosophila* (Marsh & Wieschaus, 1977), but such effects are probably due to broad metabolic changes rather than a specific determinant for the eyes.

A classic case, giving evidence for spatial patterning within the oocyte, comes from the snail *Lymnaea peregra*, where two races can be found with opposite directions of coiling of the shell (Fig. 1.10(*a*)). In experimental crosses, direction of coiling was found to be controlled by a single pair of genes, with dextral coiling being dominant, but with the direction being determined not by the genes of the snail inhabiting the shell but by the genes of its mother (Sturtevant, 1923). The difference in symmetry is in fact to be seen much earlier, at the second cleavage, and more clearly at the third cleavage (see Fig. 1.10(b)). There is now good evidence that action of the dextral gene leads to the appearance of a cytoplasmic determinant in the egg, as cytoplasmic transfers from dextral to sinistral eggs can reverse their direction of cleavage asymmetry (Freeman & Lundelius, 1982). The nature and mode of action of the determinant remain unknown, however. The stainable subcortical patches already described for this genus (p. 16) do not appear to show reversed patterns of symmetry in the two races (Ubbels *et al.*, 1969).

More recently, a systematic search has been made for genes acting under maternal control and affecting early developmental pattern. The approach has been to expose female animals to mutagenic treatments, allow them to mate with normal males and identify female-sterile mutations by their failure to lay viable eggs. Some of the mutant females are able to lay normal eggs at low temperatures, but fail at higher temperatures which wild-type animals would tolerate. This allows mutant stocks to be maintained yet also can provide material for analysis of developmental defects.

Most work has been done with *Drosophila* where genetic analysis is generally so far advanced. Hundreds of female-sterile lines have been isolated and their oocytes, eggs and early embryos studied (Wright, 1970; Bakken, 1973; Fullilove & Woodruff, 1974; Gans, Audit & Masson, 1975; Rice & Garen, 1975; Zalokar, Audit & Erk, 1975; Mohler, 1977; Bakulina, Pankova & Mitrofanov, 1984; Cheney *et al.*, 1984; Perrimon, Engstrom & Mahowald, 1984). In many cases the mutations cause visible defects in the events of oogenesis such as yolk deposition or chorion formation, and in others oviposition is affected. Most of the defects which become apparent in the early embryo are of a very unspecific nature, although, for instance, cell formation may fail in broadly restricted regions of the embryo (Rice & Garen, 1975; Perrimon, Engstrom & Mahowald, 1985) or the extent of morphogenetic movements may be affected (Zalokar *et al.*, 1975; Cheney *et al.*, 1984). Several maternal mutations cause defective germ cell formation (further discussed on p. 27), but none lead to similar defects in single somatic tissues or organs. As Zalokar *et al.* (1975) point out, this suggests that there is no mosaic of tissue-specific ooplasmic determinants, or that the genes involved are reiterated and therefore difficult to disrupt by mutation.

More important than this negative evidence, *Drosophila* maternal mutations provide positive evidence for an entirely different system of spatial determination. Many mutations are now known in which spatial

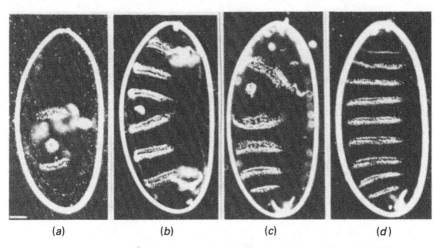

(a)                    (b)                    (c)                    (d)

Fig. 1.11. The ventral hypoderm patterns of *bicaudal* phenotypes. $(a),(b)$, symmetrical *bicaudal* embryos with $1\frac{1}{2}$ $(a)$ and $3\frac{1}{2}$ $(b)$ posterior segments duplicated. $(c)$ an asymmetrical *bicaudal* where $4\frac{1}{2}$ segments have normal polarity and $1\frac{1}{2}$ are reversed. $(d)$ a headless embryo where all the thoracic and abdominal segments are formed. Scale bar = 50 $\mu$m. (From Nüsslein-Volhard, 1979.)

pattern is disturbed in gross ways affecting much of the embryo. First, Bull (1966) isolated the *bicaudal* mutation in which some embryos showed reversal of polarity in the anterior half of the embryo, so that two partial and mirror-symmetrical abdomens were formed without any anterior structures. The effect was seen in only a low proportion of embryos from mutant mothers, but manipulation of the genetic background by Nüsslein-Volhard (1977) has produced stocks with greater penetrance, and yields can be increased further if the females used are young and temperatures high. Nüsslein-Volhard finds that there is a range of phenotypes produced, from the extreme of mirror-symmetrical ones – in which the posterior half is also affected as its segments are too few and too large – via intermediates such as embryos which have normal polarity but are headless, to normal embryos (Fig. 1.11).

Non-allelic dominant mutations with similar effects are now known (Nüsslein-Volhard, 1979; Mohler & Wieschaus, 1986), as well as mutations affecting the coordinates of development in different but equally wide-ranging ways. Embryos with double heads in mirror-image symmetry develop in eggs laid by *dicephalic* mothers (Fig. 1.12). This is associated with a distribution of nurse cells to both poles of the oocyte (Lohs-Schardin, 1982), but occurs unless both the follicle cells and the germ line bear a wild-type allele at the *dicephalic* locus (Frey & Gutzeit, 1986). Offspring of *bicoid* mothers lack head and thorax, with the abdominal segments occupying more space than usual but with only the

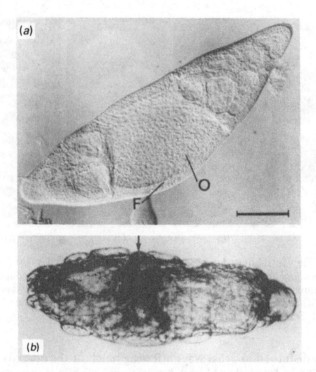

Fig. 1.12. Effects of the maternal *dicephalic* mutation. (*a*) a stage 10/11 follicle, viewed with Nomarski optics, with nurse cell clusters at both ends of the oocyte, O. F, follicle cells. Scale bar = 100 $\mu$m. (*b*) a larva with two heads of opposite polarity, that on the right having failed to invaginate. Arrow indicates level of polarity reversal. (From Lohs-Schardin, 1982.)

telson being duplicated at the anterior end (Frohnhöfer & Nüsslein-Volhard, 1986). Other maternal effect mutations, including *oskar*, have the complementary effect of deleting abdominal structures without duplicating anterior ones (Boswell & Mahowald, 1985; Schüpbach & Wieschaus, 1986; Lehmann & Nüsslein-Volhard, 1986); while still others delete structures from both poles (Schüpbach & Wieschaus, 1986). In at least some of these cases eggs can be 'rescued' from abnormal development by the early injection of cytoplasm from wild-type eggs. Anterior cytoplasm injected anteriorly restores the formation of head and thorax to the offspring of *bicoid* females (Frohnhöfer & Nüsslein-Volhard, 1986), and posterior cytoplasm injected to the prospective abdominal region restores abdominal development to the offspring of *oskar* females (Lehmann & Nüsslein-Volhard, 1986). Such work (discussed further on pp. 126 and 182) provides strong evidence that the products of these genes play essential roles in development at the anterior and posterior poles respectively.

Quite a large set of maternally acting loci is now known to affect dorso-ventral determination in *Drosophila*. A few of these act sufficiently early to change the dorso-ventral organisation of the whole follicle; one, *torpedo*, being the first maternal-effect gene shown to act exclusively in the somatic cells (Schüpbach, 1987). Products of these genes may control the expression of a second gene group which seems not to affect follicular organisation, but where a lack of maternal function leads to a dorsalis-ation of the embryo (i.e. dorsal tissues spread around the embryo at the expense of ventral ones). This effect has been most thoroughly studied for mutations of the *dorsal* locus (Nüsslein-Volhard, 1979; Nüsslein-Volhard *et al.*, 1980; Anderson & Nüsslein-Volhard, 1984b). Females bearing dominant gain-of-function alleles at another locus, *Toll*, produce ventralised offspring (Anderson, Jürgens & Nüsslein-Volhard, 1985b). The implication of these phenotypes is that the normal products of this latter gene set are required for the formation of lateral and ventral embryonic derivatives.

The effects of mutations in the *dorsal* gene set can again be reversed by injecting eggs with wild-type cytoplasm (Santamaria & Nüsslein-Volhard, 1983), or indeed with RNA purified from it (Anderson & Nüsslein-Volhard, 1984a), which indicates that at least some of the maternal information is stored as the mRNAs (see too Steward, McNally & Schedl, 1984; DeLotto & Spierer, 1986). However, in contrast to work on antero-posterior determination, there is little evidence that the gene products are localised in the dorso-ventral plane of the egg. Maternal mutations at five loci can be rescued by cytoplasm taken from, and injected into, any meridian of the egg, and the embryo will still develop with its original bilaterality which is betrayed by the form of the egg-shell (Anderson, Bokla & Nüsslein-Volhard, 1985a) although quantitative differences in rescue ability may be seen (Müller-Holtkamp *et al.*, 1985). The effects of *dorsal* mutations can only be corrected by ventral injec-tions, but again the donor cytoplasm can come from any meridian (Santamaria & Nüsslein-Volhard, 1983). Only in one case, the offspring of *Toll*⁻ females, is the dorso-ventrality of the oocyte changed. In this case the side injected with wild-type cytoplasm forms the ventral side of the rescued embryo, but even so the concentration of rescue factors seems to be as high dorsally as ventrally in the cytoplasm of the wild-type donor (Anderson *et al.*, 1985b). This has led Anderson *et al.* (1985a,b) to propose that a precursor of the *Toll*⁺ product (e.g. *Toll* mRNA) is equally distributed around the egg circumference but is activated (e.g. translated) preferentially on the ventral side. The other genes of the set could be involved in the activating process or the further production of a ventralising morphogen. A possible activation mechanism is indicated by DeLotto & Spierer's (1986) evidence that the product of one gene, *snake*, is a serine protease with a calcium-binding site and a cleavage site. In

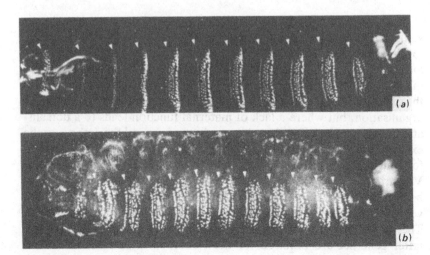

Fig. 1.13. Ventral hypoderm patterns of a normal *Drosophila* larva (*a*), and of the offspring of an *esc* mutant female (*b*). (*a*) × 80; (*b*) × 110. See text. (From Struhl, 1983.)

other situations, such as mammalian blood clotting, similar enzymes are activated by cleavage in a cascade of reactions, so a similar cascade here could amplify small differences to fix the dorso-ventrality of the embryo.

A maternal-effect mutation of a different kind, but still affecting the whole embryo, has recently been described by Struhl (1981*a*; Struhl & Brower, 1982). It is apparently a complete deletion of the *extra sex combs* (*esc*) locus. Mothers homozygous for the deletion (*esc⁻/esc⁻*) lay eggs which die as unhatched first instar larvae where all segments behind the head look like eighth abdominal segments (Fig. 1.13). The wild-type (*esc⁺*) gene must act mainly maternally, since *esc⁻/esc⁻* embryos can develop normally if the mother was heterozygous, but there is also some embryonic activity since sperm carrying two doses of the *esc⁺* gene can effect an almost complete rescue. This wholesale change of segments towards the most posterior type will be considered again in a later chapter (p. 182) as it is from the blastoderm stage of the embryo that the *esc⁺* gene product is needed. Other maternally active loci cause more localised transformations of segment type (Forquignon, 1981; Digan *et al.*, 1986) or affect the number of recognisable segments produced (Schubiger & Newman, 1982; Macdonald & Struhl, 1986). Products of some of these genes have been detected in oocytes and early embryos and in the case of *caudal* embryos they are distributed in a postero-anterior gradient by blastoderm stages (Mlodzik, Fjose & Gehring, 1985; Macdonald & Struhl, 1986; see p. 186).

The nature of these maternal effects again suggests that it is general organisation, such as polarity, bilaterality and some features of segmental pattern which are specified in the oocyte. In principle, polarity could be determined by the graded distribution of a single substance as suggested in Fig. 1.14 (from Nüsslein-Volhard, 1979). Different concentrations of the morphogen would determine the development of head, thorax and abdomen, which would thus appear in sequence from the anterior to the posterior pole of the normal egg. Changed patterns of morphogen concentration along the egg could explain the various *bicaudal* phenotypes, including their reduced total segment number. Meinhardt (1977) has proposed a model which could explain both the normal gradient form and the appearance of a second morphogen peak at the anterior pole following various kinds of interference. Nüsslein-Volhard (1979) has proposed a similar gradient system to explain dorso-ventral determination, with the high-point of the gradient midventrally. It now seems highly likely that determination mechanisms in both planes are far more complex than these models suggest, with several genes being involved in each case. None the less, the point to stress here is that maternal transcription seems not to produce a detailed 'map' in the oocyte, but rather some graded coordinate system. The positional information this supplies is then probably used to activate appropriate genes in various areas of the early embryo and it would be the zygotic transcription which further refines determination to the level of individual segments and organs (see p. 182).

Another species screened for temperature-sensitive maternal-effect mutations on a large scale is the nematode *Caenorhabditis elegans*, and the types of developmental abnormality seen are described by Hirsh (1979), Wood *et al.* (1980), Schierenberg, Miwa & von Ehrenstein (1980), Denich *et al.* (1984) and Kemphues *et al.* (1986) Mutations of the strict maternal effect genes (where maternal expression is both necessary and sufficient for development) lead in most cases to abnormalities in very early development. Defects can be seen in the eggshell or cytoplasmic ultrastructure, in the nuclear or cytoplasmic events of fertilisation, in cytokinesis or the planes, tempo or sequence of the cleavage divisions. Many of the early defects can be phenocopied by dissecting wild-type eggs from the gonad, by contact with oil or by applying pressure (Denich *et al.*, 1984), which suggests that they are of low specificity. Many of the mutations have other temperature-sensitive phases at postembryonic stages and could affect metabolic enzymes of significance in many cell types (Hirsh, 1979).

In nematode embryos, cells of different lineages cleave at different rates, so it is conceivable that maternal factors which change the cleavage rate of a lineage could indirectly affect its fate. However, where markers for differentiated tissues are available, it seems that they are developed by the appropriate lineages even following early cleavage arrest (see

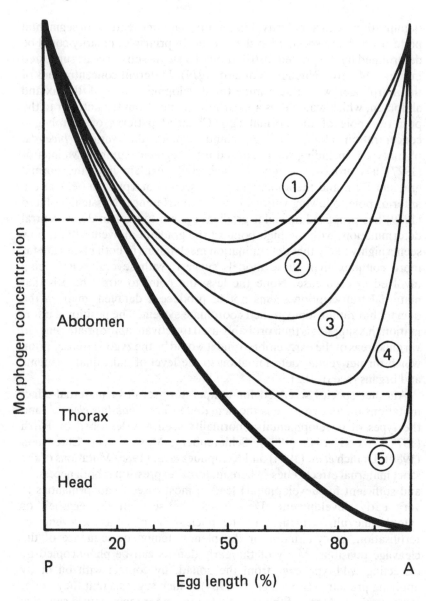

Fig. 1.14. Hypothetical gradients of morphogen concentration along the axis (posterior, P to anterior, A) of normal and *bicaudal* embryos. The dotted lines indicate threshold concentrations for abdomen, thorax and head. The heavy line shows the gradient appropriate for a normal embryo and the fine lines for *bicaudal* phenotypes, including symmetrical *bicaudal* (1, 2), asymmetrical *bicaudal* (3, 4) and headless (5) embryos. (From Nüsslein-Volhard, 1979.)

p. 118). Thus, once again a search has failed to find maternal genes responsible for specific and localised determinants in an egg.

A less intensive search in other animal species has revealed smaller numbers of maternal effect mutations. Humphrey has described five in the axolotl (for a review see Malacinski & Spieth, 1979). Females homozygous for each of these mutations lay eggs which arrest at characteristic stages with characteristic abormalities. Four seem to affect general processes such as cleavage or the fluid balance of the embryo, although in one of these cleavage is prevented only in the unpigmented vegetal half. The most interesting is the *o* (*ova deficient*) mutation (Humphrey, 1966). Eggs laid by *o/o* females do not normally develop beyond gastrulation but they can be rescued by the injection of material from the germinal vesicle of wild-type oocytes, suggesting that the $o^+$ gene product is accumulated there (Briggs & Cassens, 1966). The *o* mutation is considered further on p. 195.

In another urodele amphibian, *Pleurodeles walti*, females homozygous for the *ac* (*ascites caudale*) mutation lay eggs which become abnormal in gastrulation (see Beetschen & Fernandez, 1979). Irregular furrows appear chiefly in the animal hemisphere, but curiously this effect can be prevented simply by pricking the egg with a micropipette! The nature of the mutation is not yet clear, but although it shows some localisation in its effect it seems unlikely to involve a specific determinant.

In fact, the only maternal mutations which suggest that tissue-specific determinants may be localised in oocytes concern a single cell type – the germ cells of *Drosophila*. Several mutations are known where females lay eggs capable of developing all the somatic tissues but lacking germ cells. Of these the best known is the *grandchildless* (*gs*) mutation of *D. subobscura* (Fielding, 1967), but comparable mutations in *D. melanogaster* are described by Niki & Okada (1981) and Engstrom *et al.* (1982) who also refer to many other mutations with rather less specific effects. Other kinds of evidence support the case for the existence of specific germ cell determinants in the *Drosophila* oocyte and egg (for review see Mahowald *et al.*, 1979*a*). The area which normally forms the germ cells is at the extreme posterior pole of the oocyte and is the only area to contain polar granules (Fig. 1.15). Constituents of the polar granules apparently include RNA (Mahowald 1971*b*) and a single major basic protein (Waring, Allis & Mahowald, 1978). The posterior plasm of the oocyte is capable, after transfer to the anterior pole of a recipient egg, of inducing the formation of germ cells there (Illmensee, Mahowald & Loomis, 1976). Anterior plasm, lacking polar granules, is incapable of forming germ cells even in the offspring of *bicaudal* females when all other posterior structures are duplicated there (Fig. 1.16). All this evidence indicated that polar granules contained germ cell determinants, which Mahowald suggested might be the RNA, with the basic protein acting as a

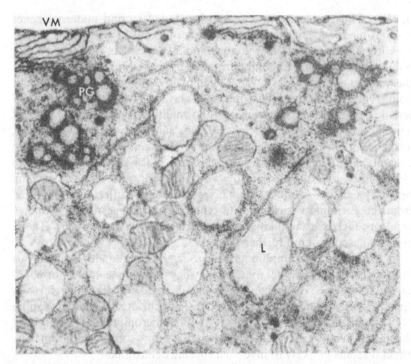

Fig. 1.15. Polar granules (PG) which have coalesced into large clusters at the posterior pole of a stage 14 *Drosophila hydei* oocyte. L, lipid droplet; VM, vitelline membrane. × 22,500. (From Mahowald, 1971*a*.)

matrix for its attachment. A problem remained, however, in that polar granules were still present in eggs of flies with *grandchildless*-type mutations, and it was rather their persistence or interaction with embryonic nuclei which was abnormal (discussed further on p. 126). By now we do know two maternal mutations which delete the polar granules, but they cause the suppression of posterior somatic structures as well as the germ cells. The loci concerned are *tudor* and *oskar* (Boswell & Mahowald, 1985; Lehmann & Nüsslein-Volhard, 1986), and their role in spatial determination has already been considered (p. 22). Their dual effects re-open the question of the link between polar granules and germ cell determination. Are these granules specific germ cell determinants whose synthesis or localisation fails where no posterior pole is designated, or are they part of the coordinate system concerned in the determination of both soma and germ line?

   The eggs of many other species in a very wide range of animal groups contain distinctive cytoplasmic structures which are segregated to the germ line during development (see Table 1 in the review by Eddy, 1975). In the anuran amphibians where electron-dense granules are seen near

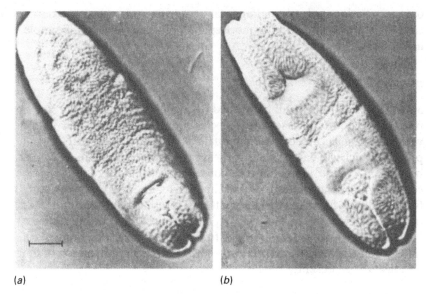

(a)                                    (b)

Fig. 1.16. Dorsal views of wild-type (*a*) and *bicaudal* (*b*) gastrulae viewed by Nomarski optics. Posterior midgut invaginates at both poles of the mutant, but only the posterior invagination contains pole cells. Scale bar = 50 $\mu$m. (From Nüsslein-Volhard, 1979.)

the vegetal pole, there is, as in *Drosophila*, circumstantial evidence for their role in germ cell determination (for review see Smith & Williams, 1979). Loss, or ultra-violet induced damage, to the vegetal cytoplasm causes reduced fertility or complete sterility (Bounoure, 1937; Buehr & Blackler, 1970), and the injection of undamaged vegetal cytoplasm can increase fertility (Smith, 1966). Smith also found that 254 nm was the most effective uv wavelength for sterilisation, suggesting that the determinant may contain a nucleic acid. Recently, however, Wylie *et al.* (1985*b*) demonstrated that embryonic cells containing these granules are not committed to germ cell formation: they can form a variety of other cell types if transplanted to new positions. Thus once again the evidence for specific germ cell determinants is now in doubt (and see p. 58). Studies of germ-line granules in nematodes are considered on p. 116.

As we have seen, genetic evidence makes it very unlikely that some oocytes bear localised determinants for specific somatic tissues. However, there are many groups for which no genetic studies are available, and in some cases other lines of evidence reviewed elsewhere in this book do indicate the existence of tissue-specific determinants. The strongest case can be made for the ascidians (see especially from p. 100) but even there the results of egg bisections (p. 52) suggest little spatial organisation of these materials. In general, it seems that oocytes have only very basic concentric and/or polar organisation. In other words, the embryo is provided with little more spatial information than the simple instruction: 'This side up'.

## Conclusions

Many materials which will be needed by the embryo are already stock-piled in the oocyte. These usually include massive amounts of yolk, of various organelles and of informational RNAs; and several mRNAs and proteins which will have specific roles in early development are known to be present. These materials are, however, distributed in very simple concentric or polar patterns within the oocyte, and there is as yet very little evidence for more specific patterns foreshadowing the different organs and tissues of the embryo. Experimental and genetic interference have also failed to produce evidence for local embryonic determinants in the oocyte, and suggest instead that only a general determination of the various axes is present.

# 2

## From oocyte to zygote

---

Several important events separate the oocyte from the zygote that is genetically the starting point of the new individual. Over the same period, ooplasmic flows commonly occur with great effects on spatial arrangements and often on developmental localisations.

### The events

Briefly, the events of this period are: 1) the completion of the maturation divisions, 2) various kinds of cytoplasmic maturation, 3) release of the oocyte from the ovary (ovulation) and 4) fertilisation. The sequence of these processes varies among different animal groups and they frequently overlap in time. They are of interest here only where they may affect spatial organisation, and so will be reviewed extremely briefly. Reference is made to reviews from which further details may be obtained.

Maturation is usually initiated hormonally, and involves widespread physiological changes (for reviews see Masui & Clarke, 1979; Meijer & Guerrier, 1984; Maller, 1985; Sardet & Chang, 1987). Rates of many activities, such as respiration and protein synthesis, usually increase, and there are often qualitative changes in the species of protein synthesised. Meiosis resumes, starting with germinal vesicle breakdown (GVBD), but often arrests again at a later stage until fertilisation. The meiotic events are accompanied by cyclic changes in the organisation of cortical microtubules, and the consistency of the general cytoplasm also seems to change. The annulate lamellae vesiculate and disperse, and shells of endoplasmic reticulum form around the cortical granules making close junctions with the surface in an apparent preparation for fertilisation. The cytoplasm also acquires the ability to induce maturation in other oocytes to which it is injected. This is ascribed to the production of a maturation promoting factor (MPF) which is apparently widespread in meiotically and mitotically dividing cells.

Most of the cytoplasmic events occur independently of GVBD or even in oocytes enucleated before the start of maturation. Such oocytes can often be fertilised but the sperm nucleus does not then synthesise DNA

31

and cleavage is usually impossible. The addition of GV material therefore has far-reaching effects, one of the added factors being the $o^+$ substance in axolotls (see p. 27). Materials added externally in the female's reproductive tract may be essential for fertilisation (jelly) or development (e.g. the albumen in birds).

At fertilisation the sperm contributes its chromosomes and centrioles, but its other organelles seem to have a limited life and significance. The events of sperm entry and zygote formation have been reviewed by Summers *et al.* (1975) and Schatten (1982). The response to fertilisation includes many changes at the egg's surface (see Vacquier, 1981; Sardet & Chang, 1987), including the discharge of the cortical granules which contribute to the fertilisation membrane, and in sea urchins to the hyaline layer over the plasma membrane. This is followed by an intense endocytotic activity, but we know little of how this enormous turnover affects the plasma membrane's properties. A metabolic activation similar to that seen at maturation in other groups occurs at fertilisation in sea urchins (although there is no qualitative change in the proteins being formed), and at ovulation or oviposition in other cases.

The sperm's nuclear membrane is rapidly remodelled after its incorporation, converting it to a male pronucleus. The chromosomes become more dispersed and in sea urchins histones characteristic of the sperm are replaced by those characteristic of cleaving embryos (Poccia, Salik & Krystal, 1981). An aster formed from the sperm centrioles appears to be responsible for the movements of both pronuclei leading to syngamy, and descendants of those centrioles will direct the formation of the mitotic apparatus through cleavage and further development.

## Ooplasmic segregation

One reason for thinking that the events of maturation, ovulation and fertilisation may have important consequences for spatial determination is that they are often accompanied or followed by visible movements and segregation within the cytoplasm (Costello, 1948). These commonly increase the polar differences in the egg, and sometimes produce clear bilateral differences too. Also over this period, though not always in close correlation with ooplasmic segregation, many eggs show great distortions in shape suggesting the action of a contractile system.

### (a) Bipolar differentiation

Simple examples of bipolar differentiation are seen in some large and yolky eggs such as those of cephalopods and fish. Here there is only a thin layer of cytoplasm around a large mass of yolk, and the germinal vesicle lies in a small cytoplasmic area at the animal pole, beneath a micropyle

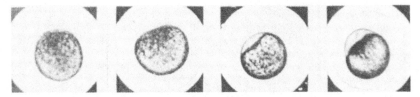

Fig. 2.1. Photomicrographs of living zebrafish eggs taken from a film. Clear plasm is seen to accumulate at the animal pole and streams joining this area are visible. (From Roosen-Runge, 1938.)

which allows sperm entry only at this point. Following fertilisation extra cytoplasm streams to the animal pole (or to points which have been activated with a needle, even in vegetal half eggs, see Sakai, 1964). In cephalopods and in some fish much of this cytoplasm comes from the interior of the egg among the yolk platelets, cytoplasmic channels allowing its flow (Roosen-Runge, 1938; Arnold, 1971; see Fig. 2.1). A counter-streaming towards the vegetal pole is also seen in fish (Roosen-Runge, 1938) and lipid globules move towards the vegetal pole (Sakai, 1964). Other teleosts have no internal cytoplasm, so the flow towards the animal pole must come entirely from the peripheral layers (Devillers, 1961). Vital staining of both cephalopod and teleost eggs after polar body formation demonstrates that the animal pole plasm has a markedly higher pH than the rest of the egg (Spek, 1933, 1934). This is probably just a confirmation of the segregation of cytoplasm from yolk (see Raven, 1938). In an acraniate, the lamprey, several layers distinguishable by their fine-structure (Fig. 2.2) have segregated at the animal pole within two minutes of fertilisation (Nicander, Afzelius & Sjödén, 1968).

The segregation of an animal pole plasm is not restricted to the groups with very large yolky eggs. Spek (1930, 1934) also saw the accumulation of alkaline material at the animal pole of polychaete and mollusc eggs which had completed their maturation divisions. In the ctenophore *Beroë* the ectoplasmic layer is recognisable in dark-field illumination because it fluoresces green (Spek, 1926), and it is also greatly enriched in mitochondria (Reverberi, 1957). Following fertilisation the material accumulates at the sperm entry point (Carré & Sardet, 1984), which in normal conditions coincides with the site of polar body formation (Freeman, 1977). Rings of contractile activity pass over the maturing egg of the barnacle *Pollicipes* which is changed from a spherical to an elongate form and finishes with the yolk segregated towards the vegetal pole (Lewis, Chia & Schroeder, 1973). Even in the almost yolkless mammalian egg an animal pole area acquires distinct surface properties during maturation (Nicosia, Wolf & Inoue, 1977), the egg having previously shown apparently complete concentric symmetry. The changes occur in this case as the meiotic spindle reaches the cortex (Longo & Chen, 1985), and have

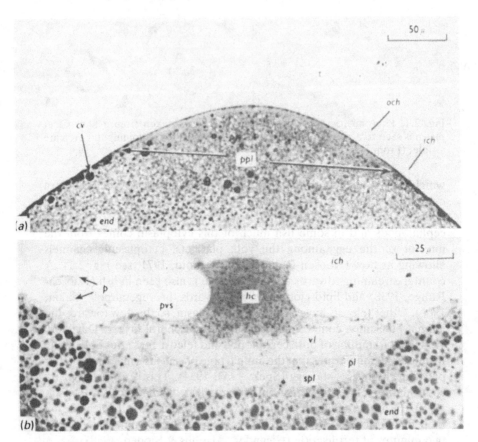

Fig. 2.2. Ooplasmic segregation at the animal pole of *Lampetra* eggs stained with toluidene blue. (*a*) unfertilised: there is a small pole plasm (*ppl*) lacking cortical vesicles (*cv*). (*b*) 2 min after insemination: 'hyaline' (*hc*), vesicular (*vl*), finely granular (*pl*) and 'spongy' (*spl*) layers can be recognised above the endoplasm (*end*). *ich*, inner chorion; *och*, outer chorion; *p*, peripheral projections; *pvs*, perivitelline space; *t*, tuft. (From Nicander *et al.*, 1968.)

parallels in a hydrozoan (Freeman, 1987) and in immature starfish oocytes (see p. 10), although they do not determine embryonic polarity in the latter case (see p. 18).

In some other animal groups both poles of the maturing egg accumulate distinctive plasms. These appear in leeches and oligochaetes following externally visible constriction movements (Schleip, 1914; Penners, 1922*a*; see Fig. 2.3). In *Tubifex*, Shimizu (1982*a*) thinks that segregation occurs in two stages – movement of yolk-free cytoplasm from the egg interior to the periphery, followed by movement of this subcortical layer towards the two poles forming accumulations of membraneous organelles there (see also Shimizu, 1986). The deformation movements

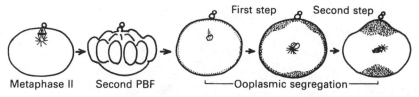

Fig. 2.3. Deformation movements and ooplasmic segregation in the *Tubifex* egg from second meiotic metaphase to first cleavage metaphase. Meridional sections except at second polar body formation (second PBF). The segregation of ooplasms (dotted) occurs in two stages as described in the text. (From Shimizu, 1984.)

occur before the first of these phases. In nematodes, constriction movements including a pseudocleavage accompany the accumulation of ectoplasm at the animal pole (Tadano, 1962) and of distinctive P granules at the vegetal pole (Strome & Wood, 1983; see also Yamaguchi *et al.*, 1983). In the polychaete *Chaetopterus*, the ooplasmic flows include the movement of germinal vesicle plasm from the animal pole to the central area and the splitting of the ectoplasm to leave a plasm beneath each pole (Eckberg, 1981). The ectoplasm is enriched for certain organelles and for mRNAs, which may all be associated with the cytoskeleton (Jeffery, 1985*a*).

During the maturation divisions of many polychaete (including *Chaetopterus*) and molluscan eggs the vegetal pole material is protruded as a quite discrete polar lobe. This material is later segregated to specific embryonic lineages (see Chapter 3), and its removal causes profound disturbances in development (see p. 51). Not surprisingly, considerable efforts have been made to recognise materials which are localised specifically to the vegetal plasm and polar lobe of such eggs. In some cases they have failed: no ultrastructural differences being found among the parts of the *Barnea* egg (Pasteels & de Harven, 1963), or between the polar lobe and the rest of the egg in *Mytilus* (Humphreys, 1964). In *Ilyanassa* (Crowell, 1964; Pucci-Minafra, Minafra & Collier, 1969) and *Dentalium* (Reverberi, 1970) double membrane vesicles with a granular content appear to be at least far more concentrated in the polar lobe than elsewhere. *Bithynia* presents a clear qualitative difference: we have already seen (p. 12) that the oocyte has an RNA-containing vegetal body, and this enters the polar lobe (Dohmen & Verdonk, 1974). In fact it occupies much of the polar lobe (Fig. 2.4), which in this species is particularly small (less than 1% of the egg volume). Electron microscope studies show that the body is composed primarily of small vesicles with a considerable number of multivesicular bodies. Dohmen & Verdonk (1974) consider the RNA in the body to be 'an obvious candidate for the role of morphogenetic factor'. Other species with small polar lobes have

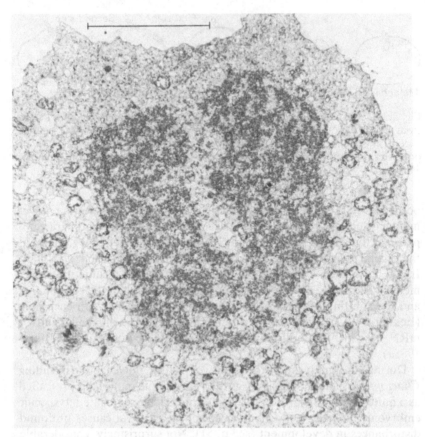

Fig. 2.4. Section through the *Bithynia* first polar lobe seen in the electron microscope. The cup-shaped mass of small vesicles in its centre is the vegetal body. Scale bar = 5 µm. (From Dohmen & Verdonk, 1974.)

since been investigated and found to contain similar organelles (Dohmen & Verdonk, 1979a), while the lobe of *Crepidula* also contains granules similar to the polar granules of *Drosophila* (Dohmen & Lok, 1975). It is therefore possible that vesicles of this sort are present in species with larger polar lobes, but have been missed by previous workers because they are dispersed in such a large volume.

The plasma membrane around the polar lobe may also have distinctive properties, indicated by the local accumulation of apparently symbiotic bacteria there in *Dentalium* (Geilenkirchen *et al.*, 1971), local folding of the surface in several molluscan species (Dohmen & van der Mey, 1977: see Fig. 2.5) and in the polychaete *Sabellaria* (Speksnijder & Dohmen, 1983), locally increased excitability of the membrane in *Dentalium* (Jaffe & Guerrier, 1981), and increased numbers of intramembraneous

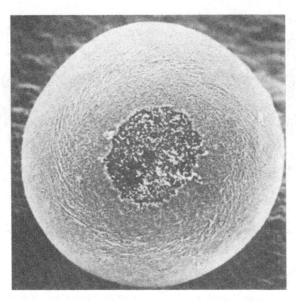

Fig. 2.5. Scanning electron micrograph of the egg of the gastropod *Nassarius* at the time of first meiotic division. In the centre is a distinct vegetal pole patch and further out is a spiral array of folds formed as the first polar lobe constricts. × 410. (From Dohmen & van der Mey, 1977.)

particles (Speksnijder *et al.*, 1985). The vegetal body of *Bithynia* may also be attached beneath the plasma membrane (see p. 55). In 1942 Dan & Dan reported a major redistribution of the surface during the second maturation division of *Ilyanassa*, amounting to a shrinkage at the animal pole and a stretching over the polar lobe. It may be during this process that the lobe membrane acquires its distinctive properties.

Cytochalasin B is able to suppress polar lobe formation (Raff, 1972) and egg deformation in *Pollicipes* (Lewis *et al.*, 1973), *Tubifex* (Shimizu, 1978*b*) and the nematode *Caenorhabditis* (Strome & Wood, 1983). It can also prevent the segregation of ooplasms in *Caenorhabditis* (Strome & Wood, 1983), the squid (Arnold & Williams-Arnold, 1974) and the zebrafish (Katow, 1983), the polar phase of segregation in *Tubifex* (Shimizu, 1982*a*) and membrane polarisation in the mouse (Longo & Chen, 1985). As these effects suggest, microfilaments are present in the constriction which separates the polar lobe (Conrad *et al.*, 1973; Schmidt *et al.*, 1980) and in the cortex of many eggs including that of the zebrafish (Katow, 1983). This cortical network contracts towards the animal pole in the mouse (Maro *et al.*, 1984; Karasiewicz & Soltyńska, 1985) and *Caenorhabditis* (Strome, 1986*a,b*). In *Tubifex* there are microfilaments cortically and subcortically, of which the latter may effect the radial phase of ooplasmic segregation while the former leads in the polar phase,

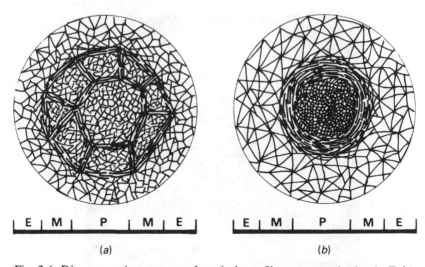

Fig. 2.6. Diagrammatic summary of cortical myofilament organisation in *Tubifex* eggs during the first (*a*) and second (*b*) stages of ooplasmic segregation. Polar (P), mid (M) and equatorial (E) zones of the cortex are indicated. (From Shimizu, 1984.)

being disrupted equatorially and contracting towards both poles (Shimizu, 1984; see Fig. 2.6). Calcium ions may be involved in these activities, although in *Tubifex* it is the earlier furrowing movements which are inducible by the calcium ionophore A23187 (Shimizu, 1978a). In fish eggs, a wave of calcium release spreads from the animal pole following sperm entry there (Gilkey *et al.*, 1978).

Ooplasmic segregation is, however, more strongly inhibited by colchicine than by cytochalasin B in *Pollicipes* (Lewis *et al.*, 1973) and *Chaetopterus* (Eckberg & Kang, 1981), suggesting that microtubules may be the important cytoskeletal element in these cases (although not in the deformation movements of *Pollicipes*). Resorption of the polar lobe is also inhibited by colchicine (Raff, 1972). Although it destroys the meiotic spindle of the mouse egg, colchicine cannot prevent the movement of the chromosomes to the cortex nor the modification of the local surface (Longo & Chen, 1985). It is the presence of the chromosomes themselves which promotes the surface changes and the thickening of the cortical microfilament band (Maro *et al.*, 1986; van Blerkom & Bell, 1986).

### (b) Lymnaea

A more complex pattern of ooplasmic segregation is seen in the maturing *Lymnaea* egg (Raven, 1948, 1963a). Altogether there are plasms at both poles and six other subcortical patches which show differential staining.

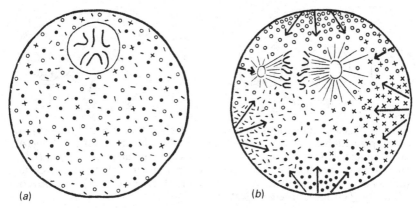

Fig. 2.7. Diagram to illustrate Raven's concept of the cortical map. (*a*) a fertilised egg with the cytoplasmic constituents randomly arranged. (*b*) the segregation of these constituents under the control of the cortex (solid arrows) which also controls the localisation of the nuclei and cleavage spindle (broken arrows). (From Raven, 1963*b*.)

All except the animal pole plasm appear during the passage of the egg through the reproductive tract, at a time of extensive amoeboid movements (see Hess, 1971). They have already been illustrated in Figure 1.8. Later the vegetal pole plasm spreads beneath the cortex and the animal pole plasm appears. Their further movements during cleavage are considered on p. 90. The animal pole plasm includes an accumulation of mitochondria which Raven & van der Wal (1964) suggest may be transported there by the action of the maturation spindle. There is a dense cytoplasmic matrix in the six subcortical accumulations, including some possibly specific kinds of small organelles (Raven, 1967).

Raven (1963*b*, 1966) sees the accumulation of these plasms as evidence of a cortical 'map' each part of which acts to selectively accumulate particular egg constituents beneath itself (Fig. 2.7). As already explained (p. 16) there is evidence that the map is 'imprinted' during oogenesis by effects of the follicle cells and the oocyte's position in the ovary. However, the nature of these apparent effects is still unknown, and we will find (p. 57) that there are still great difficulties for the whole concept of a cortical map.

### (c) Sea urchins

In sea urchins the only cytoplasmic markers allowing visual study of internal movements are the pigment granules of many species, and these are dispersed throughout the oocyte. However, a population of *Paracentrotus* from Villefranche-sur-Mer in France has eggs with a banded pigment pattern, and was studied by Boveri (1901*a*). Here the orange-red

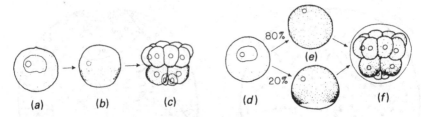

Fig. 2.8. Pigment migrations in *Paracentrotus lividus*. (*a*)–(*c*) according to Boveri, pigment has retracted to a subequatorial band in the unfertilised egg (*b*) and is almost entirely segregated to the macromeres at 16-cells (*c*). (*d*)–(*f*) Schroeder's re-investigation shows that retraction occurs only from the vegetal pole and only in a minority does it occur before cleavage. ((*a*)–(*c*) from Boveri, 1901*a*; (*d*)–(*f*) from Schroeder, 1980*a*.)

pigment seemed to have retracted, during maturation, from both poles into a subequatorial cortical band even in the unfertilised egg (Fig. 2.8(*b*)). After fertilisation, the pigment seemed to be rather exactly segregated by the early cleavages into cells of the presumptive endoderm (Fig. 2.8(*c*)). This celebrated case has been interpreted to mean that the zones separated in the egg define and perhaps determine the three germ layers (e.g. Wilson, 1925, p. 1067).

Schroeder (1980*a*) has recently reexamined this phenomenon and finds that pigment granules migrate to the cortex in all *Paracentrotus* eggs during maturation, but only 20% of the females at Villefranche-sur-Mer lay 'banded' eggs. More importantly, pigment remains in the animal half of such eggs at its previous levels, and is retracted only from the vegetal pole in a way which increases the concentration of pigment in the subequatorial band (Fig. 2.8 (*d*)–(*f*)). Similar processes occur at later stages in other sea urchin species with pigmented eggs. In *Arbacia* the pigment granules move to the cortex after fertilisation (Harvey, 1956), and the retraction of pigment from the vegetal pole occurs during the first cleavage divisions, being always complete by the fourth cleavage (Morgan, 1893; Belanger & Rustad, 1972). As Schroeder (1980*a*) points out, since pigment retraction occurs from only one pole, banding requires a physiological change only at this (vegetal) pole, and it is less likely to effect the determination of the three germ layers.

As with many examples of ooplasmic segregation already discussed, the migration of sea urchin egg pigment to the cortex and its retraction from the vegetal pole are sensitive to cytochalasin B but not to colchicine (Belanger & Rustad, 1972; Tanaka, 1981). This suggests a requirement for microfilaments and not for microtubules. A fibre system which could be responsible at least for the movement to the cortex has been described by Harris (1979). Schroeder (1980*a*) has proposed that pigment with-

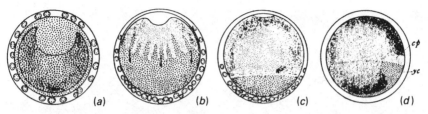

Fig. 2.9. Ooplasmic segregation in the ascidian *Styela*. *cp*, clear plasm; *yc*, yellow crescent. For details see text. (From Wilson, 1925, after Conklin.)

draws from the vegetal pole because it is embedded in a cortical matrix which is disrupted there and retracts elastically.

### (d) Ascidians

Ooplasmic segregation in ascidians converts an oocyte with almost complete concentric symmetry into a zygote with clear polar and bilateral organisation. The classical study is that of Conklin (1905a) using *Styela partita* (see Fig. 2.9). The oocyte cytoplasm of *Styela* has an outer layer recognisable by yellow pigment granules, and an inner grey yolky area. The germinal vesicle is beneath the animal pole and has clear contents which remain in place when GVBD occurs in the oviduct. According to Conklin the sperm enters near the vegetal pole and triggers ooplasmic segregation. The yellow outer material streams towards the vegetal pole followed by the clear plasm released at GVBD. This streaming occupies ten minutes. For a short time the three plasms are arranged axially in the sequence grey : clear : yellow from animal to vegetal pole. Then the clear and yellow plasms reverse their flows on one side of the egg, apparently in association with the male pronucleus as it approaches the female pronucleus. In this way they form two crescents, a clear crescent above a yellow crescent on one side of the egg axis just before first cleavage. This side will later form the posterior end of the embryo. The clear material appears to continue extending through the animal half at the expense of the grey yolky material. In two-cell embryos, approximately the animal half of the blastomeres appears clear, and there is a pale grey crescent on the future anterior side (as well as the yellow crescent posteriorly) above the deeper grey material. Conklin was also able to follow these plasms as they were distributed to different cells at cleavage, and to determine their fates in development (see p. 96). The plasms can therefore be named according to their prospective fates: the yellow crescent as myoplasm (forming muscle), the clear area ectoplasm (for ectoderm), the dark grey material endoplasm (for endoderm) and the pale grey crescent (which is least distinct and has received least further study) would be chordoplasm

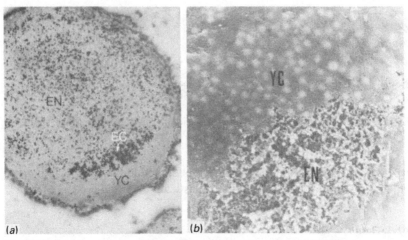

Fig. 2.10. Some components of *Styela* ooplasms. (*a*) poly(A)$^+$RNA. × 900. (*b*) scanning electron micrograph of the boundary between yellow crescent and endoplasm in a Triton-extracted egg, showing that pigment granules and surface-associated material are confined to the former. × 1400. EC, ectoplasm; EN, endoplasm; YC, yellow crescent or myoplasm. ((*a*) from Jeffery *et al.*, 1983; (*b*) from Jeffery & Meier, 1983.)

and neuroplasm (forming notochord and nervous system). Ooplasmic segregation has thus produced quite a complex bilateral organisation which relates both to the axes of the embryo and the differentiative fates of its parts. Observations of other ascidian species reveal a similar sequence of changes marked by differently coloured ooplasms (see Jeffery, 1984*a*).

The composition of the various ascidian ooplasms has been studied using cytochemistry and electron microscopy. Myoplasm of course contains the pigment granules (yellow in *Styela*), but it is also enriched to varying degrees in mitochondria (Reverberi, 1956), endoplasmic reticulum and ribosomes (Mancuso, 1964), and lipid droplets (Berg & Humphreys, 1960). The grey crescent seems relatively deficient in the larger organelles (Berg & Humphreys, 1960). The endoplasm has most protein yolk.

Jeffery is now making a more detailed molecular analysis of the plasms, using *in situ* hybridisation techniques for RNAs, two-dimensional electrophoresis for proteins and detergent-extracted eggs to examine the cytoskeleton. The high concentration of poly(A)$^+$RNAs in the ascidian germinal vesicle (see p. 11) remains with the clear ectoplasm throughout segregation (Jeffery & Capco, 1978, see Fig. 2.10(*a*)), although this is rather less than half of the egg's total poly(A)$^+$RNA (Jeffery *et al.*, 1983). Histone mRNAs distribute uniformly through the egg, while actin mRNA is always most concentrated in the myoplasm and rather less so in

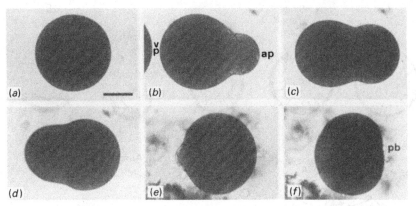

Fig. 2.11. Changes in egg shape during the 1st stage of ooplasmic segregation in *Ciona*. (*a*) unfertilised egg; scale bar = 50 $\mu$m. (*b*)–(*e*) during the 3 min after insemination a wave of constriction passes from the animal to the vegetal pole. (*f*) the first polar body (pb) is formed. (From Sawada & Osanai, 1981.)

the ectoplasm (Jeffery *et al.*, 1983). The actin mRNA seems not to code for muscle actin, so its significance as a preparation for myoplasmic fate (or even as a determinant of that fate) remains unclear. No mRNAs have been shown to be completely restricted to the myoplasm, but 15 polypeptides (some of them cytoskeletal) do appear to be so restricted, and many more are enriched there (Jeffery, 1985*b*). Most of the new mRNAs retain their normal distributions in detergent-extracted eggs, suggesting an association with the cytoskeleton, but they are not detached by colchicine or cytochalasin B (Jeffery, 1984*b*). In fact the characteristics of the internal cytoskeleton suggest that it is largely composed of intermediate filaments (Jeffery & Meier, 1983).

During the first phase of ooplasmic segregation a constriction travels along the animal–vegetal axis of the egg leaving a group of microvilli around the vegetal pole (Sawada & Osanai, 1981; see Fig. 2.11). Particles placed on the egg surface (Ortolani, 1955) and lectins bound to it (Ortolani, O'Dell & Monroy, 1977) also move towards the vegetal pole, and Sawada & Osanai (1981) could show that such particles are collected by the constriction as it passes across the egg. Test cells, which are the innermost cells of the complex ascidian follicle and are embedded in the oocyte surface (Tucker, 1942), also accumulate vegetally (Conklin, 1905*a*). At the end of this phase the surface of the animal hemisphere is particularly fragile as though depleted of some components (Sawada, 1983), and there are indeed no surface-associated materials to be seen outside the myoplasmic area of detergent-extracted eggs (Jeffery & Meier, 1983; see Fig. 2.10(*b*)). A co-ordinated vegetal movement of the myoplasm, the cortical cytoskeleton, the plasmalemma and all externally attached materials therefore appears to have taken place.

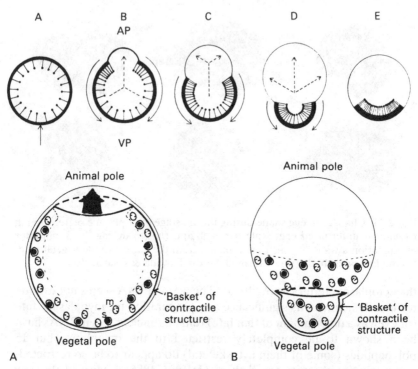

Fig. 2.12. Interpretations of the first stage of ooplasmic segregation in ascidians according to Jeffery & Meier, 1983 (above) and Sawada & Osanai, 1984 (below). See text. These groups disagree over the nature and relationships of the peripheral structures and their completeness before segregation.

It now seems certain that the active component in this movement is the cortical cytoskeleton, which contains actin (Sawada & Osanai, 1985). Actin staining is restricted to areas vegetal of the surface constriction (Sawada, 1983) and cytochalasin B prevents segregation (Zalokar, 1974). Myoplasmic and surface-associated components presumably move because they are attached to the cortical cytoskeleton, while endoplasm is squeezed towards the animal pole. The models for this process provided by Jeffery & Meier (1983) and Sawada & Osanai (1984) are reproduced as Figure 2.12. The contraction may be directed by a release of calcium ions normally starting at the sperm entry point (see also p. 38): in gradients of the calcium–ionophore A23187 myoplasm usually collects on the side where the concentration is highest (Jeffery, 1982). In fact, eggs placed between two A23187 sources can develop two myoplasmic crescents, suggesting that any point can act as a vegetal pole, and it would seem advisable, even in normal eggs, to check whether polarity is determined by sperm entry point rather than the other way round.

There has been very little study of the second stage of segregation when

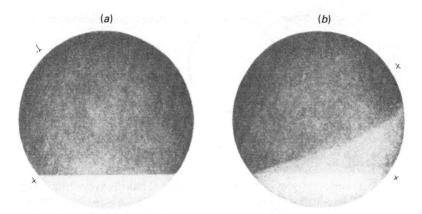

Fig. 2.13. The egg of *Rana fusca* before and after formation of the grey crescent, seen here with its mid-point on the right. (From Schleip, 1929.)

bilateral organisation becomes obvious. Conklin (1905*a*) suggested that this plane might be predetermined in the unfertilised egg. The apparent association with the movements of the male pronucleus suggests a possible role for the sperm aster. However, according to Jeffery (1982) some at least of the second stage flows occur in eggs activated by A23187 gradients, when there would of course be no sperm aster.

### (e) Amphibians

The maturation of amphibian oocytes is accompanied by movements which extensively remodel the ooplasm but have received very little attention (see Hausen *et al.*, 1985). In contrast a great deal of study has been devoted to events in the fertilised egg which usually culminate in the formation of the grey crescent. This appears between the pigmented animal and unpigmented vegetal areas some 45 minutes (*Xenopus*: Paleček, Ubbels & Rzehak, 1978) to 4–6 hours (axolotl: Benford & Namenwirth, 1974) after fertilisation (Fig. 2.13), although it is not visible in all species. It normally appears in anurans on the side opposite that penetrated by the fertilising sperm (see Roux, 1903), implying that fertilisation has a causal role in its appearance. Newport (1854) had already established a relationship between the sperm entry point and the orientation of the embryo, and it soon became clear that the grey crescent is an early marker of the future dorsal side. In the polyspermic urodele egg, a grey crescent may often still appear, and still marks the dorsal side, although this is probably unrelated to sperm entry pathways. In other species the whole pigmented area seems to move to the ventral side (Malacinski, 1984).

The first cell cycle in amphibians (and indeed many other embryos) is a

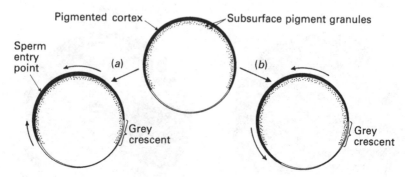

Fig. 2.14. Theories of grey crescent formation. A radially-symmetrical egg (centre) could acquire a grey crescent on the future dorsal side (*a*) by a contraction of the cortical layer towards the sperm entry point, or (*b*) by a rotation of this layer around the endoplasm.

particularly long one, and useful background information can be gained by listing the sequence of activities through this cycle. Studies mainly using cinematography of the egg reveal the following externally visible changes: 1) An 'activation wave' travels at 0.6 mm per minute away from the point of sperm entry and is probably associated with cortical granule breakdown and the raising of the fertilisation membrane (Hara & Tydeman, 1979; Takeichi & Kubota, 1984). 2) A few minutes later, a contraction towards the animal pole reduces the pigmented area of the anuran egg and is slowly reversed (Ancel & Vintemberger, 1948; Elinson, 1975). Elinson suggests that it may bring the pronuclei closer together, and Paleček *et al.* (1978) that it may further widen the perivitelline space so assisting free rotation of the egg. 3) Two further waves, possibly of contractile activity, spread again from the sperm entry point of anurans, but at only 0.06 mm per minute: the second occurs at about the time of grey crescent formation (Hara, Tydeman & Hengst, 1977). 4) Finally, waves of relaxation and contraction travel from the animal to the vegetal pole a little before, and probably as a preparation for, first cleavage (Hara, 1971; Sawai, 1982); and so are probably of less significance here.

The formation of the grey crescent obviously involves movements of the pigment granules. These are mostly in the cortical and subcortical layers of the egg, but after the second stage above, much has left the cortex for deeper locations (Paleček *et al.*, 1978). (The sperm entry point retains its cortical pigment and so its dark colour (Kubota, 1967), and later develops a clump of elongated microvilli (Elinson & Manes, 1978).) According to Merriam, Sauterer & Christensen (1983) the thickness of the subcortical layer varies from 3–7 μm at the animal pole to effectively nothing at the vegetal pole. Two theories have been put forward to explain the pigment movements (see Fig. 2.14(*a*) and (*b*)).

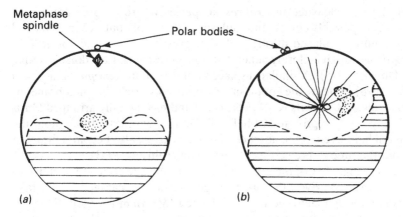

Fig. 2.15. The internal organisation of the amphibian egg (*a*) before fertilisation and (*b*) after symmetrisation. The vegetal mass of large yolk platelets is shown hatched and an area of yolk-poor cytoplasm is dotted. The pronuclei are at the centre of the sperm aster.

One is that the various surface layers of the animal half contract towards the point of sperm entry, stretching those layers on the dorsal side and raising the margin of pigment there (Bánki, 1929; Løvtrup, 1958). The other is that these surface layers of the whole egg rotate on the inner cytoplasm, raising the margin of pigment dorsally and lowering it ventrally (Ancel & Vintemberger, 1948; Elinson & Manes, 1978). Vincent, Oster & Gerhart (1986) have recently re-examined this problem by marking both the surface and subcortical cytoplasm of *Xenopus* eggs. They found evidence for both a contraction and a 30° rotation of the cortex, but considered that it was the latter which specified the bilaterality of the embryo. A thickening of the surface layers ventrally relative to those dorsally should result from both kinds of movement, and has indeed been inferred from the thickness of the layer excluding yolk platelets (Dalcq & Dollander, 1948; Hebard & Herold, 1967) and the rates at which dyes penetrate these layers (Dollander & Melnotte, 1952).

Symmetrisation also has deeper effects (Fig. 2.15). A 'vitelline wall' of large yolk platelets is seen on the dorsal side, having apparently been pulled with the overlying surface layers from more vegetal regions (Pasteels, 1964). An area of cytoplasm at the sperm entry point shows increasing rigidity which gradually extends over the animal hemisphere, apparently in association with the formation and growth of the sperm aster (Kubota, 1967; Elinson & Manes, 1978). This aster also directs the movements of the pronuclei each marked by a trail of pigment originating beneath the surface (Manes & Barbieri, 1977). Even the deepest materials in the animal half appear to be displaced: Brachet (1940) found that nuclear materials released there at maturation were localised on one side by first cleavage in *Triturus*, a species where no grey crescent is visible. A possibly equivalent ooplasm, poor in yolk but rich in RNA and glycogen is displaced during and after grey crescent formation in *Discoglossus* until it is closely beneath the grey crescent cortex dorsally (Klag & Ubbels, 1975). A similar displacement is seen in *Xenopus* (Paleček *et al.*, 1978) where the ooplasm is especially rich in membraneous organelles (Herkovits & Ubbels, 1979) and has sagitally arranged fibrils of a myosin-like protein (Ubbels, 1977). The cytoplasm throughout the animal half, even in the absence of a sperm aster, also becomes firmer at about the time of grey crescent formation (Elinson, 1983).

Surprisingly few of the events of the first cleavage cycle are sensitive to cytochalasin B. Even the precleavage contraction can apparently take place, although cleavage itself is blocked (Merriam *et al.*, 1983). Merriam's group think that actin-containing filaments are in the thin cortical layer only (see too Franke *et al.*, 1976), and that their roles include sperm engulfment and cleavage. In particular, cytochalasin B cannot prevent grey crescent formation, while colchicine can (Manes, Elinson & Barbieri, 1978). Colchicine or vinblastine also prevent the stiffening effect which spreads from the sperm entry point (Elinson & Manes, 1978), the dorsal movement of yolk-poor cytoplasm (Ubbels *et al.*, 1983) and the general cytoplasmic gelation occurring as the grey crescent forms (Elinson, 1983). For some of these events circumstantial evidence implicating the sperm aster has already been noted above, but other microtubules may also be involved since enucleated eggs with no sperm aster can still form a grey crescent (Manes *et al.*, 1978; see also Elinson, 1985). It is likely, however, that the aster at least orients grey crescent formation, since this always occurs on the opposite side of unfertilised eggs following subcortical injection of homogenised sperm (Manes & Barbieri, 1976). Another way in which grey crescent formation could be effected is by the action of a contractile system within the subcortical layer. Merriam & Sauterer (1983) have given evidence that this layer has its own contractile properties which are not sensitive to cytochalasin B. Such activities could be due to intermediate filaments, and cytokeratin is in fact found in the subsurface layers after maturation (Godsave *et al.*, 1984*b*) and soon becomes organised into a network in the vegetal half (Klymkowsky, Maynell & Polson, 1987).

There has been surprisingly little biochemical study of the effects of symmetrisation. No surface differences in the grey crescent region have been revealed by lectin binding (O'Dell *et al.*, 1974; Brachet, 1977). Some differences in the protein species of the grey crescent cortex and the animal pole cortex were reported by Tomkins & Rodman (1971), but it would be more interesting to compare dorsal with ventral cortex. Only Phillips (1982) has attempted to study the dorsoventral distribution of RNAs (in fact mainly using four-cell embryos). After cutting embryos to six sections along this axis, he found a raised concentration of total RNA in the most dorsal section, and raised concentrations for poly(A)$^+$RNAs in both the most dorsal and most ventral sections. This suggests that the most ventral section has more poly(A)$^+$RNA per ribosome than any other section, but the implications of this are unclear. In any case the differences are less marked than those along the animal–vegetal axis which are already seen in oocytes (see p. 11) and remain of a similar magnitude after fertilisation (Phillips, 1982). Animal–vegetal RNA concentration gradients are also seen, in a less simple form, if calculated on a basis of the non-yolk volume.

Some more specific differences along the animal–vegetal axis may be noted here although they are unlikely to arise during ooplasmic segregation. The great majority of maternal mRNAs seem to be rather evenly distributed along the egg axis with only a few species being enriched towards one of the poles (Carpenter & Klein, 1982; Rebagliati *et al.*, 1985). It is therefore surprising that poly(A)$^+$RNAs from either pole of an unfertilised egg, when injected to a fertilised egg, quickly distribute preferentially to the half where they originated, a result which also indicates that they can recognise specific cytoplasmic sites (Capco & Jeffery, 1981). Several proteins differ in concentration along the animal–vegetal axis (Moen & Namenwirth, 1977; Jäckle & Eagleson, 1980; Smith, 1986; Smith, Neff & Malacinski, 1986) but few are strictly localised to specific areas (Jäckle & Eagleson, 1980). Of the proteins localised in the germinal vesicle of oocytes (see p. 12) two seem to pass exclusively to the vegetal half and one to the animal half in the zygote (Jäckle & Eagleson, 1980). A sharp difference between surface membrane mobilities in the animal and vegetal halves is established at fertilisation chiefly by immobilisation of membrane lipids in the animal half (Dictus *et al.*, 1984).

## The extent of spatial determination

The visible movement of ooplasms suggests that developmentally important materials may be increasingly localised when many eggs mature or are fertilised. However, as already noted for the oocyte, other experimental procedures are needed to test whether various parts of eggs do indeed become determined at this time, and whether such developmental effects correlate with the visible flows of materials. This section reviews the evidence from such procedures.

### (a) Work with egg fragments

We have already seen (p. 18) that most nucleate oocyte fragments of the nemertine *Cerebratulus* develop normally. This remains true of most

fragments made after GVBD (Wilson, 1903), but thereafter the regulative ability decreases. About half the nucleate fragments separated at first
meiotic metaphase developed normally, but most fragments produced at
pronuclear fusion were abnormal (Yatsu, 1904). The increasing failure to
regulate is not simply due to lack of time as Yatsu (1910) showed by
delaying fertilisation for varying lengths of time. The abnormalities
produced by later operations were often deficiencies of particular larval
organs, and where each fragment contained a pronucleus they often
formed complementary parts, suggesting that determinants are indeed
localised over the period of maturation and fertilisation. A small area
around the animal pole did, however, seem dispensable, and Yatsu
(1910) showed that animal and lateral fragments were dispensable even at
first cleavage, while removal of even a small vegetal area caused the
deletion of the gut and disturbances of the ciliated lobes.

Freeman (1978) has re-investigated this problem and finds that even
the ability to form apical tuft (an animal pole derivative) is greatest in
vegetal fragments during maturation. By the time maturation is completed most animal halves can also form apical tufts, and Freeman has
shown that this change can be inhibited by removing the meiotic asters,
or produced precociously by inducing asters. This suggests that
microtubules have a role in moving putative determinant factors, a role
which probably continues using the mitotic apparatus during cleavage
until all tuft-forming potential is lost from the vegetal cells and gut-
forming potential from the animal cells (see too p. 77). These changes
are not inhibited by cytochalasin B, so microfilaments seem relatively
unimportant (Freeman, 1978). Fragmentation work with *Cerebratulus*
therefore supports the view that developmentally important materials
begin to segregate during maturation and fertilisation, but this segregation is only completed during cleavage. Even then we cannot say that
specific determinants for polar organs are localised: only that the polar
properties are sufficiently defined to allow appropriate organs to form
there during further development in isolation.

The ctenophores are another animal group used in several early
fragmentation experiments, and again Freeman (1977) has re-investigated the problem. Maturation as such appears unlikely to be significant
for spatial determination as the position of the polar bodies does not
always relate with the embryonic axes. Freeman (1977) suggests that it is
sperm entry point which decides where first cleavage will begin, and that
this in turn decides embryo polarity. He believes that the data from egg
fragmentation support this conclusion. Fragments cut as maturation is
completed do seem to show an increasing failure to regulate their
development (lit. in Freeman, 1977), but this could equally be a response
to fertilisation. Even at the start of first cleavage several authors have
shown that oral halves can develop normally, but when the losses were

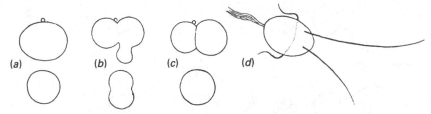

Fig. 2.16. The development of *Sabellaria* egg fragments separated horizontally during maturation (*a*). Both fragments constrict polar lobes (*b*), but only the nucleate one cleaves (*c*) and goes on to form a larva (*d*). (Redrawn from Hatt, 1932.)

from the oral region Fischel (1903) reported that larvae formed with defective comb plates. This is a potentially interesting result since the oral pole is determined by sperm entry point and we have seen (p. 33) that ectoplasm flows to this point. At later cleavage stages developmental potential certainly correlates well with the presence of ectoplasm (see pp. 110–12), but the precleavage correlation requires confirmation.

Polar lobes are particularly easy to remove from those eggs which bear them, and Wilson's (1904) classical studies with *Dentalium* laid the basis of our understanding of localisation in such cases. If the vegetal plasm (which would enter the polar lobe) was removed from an unfertilised egg, and the egg was then fertilised, Wilson found that it formed a radially symmetrical larva lacking the apical organ and all of the post-trochal region. The same effects were obtained by removing the polar lobe from first cleavage embryos, and are shown in Fig. 3.12. Wilson suggested that the vegetal pole plasm and polar lobe contain 'specific cytoplasmic stuffs' required for the apical organ and for post-trochal structures including shell gland, shell, foot, mantle folds and probably coelomoblasts. The lobeless embryo, however, still forms histologically normal ganglia, stomodaeal invagination and liver (van Dongen, 1976). Lobe removal has very similar effects in other molluscs (see Chapter 3), including *Bithynia* where the lobe is so small (Cather & Verdonk, 1974). In the polychaete *Sabellaria* it is again apical tuft and post-trochal structures which are lobe dependent (Hatt, 1932). Hatt also showed that the restriction of these potencies to the vegetal pole occurs during maturation: at the first meiotic division the vegetal one-third of the egg can be removed and the remaining animal fragment can still form a lobe and can develop to a larva with the lobe-dependent structures (Fig. 2.16). Eggs treated through the maturation divisions with cytochalasin B often show abnormalities of the lobe-dependent structures (Peaucellier, Guerrier & Bergerard, 1974).

The vegetal pole of these eggs seems to be acting as a determination centre, to which (as in *Cerebratulus*) even the potential to form apical tuft is restricted. When large amounts of material are removed from other

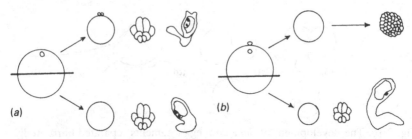

Fig. 2.17. The development of ascidian egg fragments cut before and after fertilisation. (*a*) an unfertilised *A. malaca* egg was cut in half equatorially and the halves fertilised: each produced a complete tadpole. (*b*) a fertilised egg, after 1st polar body formation, was cut subequatorially leaving a pronucleus in each fragment: the smaller vegetal fragment formed a tadpole but the larger animal one a radially symmetrical blastula at best. (Redrawn from Ortolani, 1958.)

parts of the egg, Wilson (1904) and later authors found that the vegetal fragments regulate the size of their polar lobes and form normal larvae. It is, however, possible to separate the developmental potential to form apical tuft from that for post-trochal organs, by removing only a part of the vegetal pole plasm from *Dentalium* eggs. Up to 70% of the plasm can be removed before apical tufts fail to appear, but post-trochal structures were reduced in size after smaller losses (Geilenkirchen, Verdonk & Timmermans, 1970; Verdonk, Geilenkirchen & Timmermans, 1971; and see the related data in Render & Guerrier, 1984). These authors suggest that determinants for the tuft are in the upper part of the lobe and above those for post-trochal structures. However, in Wilson's (1904) original description of lobeless development he noted that pretrochal structures are enlarged at the expense of post-trochal ones, and that with partial lobe removals the size of the latter was in 'direct quantitative relation' with the amount of lobe material remaining (Wilson, p. 30). Such results suggest that the vegetal plasm affects determination along the whole animal–vegetal axis (and see pp. 78–82).

Leopoldseder (1931) has investigated the developmental effects of removing the polar plasms from eggs of the leech *Clepsine*. Without the animal pole plasm, eggs did not cleave. Without the vegetal pole plasm, cleavage patterns were affected and mesodermal structures failed to differentiate. Work with leech embryos (Chapter 3) confirms that the polar plasms carry a very large amount of developmental information.

We have seen that ooplasmic segregation converts an almost concentrically symmetrical ascidian oocyte to a zygote first with polar and later with bilateral symmetry. Studies of the development of egg fragments show a close correlation with at least the first of these changes (Fig. 2.17). Any part of the unfertilised egg so long as it has at least half of the egg volume can, after fertilisation, develop to a normal larva (Reverberi, 1936; Ortolani, 1958; Reverberi & Ortolani, 1962). Smaller animal fragments cleave radially and can only develop to blastulae (Ortolani, 1958). Dalcq (1932), however, reported that animal fragments gastrulate with too large an ectoderm and vegetal fragments with too little ectoderm, and also claimed to have occasionally separated the future anterior and posterior plasms in such halves, and not in meridional halves where this might (at least later) be expected. All authors agree that meridional halves of unfertilised eggs develop normally. Fragments from fertilised eggs show far more localisation of developmental potential: animal fragments, whatever their size, cleave radially and form blastulae at best, while meridional or vegetal fragments can still develop normally (Reverberi, 1937; Ortolani, 1958). This would seem to correlate very well with the ooplasmic flows to the vegetal pole. Whether the second, bilateral, stage of ooplasmic segregation is accompanied by bilateral restrictions of developmental potential does not, however, seem to have been tested using egg fragments. Some other operative work with ascidian eggs may also be mentioned here. The great regulative ability of the unfertilised egg is confirmed by the development of harmonious giant larvae following the fusion of two eggs (Farinella-Ferruzza & Reverberi, 1969), and of larvae after the removal of quite large amounts of ooplasm (Abbate & Ortolani, 1961) or of cortex (Abbate & Ortolani, 1961; Cammarata, 1973). Fertilised eggs still regulate very well after the removal of cortical patches (see especially Cammarata, 1973), but after the loss of ooplasm they never do (Abbate & Ortolani, 1961), so that any localisation of determinants would appear to occur internally.

Nematode embryos, like ascidians, show a generally mosaic control of development, but here even in the fertilised egg there is little experimental evidence for localisations. It is not possible to cut fragments from nematode eggs, but cytoplasm with a part of the plasma membrane can be extruded through the eggshell, and normal worms develop when up to 20% of the egg volume has been lost (Laufer & von Ehrenstein, 1981). The distinction from egg fragmentations should be remembered, however, and these authors point out that determinants fixed to a lattice system might not be extruded. In any case vegetal losses of more than 25% of the egg volume do in fact abolish those features of the cleavage pattern characteristic of the vegetal embryonic lineages (Schierenberg, 1986), so that some polarisation of developmental potential has taken place. However, the case against complex patterns of localisation is also supported by zur Strassen's (1898) demonstration that *Ascaris* eggs which have been fused together can sometimes develop into giant worms.

Boveri (1901*b*) carried out fragmentation studies on the regulative sea urchin egg. He used the subequatorial pigment band on some unfertilised *Paracentrotus* eggs to orientate them while fragments were separated,

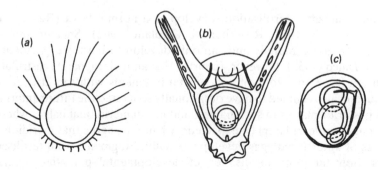

Fig. 2.18. The development of *Arbacia* egg fragments following equatorial separation and fertilisation. Animal halves formed blastulae with an enlarged apical tuft (*a*); vegetal halves gastrulated and formed plutei (*b*) or ovoid larvae (*c*). (From Hörstadius, 1937*a*.)

then each was fertilised. He found that vegetal halves could gastrulate and form all tissues but animal halves arrested as blastulae. Boveri, in effect, argued from this that the sea urchin vegetal pole acts as a determination centre: 'if the differentiation has started here, then all other regions will have their role determined by regulation emanating from that point' (translation given by Hörstadius, 1973*a*, p. 158). Hörstadius (1937*a*) confirmed these results, and contradicted other early findings, in a study using *Arbacia*: animal halves formed blastulae with an enlarged apical tuft, and vegetal halves ovoid larvae or plutei (Fig. 2.18; see also the studies of Maruyama, Nakaseko & Yagi, 1985, using *Hemicentrotus*). These are very similar results to those obtained following equatorial separations at the 8- or 16-cell stage (see p. 130), so fertilisation seems to have little effect on spatial determination. The animal half is producing less than it would within a whole embryo, while the vegetal half regulates to produce more. What is most surprising is that this result for a regulative egg is so similar to those we have seen for eggs showing more mosaic development: those of *Cerebratulus*, the polar lobe-bearing species or the fertilised ascidian egg. In fact, the only clearly different case is that of the ctenophore where a similar centre of determination could exist but at the animal (oral) pole.

It was argued earlier (p. 40) that the polarity of the sea urchin egg or early embryo is expressed in a retraction of cortical components from the vegetal pole. This has now been supported by Isaeva & Presnov (1985) who have induced a local disruption of the cortex in the *Strongylocentrotus* egg, and shown that this point usually becomes the vegetal pole of the embryo.

There have been few egg fragmentation studies designed to investigate spatial determination in vertebrate eggs. Using fertilised goldfish eggs, Tung & Tung (1944) found that some meridional halves form embryos while others form undifferentiated vesicles. Latitudinal cuts removing varying amounts of yolk showed that more than half the yolk was required before an embryo could form (Tung, Chang & Tung, 1945). Tung's group interpreted this as evidence that determinants are again localised in the vegetal half of fish eggs, but that they are to one side of the animal–vegetal axis there (see also pp. 277–8). Recently Gurdon *et al.* (1985c) cultivated fragments separated from *Xenopus* eggs and showed that the ability to transcribe the gene for muscle actin is restricted to the vegetal part of the egg and may be greater dorsally than ventrally. In other studies fertilised eggs have been fragmented without consideration of the orientation of the cut: a quarter of a newt egg can develop to a small but normal larva (Kobayakawa & Kubota, 1981) and a haploid half mouse egg at least to a blastocyst (Tarkowski & Rossant, 1976; Tarkowski, 1977). Development of all mouse haploid embryos does in fact fail at a later stage, but this is because nuclei from both parents are required (McGrath & Solter, 1984; Surani, Barton & Norris, 1984).

### (b) Centrifugation

Developmentally important ooplasms often contain a distinctive range of larger organelles. When these are displaced by centrifugation it is possible to test whether the developmental properties are due to such organelles or to other factors such as the undisplaced 'ground substance' or a 'map' in the surface membrane or cortex.

Most studies of this kind have indicated that the developmental properties of ooplasms do not depend on the larger organelles. Such organelles can be stratified yet most kinds of eggs can still go on to develop normally. Lyon (1907) demonstrated this for the sea urchin *Arbacia*, in which plutei form still showing disturbed pigment patterns. Morgan & Spooner (1909) used the position of the jelly canal to prove that this embryo developed with little or no change in polarity despite its abnormal distribution of pigment. In the stratified eggs of *Ilyanassa* (Clement, 1968) and *Dentalium* (Verdonk, 1968a), which both produce polar lobes, it has always been found that the lobe retains its developmental importance whatever components come to fill it. Clement (1968) even embedded eggs in gelatine and reversed their orientation before centrifuging them. From their results, these authors considered that none of the larger cytoplasmic organelles could be responsible for the lobe's determinative properties, and they favoured rather a cortical map. However, it is also possible that some larger organelles are not displaced because they are held by the cytoskeleton: certainly the vegetal body of *Bithynia* cannot easily be displaced or dispersed by centrifugation despite its large size and the lack of visible binding structures (Dohmen & Verdonk, 1974).

There are, however, other cases where development is usually abnormal following the centrifugal displacement of visible plasms. This is true of the polar plasms of *Tubifex*, but some embryos still develop normally, and Lehmann (1948) concluded that the egg does not develop as a mosaic work of determined plasms. The abnormal development of ascidians following egg centrifugation also seems not to be correlated with the displacement of organelles. In particular, Conklin (1931) was able to show that pigment granules and mitochondria are not required for the formation of muscles, but if higher speeds were used to displace the fine granular cytoplasm of the myoplasm as well, then muscle formed in a new position. Conklin (1931) obtained similar results for other ooplasms with their own distinctive fine-structure, and so concluded that it is the 'ground-substances' of the ooplasms which possess 'organ-forming' potentials.

By centrifuging many eggs in sucrose solutions it is possible first to stratify and then to separate them into halves or even smaller fragments. Such fragments completely lack certain kinds of the larger organelles and large areas of the egg surface, yet when fertilised they still frequently develop to larvae, as Harvey (1933) showed with sea urchins. The larger fragment of the ascidian egg can develop to a tadpole, and the poor development of the smaller fragment may relate with its lack of mitochondria (Reverberi & La Spina, 1959; La Spina, 1960). La Spina (1963) also separated fertilised ctenophore eggs into a larger endoplasmic fragment and a smaller one mainly composed of ectoplasm. One fragment cleaved, and was assumed to contain the nucleus, but only if it also contained ectoplasm did development continue beyond cleavage. This supports evidence given earlier showing a correlation between the presence of ectoplasm and full developmental potential in ctenophores. It would be interesting to know whether the ground substance of the ectoplasm, as well as its larger organelles, is restricted to one fragment. In any case, ectoplasm may not contain all the developmentally important information: according to Freeman & Reynolds (1973) a little yolk must also be present in fragments before photocytes will develop.

When centrifugation speeds are sufficiently increased, even the polarity of some eggs can be reversed. This has been shown with sea urchin eggs (Motomura, 1949) and with fertilised *Parascaris* eggs aligned with the vegetal pole centrifugally (Tadano, 1962; Guerrier, 1967). In both cases one effect of high-speed centrifugation is a disturbance of the ectoplasm. In normal *Parascaris* eggs this has accumulated beneath the animal pole, so that an ectoplasmic gradient may be the basis of polarity (but see also Guerrier, 1967). In amphibian eggs, where simple inversion in unit gravity can reverse polarity (see p. 57), the critical materials appear to be in the deeper cytoplasmic layers.

There is another way in which centrifugation can be used to study spatial determination in eggs. It can displace the meiotic apparatus so that instead of the usual small polar body a much larger one forms. Morrill (1963) has done this with *Lymnaea* eggs and finds that the eggs produced after this abnormal meiotic division can still develop normally. Even Raven (1966), who had proposed that the *Lymnaea* cortex has a map of developmental coordinates, calculates that these eggs have lost 27% of their normal surface (and a quarter of their content) to the enlarged polar body. In some polyclads the enlarged polar bodies are themselves fertilisable and can develop to larvae (Wilson, 1925, p. 494). At first sight these results suggest that there can be no localisation of essential determinants even along the animal–vegetal axis in these species. This may not be so, however, since as sea urchin polar bodies are enlarged (by other techniques) the cleavage plane gradually rotates until it is vertical when the polar body is as large as the 'ovum' (Lindahl, 1937). In any case, these results, and the development of centrifugally separated egg fragments, present very great difficulties for any theory of a detailed map.

### (c) Rotations and other interference with amphibian eggs

While even centrifugation in a reversed orientation cannot change the animal–vegetal polarity of many eggs, a simple inversion in unit gravity is enough to reverse the polarity of somatic development in fertilised amphibian eggs. Partial development of inverted eggs was first seen by Pflüger as long ago as 1883, and interest has been renewed in the last few years following the development of a simple method for keeping eggs permanently inverted (Kirschner & Hara, 1980). Many of the egg's internal constituents redistribute following inversion (Born, 1885), the gradient in sizes of yolk platelets reversing to various extents in different species (Chung & Malacinski, 1982). Little cortical pigment redistributes, however, so the development of inverted embryos can give information on the importance of cortical localisations. Cleavage patterns partially reverse to give fairly small unpigmented cells and a blastocoel at the new upper pole (Stanisstreet, Jumah & Kurais, 1980). The site at which gastrulation begins also frequently reverses, and some embryos, particularly if inverted before fertilisation, go on to become tadpoles with abnormal pigment patterns (Neff *et al.*, 1983). After later inversions, embryos arrest in gastrulation with the gastrular lip apparently defective in inducing ability (Chung & Malacinski, 1982), so different developmental determinants may vary in the ease with which they redistribute. In any case it is clear that most of the animal–vegetal organisation of the amphibian egg can reverse under the influence of gravity, suggesting that developmental determinants are internal and move with such inner organelles as yolk platelets, while the cortical role including that of the grey crescent cortex is quite restricted. One material

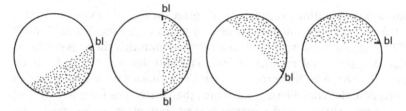

Fig. 2.19. The position of the dorsal blastopore lip (bl) in axolotl embryos developing from rotated eggs. All embryos are shown with the original animal pole upwards and the dorsal side to the right. The lip always forms at that edge of the heavy yolk (dotted area) nearest to the grey crescent position. (After Pasteels, 1964.)

which does not reverse is the germ plasm (Neff *et al.*, 1983), strengthening the case that it is determined by a mechanism totally independent of that for the embryo's somatic organisation. On the other hand, less primordial germ cells arise from the germ plasm in this new environment (Wakahara, Neff & Malacinski, 1984) or when animal pole plasm is injected to the vegetal pole of normally orientated eggs (Sakata & Kotani, 1985), suggesting a possible interaction between germ-line and somatic determination.

Because the animal–vegetal polarity of the amphibian egg is so easily reversed, Malacinski (1984) considers it to be less significant than has previously been thought. However, if we turn now to consider the egg's dorso-ventral organisation, we will find that this is far more labile and perhaps arises secondarily upon animal–vegetal polarity. Inversions of fertilised amphibian eggs have probably been used to study dorso-ventral organisation more often than animal–vegetal polarity. In normal embryos, gastrulation is first externally visible as an invagination of cells at the dorsal lip, a point on the surface near that which once formed the grey crescent. Several authors have used inverted embryos in attempts to identify the local conditions which favour the later development of a dorsal lip.

The inversion work already described demonstrates that many points on the surface have the ability to form a dorsal lip. Indeed, Penners & Schleip (1928*a,b*) found that dorsal lips could form at any point on the surface after rotations. If the eggs are also slightly compressed (which used to be the method of maintaining their inverted orientation) two lips frequently form and embryos with two dorsal axes result (Schultze, 1894). Penners & Shleip (1928*a,b*) concluded that dorsal lips tend to appear at points on the surface overlying the edge of the dense yolky mass, which is, of course, in a new position following rotation. They also suggested that it was the part of this edge nearest to the original dorsal side which later formed a lip (Fig. 2.19), and that pigment in the surface

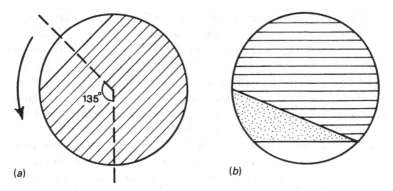

Fig. 2.20. Artificial determination of grey crescent position in *Rana fusca*. (*a*) unfertilised eggs were orientated as shown; following insemination they rotated in the direction shown by the arrow. (*b*) the grey crescent (dotted) formed on the side which was originally uppermost. The pigmented area is shown hatched. (After Ancel & Vintemberger, 1948.)

layers might also have a role. Pasteels (1938) reached similar conclusions with the exception that pigment itself was found not to be necessary, and he stressed the combination of effects of an animal–vegetal yolk gradient and a dorso-ventral cortical field. Thus observation of gastrulation allows us to make the inference that fertilised eggs show some dorso-ventral polarisation which may well be due to factors localised in the superficial layers.

The formation of the normal grey crescent involves movement of the egg's superficial layers relative to the inner cytoplasm, whether this is due to a contraction or a rotation (p. 46). Ancel & Vintemberger (1948) gave evidence that it is this relative movement which determines the dorso-ventral polarity of the egg. They showed that the position of the grey crescent can be determined at will by fertilising an egg with its vegetal pole pointing obliquely upwards (Fig. 2.20). As the perivitelline space forms, the egg rotates under gravity as shown, and, whatever the entry point of the fertilising sperm, the grey crescent will form at that edge of the yolky region which was placed uppermost. At this point gravity should have caused dense yolk to move away from beneath the cortex. This can be compared with the normal grey crescent where contraction or rotation causes the dorsal cortex to move away from dense yolk (Fig. 2.14). In fact, since eggs are usually spawned in all orientations, the redistribution of yolk under gravity could in natural conditions determine dorso-ventral polarity more often than sperm entry does. Now that methods are available for maintaining fertilised eggs in abnormal orientations it has been possible to show that smaller displacements relative to gravity (30° displacements maintained for 12 minutes) are enough to determine an egg's dorso-ventral polarity (Gerhart *et al.*,

1981). Yolk will not redistribute significantly before an egg is activated, as though the endoplasm is reorganised at this time (Neff *et al.*, 1984).

The relative movement of cortex and yolk is, however, only one of the dorso-ventral asymmetries produced following fertilisation (see p. 47). We need to know which of these are important for the developmental polarisation along the dorso-ventral axis, and some evidence is available. As already suggested for other reasons pigment cannot be a critical factor as albino *Xenopus* embryos lacking even premelanosomes develop with normal bilateral symmetry (Bluemink & Hoperskaya, 1975). The inner yolk-poor cytoplasm is not critical either, as it always moves to the side opposite sperm entry point even when egg rotation causes this side to become ventral (Ubbels *et al.*, 1983). Similarly, sperm aster position cannot be the critical factor in conditions where rotation overrides the polarising effects of fertilisation, even if it is used as a cue in normal conditions. The vitelline wall of large yolk platelets (Fig. 2.15), left by any relative movement of yolk and cortex, has been suggested as the critical factor for example by Gerhart *et al.* (1981), but this too has been queried by Neff *et al.* (1984).

The evidence above suggests that the differences between the dorsal and ventral sides of the symmetrised egg are slight and quantitative: the same materials are present on both sides though their distribution is rather different. However, until a few years ago, most authors spoke of the grey crescent as if it contained specific determinants for dorsal structures. Key reasons for this are that the grey crescent area will become the 'organisation centre' of the gastrula (Spemann & Mangold, 1924; see p. 215), and that early blastomeres lacking grey crescent cytoplasm usually fail to form dorsal structures (p. 135). By 1925, Spemann was speculating that the grey crescent already has the properties of an organiser (see also Waddington, 1956, p. 148). Here we will consider only the evidence for and against the hypothesis that the grey crescent has determinants in the uncleaved egg.

A. Brachet (1906) found that newly fertilised *Rana* eggs regulated perfectly after being pricked with a needle, but eggs pricked later sometimes showed deficiencies of dorsal organs, suggesting that some developmental fates have become determined. Any determinants may, however, be much deeper than the cortex as Pasteels (1932) found that only deep penetration by the needle caused deficiencies. There are also other possible explanations for the effects. Pricking may disturb the movements required to confer dorso-ventral polarity on the egg; and it may even be followed by mitotic abnormalities leading to aneuploidy especially when the dorsal cortex is injured (J. Brachet & Hulin, 1972). Pricking can sometimes change the dorso-ventral orientation of the embryo (Wakahara, 1986), but this could indicate quantitative rather than qualitative differences in this plane.

More positive evidence for dorsal determinants appeared to be provided by the cortical grafting experiments of Curtis (1960). Grey crescent cortex grafted into the ventral side of host eggs caused the development of a second dorsal lip and embryonic axis; while ventral cortex grafted dorsally later split the dorsal lip and caused closer double embryos. Unfortunately nobody has been able to repeat these results, although Tomkins & Rodman (1971) have shown that grey crescent cortex is more active than animal pole cortex in inducing secondary axes when implanted into the blastocoels of blastulae. Malacinski, Chung & Asashima (1980) could repeat neither Curtis' grafting work nor the implantation data of Tomkins & Rodman. They and Gerhart *et al.* (1981) suggest trivial explanations for the double axes, including the fact that Curtis maintained his operated embryos in abnormal orientations.

Grossly abnormal development affecting particularly the nervous system can also be caused by ultraviolet irradiations of the vegetal or marginal zones (Grant & Wacaster, 1972). The effect requires a higher dose than the sterilising effect of vegetal uv, is distinct from it in other ways (Thomas *et al.*, 1983), and the area of the grey crescent appears to be particularly sensitive (Malacinski, Benford & Chung, 1975). The defect has been shown to be cytoplasmic rather than nuclear and can be corrected by injections of material from normal eggs (Grant & Wacaster, 1972). This was seen as evidence for a neural determinant in the normal egg, presumably located in the grey crescent region. The action spectrum for this effect shows a peak at 280 nm suggesting a protein target (Youn & Malacinski, 1980; see too unpubl. work of Smith cited by Smith & Knowland, 1984). Irradiation has other developmental effects, including inhibition of vegetal cleavages (Grant & Wacaster, 1972; Züst & Dixon, 1975), delayed gastrulation (Grant & Wacaster, 1972) and inhibited invagination of the dorsal lip (Malacinski *et al.*, 1975), so that the damage may be unspecific. But a far more important objection to the idea of uv-sensitive dorsal determinants was discovered by Scharf & Gerhart (1980): uv-treated eggs can be completely rescued by a subsequent rotation through 90 ° (Fig. 2.21), when dorsal organs form on the side placed upwards. It does not seem credible that uv irradiation is destroying dorsal determinants which can then be re-synthesised simply by rotating the egg! Rather, the irradiation may prevent the movements that normally rearrange materials in the fertilised egg.

Further work has strengthened the old view that dorso-ventrality originates in asymmetric movements within the fertilised egg. Manes & Elinson (1980) found that the first effect of uv irradiation is to prevent the grey crescent forming, and noted that the later embryo lacks any obvious dorso-ventrality. Scharf & Gerhart (1983) could produce similar embryos also by exposure to cold (1 °C) or pressure, two physical treatments known to depolymerise elements of the cytoskeleton (Fig. 2.22).

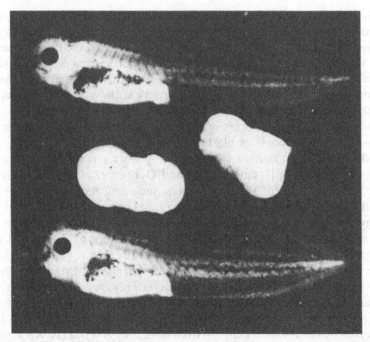

Fig. 2.21. The inhibition of dorsal development in *Xenopus* by uv irradiation and its restoration by oblique orientation. All embryos are from the same batch. *Top:* a stage 42 control tadpole. *Middle:* two aneural embryos resulting from vegetal uv followed by rearing in normal orientation. *Bottom:* a completely rescued embryo following the same uv treatment but reared at a 90° oblique angle to the 16-cell stage. (From Scharf & Gerhart, 1980.)

Embryos treated while immersed in heavy water were largely protected against the effects of uv and cold, suggesting that they act by depolymerising microtubules. Heavy water cannot, however, protect embryos from pressure effects, which may mean that other cytoskeletal elements also have a role. Oblique orientation can rescue eggs after all three treatments, and normal eggs lose their sensitivity to all three at about the same time, which is as the grey crescent is completed.

In understanding what happens during symmetrisation it has to be remembered that it is not producing bilateral organisation from nothing. The unfertilised egg has animal–vegetal organisation both in its surface-associated layers and in the endoplasm. By moving one of these relative to the other, small but apparently decisive differences could be established between the dorsal and ventral sides. Heat-shock treatments (Beetschen, 1979) and some inhibitors of protein synthesis (Gautier & Beetschen, 1983) can induce grey crescent formation in unfertilised eggs, showing that the whole mechanism for motility preexists in the egg, and suggesting that protein synthesis may even be required to prevent its

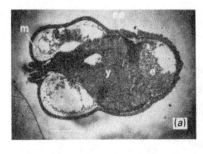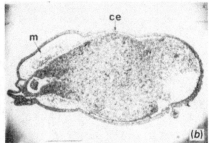

Fig. 2.22. Median sections through *Xenopus* embryos radialised by uv irradiation (*a*) or cold treatment (*b*) of the fertilised eggs. ce, ciliated epidermis; e, erythrocyte-like cells; m, mesenchyme cells; y, yolk-filled cells. Scale bar = 0.5 mm. (From Scharf & Gerhart, 1983.)

premature operation. Heat shock probably allows the bilateral reorganisation of the egg by disrupting the cytoskeleton and allowing yolk to redistribute under gravity earlier than would otherwise be possible (Beetschen & Gautier, 1987).

There is some evidence that small quantitative differences in metabolism can make the difference between dorsal and ventral pathways of development. Rearing embryos between fertilisation and blastula stages in a temperature gradient of 2 °C across the embryo can cause the warmer side to become dorsal (Glade, Burrill & Falk, 1967). Løvtrup (1958) suggested that the different permeabilities of dorsal and ventral sides would result in differences in oxygen supply, and found (Løvtrup & Pigon, 1958) that early cleavage embryos sucked into blind-ending glass tubes and left until gastrulation almost always began invagination on the side towards the open end. Later, he added to this hypothesis the effects of the vitelline wall changing the local ratio of yolk to 'light cytoplasm' (Løvtrup, 1965). Black & Gerhart (1986) speak of the relative movements activating one meridian of the egg in some unspecified way. Cooke (1986) believes that other regions of the pre-cleavage egg may also be determined according to their dorso-ventral position (see also Render & Elinson, 1986), but we should remember that uv-treated embryos are rescued when only a small part of the gastrula lip is replaced by a normal dorsal lip (Malacinski, Allis & Chung, 1974). The implications of these findings are somewhat paradoxical: the dorso-ventral differences in a grey crescent stage egg appear to be quantitative, relatively slight and reversible in temperature or other gradients, yet they seem to be responsible for the formation of the dorsal lip which apparently organises so much of the embryo (see p. 215). How real this paradox is will only be seen as we examine the dorso-ventral organisation of amphibian embryos through cleavage and blastula stages (see pp. 134 and 192).

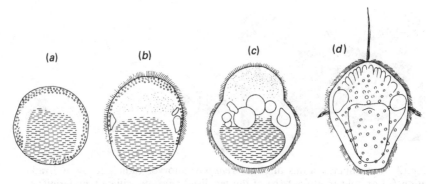

Fig. 2.23. Differentiation without cleavage in artificially activated *Chaetopterus* eggs. The eggs pass through a stage recalling a gastrula (*a*), and ciliate (*b*) to form a unicellular larva (*c*). (*d*) a normal trochophore larva. (From Kühn, 1971, after Lillie.)

### (d) Differentiation without cleavage

This chapter would be incomplete without a consideration of the phenomenon of differentiation without cleavage. Lillie (1902) described this for unfertilised eggs of the polychaete *Chaetopterus* which had been activated by potassium chloride treatment (see Fig. 2.23). Maturation divisions can take place accompanied by amoeboid movements and movements of the recognisable ooplasms as they would in a fertilised egg (more recently studied by Eckberg, 1981, and Jeffery & Wilson, 1983). Hyaloplasm derived from the germinal vesicle moves from beneath the animal pole into the egg's centre and the outermost ectoplasm divides into animal and vegetal portions. No cell division follows, but, by a stage equivalent to 16 cells, the hyaloplasm and the larger part of the ectoplasm have moved back beneath the animal pole. At the time of gastrulation these plasms gradually cover the yolky endoplasm once again. Still later, cilia develop on all parts of the surface underlain by ectoplasm, so that a single cell shows motility as well as some of the structure of a larva. It does, however, lack the apical tuft of longer cilia and is radially symmetrical.

These surprising abilities show that the egg can pass through a sequence of events mimicking normal development independently of any influence from the sperm and without being subdivided into cleavage cells. Considerable animal–vegetal organisation is established, though the lack of dorso-ventral organisation may relate with the failure to segregate the contents of the polar lobe to specific cells (see p. 78). The changes also seem to be programmed from the time of activation as they would normally be from fertilisation.

Cytochalasin B (at levels blocking cleavage of fertilised eggs) does not prevent the ooplasmic flows at maturation or pseudogastrulation (Brachet & Donini-Denis, 1978; Eckberg & Kang, 1981). In fact, colchicine is more inhibitory, allowing the flows at maturation stages but not pseudogastrulation. This suggests a role for microtubules, and cycles of aster formation were seen in activated eggs by Brachet in 1937. Despite the lack of cleavage, the chromosomes are replicated, and the replication cycle may provide the 'clock' mechanism programming differentiation (Alexandre, de Petrocellis & Brachet, 1982). Many points, however, remain to be clarified before we will understand differentiation without cleavage in *Chaetopterus*.

Similar phenomena have been seen in the activated eggs of several other species (Wilson, 1925, p. 1084) most interestingly in the eggs of anuran amphibians where development is very unlike that of annelids. Matured but unfertilised anuran eggs will, after an approximately appropriate time in culture, make pseudogastrulation movements (Holtfreter, 1943; Smith & Ecker, 1970; Baltus *et al.*, 1973; Malacinski, Ryan & Chung, 1978). The pigmented area extends into the vegetal half, a ring-shaped groove appears at about the position of the normal blastopore, and tongues of homogeneous cytoplasm extend inwards from this ring (Fig. 2.24) much as the invaginating chordamesoderm cells would normally do. Again, this shows that the unfertilised egg is capable of mimicking the events which increase the animal–vegetal complexity of the embryo. It is not clear whether any dorso-ventral organisation is established: in the absence of sperm entry or rotations there is no reason to expect it. In the case of anurans, pseudogastrulation seems to be programmed from maturation, and this is when much metabolic activation occurs in this group. Here, too, the events even occur in oocytes enucleated before maturation (Smith & Ecker, 1970) so that DNA replication cycles cannot provide the 'clock' and no transcription of nuclear genes can be required. Inhibitor studies confirm that RNA synthesis is not required, but protein synthesis and the activity of microtubules and microfilaments apparently are (Malacinski *et al.*, 1978).

## Conclusions

During the period in which an oocyte is matured, ovulated and fertilised, a variety of important changes occur, commonly including activation of many metabolic pathways as well as the addition of germinal vesicle materials, a sperm nucleus and centrioles to the ooplasm. Over this same period, waves of contractile activity pass over many eggs and their contents become increasingly segregated, due to activities of the cytoskeleton. Commonly, organelles are transported to the peripheral layers of the egg, and then towards the animal and/or vegetal poles. Sometimes dorso-ventral differences are also established.

Fig. 2.24. Pseudogastrulation in *Rana pipiens*. (*a*), (*b*) external views of stages equivalent to early (*a*) and late (*b*) gastrulation. (*c*) vertical section through a late stage showing invagination of homogeneous cytoplasmic material. (*d*) pseudogastrulation of an egg from which the GV was removed before hormonal treatment. (From Smith & Ecker, 1970.)

The development of fragments of various eggs shows that important materials do, indeed, become more localised often in correlation with the movement of ooplasms. However, rather than determinants for various organs moving to different areas, a single area seems often to become a kind of 'determination centre'. This is at the vegetal pole of many eggs but may be at the oral (animal) pole for ctenophores. Elsewhere large volumes of ooplasm and areas of cortex may be dispensable, and, if isolated, can usually form blastulae at most. The development of centrifuged eggs suggests that determination is not due to the larger organelles, but amphibian eggs are unusual as even animal–vegetal polarity can be reversed simply by maintaining the fertilised egg in an inverted position where yolk platelets (and other inner organelles?) redistribute while pigment (and other cortical elements?) do not. It is probably relations between these two domains which determine the dorso-ventral plane, with the movements which lead to grey crescent formation in fertilised eggs establishing slight differences between the dorsal and ventral sides of the egg.

Some unfertilised eggs are known to be able to mimic several aspects of embryonic development following maturation or artificial activation. This confirms that unfertilised eggs are to some extent programmed for development and spatially organised along the animal–vegetal (though perhaps not the dorso-ventral) axis.

# 3

## Does cleavage cut up a preformed spatial pattern?: the case of spiralian embryos

Cleavage is the process which divides the zygote into many cells: cells, moreover, of a more normal size, and with a more normal ratio of cell size to nuclear size. Division is usually rapid with almost all of the interphase time being occupied by replication of the DNA, and that itself occurring very fast using closely packed DNA polymerase molecules (Callan, 1973). The divisions are mitotic, using a spindle organised by a pair of centrioles originating usually from the sperm. Asters determine where cleavage will occur, but it is a band of cortical microfilaments which actually constricts the cell (for review see Rappaport, 1971). Cycles are commonly detected in nuclear, cortical and general cytoplasmic properties (see references in Satoh, 1982b). It is probably an oscillating cytoplasmic factor which ensures that all these cycles are precisely synchronised in normal cleavage (Yoneda & Schroeder, 1984). Activities which oscillate in the cleavage cycle include maturation promoting factor (MPF) and an inactivator of MPF (Gerhart, Wu & Kirschner, 1984a; see also Evans et al., 1983), and, since the artificial addition and removal of MPF can drive the mitotic cycle of Xenopus eggs (Newport & Kirschner, 1984a), they seem to play a key role in the oscillation process.

We have seen in the last chapter that developmentally important materials do become differentially distributed in many maturing oocytes and zygotes. Wilson (see 1925, p. 1069 et seq.) long ago noted that the extent to which blastomeres showed restrictions in their potency would depend upon whether cleavage planes separated areas containing different determinants. Morgan added in 1934 (p. 10) the suggestion that 'the initial differences in the protoplasmic regions may be supposed to affect the activity of the genes'.

In this and the next chapter we will consider the extent to which cleavage simply cuts up a system of determinants already organised in space in the zygote. This chapter considers the supposedly mosaic embryos of groups showing spiral cleavage patterns.

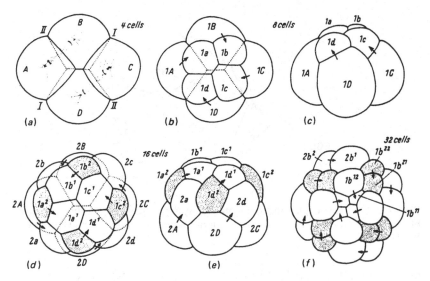

Fig. 3.1. Early cleavage stages of the snail *Trochus*. Cells of the trochoblast lineage are stippled in (*d*)–(*f*). (From Kühn, 1971, after Robert.)

## Spiral cleavage patterns

Embryos of many animal groups show spiral cleavage; molluscs (other than cephalopods), annelids, nemertines, polyclad turbellarians, echiuroids and sipunculoids provide clear examples, and other groups show apparent modifications of the spiral pattern (Costello & Henley, 1976). There are important differences in the cleavage pattern among such a large array of animals but there are important similarities too which justify considering them together.

Figure 3.1 gives the early cleavage pattern of *Trochus* as an example, with the spiral form becoming obvious at the third cleavage when mitotic spindles form obliquely and the four animal cells are formed above the furrows separating the four vegetal cells. Viewed from the animal pole, if the animal cells are formed on the clockwise side of the vegetal cells (as in Fig. 3.1) cleavage is said to be dextral, while if they are cleaved on the anticlockwise side it is said to be sinistral. The following cleavage will always be in the opposite direction, and dextral and sinistral cleavages then alternate. However, it is the direction of the third cleavage which is used in describing an embryo as dextrorotatory or laevorotatory. As already noted, *Lymnaea peregra* embryos may show either form (already illustrated as Fig. 1.10(*b*)).

Figure 3.2 shows the notation used to describe blastomeres in most spirally cleaving embryos. The first four cells are named A to D, D marking the later dorsal side of the embryo. At third cleavage each cell

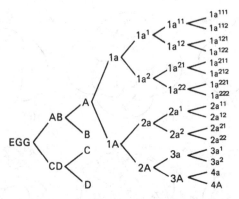

Fig. 3.2. Diagram explaining the notation system used to describe spiralian blastomeres. Only the A lineage is shown here, but the same system is used for B–D.

divides usually unequally to give an animal micromere and a vegetal macromere: 1a and 1A, 1b and 1B, etc. Further quartets of micromeres, 2a, 3a and 4a etc. are usually budded from the macromeres, whose names change to 2A, 3A and 4A, etc. as they lose this material. At the same time the micromeres undergo further cleavages to produce $1a^1$ and $1a^2$, $1b^1$ and $1b^2$, etc.

The immediate cause of the oblique cleavage planes is the oblique angle of the mitotic spindles, and it has long been observed that spindles form perpendicularly to previous spindles (the rule of Sachs: see Conklin, 1897, p. 185), providing an explanation of the alternating cleavage divisions. The positions and angles of the centrioles can in turn explain the alternating spindle angles (Costello, 1961), but this does not explain the origin of the oblique angles, particularly as the centrioles are paternally supplied but cleavage direction is maternally inherited. Mesh-cheryakov & Beloussov (1973, 1975) have presented evidence for a spiral organisation of the contractile ring which divides the egg. This is inferred from the rotation of one blastomere relative to the other during cleavage, and already at first cleavage these rotations are in opposite directions in dextrorotatory and laevorotatory species (Fig. 3.3). This will affect the orientation of the spindles in the first two cells, making second cleavage more obviously spiral with the result that only the B and D blastomeres reach the vegetal pole (see Fig. 3.1). There are further rotations at second cleavage, making third cleavage still more obviously spiral.

The case for obliquely oriented cortical microfilaments is supported by other evidence (e.g. see Fig. 2.5), but internal factors may also be important as Freeman & Lundelius (1982) were able to change the cleavage pattern of *Lymnaea peregra* eggs from sinistral to dextral by injecting cytoplasm from a dextral egg. Cell shape, including the flatten-

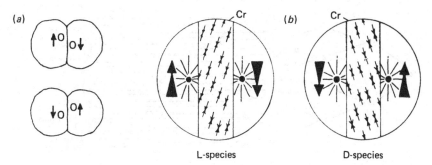

Fig. 3.3. Spiral organisation at first cleavage in molluscs. (*a*) shifts in nuclear position seen in dextrorotatory (upper) and laevorotatory (lower) species. (*b*) an explanation of these movements in terms of hypothetical 'spiral contractile rings' (cr) of microfilaments. With the orientations shown, and assuming the integrity of the ring and its binding to the plasma membrane, the nuclei will indeed move in the direction of the arrows. (From Meshcheryakov & Beloussov, 1975; (*a*) after Conklin.)

ing at cell contacts, also affects cleavage direction in spiralian (Meshcheryakov, 1978*a*,*b*) and other embryos. When cells cleave unequally, other interactions of an unknown kind seem to be involved as the aster at one pole of the mitotic apparatus is flattened and attached to the cortex (Dan & Ito, 1984; Dan & Inoué, 1987; see also pp. 114 and 128).

### Lineage plans and fate maps

If we wish to know how far cleavage cells are already determined for a specific fate, we have to know what the fate of the cells will be in normal development. This can be investigated by marking the living cells in some way and observing which structures in the embryo later bear the marker. These results can be represented on a lineage plan such as that for *Nereis* shown as Figure 3.4(*a*). Very similar lineage plans are found in many other spiralian embryos, including some from other phyla. In particular, the 1a$^1$–1d$^1$ quartet commonly contribute to the apical tuft, and 1a$^2$–1d$^2$ to the ciliated cells of the prototroch. Many of the micromeres form general larval ectoderm, but specific cells of the second or third tier often form larval mesoderm. Specific cells of the D quadrant are set aside to be used later in constructing the ectoderm and mesoderm of the adult, and are known as teloblasts. The ectoteloblasts are usually derivatives of the 2d cell, while mesoteloblasts arise from 3D or 4d. In molluscs, 4d also contributes to the intestine and is known as the mesentoblast cell. The macromeres, and commonly the 3a to 3c micromeres, form endoderm. Each cell in the spiralian embryo has a precise fate, but it is clear that homologous cells have different fates in different species. Another way of portraying the fates of cells is to mark them onto a drawing of the embryo

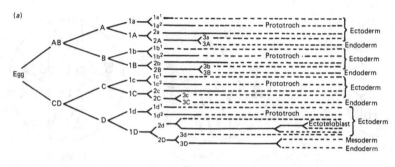

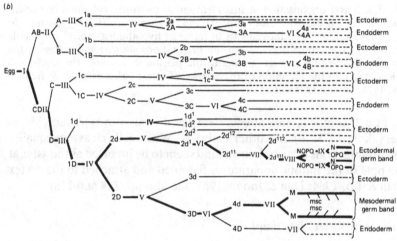

Fig. 3.4. Diagrams of the cell lineage of *Nereis* (*a*) and *Tubifex* (*b*). Further division cycles are omitted in the parts indicated by dashed lines. In (*b*), Roman numerals denote the number of cleavages and the teloblast lineages are indicated by heavy lines. msc, mesoderm stem cell. ((*a*) after Wilson, 1892; (*b*) from Shimizu, 1982, after Penners.)

to make a fate map. This device has been used widely for other types of embryos (see later chapters). Anderson (1966, 1973) has applied it to spiralian blastulae as shown in Figure 3.5. His system makes clear the basic similarity of the maps for different species, as the relative positions of the different prospective areas is usually conserved. To show the specific cells with their different fates in different species would, according to Anderson, obscure this basic similarity.

In leeches and oligochaetes there is no larval stage, the large egg developing directly to a small adult. In such cases the non-teloblastic micromeres contribute to the presegmental ectoderm, including the stomodaeum (Schleip, 1914; Penners, 1922*a*) and supraoesophageal ganglion (Weisblat, Kim & Stent, 1984), and to extraembryonic ectoderm (Anderson, 1973). Endoderm still develops from the macromeres,

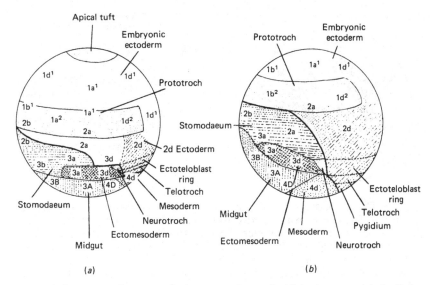

Fig. 3.5. Fate maps for two polychaete species at the blastula stage. (*a*) *Podarke obscura*. (*b*) *Scoloplos armiger*. (From Anderson, 1966, based on earlier data.)

but almost all else arises from the teloblasts (see Fig. 3.4(*b*)). The further development of teloblasts has been followed in detail, and Fernández (1980) has proposed a simpler notation for describing it in leeches. The first pair of mesoteloblasts are called the M cells. Another pair of large cells (derived from 2d) are called the NOPQ cells, and each divides to give an N, O, P and Q ectoteloblast. Each teloblast (M–Q on each side) begins a long series of unequal cleavages producing a string of m–q cells called a germinal bandlet (Fig. 3.6). As this figure shows, the five bandlets on each side unite to a broader germinal band, and these in turn coalesce on the ventral (originally vegetal) side of the embryo as a germinal plate. The continuous germinal plate later becomes visibly subdivided in the first indication of the segmentation of the adult. Probably one m bandlet cell acts as a stem cell for all the mesoderm on one side of a segment (Fernández & Stent, 1980; see also Devriès, 1973, for the oligochaete *Eisenia*). The segmental ectoderm will derive from cells of the n, o, p and q bandlets, and it is in their neural derivatives that the greatest precision of a spiralian fate map is revealed. Weisblat *et al.* (1984) have shown that a specific 'kinship group' of neurons in each segment is descended from each ectoteloblast, which also contributes a characteristic territory of the epidermis (Fig. 3.7). Thus a stereotyped cleavage pattern in the early embryo separates teloblasts each of which will make a very specific contribution to each segment along the adult body.

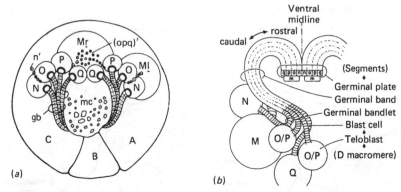

Fig. 3.6. The teloblast system in leech development. (*a*) dorsal view showing the five pairs of teloblasts which derive from the D lineage and give rise to the germinal bands (gb). The micromere cap (mc) and other cells of the N and OPQ lineages, n' and opq', are also shown. (*b*) a detailed explanation of the teloblast system also showing that the germinal bands of the left and right sides unite in the ventral midline. ((*a*) from Fernandez & Stent, 1980; (*b*) from Weisblat *et al.*, 1984.)

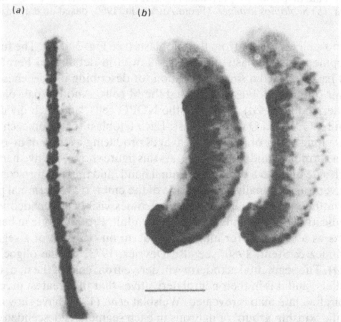

Fig. 3.7. The fates of specific teloblasts in the leech *Helobdella*. The teloblasts were injected with a lineage tracer, and some anterior segments are unlabelled because the blast cells had formed before the injection. (*a*) ventral view following injection of the left N teloblast: most labelled cells are in the ipsilateral ganglia of the ventral nerve cord. (*b*) lateral view following injection of the right Q teloblast: most labelled cells are in the dorsolateral epidermis and in neural cells outside the nerve cord. Scale bar = 200 μm. (From Weisblat *et al.*, 1984.)

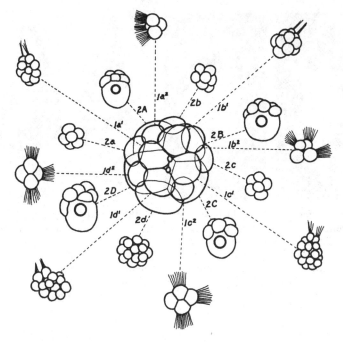

Fig. 3.8. The developmental patterns shown by each cell in the 16-cell *Nereis* embryo following its isolation and culture. (From Costello, 1945.)

### Mosaic development after separation of blastomeres

The concept of mosaic development arose from experiments involving blastomere destruction or isolation, in which the surviving or isolated cells formed just that part of the embryo which would have been their fate within the whole. In such cases Conklin (1897) said that cleavage was 'determinate': while cases where each blastomere could form a complete though miniature embryo showed 'indeterminate' cleavage. Heider (1900) used the terms 'mosaic' and 'regulative' eggs.

Costello's (1945) studies, where all the first 16 cells of *Nereis* embryos were isolated, are often cited as an example of mosaic development (see Fig. 3.8). He separated cells at each cleavage from the first to the fourth and reared them in isolation. Certainly the early cleavage patterns of the isolates were very like *in vivo* patterns particularly for $1a^2–1d^2$ (cleaving only twice) and 2A–2D (cleaving off micromeres in a cap). The top two tiers of cells also produced cilia of kinds apparently relating with their normal fate (apical tuft for $1a^1–1d^1$, and prototroch for $1a^2–1d^2$), and the macromeres began gastrulation. No isolate, however, differentiated such larval structures as prototrochal pigment, anal pigment or eye spot, nor any teloblastic structure. In other words, development is mosaic so far as

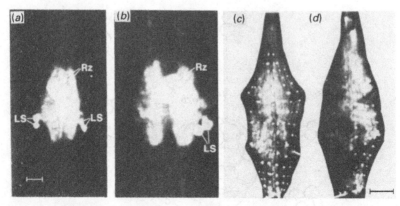

Fig. 3.9. Effects of teloblast ablations in the leech *Helobdella*. (*a*),(*b*) fluorescence photomicrographs of glyoxylic acid-stained ganglia from a control (*a*) and following the ablation of the left N teloblast (*b*). In (*a*) there are a pair of Retzius cells (Rz) and two pairs of lateral cells; in (*b*) one member of each pair is absent. (*c*),(*d*) patterns of iridescent cells in the dorsal skin of a control (*c*) and following the ablation of a Q teloblast (*d*), when such cells are absent from the operated side. Scale bars: in (*a*),(*b*) = 20 μm; in (*c*),(*d*) = 1 mm. (From Blair, 1983.)

it proceeds in isolates, but it does not proceed far enough to provide a full test of the hypothesis.

Horstadius (1937*b*) has provided another test of mosaic development using 16-cell embryos of the nemertine *Cerebratulus*. He first established the normal fates of cells using vital staining, and then compared this with the development of horizontally isolated tiers of cells. With the exception of the macromeres (which failed to develop), all isolates formed the structures expected from their normal fates including, for example, the correct proportions of the ciliated band.

However, far more impressive evidence for mosaic determination has now been given for the leeches. Mori (1932) showed long ago that the germinal band formed no ectoderm if the 2d cell was destroyed and no mesoderm if 3D was destroyed. (Penners had presented similar evidence for the oligochaete *Tubifex* in 1926.) Now Blair (1983) has been able to ablate just the N, OP (precursor of O and P) or Q cell and show that apparently only that particular kinship group of neurons expected from each cell's normal fate fails to develop in each segment (Fig. 3.9).

It is thus clear that in 'mosaic' embryos cleavage produces cells with different properties, and that by late cleavage this allows cells often to develop quite normally in isolation. This is obviously a great advance in organisation over the egg even after the completion of ooplasmic segregation. There it was found that development of fragments correlated poorly with normal fates, and that the potential to form most structures –

(a)

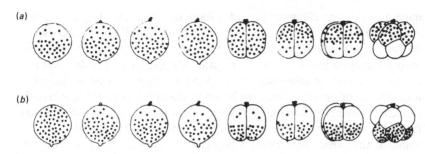

(b)

Fig. 3.10. Diagrammatic representation of the distribution of factors specifying apical tuft (*a*) and gut (*b*) formation through maturation and the first three cleavages in *Cerebratulus*. (From Freeman, 1978.)

even animal pole derivatives – was often greatest in vegetal fragments (see p. 49). It seems unlikely that cleavage is merely cutting up a preformed system of determinants to isolate each type of determinant in an appropriate cell. Another possibility is that the system of mitotic spindles, etc. used for cleavage takes an active role in redistributing materials before separating them in different cells. To test such ideas, blastomere separations and the removal of cytoplasmic areas have to be carried out at the earliest cleavages.

An investigation of this kind has been made with the nemertine *Cerebratulus*. The first two cleavages are vertical and early work indicated that there is no significant localisation of materials into any of the four cells they produce. In isolation such cells cleave as part of an embryo, but still regulate to form pilidium larvae (Wilson, 1903; Yatsu, 1904, 1910; Hörstadius, 1937*b*). Even at the eight-cell stage, after a horizontal cleavage, isolates including animal and vegetal cells can regulate (Yatsu, 1910), confirming that there is little separation of important materials in the dorso-ventral or left–right axes (or that isolates can re-form any missing materials after separation). Separating the animal quartet from the vegetal quartet at the eight-cell stage, however, reveals that materials have been segregated along the animal–vegetal axis. Zeleny (1904) and Yatsu (1910) found that only the animal cells formed apical organ and only the vegetal cells a gut, while at the 16-cell stage only the animal four cells formed apical organ.

To study the acquisition of these polar differences requires equatorial cutting of embryos before the third cleavage separates them. Freeman's (1978) study, already referred to on p. 50, has now provided such data (see Fig. 3.10). Animal halves show a continuous fall in their ability to form gut as separations are made later between the one- and eight-cell stages. Vegetal halves show a similar decrease in their ability to form apical tuft, though it appears to fall particularly steeply between the two- and four-cell stages. Such results suggest that determinants are actively moved during the cleavage period. Freeman (1978) also found that this localisation could be inhibited if early cleavage spindles and asters were suppressed by ethyl carbamate treatment, though the effect was less marked than he had found at maturation stages. He concluded that the microtubules of the asters have 'some kind of triggering role' in the polar localisation of determinants, while microfilaments seem of less significance as cytochalasin B does not inhibit localisation although it prevents cleavage. Freeman's studies (1978, 1979) appear to have revealed at least some of the basic features of a segregation system which is probably capable of extension to the 16-cell stage where all four tiers of cells along the animal–vegetal axis have different developmental properties appropriate to their normal development within the whole embryo.

This segregation system is, however, not adequate to explain the development of isolates in many other spiralian embryos. For one thing, determinants do not seem to pass to the newly formed micromeres of the acoel *Childia*, as they form only ciliated hollow spheres after isolation (Boyer, 1971). A far more important difficulty arises because of differences among the first four cells according to their dorso-ventral position, and it is this difference, apparently lacking in *Cerebratulus*, which must now be studied.

### The role of the D quadrant: the first two cleavages

#### (a) Polar lobes

The first two cleavages are often unequal in spiralian embryos. In species producing polar lobes at the first one or two cleavages, this is ensured by resorption of the lobe (vegetal) material into only one blastomere after the cleavage (Fig. 3.11). The cell receiving lobe material after first cleavage is known as CD, while the AB cell gets none of the lobe material. If a lobe is produced at second cleavage, the D cell receives its contents and not the C cell. As explained in the last chapter, Wilson (1904) found that a *Dentalium* egg from which the polar lobe was removed produced an embryo lacking the apical organ and the post-trochal region with its various derivatives (Fig. 3.12). In the same paper, Wilson reported the results of blastomere isolation experiments. He claimed that AB, A, B or C blastomeres developed in essentially the same way as lobeless embryos: CD or D blastomeres, however, were able to form the lobe-dependent structures and in fact the post-trochal region was larger than normal (see Fig. 3.13). He also found that, if the polar lobe was removed at second cleavage, an embryo still failed to form post-trochal structures but an apical tuft did now develop. Wilson therefore proposed that the polar lobe formed at first cleavage passed 'specific cytoplasmic stuffs' for apical tuft and post-trochal structures to the CD

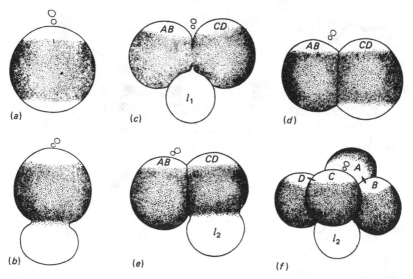

Fig. 3.11. Development through the first two cleavages in *Dentalium*. Note the lack of pigment in the polar areas and the segregation of the vegetal pole plasm via the polar lobe to CD and (although this is not shown) to D. (From Wilson, 1904.)

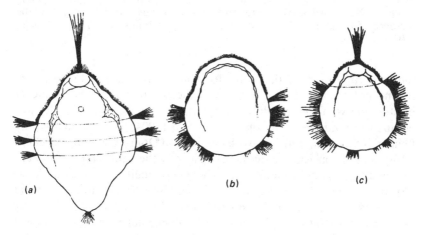

Fig. 3.12. A normal *Dentalium* trochophore (*a*) compared with those obtained after removal of the polar lobe at first (*b*) or second (*c*) cleavage. Note the loss of the post-trochal area from (*b*) and (*c*), and of the apical tuft from (*b*). (From Wilson, 1904.)

cell, and that the 'stuff' for apical tuft then left this area before the polar lobe for second cleavage formed, leaving the post-trochal 'stuffs' to be segregated to the D cell.

It may be noticed that one inconsistency remains in the above data and

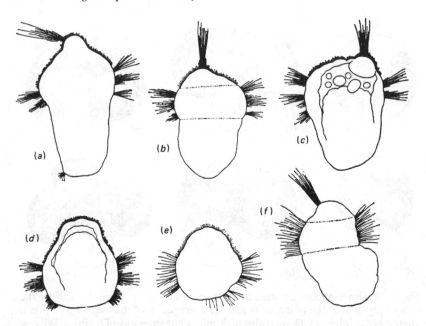

Fig. 3.13. Larvae developed from isolated *Dentalium* blastomeres. (*a*)–(*c*) CD halves; (*d*) an AB half from the same embryo as (*c*); (*e*) a C quadrant; (*f*) a D quadrant. Note particularly the relative sizes of the post-trochal region, and the presence of an apical tuft in CD halves and D quadrants only. (From Wilson, 1904.)

their interpretation. If the 'stuff' for apical tuft is not included in the second polar lobe, then it might be expected that both the C and D blastomeres would form tufts after isolation. As reported above, this was not in fact found for C cell isolates by Wilson (1904), nor indeed was it by Geilenkirchen (unpublished work cited in van Dongen & Geilenkirchen, 1974). The inconsistency can be explained if the tuft determinants are retained in the D blastomere despite being excluded from the polar lobe. However, the apical tuft is normally formed by cells of both the C and D quadrants (van Dongen & Geilenkirchen, 1974), and is produced by ABC isolates (Cather & Verdonk, 1979). Further complications have been reported for *Sabellaria* (see below), so that a more complex interpretation involving inhibitors or quantitative effects seems to be required. This interpretation should take account of the facts that tuft and post-trochal structures are differentially affected when varying amounts of material are removed from the vegetal pole of eggs (p. 52), and that the polar lobe produced at second cleavage is smaller than that at first cleavage (Render & Guerrier, 1984).

Despite such problems over detail, Wilson's interpretation is still accepted today with the 'stuffs' usually seen as determinants. The use of

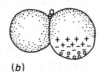  

Fig. 3.14. Render's model for the hypothetical distribution of factors affecting determination in the first two cleavages in *Sabellaria*. +, apical tuft determinants; −, apical tuft inhibitors; o, post-trochal determinants. All these factors would pass via the polar lobe at first cleavage (*a*) to the CD cell (*b*), but the tuft determinants would not enter the lobe at second cleavage (*c*) and so would enter both C and D (*d*). The other factors would be segregated to D, where the inhibitor would prevent the formation of apical tuft. (After Render, 1983.)

the polar lobe transforms a zygote which is only clearly polarised along the animal–vegetal axis into a four-cell embryo with dorso-ventral and probably asymmetric organisation as well. (Any difference between the A and C quadrants will confer asymmetry.) In this connection it is interesting that cleavages within the D quadrant occur with a different timing from those in other quadrants (see Fig. 3.4), and that lobeless *Dentalium* embryos show the same cleavage timing in all quadrants and form an essential radially symmetrical larva (van Dongen, 1976).

Results generally similar to those found with this scaphopod mollusc have been seen with other lobe-bearing embryos following lobe removal and blastomere separations. Crampton (1896), Clement (1956) and Atkinson (1971) studied the gastropod *Ilyanassa*, Rattenbury & Berg (1954) the lamellibranch *Mytilus*, and Hatt (1932), Novikoff (1938) and Render (1983) the polychaete *Sabellaria*. The greatest variations concern the control of ciliation patterns. Determinants for the apical tuft do not seem to be included in the polar lobe produced at second cleavage of *Mytilus* or *Sabellaria*. In the latter, moreover, only the C blastomere can produce an apical tuft after isolation, and this led Hatt (1932) and Novikoff (1938) to assume that tuft determinants were segregated to C though not, of course, via the polar lobe. Render (1983) has, however, shown that, if the second polar lobe is removed and C and 'D' then separated, both are capable of producing the tuft. She proposes that an inhibitor of apical tuft is present in the polar lobes and prevents tuft formation by the D blastomere (see Fig. 3.14). In this scheme the tuft inhibitor could be the same substance as the post-trochal determinant, but the tuft determinants which distribute to C and D Render feels must be a qualitatively distinct material. The present author feels that this is still unproven: a single material in the *Sabellaria* polar lobe could determine apical tuft at the low level which reaches the C quadrant and inhibit it at the higher level in the D quadrant while determining post-trochal structures. The differences between the *Dentalium* and *Sabellaria* data could be due to species-specific differences in the amount of the material retained in the second polar lobe, in the threshold required for the different responses or in the animal–vegetal distribution of the material within the D quadrant. It will be difficult to resolve such questions until the determinants have been identified. *Ilyanassa* seems to show another modification of the system in which the lobe inhibits ciliation by inappropriate cells, as lobeless embryos bear cilia over much of the pretrochal ectoderm (Cather, 1973).

Another departure from the *Dentalium* pattern, affecting even posttrochal structures, is seen in *Bithynia*. Here, the C and D cells seem almost equivalent in developmental value, either being capable of ensuring the development of all the usual lobe-dependent structures (Verdonk & Cather, 1973). It is likely that the determinants in *Bithynia* are associated with the vesicles of the prominent vegetal body (see p. 35), and Cather, Verdonk & Dohmen (1976) claimed that this did indeed disperse before second cleavage, which would explain the developmental equivalence of C and D. Dohmen & Verdonk (1979*b*) have, however, since reported that many of the vesicles from the vegetal body do pass via the polar lobe at second cleavage to the D quadrant. Thus again some clarification is required, but there is general support for the existence of determinants in polar lobe cytoplasm.

The most dramatic demonstration of the role of the polar lobe is given by experiments where its formation is suppressed so that its material is equally divided between the first two cells. Such embryos usually form two sets of the lobe-dependent structures and so develop as Siamese twins (Titlebaum, 1928; Tyler, 1930; Novikoff, 1940; Guerrier, 1970*a*; Guerrier *et al.*, 1978; see Fig. 3.15). These results have been achieved with several polychaete and mollusc species, and, following a variety of treatments including compression, temperature shocks and various chemical treatments. The only common factor is the equalisation of first cleavage, and it seems clear that this is the cause of twinning. This has many far-reaching implications. Since cells acting as D blastomeres can be induced in new positions, Guerrier *et al.* (1978) stress that dorsoventrality must be epigenetically determined, although they think that there will be a normal preference for its formation on the side opposite the point of sperm entry (see Morgan & Tyler, 1938). The drastic alterations involved in twinning suggest that many cells must have greater developmental potential than is realised in their normal fate – a characteristic of regulative rather than mosaic development. In fact, a comparison of these results with those of median and frontal constriction of amphibian embryos (see p. 134) has led Tyler (1930) and Novikoff (1940) to speculate that the polar lobe acts as an 'organiser' of the whole egg (see also Wilson, 1929). In *Ilyanassa*, development fails almost completely

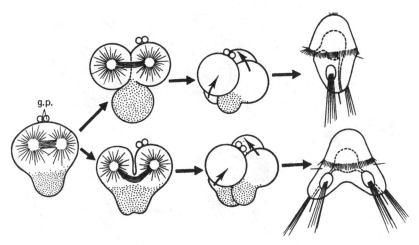

Fig. 3.15. Development in *Sabellaria* after compression perpendicular to the spindle axis during first cleavage. If the vegetal pole plasm (dotted) was segregated to one of the first two cells development was normal (upper row); if it was split between the two cells twinning occurred (lower row). g.p., polar bodies. (From Guerrier, 1970a.)

following equalisation of first cleavage (Styron, 1967), a different response but still a demonstration of the importance of the correct segregation of the polar lobe material.

In work which appears never to have been fully reported, van den Biggelaar and his collaborators obtained evidence that the 'determinants' of the *Dentalium* polar lobe do not exist free in the cytoplasm (see van den Biggelaar, 1978). Polar lobe cytoplasm could be withdrawn from one embryo and injected into a non-D blastomere of another one without affecting the development of either. If lobe membrane was also transferred, however, the recipient quadrant developed lobe-dependent structures suggesting that there is a cell-surface map of developmentally important molecules. However, it is also possible that the important materials are internal but bound to the surface by the cytoskeleton (as seems likely for the *Bithynia* vegetal body: p. 55) and so remain in the lobe when the free cytoplasm is withdrawn.

### (b) Unequal first cleavages

Some species with spiral cleavage do not form a polar lobe but the first two cleavages are unequal, so that a large D cell can still be recognised at the four-cell stage. In at least some of these species, the mitotic apparatus first forms in a central position and then moves towards one side. There is evidence that the D quadrant usually forms on the side opposite the sperm entry point, or centripetally in centrifuged eggs (Morgan & Tyler,

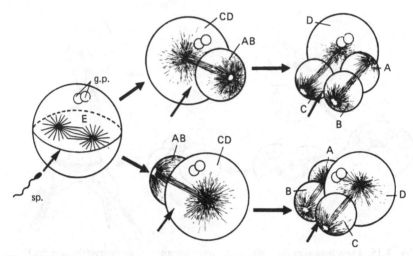

Fig. 3.16. Two alternative patterns for the first two cleavages in the lamellibranch *Pholas*. Note that A–D may be in a clockwise (upper) or anticlockwise (lower) sequence, but that D always forms almost opposite the sperm entry point (sp.), indicated by an arrow at the left. E, equatorial plane; g.p., polar bodies. (From Guerrier, 1970*b*.)

1930; Guerrier, 1970*b*), so that dorso-ventrality is again almost certainly determined epigenetically. As with lobe-bearing embryos, the only important point seems to be that cleavage must be unequal: if a cleavage is experimentally equalised, twins again result (Guerrier, 1970*b*). (If cleavage timing is also disturbed, however, development fails almost completely (Dorresteijn, Bornewasser & Fischer, 1987), providing a possible explanation of Styron's data above.) The enlarged D cell can be achieved by either of two different routes, so that the A to D sequence reads clockwise or anticlockwise (Fig. 3.16), without any apparent effect on the later development of some lamellibranchs. (With some gastropods this reversal would affect left–right asymmetry and thus the direction of coiling, see p. 19.) In the leech *Clepsine* (Schleip, 1914) and the oligochaete *Tubifex* (Penners, 1922*a*), almost all of both the animal and vegetal pole plasms is segregated to the D blastomere (Fig. 3.17). Penners (1926) killed some of the early cells of *Tubifex* and found that CD or D cells alone could regulate to develop as whole embryos even more successfully than the corresponding cells from *Dentalium*. Moreover, Siamese twins are again formed if first cleavage is equalised in *Tubifex* (Penners, 1922*b*) or *Clepsine* (Müller, 1932).

The *Nereis* embryo also cleaves unequally and it is interesting that Costello's (1945) study, so often cited as an example of mosaic development, again reveals that the development of CD or D isolates is far superior to that of other cells isolated at the two- or four-cell stage.

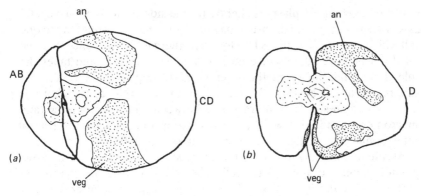

Fig. 3.17. The fate of the pole plasms (an and veg) during the first two cleavages in *Tubifex*. (*a*) at the two-cell stage; (*b*) during the separation of C and D at the second cleavage. (After Penners, 1922*a*.)

Unequal first cleavages and the use of a polar lobe seem clearly to be alternative methods of ensuring the same thing: that the vegetal pole plasm (and the animal pole plasm in leeches and oligochaetes) is segregated to one of the first four cells, polarising the embryo in the dorso-ventral axis. *Chaetopterus* in fact displays both an excentric first cleavage spindle and a small polar lobe, and the special developmental properties of the CD cell are attributable partly to the lobe and partly to other vegetal plasms in the cell (Henry, 1986).

### (c) Equal first cleavages

A D quadrant with similar developmental significance is however recognisable even in some species where the first two cleavages are equal and appear to divide all the visible plasms quite equally, e.g. in *Physa* (Wierzejski, 1905), *Patella* (van den Biggelaar & Guerrier, 1979) or *Lymnaea* (Raven, 1967). As already noted, the orientation of the spindles, however, ensures that only two cells of the four-cell embryo reach the vegetal pole. In normal development one of these will almost always take the developmental role of the D blastomere (Wierzejski, 1905): van den Biggelaar & Guerrier (1979) found one control embryo of *Patella* where this was not true. However, it remains true that any quadrant can act as D, and it seems likely that this is linked with the division of the visible plasms among the first four cells. Perhaps it is the possession of the most vegetal part of the plasm, or of the most vegetal surface, that normally gives an 'advantage' to the cells at the vegetal cross-furrow. Another problem remains, however: why do not both cells united at this furrow become D cells in a Siamese twin formation like those described above for other spiralian embryos? Raven (1963*a*) says that the vegetal pole plasm is slightly to one side of the animal–vegetal axis in *Lymnaea*, in which case it may be segregated preferentially to one cell which normally becomes D. However, there is also evidence that one quadrant achieves dominance during later development and takes the role of D: at the same time the potential of the opposite quadrant seems suppressed and it becomes B. According to this evidence the choice is made as a result of interactions between the micromeres and macromeres, and so it will be considered in a chapter on the role of intercellular interactions in determination (see p. 200).

As might be expected from the discussion above, *Lymnaea* shows greater regulative ability after cell deletions and isolations than other early spiralian embryos. Normal development is possible after the deletion of one of the first two cells, or of one or two of the first four; even at the eight-cell stage one or two macromeres or up to three micromeres can be deleted without preventing development (Morrill, Blair & Larsen, 1973). In these species there may be some predisposition for one of the first four cells to act as D, but it is certainly not an irrevocable determination.

### The role of the D quadrant: later stages

#### (a) Lobe-bearing species

Wilson (1904) was able to follow the fates of the *Dentalium* polar plasms through several cleavages. While the vegetal pole plasm was passed via the polar lobes to the D cell, the animal pole plasm was divided among the first four cells. In the A–C quadrants the first tier of micromeres received unpigmented animal pole plasm, and the second and third tiers primarily animal pole plasm with usually some pigmented cytoplasm as well. In the D quadrants, 1d again received animal pole plasm, but in the 1D macromere both polar plasms mixed and presumably both contributed to the 2d, 3d and 4d micromeres all of which are usually unpigmented. Experimental embryology has now provided data on the role in development played by these cells, although most investigators have used *Ilyanassa* where the polar plasms are unfortunately never recognisable. Clement (1962) developed a cell deletion system for testing the importance of all the d micromeres in *Ilyanassa*. He removed the D macromere from embryos at successive cleavage stages (Fig. 3.18). At the four-cell stage, embryos lacking D showed the expected 'lobeless' characteristics. They formed no external shell, foot, heart, intestine or usually eyes, the stomodaeum was everted and the velum and internal shell material was unorganised. Deletion at the eight-cell stage, leaving an embryo which also has the 1d micromere, allowed very little improvement in the later development; but after 2d had also been formed small external shells

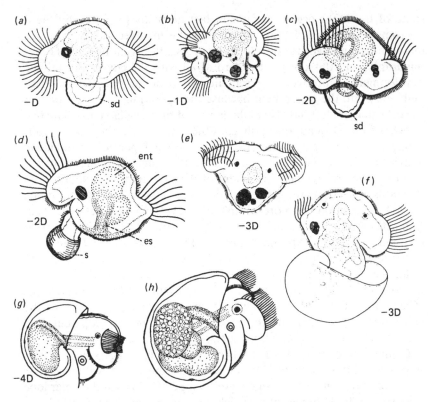

Fig. 3.18. The development of *Ilyanassa* embryos following the removal of the D macromere at various cleavage stages as indicated on the figure. *h* is a normal veliger. The variability of development following the deletion of 2D and 3D is indicated by illustrating pairs of partial larvae. ent, endoderm; es, oesophagus; s, shell; sd, everted stomodaeum. Internal shell material is hatched. (From Clement, 1962.)

sometimes developed. Embryos with the first three d micromeres could produce far more organs: the velum, eyes and shell often being well formed (though the shell was not always external) and a foot sometimes present. The addition of 4d allowed the formation also of heart and intestine, and the embryos could in fact be normal, though small, despite the loss of the 4D macromere. This gradual addition of organs might suggest that the micromeres 2d to 4d, as they are produced, are provided with determinants for certain organs: in other words that the results accord well with the concept of mosaic cleavage. One could imagine that determinants for each organ are present in the vegetal pole plasm which is passed to the 2d to 4d cells.

This interpretation could be literally true for heart and intestine which are thought to be largely derivatives of the 4d micromere. However, some of the other organs are not in fact normally produced by the d micromeres, so an inductive action must be involved. Cells of the D quadrant are inducing the development of specific organs by cells of other quadrants, and this does not accord at all with the concept of mosaic development. Moreover, as Clement (1962) points out, it introduces ambiguity into the experiment because the operations differ not only in the cells that are left but also in the length of time available for inductive effects of the D quadrant upon the other cells. Thus the extra organs produced at each stage may be due to a longer period for inductive activity by the D macromere rather than the inductive influence of the extra micromere. In fact, the extent of development possible after 3D deletions depends on the amount of time allowed since 3d formation: the later the operation the more complete is development (Clement, 1962; Labordus & van der Wal, 1986). Cather & Verdonk (1979) have carried out similar deletion experiments on the D quadrant of *Dentalium*, and also find evidence for both mosaic and inductive effects. In contrast, no inductive effects of the animal pole plasm were seen when the first tier of micromeres were individually deleted from eight-cell *Ilyanassa* embryos (Clement, 1967). The resulting embryos either showed a loss of the structures expected from the fate map (mosaic development), or regulated to develop normally.

Clement (see 1971, 1976), Cather (1967, 1971) and Atkinson (1971) have carried out a variety of other deletion experiments with *Ilyanassa* embryos, revealing a surprisingly great range of effects of the polar lobe material or the D quadrant to which it is segregated. These include: 1) segregation into the d micromeres of factors for heart and intestine (in 4d), the left half of the foot (in 3d) and part of the shell (in 2d); 2) inductive action needed for the formation of the left eye (by 1a and 2a), right eye (by 1c), organised velum and statolith (by 1a to 3a for the left and 1c, 2c and 3b for the right), some of the shell (by 2c) and the right half of the foot and its statocyst (by 3c); 3) an inhibitory influence suppressing shell formation by inappropriate blastomeres, since any quadrant can form internal shell after isolation (Cather, 1967; and see the evidence for inhibitors of ciliation on p. 82); 4) allowing the morphogenetic movements required to give the normal form to organs and the contacts between organs, even for those structures which can be produced by lobeless embryos (Atkinson, 1971); and 5) causing the D quadrant to show a unique pattern of cleavage rate and blastomere sizes.

Two other effects already described from lobeless *Dentalium* development should probably also be included: lobe material is required 6) to give the normal proportion along the animal–vegetal axis exemplified by the relative sizes of pretrochal and post-trochal regions (Wilson, 1904); and 7) to allow any dorso-ventral polarisation of the embryo (van Dongen, 1976). So great is the latter effect that all four quadrants of lobeless embryos develop like normal B quadrants, most obviously by forming stomodaeal structures (see too Verdonk & Cather, 1983); and this may well be true of *Ilyanassa* too (Atkinson, 1971). Evidence for polar interactions in *Ilyanassa* includes the formation of internal shell by almost any combination of animal and vegetal material, including combinations of ectoblasts with the polar lobe (Cather, 1967). The effects on the two developmental axes may be a necessary consequence of all the more specific effects listed before them, but as lobe-bearing species change from concentrically organised oocytes (p. 18) through the acquisition of polarity (p. 51), dorso-ventrality and asymmetry (pp. 78 *et seq.*) the lobe material seems always to be intimately involved.

The role in development of the D quadrant is thus clearly much more extensive than its presumptive fate in *Ilyanassa* and apparently other lobe-bearing species. As recognition of this fact has grown the description of their development as 'mosaic' has been increasingly criticised. As long ago as 1929, Wilson speculated that the lobe material could act as an 'organising centre' comparable with the grey crescent area in amphibians. None the less the other quadrants should not be thought of as devoid of all specific properties, a point which Clement (1976) illustrates by showing that deletion of 2c leads to the development of embryos lacking a pulsating larval heart (as with 4d deletion), and with eversion of the stomodaeum and other disturbances of the digestive tract.

### (b) Species with unequal first cleavages

Of the species where an enlarged D cell is produced by two simple but unequal cleavages, most study has concerned the directly developing oligochaetes and leeches. They provide at the same time good examples of both mosaicism of development and the dominance of the D quadrant. This is because so much of the adult structure normally arises from the D lineage.

Observations of the polar plasms again suggest that they have a leading role in developmental determination. The two pole plasms usually mix at an early stage, and are then transmitted from the D blastomere to the teloblasts. In *Tubifex* Inase (1967) claims that it is animal pole plasm which enters 2d and thus the ectoteloblasts, and vegetal pole plasm that enters 4d and thus the mesoteloblasts. Other authors have been unable to distinguish the two plasms once they mix, and Fernández (1980) speaks of the mixed polar plasms as teloplasm in the leech *Theromyzon*. It is segregated into the M to Q teloblast pairs, each teloblast being polarised with yolky cytoplasm at one pole and teloplasm at the other (Fernández & Stent, 1980). Thus, when stem cells are produced, in the series of unequal cleavages which build up the germinal bands, they receive only teloplasm (Fig. 3.19).

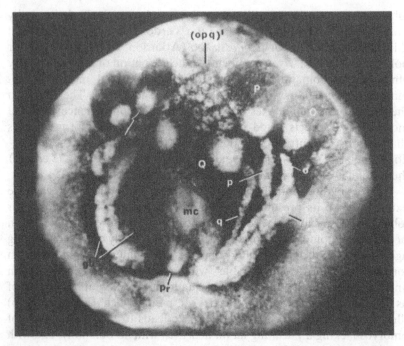

Fig. 3.19. The fate of the teloplasm in the leech *Theromyzon*, as seen in dorsal aspect. Yolk has been extracted leaving the teloplasm showing by its whiteness. It is at one end of the teloblasts (O, P and Q) and is passed via a fenestra (arrow) to the bandlet cells (n, o, p, q) making up the germ bands (gb). The N teloblasts lie beneath the bandlets. mc, micromere cap. × 60. (From Fernández & Stent, 1980.)

We have already seen that the stem cells from each teloblast make a fairly exact contribution to the structure of the adult leech (p. 73), and that deletion of a teloblast leads to specific losses from that structure (p. 76). This extreme mosaicism of development (although see p. 203) suggests that the polar plasms must have developmental determinants. These could be of qualitatively different kinds which would then have to be differentially segregated to enter different teloblasts, but this remains speculation at present. Devriès (1973) has claimed that at least the early stages of teloblast and bandlet formation can occur normally in *Eisenia* half embryos which lack one of the polar plasms.

### (c) Species with equal first cleavages

In *Lymnaea*, where the egg and its visible plasms are equally divided at the first two cleavages, the later behaviour of the plasms has been described by Raven (1948, 1970, 1974). The polar plasms fuse before third cleavage and the greater part of them passes into the micromeres. The six other subcortical accumulations of stainable cytoplasm also fuse together but pass entirely to the four macromeres at third cleavage. A part of their material is segregated into the second and third sets of micromeres, but much accumulates at the vegetal pole where it condenses to form very distinct granules. At the 24-cell stage there is a three-hour pause in cleavage, and during this these 'ectosomes' move inwards along the inner surfaces of the macromeres. It is at this stage that the 3D macromere becomes recognisable, and it is there that the ectosomes move farthest, reaching the centre of the embryo where 3D makes close contacts with the micromeres. Their further fate is uncertain and will be discussed later (p. 201).

Observations of cell behaviour also show the D quadrant attaining dominance at the 24-cell stage. Until then all quadrants have behaved identically, and, when the 24-cell stage starts, all blastomeres appear to extend inwards gradually obliterating the blastocoel. Almost always one of the macromeres united at the vegetal cross-furrow attains the most central position and forms contacts with micromeres from the animal pole (Fig. 6.18). This cell becomes the 3D macromere in *Lymnaea* (Raven, 1974) and *Patella* (van den Biggelaar, 1977a). It is apparently only at this late stage that one quadrant becomes determined to act as D. In *Patella* a macromere at the vegetal cross-furrow can be deleted and another quadrant will usually go on to become D, even if it was not at the cross-furrow (van den Biggelaar & Guerrier, 1979). Similarly, it is usually possible to decide which macromere of *Patella* or *Lymnaea* embryos will become 3D by deleting a specific micromere from the animal pole (Arnolds, van den Biggelaar & Verdonk, 1983). Finally, it is possible to produce partially radialised *Lymnaea* larvae (i.e. to inhibit dorso-ventral polarisation) by various injurious treatments of the cleaving embryo, including the use of lithium ions (see Raven, 1976), heat shocks (Visschedijk, 1953) and sodium azide (Camey & Geilenkirchen, 1970). It was once thought that this abnormal development was caused by damage to a 'cortical map'. Clearly, however, dorso-ventrality arises epigenetically and there is no relevant map, so the effects are probably due to interference with the interactions which should establish D dominance.

There is clear evidence that an inductive interaction occurs between the 3D cell and the micromeres at the 24-cell stage, and that the further development of both cell types is affected (for further details see p. 200). As noted earlier the *Ilyanassa* D quadrant exerts its main inductive effect during the same cleavage stage (Clement, 1962). What is unlike the development of other spiralians already discussed is that the D quadrant is a target as well as a source for inductive interactions. Once the D quadrant is determined in this way it does, however, have a dominant developmental role comparable in many ways with that seen in lobe-

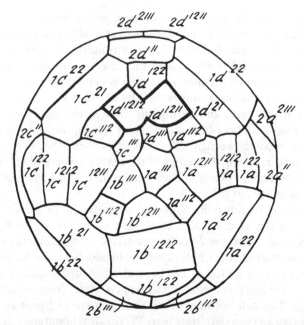

Fig. 3.20. The beginning of bilateral cleavage in the *Lymnaea* head region. In the dorsal arm of the molluscan cross the $1d^{121}$ cell (heavy outline) is dividing parallel to the axis of the arm; in the other three arms the corresponding cells have divided perpendicularly to the arm axis. (From Raven, 1976, after Verdonk.)

bearing species. Seven types of effect were listed there (p. 88), mainly from work with *Ilyanassa*, so it is interesting to consider the evidence for analogous effects in *Physa*, *Lymnaea* or *Patella*: the D lineage still 1) gives rise directly to many important larval and adult structures with the 4d cell for instance still acting as a mesentoblast stem cell (e.g. Wierzejski, 1905; van den Biggelaar & Guerrier, 1979); 2) probably acts inductively on other quadrants: this could explain the production of eyes, etc, by abnormal lineages following micromere deletions (Morrill *et al.*, 1973); 3) may inhibit especially the B quadrant from expressing its potential to act as D (although an alternative is that the B pathway is a sort of 'ground state', and a positive activation is required to achieve D status: see Arnolds *et al.*, 1983); 4) has not apparently been shown to affect morphogenesis and the form of organs; 5) certainly shows a unique pattern of cleavage positions and timing from the 24-cell stage (Wierzejski, 1905; van den Biggelaar, 1971, 1977*a*); 6) probably has smaller effects on animal–vegetal determination than in lobe-bearing species: the possibility that radialised embryos are also disturbed in this axis is discussed by Raven (1948, 1976); and 7) is apparently required for the dorso-ventral polarisation of the embryo.

The D quadrant displays its unique cleavage pattern even at late stages, and provides the first sign of dorso-ventrality in the *Lymnaea* head region, which arises from a 'molluscan cross' of micromeres (Verdonk, 1968*b*; see Fig. 3.20). In the dorsal arm of the cross the 1d$^{121}$ cell divides in parallel to the axis of the arm and the two daughters do not divide again. In the other three arms of the cross, the equivalent cell divides perpendicularly to the axis of the arm, and the daughters may divide further to form small cells difficult to distinguish from the tentacular field cells beside them.

### The fine-structure and biochemistry of spiralian development

Having just considered the special developmental properties of the D quadrant, it is interesting to start this section with evidence of its fine-structure. Accumulations of mitochondria have been reported in the d micromeres of *Dentalium* (Reverberi, 1958), and the 4d cell of the slug *Arion* (Sathananthan, 1970) and of *Tubifex* (Lehmann, 1958). In *Ilyanassa*, mitochondria are abundant in all micromeres while the 4d cell is marked by a particular accumulation of lipid (Clement & Lehmann, 1956). In *Succinea* (Jura, 1960) and *Lymnaea* (Raven, 1966), 4a, 4b and 4c seem more glycogen rich than 4d. Lehmann (1958) also stressed the fine-structural differences between 2d and 4d. Since developmentally important materials may be anchored beneath the surface (see Chapter 2), it is worth noting that the specialised vegetal surface on many lobe-bearing eggs is segregated to the D cell (Dohmen & van der Mey, 1977) and that the distribution of cortical microtubules in *Lymnaea* may correlate with that of the subcortical plasms (Morrill & Perkins, 1973).

The biochemical features of early spiralian embryos are generally similar to those known for other animal groups (for reviews see Kidder, 1976; Collier, 1983*b*). The early cleavage cycles are rapid, leaving little time for RNA synthesis, and the RNAs which are produced (at least those transcribed from reiterated genes) seem to be the same species as were provided maternally. It is possible that maternal transcripts play a more extended role in spiralians than in more regulative embryos as *Ilyanassa* gastrulates, and *Lymnaea* forms abnormal larvae, when transcription is very strongly inhibited (Newrock & Raff, 1975; Morrill, Rubin & Grandi, 1976). Studies of protein synthesis patterns also reveal that the major changes are under maternal control, although zygotic transcripts appear to modulate the translation of maternal ones.

Most attempts to study the biochemical basis of developmental phenomena in spiralians have concerned the polar lobe of *Ilyanassa*. Lobeless embryos have the same cell number and DNA content as controls until very late stages (Collier, 1975), but already transcribe much less RNA by the 25-cell stage (Collier, 1977). By postgastrula stages the

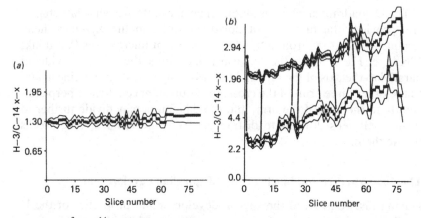

Fig. 3.21. [$^3$H]:[$^{14}$C] label ratios in *Ilyanassa* proteins electrophoresed in one dimension. (*a*) comparing two groups of control embryos labelled for 100 min in the second to third quartet stage. (*b*) two experiments in which [$^{14}$C] labelled AB-derived embryos were compared with [$^3$H] labelled CD-derived ones. Acceptable error limits were set as 10% in the upper trace on (*b*) and 15% elsewhere. Vertical lines indicate landmark differences which are repeatable and statistically significant in one or both experiments in (*b*). (From Donohoo & Kafatos, 1973.)

transcripts of normal and lobeless embryos are qualitatively different (Davidson *et al.*, 1965; Koser & Collier, 1976), but no qualitative differences could be found at the earlier stages when determination is probably occurring. In contrast, the profile of newly synthesised proteins, as seen in one-dimensional gels, appears to be affected much earlier by the lobe, as profiles for the progeny of the AB and CD cells are clearly different (Donohoo & Kafatos, 1973; see Fig. 3.21). Such profiles for normal and lobeless embryos also diverge even if transcription is more than 90% inhibited (Newrock & Raff, 1975), which strongly suggests that the differences are due to the translation of maternal mRNAs. However, the differences are not seen if the proteins are compared after two-dimensional separations (Brandhorst & Newrock, 1981; Collier & McCarthy, 1981). The data from one- and two-dimensional separations can most easily be reconciled if the polar lobe and the rest of the egg contain largely the same mRNA species but at different relative con-centrations.

There is circumstantial evidence to link maternal RNAs with important developmental phenomena in other spiralians, as RNA concentrations are found in the polar lobes of *Bithynia* and several other species (Dohmen & Verdonk, 1979a; Jeffery & Wilson, 1983) and in the sub-cortical accumulations and ectosomes of *Lymnaea* (Raven, 1974; see also Minganti, 1950). Although the latter are divided equally among the four macromeres, their behaviour in the 3D cell is distinctive and suggests a role in developmental interactions. Arnolds (1982*a,b*) has described a maternal-effect mutation which apparently causes the deletion of the ectosomes from *Lymnaea* embryos: at the 24-cell stage all the macromeres behave identically and the further dorso-ventral determination of the embryo is highly abnormal (see also p. 202). At present, therefore, maternal mRNAs appear to be our best candidates for the factors which confer determinative properties on the D quadrant, but we should remember that other biochemical aspects of spiralian development have been little studied. Arnolds (1982*b*) suggested that an undefined change in cell membrane properties might be a more important feature of his mutant embryos.

## Conclusions

This chapter considers determination in embryos with spiral cleavage patterns. Each of their cells often makes a precise contribution to the larva or even the adult, and ablation or isolation studies show that by late cleavage stages they may be determined for their specific fates. However, this does not mean that cleavage simply cuts up a preformed pattern inherited from the zygote. In *Cerebratulus* there is evidence that determinants are moved along the animal–vegetal axis during the first cleavages. In a large number of other species, one of the first four cells, the dorsal one D, receives at least the vegetal pole plasm, and after isolation it can form far more than it would normally do while the other three cells form less than they should. During further development the D quadrant takes a dominant role affecting the development of the other quadrants so profoundly as to almost 'organise' the whole embryo. A last group of species including *Lymnaea* shows a fairly equal division of plasms and developmental potential at the first two cleavages, but in later development one quadrant (again called D) attains dominance and shows a range of 'organising' effects. Maternal RNAs are at present the best candidates for the developmentally significant materials in the D quadrant.

# 4

## The limits of mosaicism in non-spiralian cleavage

The last chapter showed severe limitations in the concept of 'mosaic' development as it applies to spiralian embryos. It did, however, demonstrate differential segregations occurring at cleavage there, and seen for visibly distinct plasms, some macromolecules and determinants inferred from experimental studies. To what extent are differential segregations and mosaicism general phenomena of cleavage? This chapter will present the available evidence for other animal groups in an order chosen simply for convenience.

### Ascidians

Although they are a deuterostome group quite closely related to the vertebrates, it has long been recognised that the ascidians provide one of the best examples of mosaic development. We have, in fact, seen in Chapter 2 that this is foreshadowed in the fertilised egg by probably the clearest example of ooplasmic segregation into a bilaterally symmetrical pattern, accompanied by restrictions in the development of egg fragments. Figure 4.1 shows some stages in the segregation of the ooplasms by the early cleavages in *Styela*. Conklin (1905a) was able to show that the fates of the blastomeres correlated closely with the type of ooplasm each received. His fate map was later modified first by Ortolani and most recently by the lineage tracer studies of Nishida & Satoh (1983, 1985; Nishida, 1987; see also Zalokar & Sardet, 1984) from which the maps and lineage plan of Figure 4.2 are taken. Although the limits of the visible plasms are difficult to define, it seems broadly that cells receiving yellow plasm form mesoderm, dark grey plasm endoderm, pale grey plasm notochord and nervous system and clear plasm epidermis. Conklin (1905a) believed that the prospective areas for tissues such as muscle and notochord were already confined to single pairs of cells at the eight-cell stage, and the fact that they are not greatly complicates our interpretation of cell isolation and ablation studies, as we will see below.

Ascidian embryos develop extraordinarily rapidly: *Ascidia malaca* for example forming a swimming larva within seven to eight hours of

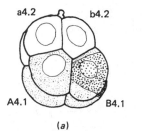 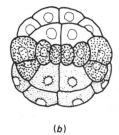 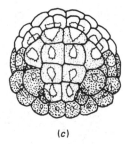

(a)  (b)  (c)

Fig. 4.1. Segregation of some of the visible plasms during *Styela* cleavage. (*a*) 8-cells in lateral view showing the notation used for the cell pairs. (*b*) 44-cells, posterior view. (*c*) 74-cells, vegetal view. Coarse dots, myoplasm; fine dots, endoplasm. (After Conklin, 1905*a*.)

fertilisation. This can allow very little time for the interactions between cells which would be required for regulative development. Thus mosaicism is probably related with speed of development at least in this group (see Conklin, 1905*b*).

First cleavage is in the plane of bilateral symmetry of the visible plasms, and so divides all plasms equally. Despite this even the daughters of first cleavage are unable to regulate to form whole embryos in ablation or isolation studies. In the first study of development in any embryo after killing a blastomere, Chabry (1887) showed that such a half embryo went on to form a lateral half larva with very little regulation. Driesch (1895) and Crampton (1897) claimed greater regulative ability, but Conklin (1905*b*) confirmed the halving of the larval form right down to the numbers of notochordal cells or of rows of muscle cells. He and Cohen & Berrill (1936), who isolated blastomeres rather than killing one of them, showed that the main regulation was in slight changes of cleavage direction or cell movements, allowing for instance the epidermis to cover the embryo on the injured side. Nakauchi & Takeshita (1983) have recently followed half larvae through metamorphosis when in fact some regulate and produce normal adults. In view of the very limited regulation in half embryos, it is surprising that when two two-cell (or even eight-cell) embryos are experimentally fused together they can sometimes develop to one large but normal embryo (von Ubisch, 1938; Reverberi & Gorgone, 1962). This must surely require that the blastomeres can either rearrange or regulate their fates far more extensively than the work with half embryos indicates.

Second cleavage divides the ascidian embryo into what are called, from their fates, anterior and posterior halves; and third cleavage separates animal and vegetal parts. The eight-cell embryo is thus made up of four cell pairs each with a distinctive ooplasmic make-up and distinctive fates in the larva. After killing some cells at the four- or eight-cell stage,

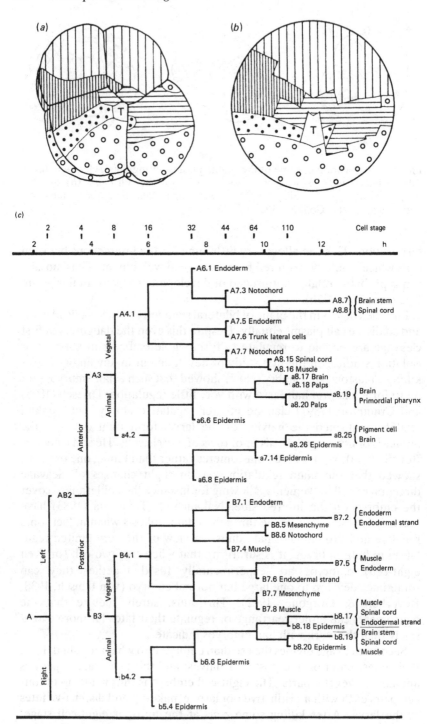

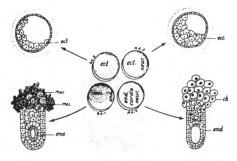

Fig. 4.3. Development of cell pairs isolated from the eight-cell ascidian embryo: see text. ch or corda, notochord; ect, epidermis; end, endoderm; mes, mesenchyme; mus, muscle; neur, neural tissue. (From Reverberi & Minganti, 1947*b*.)

Conklin (1905*b*) reported that anterior, posterior and vegetal 'halves' formed the tissues expected from their make-up. Animal 'halves', however, form only undifferentiated ectodermal cells (Conklin, 1905*b*; Rose, 1939), sometimes accompanied by unorganised neural tissue (Tung, 1934; Reverberi & Minganti, 1947*a*). Reverberi & Minganti (1947*b*) isolated all four cell pairs from eight-cell embryos, obtaining the results shown in Figure 4.3. The tissues formed by the posterior vegetal cell pair (B4.1) were those expected from the fate maps of the time, and the anterior vegetal pair (A4.1) also formed endoderm and notochordal cells as expected. However, the A4.1 pair failed to form the expected spinal cord tissue, and the undifferentiated ectoderm formed by both animal cell pairs (a4.2 and b4.2) implies particularly a failure by the a4.2 pair to form brain and sense organs. Moreover, the recent cell lineage work cited above suggests that both the A4.1 and the b4.2 pairs normally produce a little muscle but it was not seen in most progeny of these pairs in Reverberi & Minganti's (1947*b*) isolates. Neither did the B4.1 pair produce any notochord.

This is one of the best examples of mosaic development known at such a stage, but even here all four pairs of cells apparently fail to produce some of the structures expected from their lineages. There are several possible explanations for this. Small numbers of differentiated cells may

Fig. 4.2. Fate maps and a lineage plan for the ascidian *Halocynthia roretzi*. (*a*), (*b*) fate maps in lateral view with the posterior side to the right, for the eight-cell stage (*a*) and blastula (*b*). Spaced vertical hatching, epidermis; close vertical hatching, neural tissue; horizontal hatching, muscle and mesenchyme; filled circles, notochord; open circles, endoderm; T, trunk lateral cells. (*c*) lineage plan for the left half of the embryo: divisions occurring after a blastomere becomes tissue-restricted are not shown. ((*b*),(*c*) from Nishida, 1987, (*b*) redrawn; (*a*) original, constructed from data in Nishida, 1987, for comparison with Figs 4.1(*a*) and 4.3.)

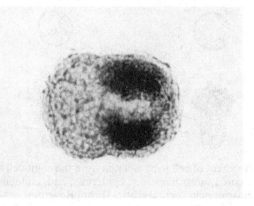

Fig. 4.4. A 9 h *Ciona* embryo stained to show acetylcholinesterase activity. The enzyme is present at high level in the muscle lineage and at low levels elsewhere. (From Whittaker, 1973*a*.)

have been missed in the analysis of the isolates, particularly when that cell type was not expected from the fate maps of the day. The culture conditions used may have been inadequate for the expression of certain potentialities even if they are determined within the cell. Even in ideal conditions such expression might fail because the number of contiguous cells determined for the same fate is below a certain threshold. However, in at least some cases it appears that fates are not adequately determined in the cells and require an inductive interaction with other cells. Reverberi & Minganti (1947*b,c*) themselves gave evidence for several such interactions after culturing cell pairs from eight-cell embryos in various combinations. In the anterior half of the embryo the induction appears to be mutual between the animal a4.2 and vegetal A4.1 lineages, causing brain development in the former and spinal cord development in the latter. The induction of brain by the A4.1 lineage had in fact already been described by Tung (1934) and Rose (1939). Reverberi & Minganti (1947*b,c*) believed that other inductive and inhibitory interactions were also involved because the presence or absence of the posterior blastomeres also seemed to affect brain determination.

Because ascidian development is so rapid, the various cell lineages form differentiated products after only a very short time. The formation of enzymes with moderate or high tissue-specificity has been a particularly useful marker for studies on determination (see Fig. 4.4). Tyrosinase activity appears in the presumptive sense organs of neurulae (Minganti, 1951), alkaline phosphatase activity increases specifically in the postneurula endoderm, although a lower level is always present in all cells (Minganti, 1954*a*), and acetylcholinesterase activity appears in presumptive muscle cells of neurulae (Durante, 1956). This last activity is also present at low levels at preneurula stages, and after hatching is present in the brain too (Meedel & Whittaker, 1979). Many of the tissues also form structural markers recognisable in the electron microscope: Crowther & Whittaker (1984) show that muscle cells, the pigment and sensory cells of the brain, notochordal cells and the test material secreted by epidermal cells can all be recognised in this way.

These markers provide new ways of testing the developmental potential of isolated blastomeres. They provide confirmation that the B4.1 pair is determined for muscle cell formation, as the progeny of B4.1 isolates develop acetylcholinesterase activity (Whittaker, Ortolani & Farinella-Ferruzza, 1977), myofilaments and myofibrils (Crowther & Whittaker, 1983). The formation of notochord by A4.1 isolates has also been ultrastructurally confirmed (Crowther & Whittaker, 1984). Isolated animal halves from eight-cell embryos were shown above to form undifferentiated ectoderm, but ultrastructural study demonstrates that they differentiate sufficiently to secrete a test (Mancuso, 1974; Crowther & Whittaker, 1984). Other results with these markers have however been confusing. While only vegetal cells form histologically recognisable gut after isolations at the eight-cell stage, both halves form alkaline phosphatase and it forms earlier than in control embryos (Minganti, 1954*b*). Animal halves apparently do not form the ultrastructural markers for brain pigment or sensory cells (Crowther & Whittaker, 1984), but there is disagreement about whether the cells form tyrosinase. Minganti (1951) found that neither half produced the enzyme after eight-cell isolations, while a review by Whittaker (1979*d*) claims that the a4.2 cell pair can produce it autonomously. These studies all support the conclusion that neural and sensory structures require induction; but Whittaker's work would suggest that the lineage is determined to form some tissue-specific proteins, and that induction may be needed only to organise these proteins into structures.

Molecular markers have also been used to test determination in lineages making minor contributions to a particular tissue. Whittaker *et al.* (1977) detected no acetylcholinesterase activity in embryos developing after the removal of the B4.1 pair. New studies were undertaken once it was known that both the A4.1 and b4.2 lineages normally contribute to muscle, and it is clear that these lineages can sometimes, and in some species, develop molecular markers of muscle autonomously (Deno, Nishida & Satoh, 1985; Meedel, Crowther & Whittaker, 1987; Nishikata *et al.*, 1987a). Nishikata *et al.* (1987a) suggest that these lineages bear muscle determinants, but that differentiation fails if they do not reach a threshold concentration. Meedel *et al.* (1987), however, believe that in *Ciona* muscle may be determined mosaically within the B4.1 lineage but inductively elsewhere, and they suggest that a similar difference may exist between the A4.1 and non-A4.1 contributions to notochord.

Further work should resolve at least some of these difficulties, but at present we can add little to the conclusions of Reverberi & Minganti (1947*b*). Cells of the vegetal half are determined for the formation of muscle, notochord and endoderm, presumably due to their possession of myoplasm, chordoplasm and endoplasm respectively. They are also presumed responsible for the different determination of anterior–vegetal and posterior–vegetal cells, although the precise extent of the three tissue types may still not be determined at the eight-cell stage. Animal halves show sufficient epidermal determination to produce a test. Interaction between the halves is required before either can form neural or sensory structures.

What is the nature of the determinants? Conklin (1905*a,b*) concluded that they were materials in the ooplasms, and recently new kinds of evidence for this have been presented. Tung *et al.* (1977) caused nuclei to populate cytoplasmic areas other than those they usually enter, and showed that it is not the nuclei which decide the different developmental fates. Whittaker (1982) has experimentally changed *Ascidia* cleavage planes so that some myoplasm enters the b4.2 cells; only after this manipulation was acetylcholinesterase activity detected in the b4.2 lineage. Work reviewed in Chapter 2 suggests that the determinants are not in a cortical map (see p. 53) nor in the larger organelles of the ooplasms (see p. 56). Deno & Satoh (1984) give some positive evidence that the determinants are internal, since enzyme activities sometimes appear in extra cells following cytoplasmic transfer. The possible role of mitochondria in muscle determination has been further investigated by Whittaker (1979*b,c*) using ascidian species which never develop a larval tail. In species with a tail, mitochondria are segregated to the muscle lineage of the tail (Berg, 1956; Reverberi, 1956). The species which lack a tail show no such segregation, but in some cases acetylcholinesterase still appears in the appropriate lineage. According to Whittaker (1979*c*) this shows that muscle determinants are not located in the mitochondria. This may be overstating the case, since no 'tail' myofibrils ever develop in these ascidians (Whittaker, 1979*b*), but the conclusion is also supported by egg centrifugation studies (p. 56) and is probably correct. The determinants are probably in the 'ground-substances' of the various ooplasms as Conklin (1931) concluded. They may be associated with the different cytoskeletal domains recognised by Jeffery's group (p. 43), and which persist at least to the 32-cell stage (Jeffery & Meier, 1983). Attempts to identify determinant molecules are still in their infancy (Nishikata *et al.*, 1987*b*), but we already have some information about how they operate.

A new way of investigating how determinants act was provided by the well-known studies of Whittaker (1973*a*, 1977) in which cleavage was arrested by a variety of drugs but tissue-specific enzyme activities still

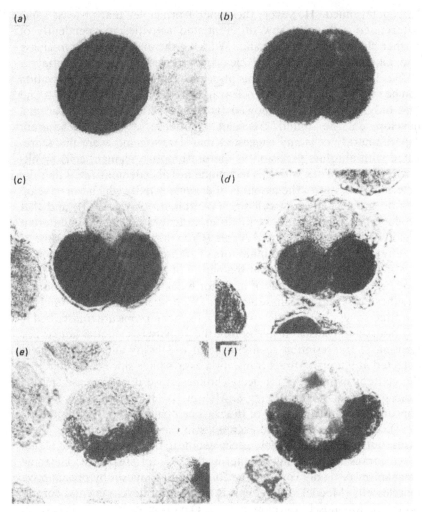

Fig. 4.5. Acetylcholinesterase activity at 12–14 h in cleavage-arrested *Ciona* embryos. The number of cells showing activity is one at 1-cell (*a*), two at 2-(*b*), 4-(*c*) or 8-cells (*d*), four at 16-cells (*e*) and six at 32-cells (*f*). (From Whittaker, 1973*a*.)

appeared at the usual times. Whittaker claimed that the maximum number of cells forming these enzymes was always appropriate for the lineage. For example, acetylcholinesterase activity developed in both cells of arrested two-cell embryos, in two cells of arrested four- or eight-cell embryos, etc. (see Fig. 4.5). There is a difficulty here, since Nishida & Satoh (1983, 1985) now claim that the muscle lineage is wider than was thought, and it would seem that the precise cell numbers involved should be redetermined. However, the general principle, that at least some determined cells can show differentiated activities independently of further cleavages, remains valid. Work on the expression of tyrosinase and alkaline phosphatase in cleavage-arrested embryos (Whittaker, 1973*a*, 1977) also supported this principle, although again a reservation can be made over detail. After cleavage arrests Whittaker (1973*a*) found that only two cells ever showed tyrosinase activity, while in normal development he found transient tyrosinase activity in adjacent supernumerary cells and suggested that determinants were not segregated with absolute precision to the presumptive pigment cells (Whittaker, 1973*b*). Work with this technique has also demonstrated that the determinants control the amounts of enzyme activity which can develop despite great changes in cell size (Whittaker, 1979*a*, 1983), and that products characteristic of several different tissues can appear in different regions of a cleavage-arrested zygote (Crowther & Whittaker, 1986).

Whittaker has also used inhibitors of transcription and translation to give further information about the nature of the determinants and their mode of action. Both kinds of inhibitor block the appearance of acetylcholinesterase and tyrosinase activities (Whittaker, 1973*a*). The timing of the effects suggests that mRNAs for the former enzyme are transcribed at 5–7 hours and translated just before the activity is seen at 8 hours. For tyrosinase, transcription seems to be at 6–8 hours and the enzyme is detected at 9 hours. Direct translation assays have since confirmed that the content of mRNAs for acetylcholinesterase does increase sharply from preneurula stages (Perry & Melton, 1983), and that this mRNA appears only in the progeny of the B4.1 cell pair (Meedel & Whittaker, 1984). The nature of the determinants in the egg is not clear in these cases, but at least one of their effects seems to be the induction of tissue-specific transcription in nuclei entering the appropriate plasms. The same mechanism probably controls the formation of myosin by presumptive muscle cells (Meedel, 1983). This is the type of developmental control proposed for mosaic development by Davidson & Britten (1971) and considered further on p. 317.

The appearance of alkaline phosphatase activity in the endodermal lineage, at 5–7 hours, can be prevented by translational inhibitors but not by transcriptional ones even if these are continuously applied from fertilisation (Whittaker, 1977). This suggests that there is maternal RNA for the enzyme in the egg, and that either the mRNA or factors allowing its translation are localised in the endoplasm. Bates & Jeffery (1987) have queried these conclusions since only nucleate fragments of ascidian eggs could go on to form the enzyme, but the role of the nucleus could rather be in controlling the time of translation, via counting of replication cycles (see later). In any case Jeffery *et al.* (1986) confirmed directly that mRNAs for another tissue-specific protein, muscle actin, are present in the egg, although unfortunately they did not test whether anucleate egg fragments could go on to form this protein. If alkaline phosphatase is indeed maternally coded, its production by isolated animal half embryos (Minganti, 1954*b*) would indicate that it is not the mRNA which is localised. This result also indicates that the system controlling formation of this enzyme involves interactions between the halves, although such animal isolates do not develop as endoderm so the determination of a tissue's enzymes and its morphology are separable here.

Whittaker's work with cleavage-arrested embryos raised the question of the timing controls for tissue-specific syntheses: clearly the counting of cleavages was not used. However, DNA is still replicated cyclically in the non-dividing cells (Satoh & Ikegami, 1981*b*), and this too must be halted to test whether replication cycles are counted as a timing control. Satoh & Ikegami (1981*a,b*) achieved this with aphidicolin, and found that 64-cell embryos did not then produce acetylcholinesterase; such embryos have completed six cleavage cycles and are in the seventh DNA replication cycle. When Satoh & Ikegami added aphidicolin at a succession of later stages they found that an increasing number of cells in the muscle lineage produced the enzyme, and it was always those in the eighth replication cycle (or later) which did so (Fig. 4.6). This and later evidence (see Mita-Miyazawa & Satoh, 1986) strongly suggests that the eighth cycle is 'quantal' for acetylcholinesterase appearance, i.e. that nuclei then become able to react (directly or indirectly) to muscle determinants, probably by transcribing the genes required for muscle development. In the same way, the eighth cycle appears to be quantal for tyrosinase production in presumptive sensory cells, and the sixth cycle for alkaline phosphatase in the endoderm (Satoh, 1982*a*). For the first two enzymes it has already been shown above that transcription occurs at approximately (although not exactly) the time of the quantal cycle. If the mRNA for alkaline phosphatase is already present in the egg, the quantal cycle would seem to play a different part here, e.g. it could begin a process leading to the later translation of this mRNA. In any case, we know more about the action of determinants in the case of these tissue-specific enzyme activities in ascidians than we do in almost any other instance of determination.

It is still not clear how far the mechanisms of determinant action described above apply even to other aspects of early ascidian determination. We have already seen that controls over tissue-specific enzyme synthesis can sometimes be separated from those for tissue fine-structure. The morphogenetic events which decide early embryo form may be controlled in yet another way. In a general review of timing controls in early development, Satoh (1982*b*) suggests that replication cycle counting may be used to time differentiation, while a cytoplasmic cycle may time morphogenetic events like the onset of gastrulation. This book is

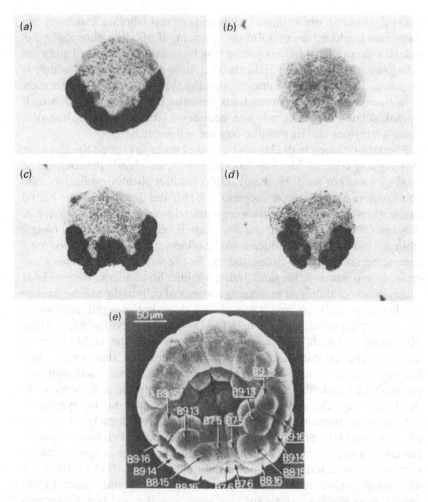

Fig. 4.6. The dependence of acetylcholinesterase (AChE) development on DNA replication in *Halocynthia*. AChE develops in the muscle lineage if cleavage is arrested at 64-cells with cytochalasin (*a*), but not if aphidicolin is also present to block DNA replication (*b*). If arrest is at the early gastrula, AChE again develops in the muscle lineage with cytochalasin (*c*), but in only a part of this lineage with aphidicolin (*d*). (*e*) is a scanning electron micrograph of an early gastrula in which 16 presumptive muscle cells are labelled. Two cell pairs at the midline are in the seventh cell generation while the more lateral ones are in the eighth or ninth generation. It is apparently the latter cells which are labelled in (*d*). (From Satoh & Ikegami, 1981*a*.)

concerned with spatial determination, and from that standpoint both the aspects considered by Satoh (1982*b*) will require local activities and therefore localised information. Some common control must surely be needed to ensure that differentiation and morphogenesis occur harmoniously in the embryo, but, beyond this, we should remember that different mechanisms may operate.

At present, our knowledge of all other aspects of early ascidian determination is so sketchy that any conclusions must be speculative. However, it has already been noted (p. 19) that ascidian androgenetic hybrids (with a nucleus from one species in the cytoplasm of another) show the morphology of the species donating the cytoplasm even as larvae (Minganti, 1959a). Normal interspecific hybrids also show maternal characteristics to stages well beyond those usual in other animal groups (Minganti, 1959b). Thus species-specific details of morphogenesis seem to be under maternal control and it would be interesting to know how far this is true of morphogenesis generally. It may be noted here that many early cytochemical, biochemical and inhibitor studies failed to find any evidence for RNA synthesis in early ascidian development (see Lambert, 1971). It is now clearly established that RNA synthesis does occur even at early cleavage stages (Smith, 1967; Meedel & Whittaker, 1978; and see studies above for specific mRNAs), but the difficulty in establishing this suggests that new transcription is at a relatively low level, and has relatively low significance, in early ascidian development.

Other aspects of early ascidian development have received very little study. Respiratory metabolism is of some interest as mitochondria are localised so early, and oxidation–reduction indicator dyes reveal other spatial differences in the embryo (Child, 1951). However, evidence has already been given that the mitochondria are not required for muscle determination, and there is apparently no difference in the oxygen consumption of anterior and posterior blastomeres (Holter & Zeuthen, 1944). Many mitochondria will be required in the larval tail muscle, and the most efficient way of providing them may be to localise them to the presumptive muscle area in the zygote, particularly as development is so rapid. The association between muscle determinants and mitochondria would then be topographic but probably not causal. For other aspects of metabolism we do not even have the topographic information.

## Amphioxus

Amphioxus is thought to have a close evolutionary relationship with the ascidians, and its resemblance to the ascidian larva is the main evidence of this. Conklin's (1932) account of development makes the extent of the similarities particularly clear. Both the ooplasmic flows and the early cleavages occur in a similar pattern, although the plasms are more difficult to distinguish than in many ascidians. Fate maps have been produced for *Branchiostoma lanceolatum* (Conklin, 1933) and *B. belcheri* (Tung, Wu & Tung, 1962a), and again show a general similarity with ascidians. The chief difference is that the mesodermal crescent

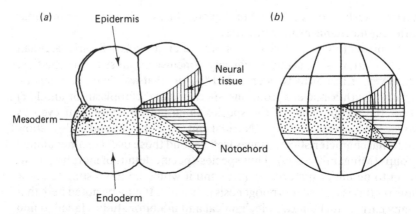

Fig. 4.7. Fate maps of the *Branchiostoma belcheri* embryo at the 8- (*a*) and 32-cell (*b*) stages. (Redrawn from Tung *et al.*, 1962*a*.)

apparently spreads much further anteriorly, at least in *B. belcheri* (Fig. 4.7).

The development of isolated blastomeres shows relatively slight, though interesting, differences from comparable ascidian work. All authors are agreed that a half embryo separated at the two-cell stage can regulate to a small but normal larva (Wilson, 1893; Conklin, 1933; Tung, Wu & Tung, 1958). This demonstrates greater regulative ability than in ascidians, but Conklin (1933) stressed that all the plasms are still present and suggested that their positions were adjusted but their fates did not change. The same may apply to isolates from the four-cell stage. All three studies cited above found quarter-sized embryos with a more complete representation of larval tissues than is seen in ascidians, but this may be explained by the differences in the fate map particularly as mesoderm spreads so far anteriorly. It is notochord and neural tissue which are usually lacking from such embryos (Tung *et al.*, 1958). Some regulation to half-sized larvae was also reported by Tung *et al.* (1958) after a separation along the second cleavage plane, but again this need not mean a change of fates as they found that some eggs separate anterior and posterior halves at first cleavage and left and right halves at second cleavage. Following horizontal separations of blastomeres at the eight-cell stage or later there is little evidence of regulation. Tung, Wu & Tung (1960*a*) separated all four tiers of cells at the 32-cell stage and found that they developed largely in accordance with the prospective fates shown in Figure 4.7. The main exception (as in ascidians) was the failure of almost all isolates from the animal half to form neural tissue (see also Conklin, 1933; Tung, Wu & Tung, 1962*b*).

When these various tiers of cells are recombined in abnormal ways, however, a surprising degree of regulative ability is revealed (Tung *et al.*,

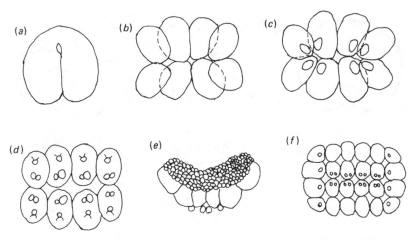

Fig. 4.8. Early development in ctenophores. (*a*) 1st cleavage; (*b*) 8-cells; (*c*) 16-cells; (*d*) 32-cells; (*e*),(*f*) stages in the cleavage of the oral micromeres. (*a*),(*e*) lateral views; (*b*)–(*d*) aboral views; (*f*) oral view. (Redrawn from Schleip, 1929; Reverberi, 1971; Freeman & Reynolds, 1973.)

1960*a*). This indicates that determination is also affected by interactions along the animal–vegetal axis, a topic for a later chapter (see p. 198).

## Ctenophores

The ctenophores are another of the animal groups which apparently show great mosaicism at cleavage stages, although they have been much less studied than ascidians. Their normal cleavage pattern (see Reverberi, 1971) is unusual and is illustrated as Figure 4.8. The first two cleavages are vertical, the only notable feature being that the furrows only cut in from one side, the future oral pole. It has been shown earlier that this is where polar bodies are usually given off, but in ctenophores the poles are described as oral and aboral rather than animal and vegetal respectively. Another unusual feature is that embryos float with their aboral surface uppermost. Third cleavage is oblique forming a curved oval plate of cells, with four slightly smaller external E cells rather above four inner M cells. At fourth cleavage, all eight cells bud off small micromeres (called $e_1$ and $m_1$ cells) at the aboral (upper) pole, and at fifth cleavage a second micromere set ($e_2$ and $m_2$ cells) is formed there while the first set also divides. A third set is then formed by the E macromeres only and is called $e_3$. All macromeres divide once equally before budding off micromeres at the oral (lower) pole, with this time the M macromeres forming two micromere sets and the E macromeres only one. Further development shows a curious inversion of the animal–vegetal polarity seen in other invertebrate groups which produce swimming larvae; the apical organ

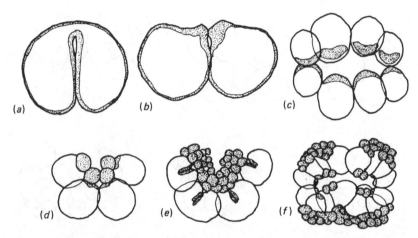

Fig. 4.9. The distribution of ectoplasm (dotted) during *Beroë* development. (*a*) during and (*b*) after first cleavage. (*c*) 8-cell. (*d*),(*e*) half embryos during formation of the aboral micromeres and, in (*e*) further cleavage of the macromeres. (*f*) formation of oral micromeres from the M macromeres. (After Spek, 1926.)

arises at the aboral pole and the blastopore at the oral pole. Because of the way the embryo floats these structures do, however, appear in their normal orientations relative to gravity.

Fate mapping of the early ctenophore blastomeres is still incomplete and sometimes contentious. Reverberi's group have followed cell fates with coloured chalk granules and believe that the aboral micromeres form ectoderm, the macromeres endoderm and the oral micromeres mesoderm (Reverberi & Ortolani, 1963; Ortolani, 1964; Reverberi, 1971). More specifically they find that $e_1$ micromeres form the rows of larval comb plates, $m_1$ the apical organ, $e_2$ general ectoderm and $m_2$ stomodaeum. The oral micromeres of the E lineage apparently contribute muscle, and those of the M lineage mesenchyme. Freeman & Reynolds (1973), however, suggested from indirect evidence that some of the oral micromeres of the M lineage give rise to photocytes, light-producing cells found near the comb plates of the larva.

In Chapter 2 it was suggested that those ctenophore egg fragments showing the greatest developmental potential are those with most ectoplasm. It is therefore interesting to follow the partition of ectoplasm during cleavage (Yatsu, 1912; Spek, 1926; and see Fig. 4.9). During the first two cleavages large amounts of the ectoplasm can be seen to move with the cleavage furrow, with the result that all visible ectoplasm is in clearly defined areas at the aboral ends of the cells at the four-cell stage. At third cleavage both the E and the M cells receive ectoplasm. At fourth cleavage almost all of it enters the $e_1$ and $m_1$ micromeres, and of the small amount remaining in the macromeres most enters the $e_2$ and $m_2$ cells at fifth cleavage. The rest moves with the cleavage furrows that divide the macromeres equally and is then passed to the oral micromeres. The latter thus contain ectoplasm and a little yolk, a composition which Freeman & Reynolds (1973) think favours photocyte formation (see p. 56). It is tempting to speculate that the early movement of ectoplasm along the oral–aboral axis may relate with the later inverted nature of development relative to other groups. As to its more specific role in development it clearly enters cells with important fates in development, but cells receiving similar amounts of ectoplasm show different fates. Reverberi (1957) has shown that this is true specifically for the mitochondria in the ectoplasm, as they enter both the $e_1$ and $m_1$ cells.

Cell isolation studies indicate a fairly strict mosaicism of development. One of the first two cells forms an apparent half-larva with half the number of comb plate rows, etc. (see Driesch & Morgan, 1895). Despite earlier confusion, Farfaglio (1963) seems to have established that each E cell isolated at the eight-cell stage forms two rows of comb plates while each M cell forms none; and that after fourth cleavage the $e_1$ micromeres form the comb plates and the E macromeres none. Deletion work by Ortolani (1964) has confirmed the mosaic determination of $e_1$ cells for comb plates, but they may also act inductively on the $e_2$ and oral micromeres (see Martindale, 1986).

Work with the M lineage is less clear, although its failure to form comb plates at least confirms that the possession of ectoplasm, or its mitochondria, is not a sufficient condition for this differentiation. Ortolani (1964) believed that her $m_1$ deletions confirmed their mosaic determination for apical organ, but Farfaglio (1963) found that possession of the M macromere is sufficient for apical organ formation, while $m_1$ deletions affected oesophageal development. There appear to have been no isolation or deletion studies of the oral micromeres. Freeman & Reynolds (1973) have shown that the ability to produce photocytes is segregated to the M cells of eight-cell embryos, and remains in the M macromeres as the two sets of aboral micromeres are budded off. However, this is really their main evidence for suggesting that photocytes arise from the oral micromeres of the M lineage, so it would be a circular argument to claim that this demonstrates mosaicism.

Freeman (1976a, 1977) has also carried out studies in which only parts of the early blastomeres were removed. The embryos still showed a great capacity to regulate and develop the larval structures scored (comb plates and photocytes). However, at the two-cell stage, it was the aboral cytoplasm which seemed most important for the formation of comb plates. The available evidence for the zygote suggested that the aboral half is dispensable (p. 50), so determinants for comb plates may move with the ectoplasm in the first cleavage furrow to collect aborally. At the two-cell stage all parts of the aboral cytoplasm still seemed equally important for comb plate formation, but, by the four-cell stage, this potential has been restricted in a second plane, to the area which will form the E cells at third cleavage. No localisation could be found for photocyte production until this potential was segregated to the M lineage at the eight-cell stage.

In work comparable with his *Cerebratulus* studies (p. 78), Freeman (1976*a,b*) has investigated the roles of the cleavage process in the two localisation movements seen for the apparent comb plate determinants, first along the oral–aboral axis and then in the plane which he describes as tentacular. Not only was localisation delayed when cleavages were reversibly inhibited, but the planes in which it occurred depended upon which cleavages still took place in the recovering embryo (Freeman, 1976*a*). The results suggested that second cleavage is required for localisation in the tentacular plane just as first cleavage is for localisation aborally (see above). In further studies, however, tentacular localisation occurred even when second cleavage was suppressed, although not if it occurred on time but in the wrong plane (Freeman, 1976*b*). Freeman (1979) concludes that determinants can be moved independently of cleavage even if the two processes are usually synchronised. In embryos recovering from cleavage inhibition the planes of cleavage are at least sometimes the same as in same-age controls, suggesting the operation of a developmental 'clock' which could also control movements of determinants. The basis of mosaicism in ctenophores therefore seems to be the active movement of factors, perhaps determinants, into specific cytoplasmic areas, followed by segregation of such areas at a cleavage. At least some of these factors are associated with the ectoplasm. Their molecular nature, mode of action and the mechanisms by which they are localised all remain unknown.

## Hydrozoans

Hydrozoans are usually classified with the ctenophores in the coelenterates. As this brief account will show, their development at cleavage stages shows some interesting similarities but also some profound differences.

The cleavages are usually radial and equal (see the review of Mergner, 1971). Teissier (1931) provides a full study for *Amphisbetia* where polarity is recognised from a yellow pigmented pole. This is the pole where polar bodies are emitted and where cleavage and gastrulation start. Later it becomes the posterior pole of the planula larva and the oral end of the hydroid, and from this fate Teissier (1931) considered it to be vegetal. In its relation with the polar bodies, however, it is clearly the equivalent of the animal pole in other animal eggs. It is probably safest to follow Freeman (1981*b*) in referring to the pole as posterior from its fate in the planula, but it must be remembered that it is on the primary, animal–vegetal, axis. Teissier (1931) claimed that all the larval endoderm formed from cells of the posterior half in *Amphisbetia*, but that other species probably had a concentric organisation from early stages with the material for prospective endoderm already internal to that for prospective ectoderm.

Polarity seems to be determined as it is in ctenophores, according to Freeman's (1981*a*) experimental studies with *Phialidium*. First cleavage is unipolar and if the furrow is caused to appear in a new position then this will become the posterior pole. If cleavage begins at two sites at the same time, both become posterior poles in partially twinned larvae, but if one site begins cleavage even three minutes later than the other it fails to affect polarity. Ostroumova & Belousov (1971) have claimed that even polarity is only fixed in later development along the axis which has by chance become longest, but their evidence for polarity changes is not decisive (see Freeman, 1981*b*).

While the blastomeres of ctenophores demonstrate mosaicism after separation, those of hydrozoans show great regulative ability. Even one of the first 16 cells can regulate to produce an apparently normal miniature planula larva (Zoja, 1895; Teissier, 1931). This means that no single area (like the yellow area of *Amphisbetia*) is essential for development, and that cleavage cells are not determined for their ectodermal or endodermal fates. One kind of positional information which was not destroyed by the isolations was polarity: Teissier (1931) and Freeman (1981*b*) showed that fragments always retained their initial polarity. Freeman (1981*b*) compared the polarised cells with magnets, an analogy which has already been offered for oocytes (p. 18). Hydrozoan embryos probably inherit their polarity from the oocyte in normal circumstances, but at cleavage stages they have not begun to determine cells according to their position along the axis. Such 'interpretation' of gradient values (see Wolpert, 1969) must occur at later stages, and must be studied by experiments on blastulae (p. 204) and gastrulae (p. 265).

## Nematodes

Observational evidence and the results of cytoplasmic extrusion, egg fusions and centrifugations (see pp. 53 and 56) suggest that the nematode zygote shows very little spatial organisation, with even polar localisation occurring at a relatively late stage. We shall see here, however, that several lineages with distinctive properties are established in the first cleavages and seem often to be determined for their fates.

The early cleavage pattern (Fig. 4.10) is unusual. Nematode embryos may be the only ones to separate the animal and vegetal poles at first cleavage, and the division is unequal so that the AB cell at the animal pole is larger than the $P_1$ cell vegetally. The position of this cleavage is

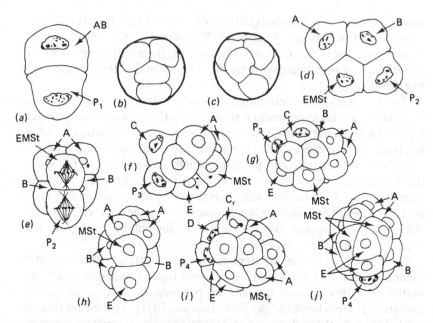

Fig. 4.10. Early cleavage stages of *Parascaris*. (*a*) 2-cell; (*b*)–(*d*) 4-cell showing the movement of $P_2$ to make contact with B; (*e*) 6-cell; (*f*) 8-cell; (*g*),(*h*) 12-cell; (*i*),(*j*) 16-cell; (*c*),(*d*) viewed from the left; (*f*),(*g*) and (*i*) from the right and (*e*),(*h*) and (*j*) ventrally. (Redrawn from zur Strassen, 1896.)

apparently determined by the mitotic apparatus which rotates in the zygote and shows inequalities in the size of its asters by anaphase (Ziegler, 1895; Nigon, Guerrier & Monin, 1960; Albertson, 1984). In *Ascaris*, second cleavage is vertical in AB and horizontal in $P_1$, forming a T-shaped four-cell stage. The most vegetal cell ($P_2$) then migrates around the next cell (EMSt) until it makes contact with one of the animal cells, producing a rhomboidal embryo with obvious bilateral symmetry (zur Strassen, 1896), a pattern reached more directly in other species. The side to which $P_2$ has migrated becomes the dorsal side, while the animal half cell which it contacts is called B and the other animal half cell is A.

The EMSt and $P_2$ cells make further cleavages in a precise pattern, to separate the stem cells of five different embryonic lineages, while a sixth lineage originates from AB. This can be seen most clearly in the early lineage plan of Figure 4.11. Thereafter the cells of each lineage divide equally and fairly synchronously, although the different lineages show different cleavage rhythms. Left–right differences become apparent in the positions of cells from early stages. At the eight-cell stage, daughters of the AB lineage are already slightly nearer the animal pole on the left side than the right side, and similar differences are soon established in

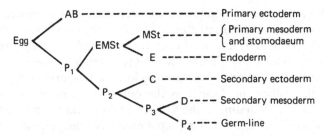

Fig. 4.11. The origin of the major lineages in the early nematode embryo. Further divisions occurring within each lineage are not shown. The main fates of each lineage are given on the right. (Original)

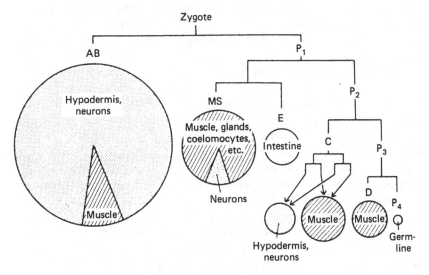

Fig. 4.12. A summary of the cell types derived from each of the early lineages in *Caenorhabditis*. Areas of circles and sectors are proportional to numbers of cells. Stippling represents typically ectodermal tissue and striping typically mesodermal tissue. (From Sulston *et al.*, 1983.)

most other somatic lineages (zur Strassen, 1951; Schierenberg, Carlson & Sidio, 1984).

Because cleavage patterns are so precise and cell numbers are relatively low, the major fates of the various lineages were established early for nematodes (zur Strassen, 1896; Boveri, 1910) and they are marked on Figure 4.11. They have been more recently confirmed for the free-living species *Caenorhabditis elegans*, where the complete lineage is now known for every cell in the newly hatched larva (Sulston *et al.*, 1983). The fates are very precise, but Figure 4.12 shows that there is only a partial correlation between the separation of lineages and the separation of fates. Some nerve and muscle cells are even sister cells separated at the final cleavage before differentiation. This raises the question of whether fates are determined by lineage (i.e. by the nature of the cytoplasm they receive) or by position (including the effects of neighbouring cells).

The best evidence for determination by localised factors is provided by the germ-line which, in *Ascaris*, is the only lineage to retain the full complement of chromatin. In the somatic lineages chromatin is eliminated at the first cleavage following separation from the germ-line, i.e. in the separation of A and B, E and MSt, etc. (zur Strassen, 1896; Beams & Kessel, 1974). This suggests that a material at the vegetal pole may protect nuclei against chromatin elimination, and the pattern of elimination seen in embryos with experimentally changed cleavage planes supported this hypothesis (Hogue, 1910; Boveri, 1910). Elimination also occurs in the $P_2$ cell following uv irradiation of the vegetal pole at first cleavage (Moritz, 1967). The P granules which become localised vegetally in the *Caenorhabditis* zygote (p. 35) are segregated to the $P_4$ cell where they associate with the nucleus, and so could act as germ cell determinants (Strome & Wood, 1982; Wolf, Priess & Hirsh, 1983; Yamaguchi *et al.*, 1983). Fertile worms can, however, develop even after large amounts of cytoplasm are extruded from the prospective germinal area (Laufer & von Ehrenstein, 1981), although the fate of the P granules was not checked (and see below).

No such localised materials have been recognised in the various somatic lineages, and there is some evidence that interactions between the cells have a role in the determination of their fates. Specifically, it has been suggested that the new contacts established at the rhomboidal four-cell stage determine the dorso-ventral organisation of the embryo (Tadano & Tadano, 1974). Before this stage, dorso-ventrality can be determined at will by centrifugation (Tadano, 1962) and the AB lineage appears to show little spatial organisation (Bonfig, 1925; Priess & Thomson, 1987). At the rhomboidal stage dorso-ventrality is fixed (Tadano, 1962) and thereafter the A and B cells make distinctive contributions particularly to ectodermal tissues (see Sulston *et al.*, 1983). Muscle-specific antigens are not formed in the A lineage if $P_1$ or EMSt are removed, or even if $P_1$ is enucleated, and they are formed instead by the B lineage if the positions of A and B are reversed (Priess & Thomson, 1987). This polarisation of the animal half by a cell from the vegetal pole makes an interesting comparison with some effects of the D quadrant in spiralian embryos (Chapter 3) and indeed with many other embryos. Bonfig (1925) and Schierenberg *et al.* (1984) have discussed the possibility that the left–right asymmetry of early cell positions also depends upon contacts with other cells. If interactions are involved here they may be mediated by specific affinities of the cell surfaces or simple steric hindrances, variables which probably also affect the finer details of cleavage pattern (Laufer, Bazzicalupo & Wood, 1980). Such effects may, however, have far-reaching consequences, as reversal of the early left–right asymmetry is followed by reversal of the asymmetric arrangements of all the internal organs (zur Strassen, 1951). Inductive effects of $P_2$ and $P_3$ on the somatic lineages have been described by Schierenberg (1986, 1987).

The only way of deciding definitively whether fates are determined by lineage or by interactions among cells is to follow the development of cells in isolation and in abnormal combinations. In nematodes this has proved difficult because of practical problems such as the presence of a thick eggshell. Pai (1928) claimed some separations of the first two cells and thought that the $P_1$ cell then regulated to produce a whole embryo, but development was not followed sufficiently far to really establish this. Stevens (1909) used uv light to arrest cleavage in some blastomeres, and recent workers have lysed or extruded specific cells. When AB is the undamaged cell it forms a blastula (Stevens, 1909) in which several of its normal derivatives fail to appear (Priess & Thomson, 1987). $P_1$ and its descendants, however, apparently develop autonomously to produce the expected tissues, gut and muscle being recognisable by the synthesis of tissue-specific molecules (Laufer *et al.*, 1980; Edgar & McGhee, 1986; Priess & Thomson, 1987). At later stages, which are really outside the scope of this book, ablation work certainly indicates a high degree of developmental mosaicism. Ablated cells are very rarely replaced and only by very similar cells, although some regulation of form may occur as cells grow, and rare cases of inductive interaction have been identified (see Sulston & White, 1980; Sulston *et al.*, 1983).

In view of this extensive mosaicism, it is surprising to find that large volumes of cytoplasm can be removed from specific areas without affecting later development (Laufer & von Ehrenstein, 1981). This was shown for a volume equivalent to the C, D and $P_4$ lineages removed vegetally from zygotes and one equivalent to D and $P_4$ from cleavage stages. As Laufer & von Ehrenstein (1981) say this could be explained by 1) lineage-specific determinants attached to a cytoskeletal system independently of other organelles and of the cell membrane, 2) the late localisation of such determinants perhaps associated with the mitotic spindles, or 3) a determination system based upon graded quantitative differences rather than specific localisations. They also note that cleavage positions are usually adjusted to give a normal ratio of cell sizes, though regulation is possible without this.

Boveri (1910) long ago suggested a determination system for nematodes involving two materials distributed in gradients decreasing from the opposite poles of the egg. Tadano (1962) has since provided evidence that egg polarity depends upon a gradient in the content of ectoplasm decreasing from the animal pole (see p. 56). Ectoplasm is thus unevenly distributed among the various cell lineages, and Tadano (1962) suggests that this affects their determination. He even has some evidence that the elimination of chromatin is promoted by ectoplasm as well as being inhibited by vegetal cytoplasm (see too King & Beams, 1938; Oliver & Shen, 1986). The classic case of determination controlled by two opposed gradients is that of the later sea urchin embryo (see Chapter 5). The operation of such a system could explain why cleavage-arrested nematode zygotes and early blastomeres (unlike those of ascidians; see p. 104) never express more than one developmental potential although they would normally contribute to several tissues (Cowan & McIntosh, 1985). Although this is not the explanation offered by the authors, the tissue which does form could be that which is specified by the morphogen ratio in the cell. Whatever the control system it displays some variability with, for example, the $P_1$ cell forming either muscle or gut markers.

In *Ascaris* the distribution of ectoplasm apparently also has a role in dorso-ventral polarisation, the first sign of which is the streaming of ectoplasm across the animal side of the EMSt cell. This causes an extension of the surface at one side of EMSt and a contraction on the other, and the $P_2$ cell moves as if attached to the contracting surface on the latter side (Tadano & Tadano, 1974). In centrifuged embryos, ectoplasm collects on the centrifugal side of the EMSt cell, and it is on this side that the surface extends while it contracts centripetally. The $P_2$ cell moves with it and the centripetal side becomes dorsal, the axis being fixed in rhomboidal four-cell embryos. What polarises the ectoplasmic movements in control embryos is unknown, although Tadano & Tadano (1974) note that the nucleus also moves.

Deppe *et al.* (1978) have noted that cleavage rates in the *Caenorhabditis* lineages relate with their position of origin along the animal–vegetal axis, being fastest for AB (animal) and decreasing through MSt, C, E and D to $P_4$ (vegetal). Some graded cytoplasmic property may thus determine the cleavage rhythm of all lineages, and, when the cytoplasm of two lineages is mixed, cleavage rhythms, and further development, are indeed affected (Schierenberg, 1984; Schierenberg & Wood, 1985). The further suggestion that the distinct cleavage rhythm of each lineage may determine its developmental fate (Deppe *et al.*, 1978) seems unlikely, however, as even embryos arresting in early cleavage develop specific markers of gut and muscle differentiation probably in the correct lineages (Laufer *et al.*, 1980; Gossett, Hecht & Epstein, 1982; Edgar & McGhee, 1986). DNA synthesis continues in cleavage-arrested embryos (Hecht *et al.*, 1982) so a system based on replication cycle timing is still possible but must explain all instances of differentiation occurring despite abnormal timing (e.g. in Schierenberg *et al.*, 1980; Denich *et al.*, 1984). It is worth noting that, in ascidians, where replication cycle counting seems used as a

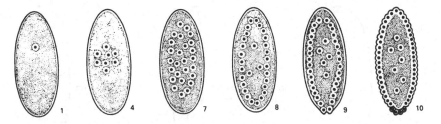

Fig. 4.13. Stages of synchronous cleavage in *Drosophila*. Numbers refer to the number of cleavage cycles undergone but not all nuclei are shown from cycle 7. The anterior end is always uppermost. (From Foe & Alberts, 1983.)

clock for determinant action (p. 105), there has been no suggestion that fates would change if the cycle length was altered.

There is an increasing body of information about genetic and biochemical aspects of nematode development. Maternal gene expression has profound effects in *Caenorhabditis* embryos (see p. 25) and even sometimes in larvae (Wood *et al.*, 1980). In fact, embryogenesis would seem to be mainly controlled by maternal gene products, sometimes supplemented by further expression of the same genes in the embryo; a few new genes are apparently activated in the early embryo but required for development only at late (morphogenesis) stages (Miwa *et al.*, 1980; Wood *et al.*, 1980; Isnenghi *et al.*, 1983). The DNA lost from the somatic lineages of *Ascaris* consists mainly of many repeats of a few related short sequences (Roth & Moritz, 1981; Tobler *et al.*, 1985): its role within the germ-line is unknown (see also Bennett & Ward, 1986). The uv sensitivity of the vegetal pole suggests a localised RNA (Moritz, 1967), and a possible gradient in maternal poly$(A)^+$RNAs in *Caenorhabditis* embryos is reported by Hecht, Gossett & Jeffery (1981). New transcripts are formed at least from the four-cell stage (Kaulenas, Foor & Fairbairn, 1969), and by about 100 cells they include collagen mRNA possibly localised in hypodermal precursor cells (Edwards & Wood, 1983). A zygotically programmed gut enzyme appears at about the same time (Edgar & McGhee, 1986).

## Insects

In the early development of insects the zygote nucleus divides repeatedly without cleavage of the cytoplasm, so producing a syncytium. The early nuclear divisions occur in the central yolk and the daughter nuclei then migrate to the outer cytoplasmic layer (periplasm) where a few are quickly segregated in the vegetal pole cells (Fig. 4.13). The remaining nuclei undergo a few further divisions peripherally at a syncytial blastoderm stage before cellularisation occurs there too. There is con-

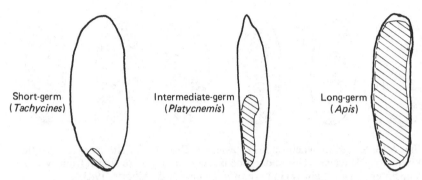

Fig. 4.14. The egg types of insects. By adjusting the magnification, these three eggs are shown with the same egg length. Eggs are classified into three types according to the proportion of this length occupied by the early germ-band (hatched). (After Krause, 1939.)

siderable evidence that determination at the syncytial blastoderm stage involves interactions throughout the embryo, and this will be considered in a later chapter (p. 168). Here we want to assess the spatial information already available at intravitelline cleavage stages, including that apparently responsible for the mosaic determination of pole cells.

The intravitelline nuclear divisions are usually synchronous and can be very rapid indeed (Agrell, 1964; Lundquist & Löwkvist, 1983). Between divisions, the daughter nuclei move apart within the yolk. It has been suggested that the zygote nucleus enters a 'cleavage centre' required for the initiation of cleavage, and that its daughters then move to a 'fountain flow initiation region' which pushes them towards both poles of the egg (Bruhns, 1974). The existence and significance of such centres, however, requires confirmation. Elements of the cytoskeleton are involved in the separation and migration of the nuclei (Zalokar & Erk, 1976; Wolf, 1980). The divisions of the nuclei seem to be activated by properties of the yolk which shows waves of change, apparently in consistency, just before the divisions (Miyamoto & van der Meer, 1982; Wolf, 1985). Wolf (1985) suggests that calcium ions are released at the wave front and prepare the nuclei for division.

As usual we have to establish the normal fates of the various parts of the egg before we can consider how far these parts are determined. A new factor enters here as only a part of the syncytium will actually contribute to the embryo, while the rest becomes extraembryonic. The former becomes recognisable as a germ band or germ anlage on the ventral side of the elongated insect egg usually at blastoderm stages of development. This germ-band can occupy virtually the whole length of the egg, only a very short length, or an intermediate proportion. In this way the development of various insect species can be classified to the long-germ, short-germ or intermediate-germ types (Fig. 4.14). When fate mapping is

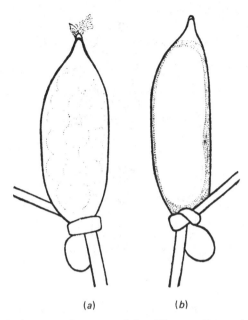

Fig. 4.15. When the posterior area of the *Platycnemis* egg is constricted off during intravitelline cleavage (*a*) no embryonic anlage forms (*b*). (From Seidel, 1926.)

carried further it is found that the germ band has the same polarity as the egg – its anterior end forms the insect's head and its posterior end the abdomen. This is enough discrimination for our present purposes, and indeed germ bands are rarely seen, and fates have rarely been checked, at intravitelline cleavage stages. The only other thing we need to know is that the pole cells will form primordial germ cells, in *Drosophila* at least as their only normal derivative (Underwood *et al.*, 1980).

The state of determination of parts of the early insect embryo has most frequently been tested by ligating the egg transversely at various levels. The parts then develop in functional isolation. When Seidel (1926) performed this experiment with eggs of the damsel-fly *Platycnemis* at intravitelline cleavage stages (see Fig. 4.15), he found that anterior fragments were unable to develop any structures at all when a relatively small posterior region was 'removed' by the ligature. If the ligature was incomplete, allowing some communication between the fragments, an embryo did form in the anterior part. Seidel (1926) saw this as evidence for a posterior 'determination centre'. Anterior pieces also fail to develop after early fragmentation of the egg of the cricket *Acheta* (see Sander, 1976), while posterior pieces as short as 16% of the egg length can form complete embryos (Sander, Herth & Vollmar, 1969). Despite the absence of a fate map, it appears probable that such posterior pieces

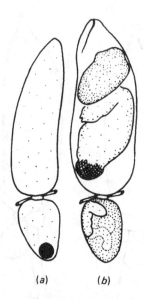

(a)        (b)

Fig. 4.16. Development of *Euscelis* egg fragments after transverse ligation during cleavage. (*a*) the ligated egg with the posterior pole material shown as a
black disc; (*b*) the typical result: a synophthalmic procephalon in the anterior
fragment, and a partial germ band of metathorax and abdomen in the posterior
one. (From Sander, 1976.)

are regulating to produce more than they would in the normal embryo.

Work with other insect species confirms the importance of the posterior region, but also shows that at least some parts of the head are
independent of posterior influences (see Sander, 1976). In the leafhopper
*Euscelis*, about a third of the egg must be removed posteriorly before all
development of anterior fragments fails, and, if eggs are ligated further
back, some anterior structures develop in the anterior fragment and
posterior structures in the posterior fragment (Sander, 1959). Following
ligations at intravitelline cleavage stages, however, the structures developed by the two fragments do not add up to a complete embryo: many of
the intermediate segments are missing (Fig. 4.16). Again Sander's (1976)
review cites many other instances of this 'gap phenomenon' in long-germ
and intermediate-germ eggs ligated at a similar stage, and it has since
been observed in *Drosophila* (Schubiger, 1976; Vogel, 1977) and a
beetle, *Callosobruchus* (van der Meer, 1979). Several authors have given
evidence that the effect is not due to cell death or other non-specific
damage at the ligature, and have shown that those structures which do
form are too large, suggesting a true change in determination. The
fragments form less than they should if they were parts of a mosaic. They
are sufficiently determined to produce appropriate terminal structures,

but specification of the intermediate segments apparently requires inter-actions between the halves, a topic for later study (p. 170).

Vogel has dissented from the view that all determination in insect eggs depends upon influences from the poles. In *Drosophila*, he suggests that the 'gap phenomenon' may be seen because the ligature prevents movements of determined areas, rather than preventing inductive inter-actions (Vogel, 1977). In *Euscelis*, he has made two ligatures and shown that the middle fragment can form some appropriate structures when isolated from both poles (Vogel, 1978). A middle fragment isolated at precleavage stages must, however, have at least 37% of the egg length before any pattern element can form, and this figure then falls sharply through intravitelline cleavage to reach 17% at blastoderm stages. This implies that influences from more polar areas at least increase the developmental potential of middle fragments, and they may be respon-sible for all that potential (see further discussion on p. 176).

Insect embryos rarely show any dorso-ventral determination even at blastoderm stages (see p. 177). One demonstration of this at cleavage stages was made by pinching *Euscelis* eggs longitudinally, when each half (including prospective dorsal and ventral ones) usually regulated to produce a whole embryo (Sander, 1971).

Evidence for the existence of anterior and posterior determinants has been obtained in entirely different ways for the long-germ eggs of chironomid midges at intravitelline cleavage stages. They were first indicated by the effects of egg centrifugation. When Yajima (1960) centrifuged *Chironomus* eggs with their anterior pole centrifugally he obtained some double abdomens with mirror-image symmetry; when the posterior pole was centrifugal double cephalons (mirror-image joined heads) were obtained. Later work has shown that the correlation between type of abnormality and direction of centrifugation is poor (see Yajima, 1983), but in *Smittia* too double cephalons are obtained when the posterior pole is centrifugal (Rau & Kalthoff, 1980). In both genera total inversion of some embryos has now been obtained following centrifu-gation with the posterior pole centrifugally (Rau & Kalthoff, 1980; Yajima, 1983). The organisation of the three abnormal types and a normal *Smittia* embryo are shown in Figure 4.17. They are strongly reminiscent of some maternal mutation effects in *Drosophila* discussed in Chapter 1, and a maternal mutation with similar effects is now known in *Chironomus* (Percy, Kuhn & Kalthoff, 1986). The structure of these abnormal forms again indicates that germ-line determination can be controlled independently of somatic determination. Whichever parts of the somatic pattern are inverted, pole cells always form at the original posterior pole (Yajima, 1970; Gollub & Sander, cited by Kalthoff, Hanel & Zissler, 1977).

These same types of abnormality can also be obtained by various local

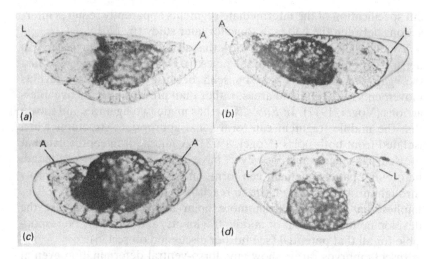

Fig. 4.17. Four basic body patterns in *Smittia* embryos: (*a*) normal; (*b*) inverted; (*c*) double abdomen; (*d*) double cephalon. A, anal papillae; L, labrum. Dorsal side always up and original anterior pole to left. (From Rau & Kalthoff, 1980.)

treatments of the poles of chironomid eggs. Some double abdomens result when the anterior pole is uv irradiated in *Chironomus* and *Smittia*; and double cephalons follow posterior irradiation in *Chironomus* (Yajima, 1964) but only very rarely in *Smittia* (Kalthoff, Rau & Edmond, 1982). However, in *Smittia*, posterior irradiation followed by centrifugation produces more double cephalons than centrifugation alone (Kalthoff *et al.*, 1982). In *Smittia*, too, double abdomens can be produced by anterior puncture of the egg alone (Schmidt *et al.*, 1975), or puncture followed by treatment with ribonuclease (Kandler-Singer & Kalthoff, 1976). All these treatments provide evidence that there are local determinants at the poles of the eggs, although it is more difficult to interfere with the posterior one in *Smittia* than in *Chironomus*. More surprising is the potential of each pole to produce structures appropriate to the opposite pole. Because of this, Kalthoff *et al.* (1982) propose that both anterior and posterior determinants are present at both poles, and that it is the ratio between them which decides the development of each half (Fig. 4.18, and see further on p. 175).

It has so far proved impossible to get such large reversals of polarity in *Drosophila* by uv irradiation at early stages. However, anterior irradiation does produce some embryos very like those offspring of *bicaudal* mothers where there is no head or thorax, the eight abdominal segments spread towards the anterior, and there is apparently some polarity reversal producing posterior spiracles at the anterior pole (Bownes & Sander, 1976). Similar effects have now been achieved by removing

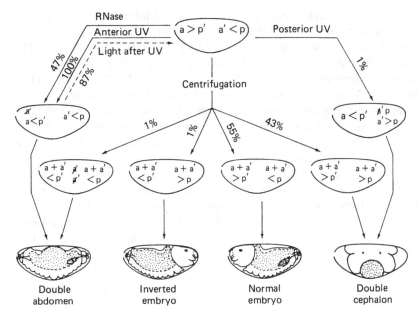

Fig. 4.18. A model of antero-posterior determination in chironomid embryos. Orientation as in Fig. 4.17. a,a', anterior determinants; p,p', posterior determinants: their relative strengths (symbolised by < or >) in each half decide determination there. Experimental treatments change this ratio by partially inactivating (indicated by letters crossed through), reactivating or redistributing the determinants. Percentages indicate the frequency of each body pattern. (From Kalthoff *et al.*, 1982.)

anterior cytoplasm, and if it is replaced by posterior cytoplasm double abdomens can result (Frohnhöfer, Lehmann & Nüsslein-Volhard, 1986), which again suggests that polar determinants have effects along the whole of the egg axis. However, removal of posterior plasm did not cause a shift of structures along the egg axis, although defects in abdominal segmentation did result (Frohnhöfer *et al.*, 1986). Attempts to assess the degree of determination elsewhere at this stage have given conflicting results (e.g. Bownes & Sang, 1974; Bownes, 1976).

A common effect of posterior damage to many insect eggs is the production of sterile adults (e.g. Hegner, 1917). Evidence has already been presented (p. 27) indicating that determinants of the germ-line are present at the posterior pole of *Drosophila* oocytes. Similar evidence exists for fertilised eggs, where uv irradiation leads to sterility, but fertility can be restored by injections of cytoplasm from the posterior pole of control eggs (Okada, Kleinman & Schneiderman, 1974a). The same pole plasm injected anteriorly can induce pole cell formation there, and if these cells are transferred to the posterior pole of new hosts they can produce fertile germ cells there (Illmensee & Mahowald, 1974). It was also suggested in Chapter 1 that determination may be due to the RNA in special polar granules, and this is consistent with its sensitivity to uv light. In the offspring of *grandchildless* mothers, most of those granules have indeed disappeared from oviposited eggs (Mahowald, Caulton & Gehring, 1979*b*). In cecidomyid flies the posterior pole plasm protects nuclei from the elimination of most of their chromosomes. Elimination usually occurs from all somatic nuclei, as it does in some nematodes, but, if pole plasm is moved to an abnormal location, the nuclei there are also protected (Geyer-Duszyńska, 1959).

The developmental role of the pole plasm is now known to be far more complicated than was thought even a few years ago. For one thing fractions from normal embryos including (polar?) granules (Ueda & Okada, 1982) and mRNAs (Togashi, Kobayashi & Okada, 1986) can restore the ability for pole cell formation to uv irradiated embryos, but fertility is not restored. This suggests that the RNA of polar granules may have only a short-term role in ensuring the precocious segregation of the pole cells. The sensitivity of pole plasm to uv irradiation decreases rapidly during early development (Graziosi & Micali, 1974). Polysomes are seen around the polar granules at cleavage stages (Mahowald, 1968) and locally applied cycloheximide inhibits pole cell formation (Okada & Togashi, 1985), suggesting that the RNA must be translated to fulfil this early role. Perhaps its product is responsible for some of the unusual features exhibited at pole cell formation, such as the early entry of the nuclei to the peripheral cytoplasm or the distinctive type of actin organisation (Warn, Smith & Warn, 1985). Certainly a delayed entry of nuclei to the pole plasm is an early effect of the *grandchildless* mutation (Fielding, 1967) and of uv irradiation (Okada *et al.*, 1974*a*). Togashi *et al.* (1986) discuss the possibility that pole plasm has a more specific role in germ cell determination, but the role could rather be in specifying the posterior pole, with its germ-line and somatic derivatives, in view of recent work with mutants lacking polar granules (see p. 28 and below).

In Chapter 1 it was suggested that the maternal genome determines not only the germ-line but also such basic features of somatic organisation as polarity and dorso-ventrality (p. 25). The relevant gene products persist into the embryo but are frequently lost by blastoderm stages (Steward *et al.*, 1984; Müller-Holtkamp *et al.*, 1985; Frohnhöfer & Nüsslein-Volhard, 1986), by which time they have presumably ensured the next steps of spatial determination (see p. 182). More particularly the evidence of Chapter 1 indicates that the product of the *bicoid* gene may act as an anterior determinant, and those of genes such as *oskar* as posterior determinants, and such polar determinants are of course expected on other grounds reviewed in the present section. Similarly the apparently even distribution of maternal gene products in the dorso-ventral direction agrees with the lability of dorso-ventral determination in the early embryo. In *Smittia*, the action spectrum for double abdomen production by uv light suggests that the targets are nucleoproteins (Kalthoff, 1973) and the effect is photoreversible by visible light (Kalthoff, 1971), during which uv-induced pyrimidine dimers in RNA are removed apparently enzymatically (Jäckle & Kalthoff, 1978). The effect of ribonuclease, cited above, also suggests the presence of RNA in the anterior determinants of *Smittia*. The lesser effects of uv light in double cephalon formation in *Smittia* (Kalthoff *et al.*, 1982), and its effects on *Chironomus* eggs (Yajima, 1983), are also photoreversible, so RNA may well be present in the determinants at both poles.

The fine structure of the egg poles has also been studied in *Smittia*. The only special materials seen localised there are in the germ-plasm posteriorly (Zissler & Sander, 1973, 1977). The anterior determinant does not seem to be associated with mitochondria (Kalthoff *et al.*, 1975) and, in centrifuged eggs, seems most concentrated in areas lacking visible organelles larger than ribosomes (Kalthoff *et al.*, 1977). From such evidence Kalthoff concludes that the anterior determinants are probably cytoplasmic maternally produced RNAs and could well be masked messenger ribonucleoproteins (mRNPs). Jäckle (1980*a*) has obtained mRNPs from *Smittia* eggs, and shown that they are indeed translationally inactive until the protein moiety is removed. Such small materials could be maintained in their positions by the microtubules of an anterior cytaster (Zissler & Sander, 1973) or other cytoskeletal elements (Kalthoff, 1979). Some macromolecules appear to be unequally distributed along the anterior–posterior axis according to electrophoretic studies of *Acheta* (Koch & Heinig, 1968) and immunological studies of *Drosophila*

## Sea urchins

Sea urchins are the best-known members of the Echinoidea, but studies of other members of the class are also considered here. The group is commonly considered to exemplify regulative development, but we will find that the limits of regulation are already obvious by the eight-cell stage.

The pattern of the first four cleavages is shown in Figure 4.19 and has been studied with mathematical accuracy, e.g. by Prothero & Tamarin (1977). The first two cleavages are vertical and equal, and the third, horizontal, cleavage is also roughly equal although the exact position of the cleavage plane varies with the species and even the batch. At the fourth cleavage the animal quartet again divides vertically and equally to produce a ring of eight mesomeres; but the vegetal quartet almost horizontally and very unequally to four large macromeres and four small

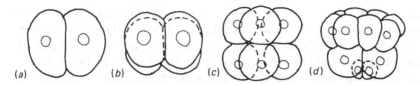

Fig. 4.19. 2-, 4-, 8- and 16-cell stages of the sea urchin *Paracentrotus* (After Boveri, 1901*a*.)

micromeres. Prothero & Tamarin (1977) found that the cell positions at the 16-cell stage were for the first time not ascribable entirely to the positions and orientations of cleavage planes. The micromeres seemed to sit in concavities of the macromeres, the mesomere ring took up an oval form which presumably could be related with dorso-ventrality, and the space between the blastomeres already seemed to be enlarging to start blastocoel formation.

The mechanisms responsible for the peculiar fourth cleavage in the vegetal half have received some special study. As Dan (1979) points out, this cleavage and the following one by the micromeres are the only ones to break the rule that each cleavage is orientated at 90° to the previous one. If eight-cell embryos are treated with detergents, the rule is obeyed and both the animal and vegetal quartet cleave vertically. Dan (1979) thus investigated the effect of the detergent on the cells of the vegetal quartet and found that it prevented a migration of nuclei towards the vegetal pole which is the immediate cause of the inequality of cleavage. Investigating further, Dan, Endo & Uemura (1983) found that the free cell surfaces of eight-cell embryos are underlain by a continuous layer of vesicles broken only by gaps near the vegetal pole (see also Uemura & Endo, 1976). The migrating nuclei and mitotic spindles point directly at these gaps (Fig. 4.20). Such cortical areas may lack microfilaments as well as vesicles, as local treatments with cytochalasin B can cause even first or second cleavage to become unequal with the small cell on the treated side (Bozhkova *et al.*, 1983). In this connection we should remember that cortical pigment, perhaps associated with the cytoskeleton, withdraws from the vegetal pole of sea urchin zygotes or early cleavage stages (p. 40). All these changes may be reflections of the same cortical reorganisation, which would be a necessary preparation for the unequal vegetal cleavages (see also Tanaka, 1981). It is interesting that a short cytochalasin B treatment after second cleavage is often followed by equal fourth cleavages (Bozhkova *et al.*, 1983).

Both classical and recent studies show that the cells of the animal half contribute only to larval ectoderm while the vegetal half contributes to all three germ layers (Hörstadius, 1935; Cameron *et al.*, 1987). The further separation of fates within the vegetal half is considered in the next chapter

Fig. 4.20. Two vegetal blastomeres of the eight-cell *Hemicentrotus* embryo. Two excentric spindles point towards gaps in the row of subsurface vesicles. × 1700. (From Dan *et al.*, 1983.)

(see Fig. 5.2) but we may note here that descendants of the micromeres form the skeleton and contribute to the coelomic sacs. Hörstadius & Wolsky (1936) could find no relationship between the first two (vertical) cleavage planes and the later dorso-ventrality of the larva, but Cameron *et al.* (1987) believe that in *Strongylocentrotus* both cleavages are at about 45° to the bilateral plane forming one ventral (oral), one dorsal (aboral) and two lateral blastomeres.

Driesch (1891, 1900) separated blastomeres after the first or second cleavage. He found that they first cleaved as partial embryos, but then closed to a hollow sphere at the morula or blastula stage, and went on to form a normal miniature larva. Work with other animal embryos had indicated that such cells are determined to become right or left and dorsal or ventral (Chabry, 1887, see p. 97; Roux, 1888, see p. 134), so Driesch's demonstration of this regulative ability was important. However, Driesch went further and described the sea urchin embryo as a 'harmonious equipotential system' in which polarity and bilaterality must be present in the smallest protoplasmic elements: he was probably the first author to compare such organisation with a magnet (see Driesch, 1908, pp. 65–6). To explain how different parts of the embryo later formed different organs, Driesch (1908) had to resort to vitalism; propos-

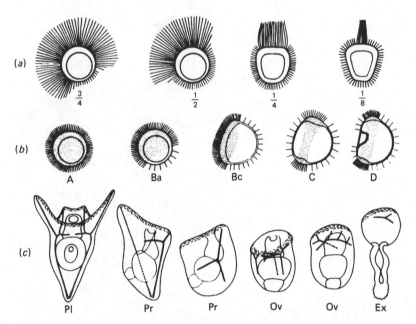

Fig. 4.21. The development of animal and vegetal halves isolated from 16- or 32-cell sea urchin embryos. (*a*) animal halves after 1 day's culture classified by the size of the animal tuft; (*b*) animal halves after 2 days' culture, from those with uniform ciliation (A) to those with a ciliated band and stomodaeum (D); (*c*) vegetal halves which may regulate to form plutei (Pl) or arrest as prism-like (Pr) or ovoid (Ov) larvae or exogastrulae (Ex). (From Hörstadius, 1973*a*.)

ing a non-material 'entelechy'. Later study was, however, to reveal at least a labile bilateral organisation in half eggs or embryos separated along the animal–vegetal axis (see especially Hörstadius & Wolsky, 1936). Sometimes one fragment showed delayed development on its left side and the other on its right side, indicating that the cut had separated prospective right and left halves respectively. In other cases both were bilaterally symmetrical, but one developed with a delay relative to the other. As will be discussed on p. 153 these were probably prospective ventral and dorsal halves with the latter showing the delay.

Far more pronounced differences are seen when animal and vegetal halves are separated. Zoja (1895) did so with 16-cell embryos and found that animal halves developed an enlarged apical tuft and failed to gastrulate, a result confirmed by Boveri (1901*b*) with the halves of unfertilised eggs (see p. 54), and one which removed the need for Driesch's entelechy. Further equatorial separations have been carried out at the 16- and 32-cell stages by Hörstadius (1935) and reveal some variability as indicated by the forms illustrated in Figure 4.21. It is particularly interesting that some batches produce mainly forms shown at

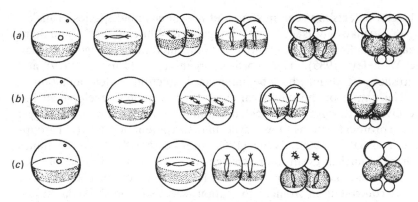

Fig. 4.22. Normal (*a*) and delayed (*b*),(*c*) cleavage patterns in *Paracentrotus*. Cleavage planes seem determined by time since fertilisation and in particular micromeres form at about the normal time even if this is at third rather than fourth cleavage. (From Hörstadius, 1973*a*.)

the left of each row in the figure, while others produce the forms at the right. Presumably third cleavage is cutting the animal–vegetal organisation at slightly different levels in such batches. Comparing the development of the isolates with their normal fates it is seen that animal halves form less than they should, overdeveloping the polar tuft; while vegetal halves at least sometimes form more than they should, regulating towards a normal larva. Both halves frequently show some dorso-ventral organisation. Their general similarity with the corresponding halves from unfertilised eggs suggests that there is little, if any, change in animal–vegetal organisation through fertilisation and the early cleavages.

Boveri (1901*b*) interpreted the great developmental potential of isolated vegetal egg fragments as evidence for a determination centre there. It is therefore interesting to find that at the 16-cell stage micromeres can self-differentiate after isolation to form a quite normal skeleton (Okazaki, 1975). They seem to be the only mosaically determined cells at this stage. The production of micromeres appears to be an important phenomenon. We have already seen that special mechanisms ensure their formation by an unequal cleavage. There is also a clock mechanism which ensures that a small vegetal area is segregated at the correct time. This can be demonstrated using shaking, development in dilute sea-water (Hörstadius, 1927), or irradiation with X-rays (Rustad, 1960) or uv light (Ikeda, 1965) to delay cleavage: micromeres still form vegetally approximately on time but at the third, second or even first cleavage (Fig. 4.22). The question thus arises as to whether embryos could develop normally if this important cleavage was equalised. Several early studies suggested that they could (Driesch, 1892; Boveri, 1901*b*; Hörstadius, 1928), but, if detergents are used to equalise the cleavage, skeletal defects are seen in some but not all echinoid species (Tanaka, 1976; Bozhkova & Isaeva, 1984; Langelan & Whiteley, 1985), and there may be a delay in the differentiation of the micromere lineage (Langelan & Whiteley, 1985). Thus cortical organisation (Tanaka, 1979) and threshold effects may both be important in micromere determination.

The nature of the biological clock for sea urchin development has received further study. Coincident with the mitotic cycle of normal eggs, some protein fractions show a cycle in their content of sulphydryl groups (review: Sakai, 1968). Ikeda (1965) showed that this cycle continues when the nuclear events and cleavage are blocked by uv irradiation, and so proposed that the sulphydryl cycle forms the biochemical basis of the clock controlling micromere formation (see also Dan & Ikeda, 1971). Other factors may also be involved, as micromere formation is delayed one cycle in half embryos separated immediately after an early cleavage and before the blastomeres have flattened against each other (Shmukler *et al.*, 1981; see also Bozhkova *et al.*, 1982).

As usual, DNA synthesis rates are high and RNA synthesis rates low during the early cleavage stages, but the latter apparently increases much faster than cell number at fourth cleavage (Wilt, 1970). Two mRNA species are also produced transiently at this stage, one coding a particular histone (Senger, Arceci & Gross, 1978). This histone enters the chromatin to higher levels in the micromeres than in the other cells, and the micromeres already show many other distinctive biochemical properties. When transcripts from unique DNA sequences are compared, the micromeres receive a smaller spectrum from the maternal pool than other cells, but they then appear to form some new species not formed by other cells (Rodgers & Gross, 1978; Ernst *et al.*, 1980). Some reiterated genes are transcribed only in the micromeres, forming RNAs not present in the maternal pool (Mizuno *et al.*, 1974). The micromere nuclei are the first to show detectable levels of U1 RNA, which is thought to be involved in mRNA processing (Nash *et al.*, 1987). There has been controversy over the relative transcription rates of the cell types. Cytochemistry (Agrell, 1958; Cowden & Lehmann, 1963) and autoradiography (Czihak, 1965a,b, 1977) suggested that micromeres were much the most active, but comparisons of separated cells indicate similar rates in the three cell types (Hynes & Gross, 1970; Hynes *et al.*, 1972; Spiegel & Rubinstein, 1972). However, transcription rates are greatly affected by cell dissociation (Hynes & Gross, 1970; Arezzo & Giudice, 1983), and in any case the concentration of the new RNAs will be highest in the micromeres because of their small size.

All three cell types form the same species of predominant proteins (Tufaro & Brandhorst, 1979), but there are clear differences in their relative synthesis rates (Senger & Gross, 1978). These differences are not seen in embryos where RNA synthesis is inhibited from the four-cell stage, so the divergence in the RNA populations being translated is due to the new transcripts (Senger & Gross, 1978).

### Starfish

Starfish are members of another echinoderm class, the Asteroidea. They have been studied much less than sea urchins, but with interesting results.

Normal development of *Asterina pectinifera* is described by Dan-Sohkawa & Satoh (1978), and the early cleavages appear to be equal and synchronous. In particular no micromeres are formed: all fourth cleavages being vertical so that the 16-cell stage has two tiers of eight cells each. The fates of some early blastomeres have been followed by Kominami (1983) using injection of intracellular tracers. This confirms that first cleavage is along the animal–vegetal plane but at random to the dorso-ventral plane. Third cleavage is equatorial and the cells of the animal half form about two-thirds of the ectoderm, while those of the vegetal half form the rest of the ectoderm, the endoderm and presumably the mesoderm.

Isolated *Asterina* blastomeres show far greater regulative ability than those of sea urchins (Dan-Sohkawa & Satoh, 1978). All of the first eight cells can form morphologically normal small bipinnaria larvae. At 16 cells the increased adhesiveness of cells makes separation more difficult, but at least seven of the cells can form larvae, and at 32 cells one cell was able to form a minute gastrula. The difference from the sea urchin is of course that animal and vegetal halves show equal developmental potential in isolation: *Asterina* is a far better candidate for Driesch's (1908) 'harmonious equipotential system'! Interactions between the cells must be required to allow them to follow different pathways in their later development (see p. 188).

### Amphibians

Amphibians are another animal group with embryos showing radial cleavage and considerable regulative ability. The first two cleavages are vertical and begin at the animal pole. Usually the first separates the future right and left halves of the embryo, and the second the dorsal and ventral halves. According to Klein (1987) this is always true in *Xenopus*. At the four-cell stage, the dorsal cell pair are often recognisably smaller than the ventral cell pair. Third cleavage is horizontal but again usually unequal so that the cells of the animal quartet are smaller than those of the vegetal quartet. Further cleavages are again seen first in the animal half, and probably continue to obey the rule of 90° rotations between successive cleavage planes. From the start of cleavage a central blastocoel

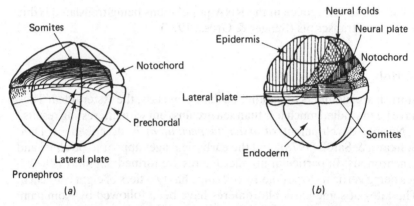

Fig. 4.23. Fate mapping studies of *Xenopus laevis*. (*a*) areas of prospective chordamesoderm at 16-cells: note particularly that each somite, except the most anterior ones, receives a contribution from both the dorsal and ventral blastomeres separated at second cleavage. (*b*) a general fate map at 32-cells. See further in text. ((*a*) after Cooke & Webber, 1985; (*b*) after Nakamura *et al.*, 1978.)

seems to be produced actively by secretions from the inner cell surfaces (Kalt, 1971*a*,*b*,).

The fates of the *Xenopus* early blastomeres have been determined by vital staining and the injection of lineage tracers, and data obtained by both methods are shown in Figure 4.23. The recent lineage tracer studies of Moody (1987*a*,*b*) and Dale & Slack (1987) show that each blastomere contributes to a greater range of tissues than is indicated in the figure, including some tissues in all three germ layers, making it impossible to express the data in a single fate map. Prospective mesoderm is more evenly split between the animal and vegetal halves than Figure 4.23(*b*) indicates, but the figures remain a useful summary for comparison with the development of embryonic isolates.

In some conditions even right or left blastomeres from the two-cell stage develop mosaically to a half larva. This happened when Roux (1888) killed one cell with a hot needle and left the other to develop next to it, and it also happens if a separated cell is reared to gastrula stages in a complete medium with serum (Kageura & Yamana, 1983). Blastomeres separated within the jelly coat however regulate to form bilaterally symmetrical larvae – both in urodeles (Herlitzka, 1897) and anurans (McClendon, 1910). Sometimes dorsal structures are overdeveloped in such larvae, perhaps because prospective ventral areas are dorsalised when they make contact with dorsal ones as the fragment heals (Cooke & Webber, 1985*b*).

There has been more confusion over the development of future dorsal and ventral halves separated during the early cleavages. The early data

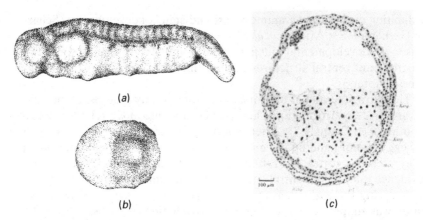

Fig. 4.24. The development of dorsal (*a*) and ventral (*b*),(*c*) halves isolated from 2- or 4-cell newt embryos. (*a*),(*b*) external views; (*c*) section through a ventral half where three distinct germ layers and dividing nuclei (Kary) can be seen; × 40. ((*a*),(*b*) from Spemann, 1938; (*c*) from Spemann, 1901.)

were quite clear; the dorsal half produced a whole embryo while the ventral half developed as a 'Bauchstuck' or belly-piece with gut and ventral mesoderm inside a covering of epidermis (Spemann, 1901; Ruud, 1925; Schmidt, 1933; see Fig. 4.24). A part of the confusion has arisen because later reviewers of this work stated that the ventral halves did not gastrulate or form any recognisable structures (e.g. Morgan, 1934; Horder, 1976): a result also experimentally obtained by Landström & Løvtrup (1974) using *Xenopus*. On the other hand, Dalcq & Dollander (1948) claimed to have obtained some normal larvae from newt ventral blastomeres. In recent studies with *Xenopus* some ventral halves formed belly pieces but others produced extensive amounts of axial tissue, while dorsal halves developed too large a head and too small a tail (Kageura & Yamana, 1983; Cooke & Webber, 1985*a*). Since second cleavage cuts across the prospective areas for many axial tissues including most of the individual somite blocks (Fig. 4.23*a*), such isolates are tending towards mosaic development. In summary, ventral halves seem to contain at least most of the 'organising' ability appropriate for their fate, though the ability to polarise these structures in the dorso-ventral direction is lower than in dorsal halves. In this connection it is worth noting that the structure of a belly piece (Fig. 4.24) is very like that of an embryo radialised by uv irradiation of the egg (Fig. 2.22; see Manes & Elinson, 1980; Scharf & Gerhart, 1983). Wilson (1929) and Tyler (1930) long ago compared the extra potential of amphibian blastomeres containing grey crescent material with that of D quadrants receiving polar lobe material in some spiralians. Parts which lack this extra potential, the uv-radialised amphibian egg and the *Lymnaea* B quadrant, have been described as exhibiting the basic programme or 'ground-state' of development (Scharf & Gerhart, 1983; Arnolds *et al.*, 1983). In fact, the ventral half probably has more developmental independence in amphibians than in spiralians where some ventral structures require induction from the D quadrant (see Chapter 3).

The extra potential of the dorsal side is apparently acquired at the time of grey crescent formation when parts of the egg move relative to one another (see Chapter 2). After that dorso-ventrality cannot be changed by egg rotation, and by the eight-cell stage no second axis results when grey crescent cortex is grafted ventrally (Curtis, 1962). There is no evidence that dorso-ventral determination normally progresses any further during the early cleavage period. However, where this determination was suppressed in the egg by uv irradiation it can be supplied by rotation after first cleavage (Scharf & Gerhart, 1980). Also in conditions where one side is 'favoured' through cleavage and blastulation by a higher temperature (Glade *et al.*, 1967) or unrestricted oxygen supply (Løvtrup & Pigon, 1958; Landström & Løvtrup, 1975), it has been claimed that it usually becomes dorsal, implying that the original dorso-ventrality can be overridden (see also p. 63). Landström & Løvtrup (1975) further claim that some ventral halves develop well-formed axes, particularly in the tail, if reared in conditions allowing access of oxygen on only one side. They suggest that grey crescent formation results in higher metabolic activity in the dorsal side, and that development of ventral blastomeres usually fails because they have no such leading side.

Although there has been so much interest in the dorso-ventral organisation of the early amphibian embryo, separations of the animal and vegetal cells at the eight-cell stage have revealed more fundamental differences (Fig. 4.25), as they have in other embryos. Early studies suggested that the quartet of animal cells form a hyperblastula with no axial organs in *Triturus* (Ruud, 1925) and *Rana* (Vintemberger, 1936). Later workers have observed small amounts of axial tissue in 30% of such isolates from *Triturus* (Grunz, 1977) and of muscle in almost 50% for *Xenopus* (Kageura & Yamana, 1983). Vegetal halves form tissues of all three germ layers including nervous tissue (Stableford, 1948; Grunz, 1977; Kageura & Yamana, 1983); sometimes almost complete embryos result (Grunz, 1977). Comparing these data with the fate map, it would seem that, as in the sea urchin, animal halves form too little and overrepresent structures from the animal pole, while vegetal halves regulate towards complete embryos. Grunz (1977) and Kageura & Yamana (1983) have, however, suggested that the results may be largely compatible with mosaic development.

Kageura & Yamana (1983, 1984, 1986) have studied the effects of various other cell deletions and combinations at the eight-cell stage. Any two animal cells can be removed with little effect on development, but, if

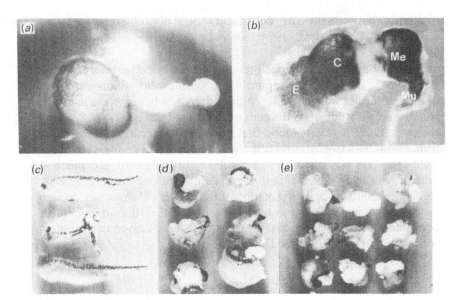

Fig. 4.25. Embryos derived from amphibian animal or vegetal cell quartets at the 8-cell stage. (*a*),(*b*) a *Xenopus* animal half after 2 (*a*) or 3 (*b*) days' culture. C, cement gland; E, epidermis; Me, melanophores; Mu, muscle. (*c*)–(*e*) *Triturus* vegetal halves develop to forms varying from nearly complete embryos (*c*) to poorly differentiated yolk-rich explants (*e*). ((*a*),(*b*) from Kageura & Yamana, 1983; (*c*)–(*e*), from Grunz, 1977.)

vegetal cells are removed, at least one dorsal vegetal and one ventral vegetal cell must be left. Such results indicate that the different developmental properties of dorsal and ventral halves are due to factors in their vegetal regions. On the other hand, the replacement of any ventral blastomere by a dorsal one can cause duplication of the dorsal axis, though the frequency and intensity of the effect is greatest when the replacement is made vegetally. By the 64-cell stage cells from the dorsal side of the vegetal octet can rescue an embryo radialised by an earlier uv irradiation (Gimlich & Gerhart, 1984; see also p. 192).

Once again cleavage is a stage of rapid cell division with possibly no nuclear transcription at all occurring (Gurdon & Woodland, 1969). Most protein synthesis must be programmed by maternal mRNAs, a few of which are known to be unevenly distributed (p. 11), and indeed some proteins are formed at very different rates by animal and vegetal blastomeres (Smith, 1986). Active movements of the various egg components (Phillips, 1985) could increase the spatial complexity. Differences in the rate of indicator oxidation suggested that smaller physiological differences also exist in the dorso-ventral plane (Child, 1948), but when specific metabolic pathways were studied no significant

138    *Non-spiralian cleavage*

differences between the early dorsal and ventral blastomeres were found (Landström & Løvtrup, 1974; Thoman & Gerhart, 1979; Sagata, Okuyama & Yamana, 1981).

The blastomeres are electrically coupled from the earliest stages (Ito & Hori, 1966). The junctions responsible sometimes allow fluorescent dyes (with a molecular weight of a few hundred) to pass to neighbouring cells, and in a recent study of all the cells of the animal half in 32-cell *Xenopus* embryos, Guthrie (1984) found that such transfer occurs far more often from dorsal than from ventral cells. When both dorsal vegetal cells at the eight-cell stage are injected with antibody to the major gap junction protein, embryos develop with axial deficiencies of varying severity recalling the effects of uv irradiation of the zygote (Warner, 1985). Gap junctional communication may thus be involved in the attainment of dorsal dominance, and its involvement in other inductive interactions is considered on pages 194 and 235.

**Mammals**

The embryos of placental mammals cleave extremely slowly and show a high degree of regulative ability: properties which place them at the opposite extreme from the ascidians. Many of their peculiarities, however, also relate with development inside the mother.

From the earliest stages, the cleavages may be slightly unequal (Abramczuk & Sawicki, 1974) and asynchronous (Lewis & Wright, 1935; Kelly, Mulnard & Graham, 1978). In the rabbit one of the first two blastomeres rotates 90° relative to the other, so that second cleavage produces a cross-formation which has also been reported in other mammalian species (Gulyas, 1975). The orientation of third cleavage in mice is such that each cell usually produces one daughter having more contacts with other cells than the other (Graham & Deussen, 1978). The midbodies of the early cleavages persist for some time, probably providing syncytial connections (Goodall & Johnson, 1984). During the eight-cell stage a further distinct event known as compaction is observed (Lewis & Wright, 1935; Ducibella & Anderson, 1975). This transforms the embryo from a collection of loosely associated blastomeres to a morphological unit where cells seem to be maximising their contacts (Fig. 4.26). This also occurs with slight asynchrony in mice, one quartet of cells compacting before the other (see Goodall & Johnson, 1984). It is a complex activity involving microvilli, microfilaments and microtubules (Sutherland & Calarco-Gillam, 1983), and thereafter the behaviour of the cells seems to depend on their position rather than their lineage (see p. 205).

Cleavage of the mammalian zygote, like that of insects, will produce cells destined for extra-embryonic as well as embryonic structures. When

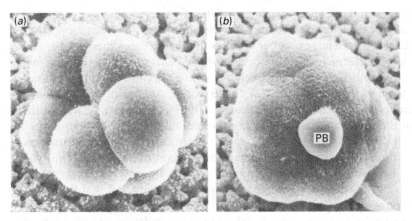

Fig. 4.26. Scanning electron micrographs of 8-cell mouse embryos before (*a*) and after (*b*) compaction. The polar body (PB on (*b*)) is not involved in the increase of cell contacts. (From Ducibella, 1977.)

we try to produce a fate-map of the cleaving mammalian embryo, these are the only two fates which we can hope to assign to different areas, and even for these no reasonably accurate map can be drawn. Wilson, Bolton & Cuttler (1972) injected drops of inert silicone fluid into cleaving embryos, either peripherally or centrally. They found that peripheral drops always entered extraembryonic pathways. In the blastocyst (see Fig. 6.21(*b*)) extra-embryonic tissues do indeed develop outside the embryonic ones, and various other studies confirm that the two prospective areas maintain their relative position with little mixing. This means that most early blastomeres contain material for both developmental pathways. According to Kelly *et al.* (1978) early-cleaving blastomeres contribute more cells to the embryo than late-cleaving ones, but this is probably because they can attain more central positions. Similarly, at the eight-cell stage Graham & Deussen (1978) found that the cells with most contacts contribute most to the embryo. Some (presumably more peripheral) blastomeres may not contribute to the embryo at all (Bałakier & Pedersen, 1982).

It is clear that single blastomeres from the two-cell stage can regulate to produce a whole blastocyst and indeed a new-born mammal (Nicholas & Hall, 1942; Seidel, 1952; Tarkowski, 1959). Even single cells from eight-cell rabbit (Moore, Adams & Rowson, 1968) or sheep embryos (Willadsen, 1981) can sometimes do so. Regulation is, however, less successful with four- and eight-cell mouse blastomeres, and an increasing proportion of the isolates form 'empty' blastocysts lacking an embryo (Tarkowski & Wroblewska, 1967). These authors proposed that the embryonic or extra-embryonic fate of cells in mammalian embryos depends on their inside or outside position, and that, with the reduced volume and cell number of later isolates, few if any cells were enclosed by the time the choice was made. This hypothesis has now been supported by the results of a large number of cell combination experiments, where cells placed externally form extraembryonic structures, and cells (or even whole embryos) which are enclosed form embryonic ones (e.g. Hillman, Sherman & Graham, 1972). Such combinations, or whole cleaving embryos fused together (Tarkowski, 1961), can integrate well in their further development, suggesting considerable regulative ability. Specifically, marked single blastomeres from eight-cell embryos, in combination with carrier blastomeres, can be shown to contribute to both embryonic and extraembryonic structures (Kelly, 1975). It seems clear that such cells are still totipotent, although it should be remembered that this has not been tested in a neutral environment as isolates are. (The distinction may be seen for the ectoderm of early amphibian gastrulae on pages 218–20).

Microscopic investigations of cleaving embryos also reveal inside–outside organisation rather than differences between cells. During the four- and eight-cell stages the nuclei move from peripheral positions to cluster nearer the centre of the embryo (Reeve & Kelly, 1983). The cytoplasm becomes polarised, with columns of organelles aligning between the nuclei and the embryo periphery in the rat (Izquierdo, 1955) and smaller differences appearing at compaction in the mouse (Reeve, 1981). The activities of some cell-surface enzymes are detectable only at points of intercellular contact from early cleavage stages (Mulnard & Huygens, 1978; Lois & Izquierdo, 1984). In compacted embryos, microvilli and a specific molecular marker are restricted to the exposed cell surfaces (Reeve & Ziomek, 1981) and this polarisation persists even in dissociated cells. Blastomeres from eight-cell embryos react to local contact (even with two-cell embryos) by polarising, and if completely surrounded by other cells they fail to polarise (Johnson & Ziomek, 1981b).

Compaction of course greatly increases the areas of intercellular contact, but the cells still polarise in its absence (Pratt, Chakraborty & Surani, 1981) and normal blastocysts can develop if compaction is prevented until the 32-cell stage (Johnson et al., 1979). This poses the problem of the normal function of compaction. It is a striking phenomenon, and one which occurs on time even in isolated blastomeres (Lehtonen, 1980) or after cleavage-arrest with cytochalasin B (Surani, Barton & Burling, 1980), suggesting that it is programmed by a developmental 'clock'. If this involves replication cycle counting then the system operates with a delay since cycles after the two-cell stage are not required for compaction (Smith & Johnson, 1985; see too Howlett, 1986). The cell surface changes include marked increases in the number and variety of intercellular junctions including the appearance of gap junctions (Ducibella & Anderson, 1975). These changes are not required to polarise the cells, but they probably are for the next step in which the daughters of these cells follow divergent developmental pathways (p. 205).

Early mouse embryos show changes in gene expression patterns which are far more radical than those we have seen in other embryos. Following fertilisation several maternally coded proteins increase transiently from minor to major products of translation (Braude *et al.*, 1979), and by the late two-cell stage large amounts of maternal mRNA and ribosomes have apparently been degraded (Bachvarova & De Leon, 1980; Flach *et al.*, 1982). New RNAs of all classes are being transcribed by the two-cell stage (Mintz, 1964; Knowland & Graham, 1972; Clegg & Pikó, 1983), and transcriptional inhibitors can sometimes block development even earlier (Golbus, Calarco & Epstein, 1973; Levey, Troike & Brinster, 1977). Development therefore comes under zygotic control much earlier than in other embryos (Whitten & Dagg, 1972; Magnuson & Epstein, 1981). Studies of rabbit embryos reveal similar patterns, although ribosomal RNA synthesis begins rather later (Schultz & Tucker, 1977).

Among the early synthetic products are proteins characteristic of cleavage stages such as histones and tubulin (e.g. Abreu & Brinster, 1978; Kaye & Wales, 1981), but other syntheses may relate with the embryo's reliance on externally supplied metabolites, with events such as compaction or with later spatial determination. Between fertilisation and the eight-cell stage many cell surface components are gained or lost (e.g. Artzt *et al.*, 1973; Johnson & Calarco, 1980; McCormick & Babiarz, 1984) and the formation of extracellular materials begins (Sherman *et al.*, 1980; Wu *et al.*, 1983). Cytoskeletal elements accumulate beneath outward facing cell surfaces (Lehtonen & Badley, 1980; Sobel, 1983). Many of these changes are probably preparations for compaction, which requires the activity of surface components such as glycoproteins (Kemler *et al.*, 1977; Surani, 1979; Bird & Kimber, 1984) and sterols (Pratt, Keith & Chakraborty, 1980), but apparently requires no further protein synthesis from the late four-cell stage (McLachlin, Caveney & Kidder, 1983; Levy *et al.*, 1986). The major glycoprotein involved in compaction is recognised by antibodies to the cell adhesion molecules of epithelial cell types (Damsky *et al.*, 1983; Peyriéras *et al.*, 1983; and see p. 208). Some components formed at, or even before, the eight-cell stage are later specific to extra-embryonic cells: this is true of two surface antigens (Searle & Jenkinson, 1978; Warner & Spannaus, 1984) and two cytoskeletal proteins (Oshima *et al.*, 1983). It has been pointed out earlier that some materials are restricted to inner or outer areas of the cleaving embryo, and it would be interesting to know whether the above components are formed only in outer areas. Pre-eight cell embryos consume little oxygen (Mills & Brinster, 1967) and require externally supplied pyruvate (Whitten, 1957), but mitochondria are morphologically reorganised at the four- and eight-cell stages (Stern *et al.*, 1971) in an apparent preparation for greater metabolic activity after compaction (see p. 208). There is a strange contrast between the lack of determination of early mammalian blastomeres and their complex biochemistry with indications that differentiation is beginning.

## Conclusions

This chapter demonstrates that embryos vary enormously in the extent to which they segregate developmental information at the earliest cleavages. Ascidians, and probably ctenophores and nematodes, provide better examples of mosaic development than the spiralian embryos; but hydrozoan, starfish and mammalian blastomeres show complete regulative ability even after several cleavages. Other groups show intermediate properties: and it is interesting that groups with a close evolutionary relationship may differ greatly in their regulative ability.

When cleavage has divided an egg in three different planes and the single blastomeres can still regulate to produce a whole embryo, it proves that no single part of the animal–vegetal, dorso-ventral or left–right axis is indispensable for development. Materials could, however, be localised in a concentric arrangement as Wilson (1925, p. 1073) suggested for hydrozoan embryos. Alternatively, materials may not be localised but the embryo may be polarised like a magnet, and this seems likely for those hydrozoans which have received most experimental study, and possible for starfish embryos. In mammals, embryonic and extraembryonic areas are concentrically arranged, but not apparently by localisation of determinants, since a later change of inside–outside position changes these fates. Mammals may be the only embryos where all spatial pattern is environmentally determined.

When developmental information is differentially segregated by the early cleavages, the vegetal half commonly receives most of it, as might indeed be predicted from the behaviour of egg fragments (see Chapter 2). The sea urchin illustrates this well, with the vegetal half often regulating towards complete development and the animal half forming too little. In amphibians, amphioxus and even ascidians and nematodes, animal halves seem to form less than they 'should', and there is independent evidence (reviewed in later chapters) that the development of all animal half tissues except epidermis requires inductive interactions with the vegetal half. A posterior determination centre has been revealed in many intermediate-germ insects, but influences from the anterior pole also seem to be required. In the long-germ chironomid egg anterior–posterior determination seems controlled from two equally important centres, one at each end of the egg.

The early segregation of determinants in a second plane is less often seen. The vegetal half of the ascidian embryo provides probably the clearest example, with determinants for larval notochord, endoderm and muscle in sequence from the anterior to the posterior side, although even for these tissues there are indications that the prospective areas and the determined areas may not be exactly the same. The dorso-ventral polarisation of amphibian embryos also seems to be largely a property of the vegetal blastomeres, and in nematodes they can determine the dorso-ventrality of the animal half. The ctenophore embryo segregates determinants in a second, tentacular, plane at least at the aboral pole and probably orally. All vertically separated halves of early sea urchin and insect embryos, however, can apparently regulate completely.

When left and right halves fail to regulate, as in ascidians and ctenophores, it is not because determinants are differentially segregated, but presumably because their distribution is sufficiently fixed to prevent redistribution into a bilaterally symmetrical pattern.

Visible plasms almost certainly provide the physical basis of ascidian localisations, and partly explain them in ctenophores and nematodes, in both of which ectoplasmic flows increase the spatial complexity of the cleaving embryo. Analysis of the nature of determination has proceeded much the farthest in ascidians, where tissue-specific structures and enzymes form after relatively short times. The mRNA for an endodermal enzyme may already be present in the egg; but other such syntheses require gene action according to a controlled programme, suggesting that specific gene regulatory agents may be segregated to appropriate lineages. In sea urchins, where special mechanisms usually ensure that micromeres are segregated vegetally at the correct time, these cells also show unique gene expression patterns. Maternal gene products, sometimes known to be inherited as RNA, have key roles in insect spatial determination, including the separation of the germ-line.

# 5

## Cellular interactions in the morula and blastula: the case of sea urchin embryos

The early cleavages usually produce a raspberry-like collection of cells known as a morula, and further divisions then produce a form at least derivable from a blastula: a hollow sphere with walls one or a few cell layers thick. This is still a period of no true growth, with the material of the zygote being subdivided amongst cells of more normal sizes. On the other hand, cleavage rates are slowing, allowing more time for transcription and other activities.

The last two chapters have shown that the early cleavages in most animal groups have separated cells with some differences in their biological properties. The next way in which developmental decisions can be made is by interactions among these cells. For instance, if one cell or area produced a labile chemical substance which diffused away from its point of origin, all the rest of the embryo's cells could be defined by their distance from this source and thus determined for various fates. There has been considerable interest in the theoretical requirements of such systems, especially since Wolpert's (1969) and Goodwin & Cohen's (1969) discussions of the subject. In this chapter we will take a classical case of interacting cells, the echinoid morula and blastula, and discuss our knowledge of the interactions at the cellular and biochemical levels.

### Normal development and fate maps

Sea urchin embryos provide a good example of the decrease in cleavage rates during the morula and blastula stages (Dan *et al.*, 1980). The daughters of the three cell types in the 16-cell embryo (see p. 127) also show differences in their cleavage cycle characteristics. There is only a slight asynchrony for mesomeres and macromeres, but micromeres always have a markedly slower cleavage rhythm than the other cells. Their first cleavage is vertical and unequal, and the small micromeres formed at the vegetal pole show the slowest cleavage rhythm of all. The micromeres show another property which is particularly relevant in considering how cells might interact. They form syncytial connections among themselves and with the neighbouring macromeres (Hagström &

144

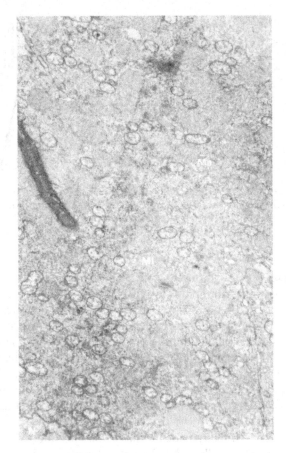

Fig. 5.1. Section through the contact between a macromere (Ma) and a micromere (Mi) in a 16-cell *Paracentrotus* embryo. N, interphase nucleus. Arrows indicate dissolved parts of the cell membrane. × 9000. (From Hagström & Lönning, 1969.)

Lönning, 1969; Schmekel, 1970). These are illustrated in Figure 5.1, and clearly would allow the transfer even of the largest organelles between the cells. Hagström & Lönning (1969) also gave evidence from time-lapse studies that organelles are indeed transferred, and described streaming, pseudopodial and pulsatile activities which may facilitate this.

The alignment of cleavage spindles tangentially to the embryo surface and the adhesion of the outer cell surfaces to the hyaline layer (see p. 32) are probably the main factors ensuring the formation of a blastula with walls one cell thick (Wolpert & Gustafson, 1961). The blastula cells show a clear polarisation in the arrangement of their organelles, the secretion of a fibrous layer over the surface facing the blastocoel, and the formation of cilia on the outer surfaces (Wolpert & Mercer, 1963). Cilia form at

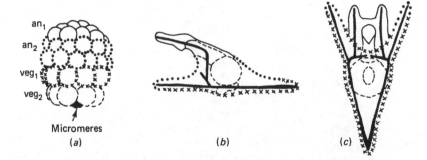

Fig. 5.2. The fates of the various cell layers in the 64-cell sea urchin embryo. (*a*) the five layers marked on the 64-cell embryo; (*b*),(c) the same notation is used to mark the fates of these layers in a pluteus larva seen from the left (*b*) or the anal side (*c*). (From Hörstadius, 1973*a*, as modified by Cameron *et al.*, 1987.)

different times on the different cell types, and in some species the micromeres do not form them (Masuda, 1979). Later, the active cilia at the animal pole are replaced by a tuft of immobile stereocilia, which probably function after hatching to detect the water surface and prevent the disruption of the whole embryo by surface tension (see Hörstadius, 1973*a*, p. 17). The hatched blastula also shows variations in the thickness of its wall which is greatest in the cells forming the apical tuft and in a ring of cells around the somewhat flattened vegetal pole. The dorso-ventral polarity of the embryo is not recognisable until gastrulation has started (see p. 253).

Hörstadius (1935, 1936*b*) made the major classical study of prospective fates by vitally staining cells in the 32- or 64-cell embryo. He distinguished two rings of cells derived from the mesomeres ($an_1$ and $an_2$), two rings derived from the macromeres ($veg_1$ and $veg_2$) and the micromeres. He found that $an_1$, $an_2$ and $veg_1$ all contributed to ectoderm, $veg_2$ to endoderm and larval muscle, and that the micromeres formed the skeleton. Each of the first three rings made a quite distinct contribution to the larval ectoderm, the stomodaeum forming from $an_1$ and the ciliary band from the margin of $an_2$ and $veg_1$. From Cameron *et al.*'s (1987) lineage tracer studies of the eight-cell embryo (see p. 128) it can be concluded that $veg_1$ must make a larger contribution to the ectoderm than Hörstadius thought, particularly at the apex of the pluteus. Pehrson & Cohen (1986) have also demonstrated that the small micromeres invaginate at the tip of the archenteron and contribute to the coelomic sacs of the larva and probably to adult tissues. Figure 5.2 takes account of these recent discoveries, though it does not indicate the differences in fate in the dorso-ventral direction which have been described by Cameron *et al.* (1987).

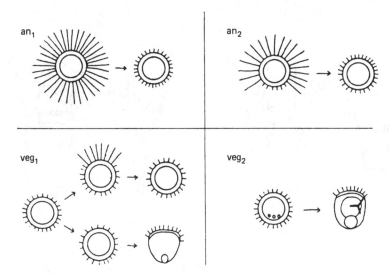

Fig. 5.3. The development of isolated cell layers from the 64-cell embryo. For the origin of these layers, see Fig. 5.2(*a*). (After Hörstadius, 1935.)

## Cell isolations and abnormal combinations

Hörstadius (1935) also followed the development of the same five rings of cells after isolation (Fig. 5.3). On the first day $an_1$ isolates formed stereocilia over their whole surface, and $an_2$ over about three-quarters of their surface. In both cases this is a more extreme differentiation of animal pole derivatives than is their normal fate, where stereocilia were formed by only part of the $an_1$ material. Veg$_1$ isolates sometimes formed a fairly normal-sized tuft rather late and (like $an_1$ and $an_2$) remained as permanent blastulae; others invaginated a small gut. These two possible pathways represent regulation either to more animal or more vegetal structures than are the normal fate of this layer. Veg$_2$ showed the greatest regulative ability producing ovoid larvae with epidermis and skeleton, i.e. both more animal and more vegetal structures than are its normal fate. As noted earlier (p. 131), micromere isolates can differentiate to fairly normal skeletons, although this was not achieved by Hörstadius.

In summary, then, isolation work showed that cells from the animal half develop in an even more extreme animal way than they would in whole embryos, while cells from the vegetal half showed greater ability both to differentiate according to their prospective fate and usually to regulate to produce other structures too. This is essentially what happens with animal and vegetal halves isolated from unfertilised eggs (p. 54) or eight-cell embryos (p. 130), although more detail is attainable as more transverse planes are separable. It makes an interesting contrast with the

148    *The sea urchin morula and blastula*

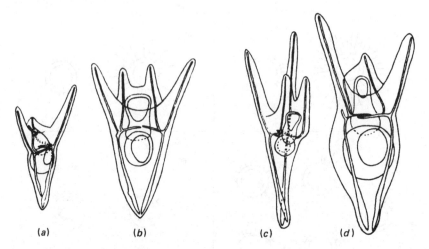

(a)          (b)               (c)         (d)

Fig. 5.4. Pairs of plutei produced from single sea urchin embryos. $(a),(b)$ the products of $an_1$ + micromeres $(a)$ and $an_2$ + macromeres $(b)$. $(c),(d)$ the products of the animal half + micromeres $(c)$ and of the macromeres alone $(d)$. (From Hörstadius, 1936$b$, 1973$a$.)

mosaic development of tiers of cells isolated from the 16-cell *Cerebratulus* embryo (p. 76).

The changes in development after transverse separations of the sea urchin embryo indicate that interactions along the animal–vegetal axis normally affect the determination of cells. Confirmation of this is provided by making abnormal combinations of the cell layers. For example, fairly normal dwarf pluteus larvae can develop from combinations of the $an_1$ layer and micromeres, or of $an_2$ with the macromeres (Fig. 5.4: Hörstadius, 1936$b$). In both cases very considerable regulation must occur and together they show that no single cell has to be present for a pluteus to form. What both combinations have is a balance of animal and vegetal cell types, so this may be a condition for normal development.

When animal and vegetal cells interact, the more obvious effect is that of the vegetal cells on the animal ones. This seems to 'moderate' differentiation, limiting the extension of the animal tuft and allowing less extreme animal cell types to form. Hörstadius (1935) has provided a semi-quantitative analysis of this effect and two of his studies may be used as illustrations. In one he followed the development of whole animal halves from 64-cell embryos alone or in combination with different more vegetal cell layers. As Figure 5.5 shows, any more vegetal material (even the prospective ectoderm of $veg_1$) was able to limit the extension of stereocilia and allow the formation of a stomodaeum and ciliated band. With $veg_2$ or the micromeres, development is more complete and the contribution from the animal half is more extensive: micromeres can induce it to

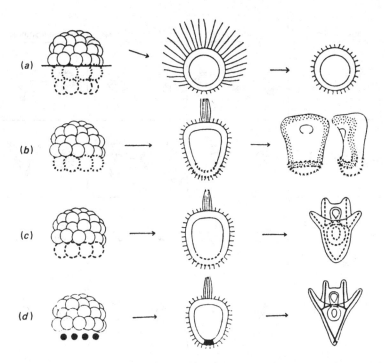

Fig. 5.5. The effects of more vegetal cells on the development of sea urchin animal halves. (*a*) the animal half alone; (*b*) with veg$_1$; (*c*) with veg$_2$; (*d*) with the micromeres. See text. (From Hörstadius, 1935, 1973*a*.)

form a fairly normal (although often small) gut. Thus in experimental circumstances the 'moderation' of animal tendencies by vegetal cells can be taken to the point where animal halves produce much more vegetal structures than are their normal fate. We can say that vegetal cells 'vegetalise' animal halves (Lindahl, 1933), in normal circumstances to diversify the kinds of ectodermal structures formed, but experimentally even to the extent of converting prospective ectoderm to endoderm. Vegetalising ability is highest in the micromeres and decreases with distance from the vegetal pole.

In the second study, Hörstadius (1935) always used micromeres as the vegetalising influence, varying both their number and the cell layer with which they were combined. Figure 5.6 summarises the results, and, to condense them still further here, they show that, of the combinations tried, four micromeres produced the most balanced development from the an$_1$ layer, two from an$_2$, one from veg$_1$ and none from veg$_2$. Lesser numbers seemed unable to completely counter the animal tendencies of the first two layers, and greater numbers shifted development too far in the vegetal direction (e.g. veg$_1$ plus four micromeres gives ovoid forms

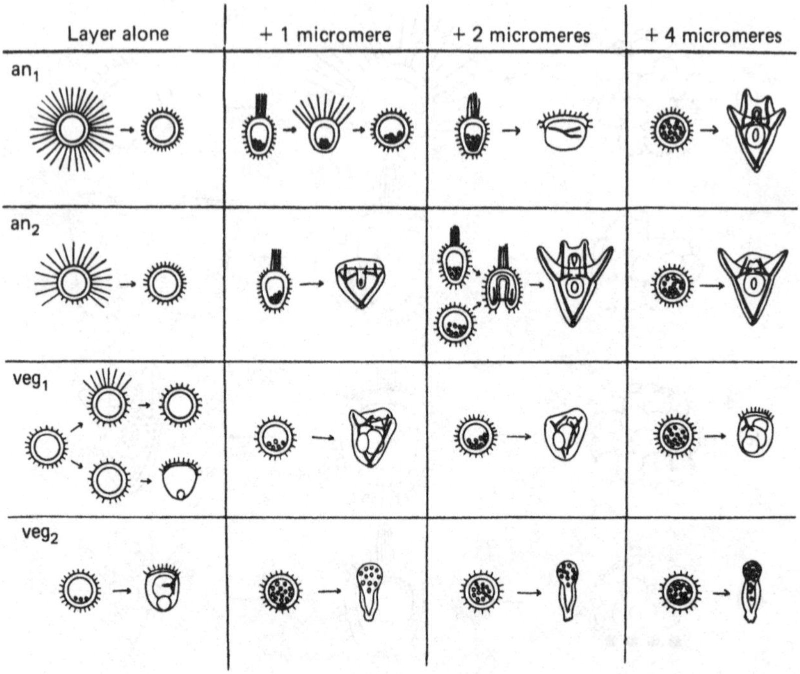

| Layer alone | + 1 micromere | + 2 micromeres | + 4 micromeres |
|---|---|---|---|
| $an_1$ | | | |
| $an_2$ | | | |
| $veg_1$ | | | |
| $veg_2$ | | | |

Fig. 5.6. The effects of added micromeres on the development of the other four cell layers from the 64-cell embryo. See text. (From Hörstadius, 1935.)

similar to $veg_2$ alone). Just as vegetalising ability decreases from the vegetal pole, a 'resistance' to vegetalisation apparently decreases from the animal pole. Since the cells of the vegetal half are susceptible to vegetalisation (e.g. $veg_1$ forming gut when combined with micromeres), the interactions are not simply between two unlike halves of the embryo.

The effect of animal cells upon vegetal ones is more difficult to demonstrate. Isolated vegetal halves usually show regulative development, rather than a more extreme vegetalisation, so no 'moderating' influence from the animal half is apparently required. In some circumstances, however, an 'animalising' effect (Lindahl, 1933), changing cell fates in an animal direction, can be detected. One probable example is in the plutei formed from animal halves plus micromeres (Fig. 5.5(d)). Hörstadius (1935, p. 331) found evidence from vital staining and counts of the primary mesenchyme cells (the normal products of micromeres) that the micromeres sometimes contributed a few cells to the tip of the archenteron. The significance of the staining data must now be in doubt, however, as Pehrson & Cohen (1986) have shown that the small micromeres always move at the tip of the archenteron and contribute to the coelomic sacs. The regulative development of meridional half

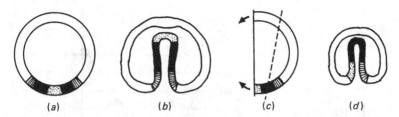

Fig. 5.7. Cell fate changes in a meridional half sea urchin embryo. (*a*),(*b*) diagrammatic representation of the relative positions of prospective oesophagus (dotted), stomach (black) and intestine (hatched) in a normal blastula (*a*) and gastrula (*b*); (*c*) when a meridional half embryo closes (arrows) a new polar axis is established (broken line). (*d*) in the resulting small gastrula the prospective endoderm cells make changed contributions to the embryonic gut. (After Hörstadius, 1973*a*.)

embryos provides another example of 'animalisation'. As noted earlier (p. 129) these cleave as half embryos before closing to a hollow sphere and developing apparently normally. In closing, the cells of the animal and vegetal poles are brought together and interact, apparently each moderating the extreme determination of the other so that new animal and vegetal centres are formed displaced from the original poles (Fig. 5.7). Prospective oesophagus (and presumably prospective skeleton, although this is not illustrated) are animalised to form more posterior gut derivatives, while the new vegetal centre forms in what were prospective stomach cells. At the same time, however, this experiment confirms that vegetalising effects are more easily obtained, as the animal centre is displaced to a far greater extent.

These experiments also demonstrate two other points. One is that skeleton determination cannot be absolutely identified with the micromeres. Not only can micromeres sometimes change their fate, but in their absence other vegetal cells can take on this fate (e.g. in Fig. 5.5(*c*)). Even skeletal determination thus seems to be a gradient-distributed property with its maximum at the vegetal pole. The other point is that the animal–vegetal interaction is so important that, if it is experimentally caused to conflict with polarity, it is the latter which is reversed. It cannot therefore be an inherent polarity which determines the animal–vegetal axis, even if the two phenomena are usually closely related.

Other operative experiments have provided evidence for interactions along the animal–vegetal axis by interfering with their transmission. Stretching an embryo along this axis is enough to cause an enlargement of the animal tuft (Lindahl, 1936), as though the vegetal cells cannot exercise their moderating influence over the greater distance. Equatorial constriction causes the two halves to develop almost as they would after

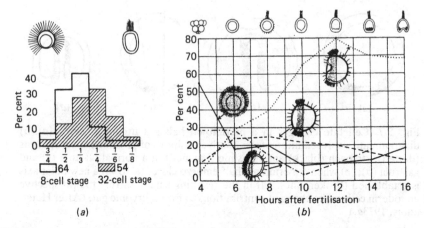

Fig. 5.8. The development of animal halves isolated at different stages: compare with the types shown in Fig. 4.21. (*a*) the size of the animal tuft after isolation at 8- and at 32-cells; (*b*) the form of 2-day isolates separated 4–16 hours after fertilisation. See text. ((*a*) from Hörstadius, 1965*a*, 1973*a*; (*b*) from Hörstadius, 1973*a*, after Lehmann.)

complete separation (Hörstadius, 1938). If two animal halves are added with normal polarity on top of a whole 16-cell embryo (Hörstadius, 1950) the total axial distance often becomes too great to be integrated into a normal larva. Two animal tufts (a partial regulation from the three animal pole regions present) and later two ciliated bands may form. Other combinations with a similar ratio of cell types in a more compact arrangement do succeed in integrating and regulating, so that it seems to be the axial distance which causes the abnormalities. The approximate balance of cell types can, however, also be important as has been demonstrated by other experiments cited above.

The studies reviewed in this section have shown that interactions along the animal–vegetal axis of the sea urchin embryo affect the determination of its constituent cells. To find out when these fates become fixed, Hörstadius (1936*a*, 1965*a*) separated animal and vegetal halves at various stages. Animal halves separated at the 32-cell stage already form smaller animal tufts than those from eight-cell embryos, showing that interaction has occurred between these stages (Fig. 5.8(*a*)). However, it is not until late blastula stages that most two-day animal isolates realise their full prospective fate by forming a stomodaeum (dotted line in Fig. 5.8(*b*)). The vegetal effect on determination in animal halves is thus exercised through the morula and blastula stages. The reciprocal animal effect on vegetal halves is more difficult to quantify, but it is only from the mesenchyme blastula that vegetal halves fail to produce an oral lobe by regulation (Hörstadius, 1936*a*). Animal halves gradually lose the ability

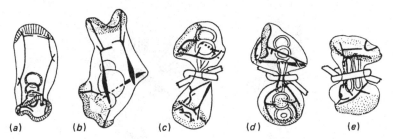

Fig. 5.9. Biventral sea urchin larvae arising from dorso-ventrally stretched (*a*),(*b*) or meridionally constricted (*c*)–(*e*) embryos. (From Hörstadius, 1973*a*; (*a*),(*b*) after Lindahl.)

to respond to implanted micromeres (Hörstadius, 1936*a*); but older vegetal halves (at least up to early gastrula stages) are able to limit the extent of the animal tuft on animal halves (Hörstadius, 1950).

As shown on page 129 there is no irreversible separation of determinants at the first two meridional cleavages, but the early development of the isolates does betray a labile bilateral organisation. The development of fragments which are apparently prospective dorsal and ventral halves is worth some further consideration here. Both can regulate successfully, but vital staining shows that both form dorsal structures at the cut surface, which means that the bilateral organisation of one fragment must have been reversed (Hörstadius & Wolsky, 1936). Furthermore, the ventral structures in one fragment developed with a considerable delay, suggesting that developmental time may have been lost while the new organisation was established. Hörstadius & Wolsky (1936) suggested that the prospective ventral half had maintained its bilateral organisation and the prospective dorsal half had undergone reversal. The same thing apparently happens when the distance between the prospective dorsal and ventral sides is increased by horizontally stretching an egg (Lindahl, 1936) or by vertically constricting an embryo (Hörstadius, 1938): in both some embryos develop with two ventral sides (Fig. 5.9). It seems therefore to be the ventral side which confers the labile dorso-ventrality on the egg and early embryo.

The progress of determination in the dorso-ventral and left–right planes can be monitored by meridional separation of embryos at various stages (Hörstadius, 1936*a*). It is certainly complete by the early gastrula when right, left, dorsal and ventral half larvae are obtained. Right and left halves separated from hatched blastulae also show little sign of regulation, but prospective dorsal sides still show a partial or complete bilateral reversal. Final determination of the dorsal side therefore seems to occur as late as that along the animal–vegetal axis.

(a)                    (b)

Fig. 5.10. Sea urchin development in ionically changed sea-water: (a) *Sphaerechinus* exogastrula after 6 days' culture in the presence of LiCl (final concentration 0.0925%); (b) *Echinus* embryo with enlarged animal tuft after 3 days' culture in artificial sea-water lacking sulphate ions. ((a) after Herbst, 1892; (b) after Herbst, 1897.)

**Chemical animalisation and vegetalisation**

The spatial pattern of development in echinoid embryos can be changed by adding various chemicals to the sea-water, suggesting that the animal–vegetal balance has been affected. Potentially such treatments could be of great value in investigating how various areas become determined and their metabolic properties thereafter. Reviews of these studies have been given by Lallier (1964), Gustafson (1965) and Giudice (1973).

The first such treatment to be discovered was the addition of lithium ions to the sea-water by Herbst (1892). This caused what would now be described as vegetalisation in which the endodermal organs were too large relative to the ectodermal ones. Where it became impossible to invaginate this endoderm into the ectoderm at gastrulation, exogastrulae were formed with a large everted gut (Fig. 5.10(a)). The opposite effect of animalisation, increasing the apical tuft and reducing the size of the gut, was first described for embryos developing in sulphate-free sea water (Herbst, 1897; see Fig. 5.10b).

A large number of other treatments with animalising effects are now known. The stronger ones include zinc ions (Lallier, 1955), trypsin (Hörstadius, 1953), iodosobenzoic acid (Runnström & Kriszat, 1952), sodium thiocyanate (Lindahl, 1936) and polysulphonated dyes (see Lallier, 1964). Treatments with vegetalising effects are fewer but biochemically interesting since they include chloramphenicol (Lallier, 1961), caffeine and theophylline (Yoshimi & Yasumasu, 1978) and cycloheximide (Yoshimi & Yasumasu, 1981). All of these treatments are highly unnatural, and an obvious danger is that they produce unspecific toxic effects. The existence of a more specific component is, however, suggested by the facts that there are two effects of such opposite types and

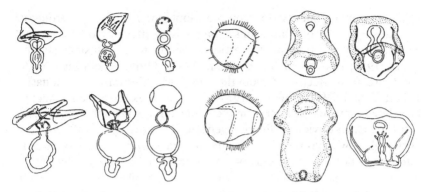

Fig. 5.11. A comparison of animalised and vegetalised embryos produced by operative (upper row) and chemical (lower row) methods. (From Hörstadius, 1973*a*, after Lindahl.)

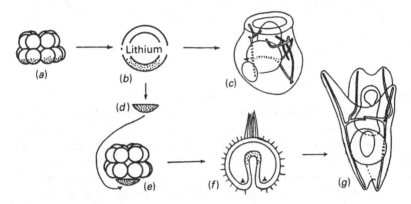

Fig. 5.12. The rescue of animal halves using lithium ions. Animal halves (*a*) reared in the presence of Li$^+$ (*b*) go on to produce ovoid larvae (*c*). If the most vegetal material from such an embryo at the blastula stage (*d*) is combined with another animal half (*e*) it can regulate to produce an essentially normal gastrula (*f*) and pluteus (*g*). (After Hörstadius, 1936*b*, 1973*a*.)

that they mimic the development of isolated animal and vegetal cells (see Fig. 5.11). Furthermore, animalisers and vegetalisers often tend to cancel out each other's effects when used in combination, or can correct the development of isolated half embryos. Lithium ions correct animal halves or even 'overcorrect' them to form exogastrulae (von Ubisch, 1925), and trypsin corrects or overcorrects vegetal halves (Hörstadius, 1965*b*). In fact, cells from the vegetal end of an animal half, after lithium treatment, are even able to act like micromeres in vegetalising other animal halves (Fig. 5.12; Hörstadius, 1936*b*).

A generalised toxic effect of these agents also seems unlikely because treated embryos live for some days after treatment at cleavage stages.

Using 3-hour treatments at various developmental stages, the strongest vegetalising effects of lithium ions or animalising effects of iodosobenzoic acid are found for treatments starting at about the 16-cell stage (Bäckström, 1953; Bäckström & Gustafson, 1953). Other vegetalising treatments are most effective at about the same stage (Fujiwara & Yasumasu, 1974a; Yoshimi & Yasumasu, 1978). In such cases the effective agents have been removed, at least from the sea-water, many hours before their action becomes obvious in the changed size of the gut, apical tuft, etc. It is as though a switch is thrown at the morula stage with the developmental effect only becoming obvious when differentiation occurs much later. This estimate of when determinative events takes place also agrees well with that deduced from operative data above. Some caution is advisable, however, as some of the agent will remain in the embryos and their coverings even when it has been removed from the sea-water. For zinc ions this internal proportion can also be effectively removed by chelating it, and then embryos are little affected by any treatment before hatching (Mitsunaga *et al.*, 1983). Zinc ions may thus act at a much later stage or any early action is reversible by chelation (see pp. 161 and 255). Even without chelation, however, embryos do not show the same early sensitive phase to zinc ions as to other agents (Lallier, 1959), and, at present, the evidence for early sensitivity to other agents still seems secure.

Many artificial animalising or vegetalising treatments also affect dorsoventral determination. Usually the result is a more radially symmetrical larva, whether the treatment was with an animalising agent or with lithium ions. This may be an inevitable consequence of the spread of one polar area, forcing lateral structures into an axial position at the opposite pole. However, this is not evident in the development of isolated animal and vegetal half embryos, where there may be a marked dorso-ventrality (see Fig. 4.21); and Lindahl (1933) has reported that dorso-ventrality can actually be heightened following treatments with lithium ions.

### Models of determination

It is obvious from the last two sections that the extent of determination within the sea urchin embryo increases greatly in the morula and blastula stages. Moreover, the results of these experimental studies put constraints upon the kinds of mechanisms which can be proposed for determination in this case. We will deal mainly with such formal considerations here, before looking at our physiological and biochemical knowledge of these stages.

When he demonstrated that the sea urchin egg is not an equipotential system, Boveri (1901b) proposed the idea of a gradient of determination centred on the vegetal pole (see p. 54). Later Okazaki (1975) confirmed that micromeres are sufficiently determined to produce a skeleton in isolation. They could also, at least in theory, act to impose an appropriate determination on their neighbouring cells, then these in turn determine the next cell group, until all parts of the embryo had been allocated a developmental pathway. However, in such a model the animal pole would appear to have only a passive role, and this is not supported by either manipulative or physiological studies. Animal halves do not fail to form animal pole derivatives but rather produce them over too large an area. In some circumstances cells from the animal pole can even change the determination of those from the vegetal pole. Dye reduction studies (see also later) suggest that the animal pole is in fact the most active area of the whole embryo at least in some aspects of metabolism. Such findings would have to be accommodated in any theory of a vegetal to animal spread of determination.

Runnström (explicitly in 1929) was the first author to suggest that influences from both poles are required to produce the normal pattern of spatial determination. Arguing largely from the effects of artificial treatments, he concluded that the poles are characterised by qualitatively different kinds of metabolism rather than being opposite ends of a single gradient. Hörstadius has since interpreted his operative data in terms of this double-gradient hypothesis, and speaks of 'hostility' or 'antagonism' between the two polar regions, either of which tends, if predominant, to suppress the expression of the other (see Hörstadius, 1973*a*, pp. 44 *et seq.*, 160 *et seq.*). Evidence that there are two qualitatively different types of metabolism comes from those cases where artificial treatments have opposite effects on the two polar regions. For example, rearing in sulphate-free sea-water (Runnström *et al.*, 1964) or at low temperatures (Hörstadius, 1973*b*) increases the animalisation of animal halves but the vegetalisation of vegetal halves. It is difficult to see how one agent could have such opposite effects at the two ends of a single gradient system, and easier to imagine that different metabolic processes are being affected in the two halves.

The two hypothetical overlapping gradients decreasing from opposite poles are often represented in diagrammatic form as in Figure 5.13(*e*). The remaining parts of the figure show how such gradients would be affected in animalised and vegetalised embryos. The two-gradient model is probably capable of explaining, at a phenomenological level, all the data on animal–vegetal determination in the previous two sections.

Dorso-ventral determination appears to be controlled in a way which is quite different, even in principle, from animal–vegetal determination. Probably only the ventral side is determined at egg and cleavage stages (see p. 153). The other sides have the potential to produce ventral structures, but this is only expressed if they are sufficiently separated from the normal ventral side by stretching, constriction or isolation. This suggests that the ventral side is the dominant zone of a single-gradient

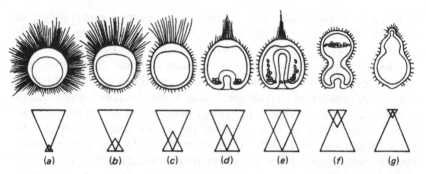

Fig. 5.13. The two-gradient theory of spatial determination along the sea urchin embryonic axis. In the normal embryo (fifth in line) animal and vegetal gradients are equal in strength and interact along the whole axis. In animalised embryos (to the left) the animal gradient is dominant, and in vegetalised embryos (to the right) the vegetal one is dominant. (From Hörstadius, 1973a, after Gustafson, 1965.)

system. In contrast when the animal–vegetal axis is extended by stretching or by adding extra cells, the apical tuft becomes enlarged or more than one tuft is able to develop (see p. 152). That is, the dorsal side reacts to a diminishing influence from the ventral side by escaping to form a new ventral centre; but the animal pole reacts to a diminishing influence from the vegetal pole by extending its own sphere of influence. It is worth remembering that the dorso-ventral polarisation of the amphibian egg probably requires the establishment of only small quantitative differences, in that case apparently favouring the dorsal side (see pp. 63 and 136).

A final consideration is that, despite their differences, animal–vegetal and dorso-ventral determination may be linked in some way, since animalising and vegetalising treatments also affect the bilateral symmetry of the embryo. This led Schleip (1929 p. 537) and Runnström (1931) to propose that the high point of the vegetal gradient was displaced slightly to the ventral side of the vegetal pole, allowing interaction to occur first on that side. Alternatively, the animal and vegetal gradients may be centred strictly on the poles, but the responsible factors may be transmitted fastest on the ventral side (Lindahl, 1936). The validity of such ideas will probably only be tested when we have some understanding of the physiological or biochemical basis of the proposed gradients.

### Approaches to a physiology of determination

Studies in the early part of this century provided several promising clues to the spatial differences in metabolism within the early sea urchin embryo. In this section we will consider these clues and the more recent work to which they have given rise, separating them somewhat artificially from a final section on modern biochemical studies primarily devoted to gene expression.

Child (1936a) produced evidence for a metabolic gradient from the different rates at which dyes become reduced (changing colour) in the different parts of the sea urchin embryo. At egg to blastula stages it was the animal pole cells which reduced the dye fastest, and the colour change spread towards the vegetal pole. Child (1916) had already described gradients in susceptibility to treatments with such agents as potassium cyanide again spreading from the animal pole. He proposed that there is a gradient in metabolic activity, decreasing from a maximum at the animal pole, throughout the stages which now appear to be crucial for determination. At later stages more complex patterns develop (see p. 258), and a more recent study suggests that even in early embryos it is the equatorial regions which are least active (Ostroumova, Belousov & Mikhailova, 1977). Changes in dye reduction patterns in animal and vegetal half embryos relate quite closely with changes in developmental properties, which makes it more likely that they do indeed reflect spatial differences of real developmental significance (Hörstadius, 1952).

This raises the question of which metabolic activities underlie the gradient in dye reducing ability. An obvious candidate would be respiration since the reductions were seen in conditions of limited oxygen supply and potassium cyanide is a respiratory poison. Respiratory rates are indeed reduced in vegetalised embryos (Lindahl, 1936), but they are not increased in animalised ones (Horowitz, 1940). Isolated animal and vegetal halves normally respire at equal rates (Lindahl & Holter, 1940), but only animal halves can increase this rate if oxidation is uncoupled from phosphorylation by dinitrophenol (de Vincentiis, Hörstadius & Runnström, 1966). This means that animal halves could respire at a greater rate if other activities required more energy; but vegetal halves would be unable to respond probably because of limiting activities within the respiratory pathways themselves. These pathways are incompletely known and apparently variable: for example, some species use glycogen and others lipids as the main energy source (Hino & Yasumasu, 1979; Løvtrup-Rein & Løvtrup, 1980; Yasumasu *et al.*, 1984). There are several indications that the hexose monophosphate shunt may be important for determination, including changed activities for one shunt enzyme (raised in animalised embryos, lowered in vegetalised ones) over the period when determination is occurring (Bäckström, 1959; but see also Broyles & Strittmatter, 1971). There are no gradients in mitochondrial numbers or sizes in early embryos (Shaver, 1957; Berg & Long, 1964), and even the difference in numbers of stainable mitochondria claimed by Gustafson & Lenicque (1952) develops rather late as if a consequence rather than a cause of determination.

Alternatively, the developmental relevance of the dye reduction work may be that native molecules are also most easily reduced in the cells of the animal pole, and there are indications that the native state of the embryo's proteins varies along the axis. Animalisation decreases their viscosity and precipitability by ammonium sulphate solutions, and makes the embryo easier to stratify by centrifugation; vegetalisation has the opposite effects (see Ranzi, 1957, 1962). Ranzi's interpretation of these data was that animalisation resulted from a breakdown of some cytoplasmic structures, while vegetalisation occurred when these structures were protected in some way. This is supported by the facts that proteases such as trypsin (Hörstadius, 1953) and denaturing agents such as urea (Pedrazzi, 1957) have animalising effects. If ions are arranged from the most strongly vegetalising to the most strongly animalising, they follow approximately the lyotropic series, which ranks ions from those which most strongly precipitate colloids to those with the strongest swelling effects on them. This series derives from the work of Hofmeister (1888) and is now known to be one of decreasing hydration of the ion (Voet, 1937; Fraústo da Silva & Williams, 1976). The link with colloids can be used to draw together many data obtained with many embryonic groups (Ranzi, 1957, and see p. 227). The disadvantages are that the mechanisms and target molecules are still ill defined. Is the stability of certain specific proteins crucial for animal–vegetal determination and, if so, which are the critical proteins and what does their breakdown involve?

### The biochemistry of the morula and blastula

As a source of rapidly dividing unspecialised cells early sea urchin embryos have been a favourite subject for biochemical study. Relatively few of these studies have been concerned with spatial determination, and whole embryos have usually been homogenised so that it is impossible to assess the contribution of the different embryonic areas. The stages studied are sometimes not stated with any precision, and are often not those which are critical for determination. In the following brief account, data of possible relevance to determination mechanisms are stressed.

Cleavage rates decrease during blastula stages but except for the micromere lineages all cells cleave at the same rate (Okazaki, 1975) or macromere descendants cleave first in a mitotic wave passing towards the animal pole (Agrell, 1964; Parisi et al., 1978). Overall cleavage rates are decreased by lithium ions and increased by animalisers (Hagström, 1963), a result which cannot easily be linked with the normal spatial pattern.

As in most embryos, transcription rates rise rapidly as cleavage rates fall, and the increases in cell number and interphase time may provide a sufficient explanation for the rise (see Peters & Kleinsmith, 1984). The contribution of mitochondrial RNA synthesis is still relatively high but falling (Cabrera *et al.*, 1983). The spectrum of transcripts is unusual, being dominated by heterogeneous RNAs with histone mRNAs being prominent (Kedes *et al.*, 1969). Ribosomal RNA synthesis makes a very minor contribution, and was, for a long time, thought to be absent, but probably occurs at least in blastulae at similar rates per nucleus as in older embryos (Surrey, Ginzburg & Nemer, 1979). 5S and transfer RNAs are synthesised from at least the 32-cell stage (O'Melia, 1979), and small nuclear RNAs from the morula (Nijhawan & Marzluff, 1979). Most of the newly formed mRNAs are probably species which were also supplied maternally (Glišin, Glišin & Doty, 1966; see also Davidson, 1986), but some qualitatively new transcripts do form (Hynes & Gross, 1972; Hough-Evans *et al.*, 1977).

Some of the new transcripts in the 16-cell embryo are formed preferentially or entirely by the micromeres (see p. 132), but there is no other direct evidence of localised transcription events during the period when determination is occurring. Both animalised and vegetalised embryos form 5S and transfer RNAs at normal rates (O'Melia, 1984). After hatching, several prospective areas begin the transcription of specific genes (Angerer & Davidson, 1984, see also p. 256), but this is probably the beginning of differentiation. Of these transcripts, two are produced from the morula–blastula transition (Nemer, 1986) and that coding metallothionein a few hours later (Nemer *et al.*, 1984), and all of these are later characteristic of larval ectoderm. Their synthesis is depressed by lithium ions and stimulated by zinc ions which may mean that they are formed by prospective ectoderm even at early stages. Other transcripts formed from blastula stages include those coding two cytoskeletal actins (Shott *et al.*, 1984) and a calcium-binding protein (Nemer, 1986) and will later characterise just the aboral ectoderm. Their synthesis, however, appears to be depressed by both lithium and zinc ions, and Nemer, Wilkinson & Travaglini, (1985; see also Nemer, 1986) suggest that in this case zinc ions are acting to arrest development while conditions are unfavourable rather than to change cell fates.

It is difficult to assess the developmental significance of early embryonic transcription. It is not required for cleavage, blastulation or ciliation as these occur following enucleation (Harvey, 1936), rearing in actinomycin (Gross, 1964) or in cordycepin (Spieth & Whiteley, 1980). However, transcripts produced in the early stages may affect determination patterns or later developmental events. This possibility has been tested by inactivating nuclei at various stages, but the work is subject to many uncertainties. False positives (indicating 'developmental' transcripts where there are none) may result from unspecific effects of the treatments or residual drug action after a pulse treatment. False negatives (missing the developmental role of early transcripts) may result if markers of early development are inadequate or later transcription can make up for that missed during a pulse treatment. These limitations should be remembered in considering the following data. Work with X-rays (Neyfakh, 1964), actinomycin (Giudice, Mutolo & Donatuti, 1968) and cordycepin (Spieth & Whiteley, 1980) indicates that prehatching transcripts have roles in establishing embryonic polar organisation and in later development through mesenchyme blastula and gastrula stages. Transcription between the morula and hatching stages is also required to allow a rise in protein synthesis rates in the mesenchyme blastula (Krigsgaber & Neyfakh, 1972). Most surprisingly, gastrulation is blocked by bromodeoxyuridine (BUdR) treatments before the 16-cell stage but rarely by later treatments (Gontcharoff & Mazia, 1967). The BUdR is incorporated to DNA, and Czihak (1978; Schreuer & Czihak, 1978) calculates that the percentage gastrulating is inversely related with the number of DNA strands incorporating it in the 16-cell embryo. This would indicate that transcription in the 16-cell embryo is required for gastrulation. It seems likely that prehatching transcripts do have a developmental role, which could be to 'top up' a diminishing maternal pool or to provide qualitatively new information.

There is some evidence that new transcription is especially important for the vegetal tissues, but it suffers from the same low specificity of treatments and ambiguity of developmental markers noted in the previous paragraph. Runnström (1964) found that polyspermic embryos with chromosomal abnormalities showed most developmental disturbance in the vegetal tissues. Using eight-cell embryos, Hörstadius, Lorch & Danielli (1953) found that four enucleated vegetal cells were unable to limit the spread of the apical tuft over the remaining nucleated cells, while four enucleated animal cells could still improve the development of nucleated vegetal halves. Treatment of 16-cell embryos with analogues of RNA bases seems to particularly affect the archenteron (Czihak, 1965a,b), and the gastrulation-blocking effects of BUdR (see above) and actinomycin (Yoshimi & Yasumasu, 1978) have been interpreted in the same way. Studies of local effects of actinomycin have produced conflicting data, however, and it should be noted in particular that micromeres from actinomycin-treated embryos can still act inductively upon untreated animal halves (Giudice & Hörstadius, 1965).

In considering the possible involvement of transcription in the early development of vegetal tissues, it should be remembered that micromeres probably produce more RNA (and certainly have a higher density of new RNA) than other cells, that the spectrum of RNAs is unlike that in other cells, and that micromeres form syncytical connections with other cells (see pp. 132 and 144). In this situation it would not be surprising if RNA were transferred to other cells and affected their

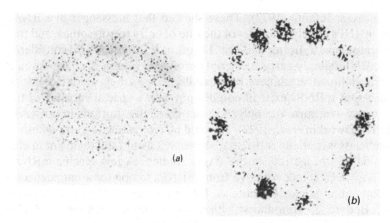

Fig. 5.14. Autoradiographic evidence for RNA transfer from sea urchin micromeres: (*a*) an animal half after 2 h contact with uridine-labelled micromeres: stained with Ehrlich's haematoxylin; (*b*) a normal blastula of the same age which was allowed to take up labelled uridine at the 16-cell stage: unstained. Note the uniform distribution of label in (*a*) and its concentration over nuclei in (*b*). (From Czihak & Hörstadius, 1970.)

development. Czihak in particular has suggested that this is the basis of the micromeres' vegetalising activity. Czihak & Hörstadius (1970) transplanted uridine-labelled micromeres onto unlabelled animal halves. Very large amounts of the label were later seen autoradiographically in macromolecules in the cytoplasm of the animal half cells (Fig. 5.14(*a*)). The work incorporated controls against the possibility that transfer occurred as free uridine followed by transcription within the animal half: transfer still occurred if the operated embryos were reared in excess 'cold' uridine, while if they are reared in labelled uridine it is the nuclei which show the main labelling (Fig. 5.14(*b*)). Despite this, Spiegel & Rubinstein (1972) suggest that the transfer is non-specific.

Newly synthesised RNA leaves the nuclei and enters polysomes later in lithium-vegetalised embryos than in controls (Runnström, Nuzzolo & Citro, 1972), suggesting that the control of post-transcriptional events could be of importance for development. Those messenger-like RNAs which are in the cytoplasm but not in polysomes are complexed with proteins in ribonucleoproteins. Such bodies, commonly called mRNPs, were first described in early embryos including those of the sea urchin (Spirin, Belitsina & Aitkhozhin, 1964; Spirin & Nemer, 1965). Conceivably, the possession of mRNPs could determine the fate of a cell, while later translation of the message would lead to differentiation (see Tyler, 1967). However, ideas of a general storage role for mRNPs now seem to be ruled out following kinetic studies of the passage of RNAs through such bodies and through polysomes (Dworkin & Infante, 1976; Dworkin,

Rudensey & Infante, 1977). These showed that messenger-like RNAs enter mRNPs at only a quarter of the rate of entry to polysomes, and that the former have a shorter half-life. Dworkin & Infante (1976) considered that mRNPs simply carry transcripts produced in excess of the amount the translational machinery can handle. Later, they modified this to suggest that mRNPs exist in equilibrium with a special subclass of the polysomes – because the poly(A) tract gradually shortens in poly(A)$^+$ mRNPs (Dworkin *et al.*, 1977). It would be interesting to know whether this subclass is spatially restricted, although it would still seem not to play a significant storage role at the stages studied. A few specific mRNAs have now been shown to move from mRNPs to polysomes at particular developmental stages (Infante & Heilmann, 1981; Baker & Infante, 1982; Bédard & Brandhorst, 1986*a,b*). Different mRNAs will also compete for the limited translational machinery, and the spectrum of proteins formed will depend on their differing efficiencies, which are highest for histone mRNAs and lowest for other poly(A)$^-$mRNAs (Dworkin & Infante, 1976). As would be expected there are more low-efficiency messages in the mRNPs than in the polysomes (Rudensey & Infante, 1979).

According to the data above, protein synthesis rates in the early embryo are not decided by the amount of mRNA (see also Neyfakh, 1971). The limits may be set by inhibitors attached to the ribosomes and lost gradually during development (Hille, 1974), or by the activities of initiation factors (Nemer & Surrey, 1976; Hille *et al.*, 1981). During the first eight hours of development the suppression of transcription by more than 90% using actinomycin D also has the paradoxical effect of increasing protein synthesis (e.g. Gross, 1964). This could of course be a non-specific effect of the drug, but it is not seen in the data of Sargent & Raff (1976) where activated anucleate merogones were treated with actinomycin D. It may therefore be that some embryonic transcripts regulate translation rates, probably inhibiting the translation of maternal mRNAs.

Early work by Markman (1961*a,b*) suggested a gradient-wise decrease in protein synthesis rates from the animal to the vegetal pole of early sea urchin embryos, but Berg (1965, 1968*b*) found no significant differences between the rates in animal and vegetal halves of another species. He did, however, find that rearing in lithium ions caused a decrease in protein synthesis rates before hatching proportional to the degree of vegetalisation (Berg, 1968*a*). Several other vegetalising treatments inhibit protein synthesis and indeed some are known primarily as translation inhibitors (Fujiwara & Yasumasu, 1974*a*; Yoshimi & Yasumasu, 1978, 1981). Yoshimi & Yasumasu (1979, 1981) suggest that vegetalisation results when chromatin proteins are formed more slowly than DNA, allowing extra RNA synthesis on the naked DNA.

The chromatin proteins certainly change qualitatively during early sea urchin development. The core histones and histone H1 undergo a particularly clear programme of change, mainly involving histone variants each coded by a different gene. Special 'cleavage-stage' variants are formed from fertilisation to the morula, other 'early' variants from about four-cells to the hatched blastula, and 'late' variants usually from about hatching (Ruderman & Gross, 1974; Cohen, Newrock & Zweidler, 1975; Newrock *et al.*, 1978; Childs, Maxson & Kedes, 1979; Harrison & Wilt, 1982). Later variants enter chromatin as soon as they are formed; earlier ones are not displaced but soon become only minor constituents in a changing chromatin. Presumably as a result, chromatin organisation and properties change (Arceci & Gross, 1980*b*), and the major developmental significance may be to allow tighter packing of the DNA as nuclei and metaphase chromosomes become smaller (Harrison & Wilt, 1982). One specific H1 variant is particularly prevalent in the micromeres (see p. 132): this may result simply from differences in cell cycle time, but could still be of great significance for determination (Senger *et al.*, 1978). There is no other evidence that changes in the histones affect determination, and certainly all three cell types isolated from 16-cell embryos can undergo the later changes autonomously (Arceci & Gross, 1980*a*). Gineitis, Stankevičiute & Vorob'ev (1976) have compared the chromatin composition of normal, animalised and vegetalised embryos, but the histone variants were then incompletely separated. Cognetti, Settineri & Spinelli (1972) reported some qualitative changes in the non-histone chromatin proteins following the 16-cell stage.

The relative rates at which other proteins are formed change markedly during development to hatching, mainly by the programmed translation of various maternal mRNAs (Terman, 1970). However, determination is accompanied by very few qualitative changes in the species of proteins synthesised by the sea urchin embryo. Even two-dimensional gel separations of the new polypeptides reveal few changes before hatching (Brandhorst, 1976; Bédard & Brandhorst, 1983). Posthatching changes largely restore the synthetic profile seen in oocytes, so that of the early embryo is atypical and may relate with the requirements of rapid cell division (Grainger, von Brunn & Winkler, 1986). Micromeres isolated from 16-cell embryos also show few changes at these early stages although, of course, these few may be very important for their later differentiation (Harkey & Whiteley, 1983). Berg (1968*a*) could see no qualitative differences between the proteins formed by normal and lithium-treated embryos, and this remains true in the two-dimensional separations of Hutchins & Brandhorst (1979). The effects of zinc treatment have been checked only at posthatching stages and in one-dimensional gels – when significant pattern changes are seen (Carroll, Eckberg & Ozaki, 1975).

Some of the most obvious changes occurring at morula and blastula stages involve the shapes and adhesiveness of the cells. Early cleavage stages probably rely to a great extent on the extracellular hyalin layer to keep the cells together, but thereafter cell-surface materials such as glycoproteins play an increasingly important part (see Watanabe *et al.*, 1982). Other materials present in the egg are secreted from the cells' inner surfaces during early development to form a basal lamina (Wessel, Marchase & McClay, 1984). Fibronectin and laminin may be present in all the extracellular materials (Spiegel, Burger & Spiegel, 1980, 1983). There is evidence that the cell-surface properties affect biochemical activities within the cells (Vitorelli *et al.*, 1980), and that the surface properties in different embryonic areas are sufficiently different to allow cells to 'sort out' following random aggregation (Spiegel & Spiegel, 1978). Only the micromeres of 32- or 64-cell embryos are agglutinated by concanavalin A (Roberson & Oppenheimer, 1975), the patterns of its binding indicating a greater lateral mobility of membrane components than in other cell types (Roberson, Neri & Oppenheimer, 1975). Lithium ions seem to partially dissociate cells (Hagström, 1963), an effect also noted for other types of embryo, but concanavalin A, which has a similar effect, animalises sea urchin embryos (Lallier, 1972). Lithium ions also reduce, and zinc ions increase, the precipitability of hyalin by the calcium ions in sea-water (Timourian & Watchmaker, 1975). The latter authors in particular suggest that hyalin may play a role in normal animal–vegetal determination. At gastrulation, cell-surface properties along the animal–vegetal axis certainly become very important indeed (see p. 254), and some components secreted then by primary mesenchyme cells are apparently already present within cells of the micromere lineage in blastulae (Sugiyama, 1972).

We know less about other aspects of the biochemistry of the morula and blastula, and about their significance for development. There are changes in polyamine levels (Manen & Russell, 1973), which may have developmental significance since inhibition of their synthesis arrests embryos as animalised late blastulae (Brachet *et al.*, 1978). Yoshimi & Yasumasu (1978) have noted that several vegetalising agents are inhibitors of cyclic AMP phosphodiesterase, as well as being translational inhibitors. However, Nath & Rebhun (1973) did not report any disturbance of development when embryos were reared in the presence of cyclic AMP, despite enormous increases in its internal concentration. Interestingly, when attempts were made to purify 'natural' animalising and vegetalising materials from sea urchin eggs, the greatest success was achieved with an animalising fraction which may be a nucleotide (Josefsson & Hörstadius, 1969; see also Fujiwara & Yasumasu, 1974*b*).

## Conclusions

Early echinoid embryos provide the classical example of spatial determination resulting from interactions occurring throughout the embryo. Cells of the animal half in isolation produce only the most extreme animal pole derivatives, but more vegetal cells are able to 'vegetalise' them so that they produce the full range of ectodermal structures and even endoderm. Isolated cells of the vegetal half show a far greater self-differentiative and sometimes regulative ability: only rarely can more animal cells be shown to have an 'animalising' effect upon them. Various artificial treatments can animalise or vegetalise whole embryos, and there is evidence that they do so by affecting the normal axial determination system. Animalising and vegetalising activities can be represented by a double-gradient model, and they normally fix cell fates along the animal–vegetal axis during morula and blastula stages. Determination in the dorso-ventral plane is also finalised in blastulae, and probably begins as a labile determination of the ventral side which acts to prevent any other side from becoming ventral. Such a system has many different properties from that controlling animal–vegetal determination.

Early studies revealed two kinds of physiological gradient along the animal–vegetal axis of early sea urchin (and many other) embryos. First, the ability to reduce vital dyes decreases from the animal to the vegetal pole, and may indicate an ill-defined gradient in metabolic activity. Secondly, the native state of proteins seems to vary from a more denatured free form at the animal pole to a more cross-linked and organised one at the vegetal pole. Before dismissing such differences as hopelessly vague, the modern biochemist should reflect that sophisticated molecular techniques are only now beginning to identify differences along the early axis. Transcription is required if polarity is to be expressed, but the suggestion of a particular involvement with vegetalisation, with the possible transfer of RNA as a vegetalising agent, remains contentious.

# 6

## Interactions at morula and blastula in
## other embryos

The last chapter presented the evidence that the determination of cells in the sea urchin morula and blastula is affected by interactions occurring throughout the embryo. The present chapter will consider evidence for similar interactions occurring at the equivalent stages in other animal embryos.

### Insects

As has already been made clear (p. 119), the nuclei of insect eggs at first divide without cleavage of the cytoplasm. The stages considered here (Fig. 6.1) are those when most nuclei enter the peripheral cytoplasm (the syncytial blastoderm), when cleavage furrows appear between them (the cellular blastoderm) and when the future embryonic area becomes recognisable (germ band stages). Despite the great morphological differences from the sea urchin, developmental interactions apparently determine cell fates in a similar way, and the mechanisms involved have been subjected to genetic analysis particularly in *Drosophila*.

Cycles of nuclear division are lengthening during these stages (e.g. see Foe & Alberts, 1983), just as cleavage cycles do in the equivalent stages of other embryos. At syncytial blastoderm stages they occur as parasynchronous waves, starting in positions which vary with the species and even sometimes amongst individuals (Miyamoto & van der Meer, 1982; Foe & Alberts, 1983; Lundquist & Löwkvist, 1983). The divisions become completely asynchronous during cellular blastoderm stages, despite the fact that the nuclei are not completely separated by the new membranes. The ingrowth of furrows apparently uses a single network of microfilament bundles extending throughout the cortex of the blastoderm (Fullilove & Jacobson, 1971; Warn, Bullard & Magrath, 1980). Internally the 'cells' remain syncytially connected with the yolk, and thus indirectly with one another until gastrula stages (*Drosophila*: Rickoll, 1976) or even later. Despite the integrated nature of the furrowing process, cellularisation is first completed on the ventral side of the *Drosophila* embryo (Anderson & Nüsslein-Volhard, 1984*b*).

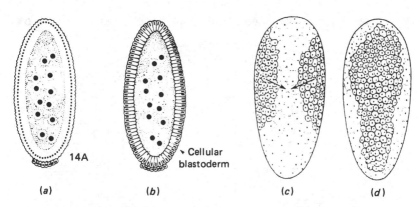

(a)                (b)                (c)                (d)

Fig. 6.1. Blastoderm and germ anlage stages of insect development. (a),(b) *Drosophila* embryos in the fourteenth cleavage cycle before (a) and after (b) cellularisation. (c),(d) in many insects a germ anlage then arises by aggregation of cells in two lateral areas (c) and their fusion on the future ventral side (d). ((a),(b) from Foe & Alberts, 1983; (c),(d) from Seidel, 1936.)

The fates of specific areas in the *Drosophila* blastoderm have been ascertained by direct observation, the effects of local damage and various kinds of genetic analysis (Poulson, 1950; Janning, 1976; Lohs-Schardin *et al.*, 1979; Struhl, 1981*b*; Hartenstein, Technau & Campos-Ortega, 1985; Jürgens *et al.*, 1986; Lawrence & Johnston, 1986). Each method has its drawbacks, but there is good agreement over most features shown on the fate map of Figure 6.2. Along the egg, each segment is represented by a transverse stripe three or four cell diameters wide. These cells apparently form all the ectodermal and perhaps all the somatic mesodermal structures of the larva and of the adult (the progenitors of the latter being set aside throughout larval development as the imaginal disc cells). The prospective areas for most different tissue types are arranged around the egg circumference. From the dorsal to the ventral side these areas will form extraembryonic tissues, dorsal ectoderm, ventral ectoderm (including the ventral nerve cord) and somatic mesoderm. The future ectoderm thus originates as separate right and left halves. Cells destined for the anterior and posterior endoderm are localised near the poles, together with prospective stomodaeum and presegmental head structures anteriorly, and proctodaeum and Malpighian tubules posteriorly.

The arrangement of prospective areas is probably similar in other long-germ insects. In intermediate- and short-germ embryos the prospective embryonic areas make up a far smaller proportion of the blastoderm (see p. 120 and Fig. 4.14), and a posterior part of the embryo cannot be mapped as it proliferates later from a small budding zone. This applies to the posterior abdomen of intermediate-germ embryos but to all structures behind the procephalic area in short-germ ones (see Sander, 1976).

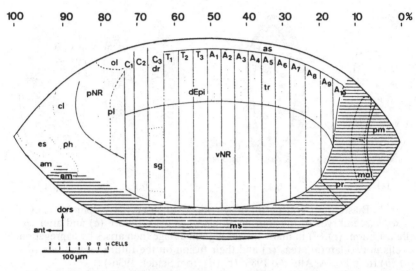

Fig. 6.2. Fate map of the *Drosophila* blastoderm. Hatched areas will invaginate at gastrulation. am, anterior midgut; as, amnioserosa; cl, clypeolabrum; dEpi, dorsal epidermis; dr, dorsal ridge; es, oesophagus; mal, Malpighian tubules; ms, mesoderm; ol, olfactory lobes; ph, pharynx; pl, procephalic lobe; pm, posterior midgut; pNr, procephalic neural region; pr, proctodaeum; sg, salivary gland; tr, trachea; vNr, ventral neurogenic region; C1–C3, gnathal segments; T1–T3, thoracic segments; A1–A10, abdominal segments. (After Hartenstein *et al.*, 1985.)

In Chapter 4 we found that determination was incomplete during preblastoderm stages. By blastoderm stages it has usually progressed, and in *Drosophila* every cell in the ectodermal area may be determined to the level of an individual segment. The clones arising from such cells seem always to be restricted to single segments (Wieschaus & Gehring, 1976), and small numbers of cells seem to develop according to their segment of origin even when transplanted to abnormal sites (Simcox & Sang, 1983). The defects resulting from local damage are not always so precisely localised however (Bownes & Sang, 1974). In other insect species, determination is far from completed at blastoderm stages: in some it has hardly begun. Even in such cases, however, current evidence suggests that the increase in spatial determination will require interactions occurring throughout the embryo.

One effect of incomplete determination in preblastoderm stages is the 'gap phenomenon' – in transversely ligated eggs the structures of several segments fail to form while terminal segments differentiate and occupy too much space (pp. 122–3). Following ligation at cellular blastoderm stages, however, this gap has shrunk to one segment or less in *Drosophila* (Schubiger, 1976; Vogel, 1977), and other dipteran species (see Table 6.1). In other insect embryos included in the table, the size of the gap also

Table 6.1 *Average numbers of body segments which fail to be formed after fragmentation at various stages ('gap phenomenon')*

| Species | Stage at fragmentation | | | | | Numbers represent |
|---|---|---|---|---|---|---|
| | Cleavage | Syncytial blastoderm | Cellular blastoderm | Germ anlage | Germ band | |
| *Acheta* | –* | – | – | 3 | | Median value |
| *Euscelis* | (9)  (5–6) | (3–4) | 2–3 | 2 | 0 | Median value |
| *Necrobia* | 8 | 6 | 5 | 2 | | Median value |
| *Bruchidius* | (8) | 7–8 | 5 | 3 | | Mean |
| *Callosobruchus* | 7–8 | 7–8 | 1–2 | 1 | | Mean |
| *Apis* | | Greater | | Smaller | | Estimate |
| *Smittia* | 7 | 4 | 1 | | | Median value |
| *Protophormia* | 6 | 4 | 2 | 1 | 0 | Mean |
| *Drosophila* | 3–4 | 2 | 0–1 | | | Mean |

* Only one fragment of the egg will form a pattern.
( ), Figures based on a few cases, because in most eggs only one fragment will develop to a stage permitting analysis of the pattern formed. (After Sander, 1976, where original sources can be found. More recent data have been added for *Callosobruchus* (from van der Meer & Miyamoto, 1984) and *Drosophila* (from Vogel, 1977).)

clearly decreases the later the ligation is made, but there is still a gap of several segments at late blastoderm stages. Gaps are usually smaller or non-existent following partial or temporary constriction of eggs. All this suggests that the determination of intermediate segments requires interactions between factors emanating from the poles occurring and sometimes even being completed during blastoderm stages.

When the *Acheta* egg is transversely fragmented at intravitelline cleavage stages, the anterior fragment forms too few, if any, embryonic structures, and the posterior fragment too many (see p. 121). Eggs fragmented at later stages show a gradual increase in the structures produced anteriorly and a decrease for those formed posteriorly. The former effect can be explained as a forward transfer of developmental information from a posterior instructing centre (see Sander, 1976). This would seem very like the moderating effect on animal halves due to a vegetal determination centre in sea urchin embryos. The reduction in the potential of posterior fragments also recalls sea urchin vegetal halves which sometimes regulate to whole embryos at early stages but lose this ability later, apparently as stable developmental gradients determine cells along the egg axis. Sander (1976) does not propose that there is any anterior gradient factor in *Acheta* (although see below), but even in the sea urchin the effects of the animal gradient are relatively difficult to demonstrate. A similar system may operate in some other species where the posterior pole must be present to allow any development, as in the intermediate-germ *Platycnemis* (p. 121) and the short-germ *Tenebrio* (Ewest, 1937). Mee (1986) has fragmented the *Schistocerca* egg when segmentation has already begun and shown that anterior fragments rarely add more than one further segment, but she offers a different interpretation of this result.

In *Euscelis* there is better evidence that two polar factors are involved in anterior–posterior determination. Following early transverse ligation both fragments seem to produce fewer structures than they would in normal development, although the absence of a fate map makes this uncertain. When ligations are made at later stages both fragments produce an increasing number of segments (Sander, 1959, 1976; see Fig. 6.3). The increasing potential of posterior halves is particularly interesting as it was not seen in *Acheta*, and apparently shows that anterior areas can influence determination more posteriorly (see also Vogel, 1982*a*). This influence can be brought prematurely into more posterior areas by centrifuging the egg (Sander, 1976). It is also noteworthy that the most posterior egg region of *Acheta* or *Euscelis* is unable to develop in isolation despite apparently being highly determined. It seems possible that, as in amphibians (p. 189), this region may require an activating influence from more anterior regions before it is able to differentiate.

The influence of the polar regions on the determination of intermediate

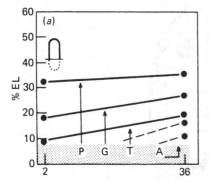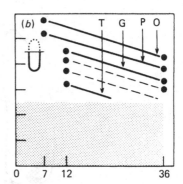

Fig. 6.3. The development of anterior (*a*) and posterior (*b*) egg fragments of
*Euscelis* separated at different stages between cleavage (2 h) and the early germ
anlage (36 h). Ligation position is given as %EL (egg length) where the posterior pole is 0 and the anterior pole is 100. P, procephalon; G, gnathocephalon;
T, thorax; A, abdomen; O, extraembryonic development. The broken curves
indicate individual thoracic segments. In (*a*) with fragmentation levels above a
given curve pattern, elements anterior to the body region indicated will usually
be formed. The dotted region cannot be analysed. In (*b*) with fragmentation
levels below a given curve pattern, elements posterior to the body region indicated will usually be formed. Fragments comprising only the dotted zone did not
differentiate. (From Sander, 1976.)

segments has been most directly demonstrated for the posterior pole of
*Euscelis*. Sander (1960) succeeded in grafting the posterior polar region
onto an anterior fragment, and showed that this was then able to form a
complete embryo (Fig. 6.4). Such anterior fragments would 'alone' (after
constriction) form only a few anterior segments, so that these results
suggest that at least two-thirds of the embryo requires instruction from
the posterior pole. The posterior pole plasm (which includes a mass of
symbiotic bacteria almost engulfed by the egg) shows other interesting
properties in experimental treatments. Transplanted to the anterior end
of a posterior fragment it can cause partial or complete reversals of
polarity (Fig. 6.5). Posterior fragments lacking this plasm are, however,
little affected (Fig. 6.4), which suggests that the posterior factors affecting development are not confined to the immediate polar area.

These findings again recall those with the sea urchin system, with the
posterior pole, like the micromeres, being capable even of causing
polarity reversals. The anterior pole seems, however, to have greater
effects on determination than the sea urchin animal pole cells. Sander
(1960, 1961) has some evidence that the two polar influences act antagonistically, and has developed a two-gradient model of the spatial
determination system in *Euscelis*, which is capable of explaining the gap
phenomenon and the effects of transplanted posterior poles (Fig. 6.6).
According to Meinhardt (1977) only a single, posterior, gradient centre

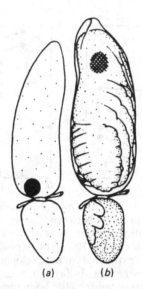

(a)                    (b)

Fig. 6.4. The development of *Euscelis* egg fragments after transposition of the posterior pole material to the anterior fragment. (*a*) the operated egg; (*b*) typical result: a complete embryo in the anterior fragment and a partial germ band of two thoracic segments and the abdomen in the posterior fragment. Compare with Fig. 4.16. (From Sander, 1976.)

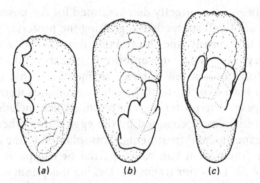

(a)              (b)              (c)

Fig. 6.5. Polarity reversal in *Euscelis* after transposition of the posterior pole material to the anterior region of posterior fragments. (*a*) a control fragment with a partial germ band of two gnathal segments, thorax and abdomen; (*b*),(*c*) fragments after the transposition with reversed partial germ bands consisting of abdomen and two or one thoracic segments respectively. (From Sander, 1976.)

may be required; while Vogel (1978, 1983) proposes the existence of extra centres at intermediate levels (see also p. 123) interacting over short distances. Other data from Vogel (1982*b*) seem irreconcilable with any gradient theory, unless the two polar areas are able to effectively cancel each other out by their mutual antagonism. In any case most data

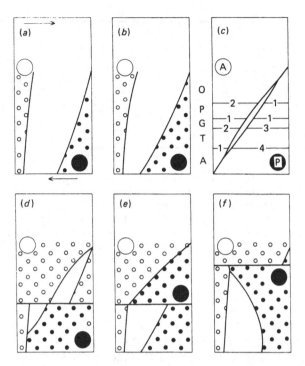

Fig. 6.6. A double-gradient hypothesis to explain the *Euscelis* data. Open circles represent the anterior gradient and filled circles the posterior one: in each case the large circle represents the source. (*a*)–(*c*) normal development from oviposition to germ anlage, when the ratio of the two gradients (a/p ratio), which varies continuously along the axis, is assumed to decide which developmental pathway will be followed (letters as in Fig. 6.3); (*d*) after early fragmentation 'anterior' substance accumulates in the anterior and 'posterior' substance in the posterior fragment causing a discontinuity in the a/p ratio and so a gap in the segmental series; (*e*) after transfers of posterior pole material and then fragmentation the anterior fragment has sources of both kinds and so can form a complete axis; (*f*) after transfers of posterior pole material within a posterior fragment the direction of the posterior gradient and of the a/p ratio reverse giving a reversed partial germ band. (From Sander, 1976, where further details may be found.)

are explicable by a two-gradient model very similar to that proposed for sea urchins (p. 157) and Fig. 5.13).

The influence of the anterior polar area is still greater in many long-germ insect embryos, with the two poles seeming about equivalent in importance for determination. The beetle *Bruchidius* provides a good example, as the thorax usually appears about half-way along the egg but develops at far more anterior levels when the anterior influence is lost and at more posterior levels when the posterior influence is lost (Jung, 1966: see Fig. 6.7). In chironomids the development of the anterior and posterior egg halves may depend on the ratio of anterior to posterior

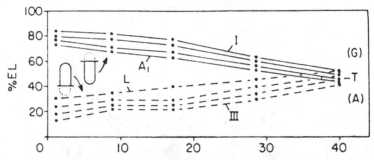

Fig. 6.7. The development of anterior and posterior fragments of *Bruchidius* eggs ligated at different stages up to the early germ anlage. Upper solid lines show the size of posterior fragments required for the formation of partial germ bands beginning with first (I), second and third thoracic segments and first abdominal ($A_1$). Lower broken lines show the size of anterior fragments required for the formation of partial germ bands terminating with labial (L), first, second or third (III) thoracic segments. G, T and A are territories for gnathocephalon, thorax and abdomen respectively at the early germ anlage stage. (From Sander, 1976, after Jung.)

determinants localised at each pole (Kalthoff *et al.*, 1982: see p. 124 and Fig. 4.18). Kalthoff *et al.* (1982) note that this is compatible with double-gradient models, although they think that the graded information is used to take a binary decision between anterior and posterior developmental pathways. The *Bruchidius* data above could not, however, be explained in this way. Factors spreading from the two poles have important effects in *Drosophila* too, as is seen when polar plasms are deleted or transferred (Frohnhöfer *et al.*, 1986; see p. 124) or when wild-type anterior plasm rescues *bicoid* mutants (Frohnhöfer & Nüsslein-Volhard, 1986). The apparent antagonism of the two polar influences is particularly clear in the latter case where anterior plasm injected at more posterior levels seems to be increasingly subject to a suppressive effect from the posterior pole.

Evidence considered in Chapters 1 and 4 indicates that maternal factors provide a basic 'lay-out' for the early insect embryo. Sander (1976) points out that eggs where the anterior pole affects determination arise from meroistic ovaries, so anterior determinants may be received from nurse cells which communicate with the oocyte at its anterior pole. It is probable that all the polar factors considered above are, at least initially, under maternal control, and their interaction may thus begin even before laying. This could explain how middle sections, isolated from both poles, can from the earliest stages form a few segmental structures (Vogel, 1978, see p. 123), in which case no third reference point need be postulated. A thoracic or other intermediate region often plays a leading role in development, being first to show commitment to the single segment level in *Drosophila* and first to show overt differentiation in many insects. It has been argued that this 'differentiation centre' also affects the development of other areas (see, e.g. Schnetter, 1934; Haget, 1953), but Sander (1976) finds the case unconvincing (see also p. 259).

The dorso-ventral axis seems to be determined at relatively late stages in many insect embryos. At first, bilaterally symmetrical embryos can develop when one lateral half of the system has been destroyed, but this regulative ability is lost before cellularisation in *Dermestes* (Küthe, 1966) and at about the time that the germ anlage becomes visible in *Leptinotarsa* (Haget, 1953) and *Gryllus* (Sauer, 1961). It remains in modified form even in the gastrula of *Tachycines* (Krause, 1952; see p. 259). No instructing regions comparable with the polar influences have yet been implicated in dorso-ventral determination (Sander, 1976). Some labile dorso-ventral organisation, however, presumably exists, and is probably controlled by maternal factors in view of the developmental effects of maternal mutations (see p. 23). As in sea urchins, the ventral side could act as a dominant zone: it has already been noted that cellularisation is first visible ventrally.

By the first division after the blastoderm stage the prospective thoracic cells of *Drosophila* seem to be determined to the level of intrasegmental compartments. This was discovered in studies of flies carrying a mutation which is only expressed following a mitotic recombination event. Such an event can be induced by irradiation of the embryo, and thereafter the daughters of the cell where recombination occurred can be recognised and followed through their further development. Clones induced just after the blastoderm stage can contribute only to the anterior or posterior half of some segments (García-Bellido, Ripoll & Morata, 1973; Lawrence & Morata, 1977). For example, mesothoracic clones originating at this stage can contribute to both wing and leg structures but not to the anterior and posterior compartments of the wing (see Fig. 6.8). The compartment boundary is respected even where the normal and mutant cells divide at different rates so that one part of a segment grows faster than the other. Similar evidence for clonal restriction of cells at the single segment level is available for blastoderm-stage embryos of the bug *Oncopeltus* (Lawrence, 1973). The anterior–posterior boundary is so important in *Drosophila* that in some circumstances the fundamental unit appears to be the 'parasegment', i.e. the posterior compartment of one segment plus the anterior compartment of the next one (Martinez-Arias & Lawrence, 1985; and see Fig. 6.11 below).

The discovery of compartments in development has shed new light on the homoeotic mutants of *Drosophila* which have been known for many years. In these mutants a single gene change transforms the cuticular structures of one compartment into those of another compartment, e.g. in *bithorax* structures of the anterior mesothorax appear in place of those

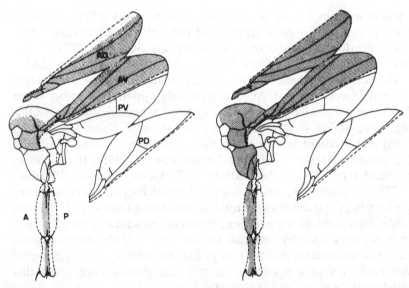

Fig. 6.8. Evidence for compartments in *Drosophila* development. The descendants of cells genetically marked at the blastoderm are seen here in the cuticular structures of adult second thoracic segments. Wings and legs are displayed as if opened at the anterior–posterior compartment boundary (broken lines): note that there are many marked cells at this boundary but none cross it. A, anterior; P, posterior; D, dorsal; V, ventral. (From Steiner, 1976.)

of the anterior metathorax (Lewis, 1963; see Fig. 6.9(*a*)). The effects can often be phenocopied by subjecting wild-type embryos to abnormal conditions during defined developmental stages, e.g. *bithorax* effects can be phenocopied by ether treatment or heat shock particularly at the syncytial blastoderm stage (Gloor, 1947; Maas, 1949; Capdevila & García-Bellido, 1974; see Fig. 6.9(*b*)). This suggests that the loci concerned normally function in making choices between developmental pathways. The transforming effects of homoeotic mutations have usually been checked in adult cuticular structures, but probably all larval and adult ectodermal (including neural) cells of a compartment are affected in the same way (Lewis, 1978; Jiménez & Campos-Ortega, 1981; Lawrence, 1985). The same genes may define fates in the somatic mesoderm to the level of single parasegments, but probably have no role in the determination of the unsegmented tissues arising near the poles of the blastoderm (Lawrence, 1985). Current studies seem to be uncovering a system capable of translating the quantitative polar differences of the preblastoderm embryo into the precisely determined pattern of the blastoderm.

An important part in this system is played by a complex of genes which includes *bithorax* and is called the bithorax complex or BX–C. This

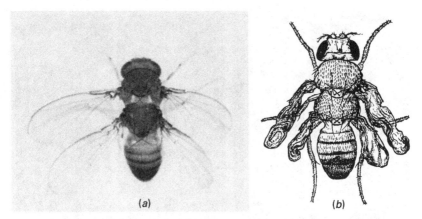

Fig. 6.9. Four-winged flies: the transformation of metathorax to mesothorax in *Drosophila*. In (*a*) the transformation was caused by mutations within the bithorax complex (BX–C), and in (*b*) by ether treatment of the early embryo. ((*a*) courtesy of Professor E. B. Lewis; (*b*) from Gloor, 1947.)

complex, on the right arm of the third chromosome, contains several genes which in mutant form produce homoeotic transformations of thoracic and abdominal segments, and has been studied over many years particularly by Lewis (1963, 1978; see also Morata, Sánchez-Herrero & Casanova, 1986). Its domain of action covers the whole embryo posterior of the anterior–posterior compartment boundary in the mesothoracic segment. Embryos homozygous for a deficiency of the whole complex die by the first larval instar, but all segments in the domain have the same form (Fig. 6.10). Lewis (1978) thought that each of these segments developed as a mesothorax, but further work has shown that only the anterior compartments have mesothoracic characteristics while the posterior ones are determined as prothoracic (Morata & Kerridge, 1981; Hayes, Sato & Denell, 1984). This shows that the ground-state of the domain affected by the BX–C is the fourth embryonic parasegment, i.e. posterior prothorax plus anterior mesothorax.

Mutations affecting smaller parts of the BX–C have more localised effects on the determination of segments or compartments. As already described above, the *bithorax* function itself affects the determination of anterior metathorax. This function is controlled by a region of the BX–C near its centromere–proximal end, and there is a general correlation of position within the complex and position along the body of the visible effect, with more centromere–distal regions affecting successively more posterior parts of the body. It now seems established that the functions are hierarchically organised, with the first subdivision of the BX–C being into three major genes, *Ultrabithorax*, *abdominal-A* and *Abdominal-B* (Sánchez-Herrero *et al.*, 1985). Figure 6.11 shows their arrangement,

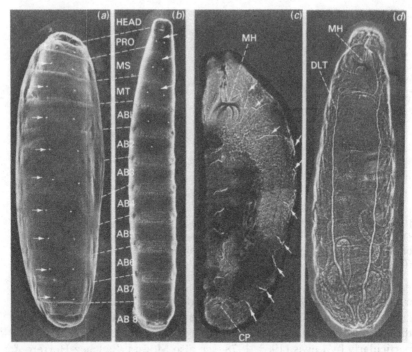

Fig. 6.10. Cuticular and tracheal patterns in *Drosophila* first instar larvae homozygous for a deletion of the BX–C (*a*),(*c*) compared with the wild type (*b*),(*d*). (*a*),(*b*) scanning electron micrographs of cuticular patterns: note that ventral setal bands of thoracic type as well as Keilin's organs (arrows) are also present in the abdominal area of the mutant; (*c*),(*d*) phase contrast views of whole larvae. Note separate tracheal trunks in each segment of the mutant (arrows) while these are joined in a continuous dorsal longitudinal trunk (DLT) in the wild type. Tiny chitinised plates (CP) posteriorly in the mutant resemble rudiments of the mandibular hooks (MH) suggesting a head-like transformation. PRO, MS and MT, thoracic segments 1–3; AB1–AB8, abdominal segments 1–8. (*a*) × 135; (*b*) × 50; (*c*),(*d*) × 100. (From Lewis, 1978.)

with *Ultrabithorax (Ubx)* controlling development in parasegments five and six, and the domains of *abd-A* and *Abd-B* meeting at about the boundary between the fourth and fifth abdominal segments.

Further progress in understanding how the BX–C operates has been made mainly for the *Ubx* region. Strong *Ubx* mutations transform parasegments five and six (i.e. a total of four anterior or posterior compartments) to the form of parasegment four. This is the sum of the effects seen when mutation affects the separate functions within the *Ubx* region: each function controls the identity of a single compartment. *Ubx* activity requires chromosome continuity over about 75 kilobases (kb) of the DNA, and it has been suggested that the primary transcript from this region may be processed in different ways to produce RNAs for the four

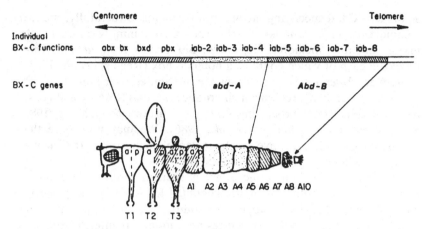

Fig. 6.11. The genetic and anatomical domains of the BX–C. The relative positions of known and postulated genetic functions along the third chromosome are shown at the top. These can be assigned to one of three major BX–C genes, each of which is responsible for the characteristic pattern of a specific anatomical domain as shown. T1–T3, thoracic segments; A1–A10, abdominal segments; a, anterior compartment; p, posterior compartment. (From Sanchez-Herrero *et al.*, 1985.)

*Ubx* functions (Bender *et al.*, 1983). This has been supported by the discovery of at least three RNAs (see Beachy, Helfand & Hogness, 1985) and three proteins (White & Wilcox, 1984) deriving from the 75 kb region and sharing a part of their sequence. A further 20 kb of the *Ubx* region, in the area of the *bithoraxoid* function, also has a role in determination (Bender *et al.*, 1983). It may regulate the expression of the 75 kb region by affecting the processing pathway taken by its primary transcript in the various compartments (Beachy *et al.*, 1985; Lipshitz, Peattie & Hogness, 1987).

There is some evidence that the *abd-A* and *Abd-B* genes may operate in a comparable way to *Ubx*. Each controls a complex of overlapping genetic functions (Karch *et al.*, 1985) probably using multiple transcription units (see Beachy *et al.*, 1985). All three include a homoeo box, a short DNA sequence which is highly conserved and present in other homoeotic genes (McGinnis *et al.*, 1984; Scott & Weiner, 1984; Regulski *et al.*, 1985).

One limitation of the compartment-specification system so far described is that it applies only to the mesothoracic and more posterior segments. There are, however, other homoeotic mutations which cause transformations between head and thoracic structures, and many again involve genes located in a small region of the right arm of chromosome three. Kaufman, Lewis & Wakimoto (1980) have called this the antennapedia-complex (ANT–C), and suggest that it acts in much the same way as the BX–C but specifying the anterior segments. Structurally, the two complexes seem very similar with the ANT–C containing several separate major functions including *Deformed, Sex combs reduced* and *Antennapedia* each of which contains a homoeo box (Regulski *et al.*, 1985). Moreover, *Antennapedia* mutations are distributed over more than 100 kb of the genome in a region which produces several RNAs possibly as a result of differential processing (Garber, Kuroiwa & Gehring, 1983; Scott *et al.*, 1983; Schneuwly *et al.*, 1986). We may note here that homoeotic gene clusters have also been described in a silkmoth (Tazima, 1964) and a flour beetle (Beeman, 1987).

Another kind of homoeotic effect is seen in *Drosophila* embryos homozygous for mutations at the *engrailed* locus. Only the posterior compartments of segments appear to require the *engrailed* product, and in its absence they develop structures appropriate to anterior compartments (García-Bellido & Santamaria, 1972; Morata & Lawrence, 1975; Kornberg, 1981). Posterior and anterior compartments seem thus to be characterised by the presence and absence, respectively, of *engrailed* gene expression. It is noteworthy that the *engrailed* region also possesses a somewhat diverged homoeo box sequence (Fjose, McGinnis & Gehring, 1985; Poole *et al.*, 1985).

By postblastoderm stages all the body segments and their anterior and posterior compartments could thus be characterised by a unique pattern of expression of the homoeotic genes. This, however, requires that cells are able to recognise their position along the egg axis with considerable precision. Maternal gene action of course provides quite a lot of this information in the form of polar determinants and presumed gradient-distributed factors (see pp. 24 and 126). If such factors include activators or repressors of the BX–C or ANT–C genes they could directly control the changing patterns of expression of these complexes along the embryonic axis. It will be very interesting to check these patterns in the offspring of *bicoid, oskar* and other such mutant females (see pp. 21–2). Offspring of *esc⁻/esc⁻* females develop with eighth abdominal structures in all segments behind the head (p. 24), a phenotype which suggests that the *esc* product acts as a repressor of the BX–C (Struhl, 1981*a*). Current evidence indicates, however, that it acts too late to decide initial specifications and may, instead, reinforce these at postgastrula stages (Struhl & Akam, 1985). Many other genes are known to have such maintenance-type effects on BX–C or ANT–C expression (Lewis, 1978; Capdevila & García-Bellido, 1981; Forquignon, 1981; Ingham, 1984; Jürgens, 1985; Breen & Duncan, 1986). The genes of the BX–C and ANT–C also seem to affect each other's expression with those acting posteriorly repressing those acting more anteriorly (Hafen, Levine & Gehring, 1984*b*; Struhl & White, 1985), but again such interactions occur relatively late.

The genes so far considered are of course those involved in producing

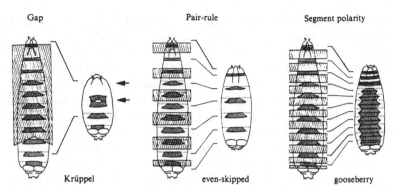

Fig. 6.12. Semi-schematic drawings indicating the regions deleted from normal pattern in three *Drosophila* segmentation mutants. Dotted regions indicate ventral setal bands, dotted lines segmental boundaries and hatched bars the regions missing in mutant larvae. Transverse lines connect corresponding regions in normal and mutant larvae. The arrows indicate planes of polarity reversal in *Krüppel*. (From Nüsslein-Volhard & Wieschaus, 1980.)

differences among the segments along the insect body. But the production of repeated patterns is probably a still more fundamental property of segmented animals and this too is controlled by quite a large number of genes in *Drosophila*. Some information is again provided by maternal gene action (see p. 24), and this seems to be refined by several stages of zygotic gene action (Nüsslein-Volhard & Wieschaus, 1980; Meinhardt, 1986; see Fig. 6.12). Mutations in three genes lead to development with large gaps anteriorly (*hunchback*), centrally (*Krüppel*) or posteriorly (*knirps*) in the segmental series, while further genes may have comparable roles in the terminal regions (see Meinhardt, 1986). *Krüppel* mutants also show local reversals of polarity suggesting interference with the transmission of maternal polarity signals (Wieschaus, Nüsslein-Volhard & Kluding, 1984). Mutations in a second set of genes cause the subdivision of the axis into half the number of double-sized segments. Such 'pair-rule' genes are presumed to control segmentation by acting in repeating units two segments wide, while further genes would act at the level of individual segments (Nüsslein-Volhard & Wieschaus, 1980). One of the pair-rule genes, *fushi tarazu*, is located within the ANT–C and actually contains one of the homoeo boxes there (see Laughon & Scott, 1984). It is now clear that there are extensive interactions between the segmentation and the homoeotic genes (e.g. Ingham, Ish-Horowicz & Howard, 1986), and the gap genes and *fushi tarazu* may provide the first functional link between the maternal gradients and the ANT–C and BX–C (Frohnhöfer & Nüsslein-Volhard, 1986; Ingham & Martinez-Arias, 1986; Lehmann & Nüsslein-Volhard, 1986).

The zygotic activity of homoeotic and segmentation genes would thus seem well capable of transforming maternal spatial information, often of a quantitative kind, into the 'genetic addresses' specifying segments and compartments along the embryonic axis. In particular, the normal function of the homoeotic genes would be the selection of one among many possible developmental pathways, and García-Bellido (1975) has thus called them selector genes. To actually follow those pathways would, however, require the activation of still further genes coding for the compartment-specific structures. In this connection it is noteworthy that homoeo box sequences (Laughon & Scott, 1984; Desplan, Theis & O'Farrell, 1985) and other sequences in *Krüppel* (Rosenberg *et al.*, 1986) and *hunchback* (Tautz *et al.*, 1987) encode polypeptide sequences which should have DNA-binding activity. Work with monoclonal antibodies also shows that the *Ubx* proteins are concentrated in nuclei (White & Wilcox, 1984; Beachy *et al.*, 1985). Such properties are consistent with a role of the homoeotic gene products as gene-regulatory agents.

Ventro-dorsal patterns are probably also determined at blastoderm stages but we know much less about the genes involved in this case. As already explained (p. 23) the injection of wild-type cytoplasm can polarise early embryos produced by *Toll⁻* females, but if the injection is delayed beyond syncytial blastoderm stages the rescue is incomplete with lateral elements being particularly difficult to restore (Anderson *et al.*, 1985*a*). Several zygotically active loci affect dorso-ventral patterns and appear to interact with the maternal information (Simpson, 1983). One, *zerknüllt*, is located in the ANT–C and contains a diverged homoeo box, and its transcripts are localised in dorsal areas of the syncytial blastoderm (Doyle *et al.*, 1986).

The coordinate system which decides the pattern of selector gene expression presumably develops in the periplasm. In *Leptinotarsa* large proportions of the yolk and of the nuclei can be removed at intravitelline cleavage stages and whole embryos still develop (Schnetter, 1965). At the same stage in *Drosophila* there is evidence that the nuclei have not been determined according to their position along the egg axis (Okada, Kleinman & Schneiderman, 1974*b*). The best positive evidence that the periplasm can elaborate and maintain a complex coordinate system independently of its surroundings is provided by those experimental cases in *Callosobruchus* where defined strips along the axis develop with an opposite polarity from the areas on either side (van der Meer, 1984).

In many insect species, RNA synthesis begins in the nuclei as they enter the periplasm (Lockshin, 1966; Sabour, 1972; Lamb & Laird, 1976) or at least by cellular blastoderm stages (Schmidt & Jäckle, 1978). Only the pole cell nuclei show a delayed activation (Lockshin, 1966; Lamb & Laird, 1976; Zalokar, 1976). In this connection it is interesting that certain proteins, which normally combine with nuclear RNAs, enter all other nuclei (including those still in the yolk) distinctly earlier than they

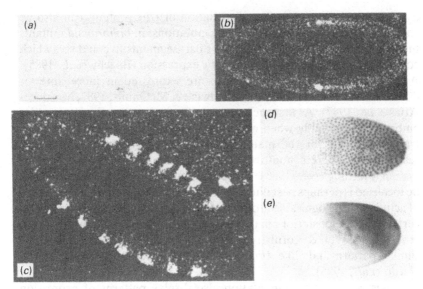

Fig. 6.13. The distribution of some gene products implicated in early spatial determination in *Drosophila*. Egg anterior to the left and dorsal at the top. mRNA distributions are visualised in bright (*a*) or dark field (*b*),(*c*) after *in situ* hybridisation of appropriate probes and autoradiography. *cad* protein is visualised immunohistochemically. (*a*),(*b*) *Ubx* mRNA in the late cellular blastoderm: scale bar = 50 μm; (*c*) *ftz* mRNA in a 3.5 h cellular blastoderm; (*d*),(*e*) *cad* protein in the twelfth (*d*) and fourteenth (*e*) nuclear cleavage cycles: note that the protein decreases anteriorly and ventrally between these stages. ((*a*),(*b*) from Akam & Martinez-Arias, 1985; (*c*) from Hafen *et al.*, 1984*a*; (*d*),(*e*) from Macdonald & Struhl, 1986.)

enter pole cell nuclei (Dequin *et al.*, 1984). All gene classes become competent for activation as nuclei enter the periplasm (Edgar & Schubiger, 1986) and the new transcripts include mRNAs (Lamb & Laird, 1976) among which, as in other embryos, are those coding histones (Anderson & Lengyel, 1980). Inhibitor studies with *Leptinotarsa* indicate that the new transcripts are required through and beyond gastrulation (Maisonhaute, 1977).

Recent work has provided direct confirmation that the homoeotic and segmentation genes are indeed expressed in the *Drosophila* embryo in highly specific spatial arrangements (see Fig. 6.13). *Ubx* transcripts appear in the BX–C domain at syncytial blastoderm stages, and following cellularisation attain high levels in what is probably parasegment six and lower levels elsewhere in the BX–C domain (Akam & Martinez-Arias, 1985). A similar distribution is seen for *Ubx* proteins at later stages of development (White & Wilcox, 1984; Beachy *et al.*, 1985). *Ubx* expression thus occurs in the areas expected from our current understanding of the BX–C except that higher activity would be expected in parasegment five (compare Fig. 6.11). The distribution of *Ubx* proteins can also be changed in predictable ways by some mutations: in *bithoraxoid* mutants the changes include equal labelling of parasegments five and six, which could indicate a failure to regulate *Ubx* expression (Beachy *et al.*, 1985). As is also expected ANT–C genes are expressed in more anterior domains (McGinnis *et al.*, 1984; Chadwick & McGinnis, 1987; Martinez-Arias *et al.*, 1987). At first such patterns are not very precise but in older embryos, presumably when regulatory interactions are complete, high transcript levels for the major genes of the BX–C and ANT–C are restricted to discrete non-overlapping regions of the central nervous system (Harding *et al.*, 1985). The precision with which transcription can be localised is perhaps best illustrated by *engrailed*, where fourteen bands of activity are recognisable around the cellular blastoderm and are almost certainly in the posterior compartments of the future segments (Ingham *et al.*, 1985*b*; Weir & Kornberg, 1985). The *engrailed* protein is found in a similar pattern and, like *Ubx* proteins, is concentrated in nuclei (Di Nardo *et al.*, 1985).

Genes controlling segmentation also display patterns of expression which are mostly consistent with their expected roles. We should recall here that the products of the maternal-effect gene *caudal* exist in a postero-anterior gradient by syncytial blastoderm stages (p. 24), probably due to the progressive loss of transcripts from the anterior end (Mlodzik *et al.*, 1985). Mlodzik *et al.* isolated this gene by its possession of a diverged homoeo box sequence, and Macdonald and Struhl (1986) demonstrated that its protein product again becomes localised in the nuclei. The first main band of *Krüppel* transcription is within the central area where segmentation fails in *Krüppel* mutants, but its full expression pattern is more complex (Knipple *et al.*, 1985; Gaul *et al.*, 1987). Transcripts of *hunchback* are found more anteriorly than those of *Krüppel* (Jäckle *et al.*, 1986). The pair-rule genes seem at first to be quite randomly expressed, but by cellular blastoderm stages there are seven broad bands of their transcripts (e.g. Hafen, Kuroiwa & Gehring, 1984*a*; Weir & Kornberg, 1985; Ingham, Howard & Ish-Horowicz, 1985*a*) and of *fushi tarazu* protein in nuclei (Carroll & Scott, 1985). Ingham *et al.* (1985*a*) suggest that *fushi tarazu* is expressed in the even number parasegments and *hairy* in the odd ones. The role of *Krüppel* has also been demonstrated in other ways: injection of the apparent wild-type transcript to the middle region of preblastoderm mutant embryos can effect a partial rescue (Preiss *et al.*, 1985); while conversely injection of antisense RNA to wild-type embryos produces mutant phenotypes (Rosenberg *et al.*, 1985).

As in sea urchins, some mRNAs appear to be surplus to the translational capacity of the early *Drosophila* embryo (Lovett & Goldstein, 1977), and different mRNA species may be translated with different efficiencies in *Smittia* (Jäckle, 1980*b*). In beetle embryos, the entry of the nuclei to the periplasm coincides with an increase in protein synthesis (Lockshin, 1966) and may control qualitative changes in the synthesis pattern (Küthe, 1972). There are minor qualitative changes at this stage in *Drosophila* too (see Sakoyama & Okubo, 1981; Savoini *et al.*, 1981), and synthesis levels for several proteins fall dramatically before cellularisation (Summers, Bedian & Kauffman, 1986). In *Smittia*, some newly formed proteins are formed only by anterior or posterior parts of the blastoderm (Jäckle & Kalthoff, 1980, 1981). Of these, one protein was found to reliably predict the future position of an abdomen, and another one of a head, even in experimentally produced double abdomens and double cephalons. The posterior indicator protein is formed by all regions of preblastoderm embryos, when it is probably maternally coded, but its later localised production seems to be under zygotic control.

## Other arthropods and other echinoderms

There have been few experimental studies of early development in arthropods other than insects or echinoderms other than echinoids. Nevertheless, evidence has been presented which suggests that developmental interactions occurring in blastoderm or blastula stages may be as important as in their better-studied relatives. This section focusses on that evidence.

Green (1971) has reviewed studies of crustacean development. As in insects the eggs are typically centrolecithal and the nuclei divide and migrate to the superficial cytoplasm, although the cytoplasm is also completely cleaved in species with small eggs. A germinal disc becomes distinct on the ventral side, and it is there that gastrulation and embryo formation occur. In *Megaligia*, no embryo develops if cells in the centre of the disc are destroyed, and even before the disc appears the ventral side may be determined to form embryonic structures (Kajishima, 1952). However, within the disc area considerable regulation is still possible, as whole embryos can form in the fragments following meridional constrictions.

Among the arachnids, the germinal disc area again seems to be critical for development, but multiple embryos can still be obtained following experimental treatments at blastoderm stages. Sodium bicarbonate treatments of the horseshoe crab *Tachypleus* produce this effect, possibly by weakening the adhesions of cells and allowing each cell mass from the germinal disc area to form an embryo (Itow & Sekiguchi, 1979). Centrifugation has very similar effects on some spider blastulae (Sekiguchi, 1957). If only a small proportion of germinal disc cells are required to form an embryo, then interactions among these cells must normally be required to ensure that different disc cells form different structures. Kautzsch (1910) long ago concluded that spider development is regulative, and it was later shown that spatial patterns are displaced by treatments which animalise or vegetalise sea urchin embryos. Lithium treatments lead to an overdevelopment of the vegetal endomesoderm as well as a diminution of the prosoma (Ehn, 1963*a*) which is reminiscent of microcephaly in vertebrates. The main effect of iodosobenzoic acid is a radialisation of the embryo which suggests action on dorso-ventral rather than animal–vegetal gradients, but the prosoma is also enlarged (Ehn, 1963*b*).

Turning to the echinoderms, some starfish blastomeres are able to form whole larvae even when isolated at the 16-cell stage (see p. 133), and it would seem that interactions are required to allocate different fates to the various cells. The transfer of ions between the cells becomes possible at the 32-cell stage of *Asterias* (Tupper, Saunders & Edwards, 1970) although fluorescent dyes are still not transferred (Tupper & Saunders, 1972). Dan-Sohkawa has described the way in which *Asterina* blastomeres acquire increased adhesiveness during morula and blastula stages, and considers blastulation as an active process transforming a loose collection of cells to an integrated individual where cells become connected by septate desmosomes (see Dan-Sohkawa & Fujisawa, 1980).

Several other observations point to the importance of the blastula stage in *Asterina* for later development. The embryos become susceptible to a vegetalising effect of lithium ions which can increase the proportion of cells entering mesendodermal pathways by 30% (Kominami, 1984). Kominami thinks that this may be due to increased cell adhesion in the treated embryos allowing developmental interactions to spread further, and believes that concanavalin A treatments decrease both the adhesions and the proportion of cells entering endodermal pathways (Kominami, 1985). (However, these agents have equivalent effects on animal–vegetal determination in sea urchin and other embryos, where lithium ions are frequently noted to decrease cell adhesions.) *Asterina* also passes through a peculiar 'wrinkled blastula' stage in which cell sheets sink into the interior of the embryo (e.g. see Komatsu, 1976). It is possible that this facilitates interactions among the cells of an embryo larger than that of the sea urchin. Finally, transcription in blastulae seems to control gastrulation in *Asterias* (Barros, Hand & Monroy, 1966).

**Amphibians**

Cleavages continue in the morula and blastula stages of amphibian development, and until the midblastula the length of the cleavage cycles is effectively constant although each vertical cleavage starts in the animal half and takes some time to reach the vegetal pole. After the midblastula all cleavage cycles lengthen, and those of the vegetal cells begin to lag increasingly behind those of the animal cells (Signoret & Lefresne, 1971; Hara, 1977). In *Xenopus* these changes begin after the twelfth cleavage, when transcription also begins and all but the largest cells attain motile properties forming blebs and pseudopodia (Newport & Kirschner, 1982). This suggests an important switch in the developmental programme, perhaps as a preparation for gastrulation, and Newport & Kirschner (1982) have termed it the midblastula transition (MBT).

Other morphogenetic events occur at these developmental stages. The blastocoel expands by the addition of mucosubstances secreted by the surrounding cells (Kalt, 1971), and by early blastula stages an extracellular material covers many cells and glycogen-like granules are present in the intercellular spaces (Komazaki, 1983). Movements foreshadowing gastrulation also begin, with cells moving away from the animal pole (compare Figs 4.23 and 7.2) and vegetal material sinking into the embryo first along cleavage furrows or towards the nuclei and later by the ingression of whole cells (Schechtman, 1935; Harris, 1964; Kobayakawa, 1985).

In all pregastrula stages prospective ectoderm, mesoderm and endoderm are present in sequence from the animal to the vegetal pole (see Fig. 4.23), and there are already marked differences in developmental properties along this axis at early cleavage stages (p. 136). When isolated at blastula stages animal pole fragments, or even the whole animal half, form only epidermis, and vegetal pole fragments seem already determined for endoderm (e.g. see Nieuwkoop, 1969a). The differentiation of the latter however appears to require a promoting influence from the mesoderm (Okada, 1953, 1954) much as the differentiation of insect posterior fragments may require activation from more anterior levels (see p. 172). The endoderm of early blastulae cannot even form a blastopore groove in isolation, and requires an influence from the mesoderm before it can do so (Doucet-de-Bruïne, 1973). Later in development even the kind of endodermal structure produced can be shown to depend on the kind of mesoderm present (see p. 251). Although large isolates from the vegetal pole show an apparently complete determination, the fate of individual cells can be changed by exposing them to a new environment. Cells from the vegetal pole of one blastula transplanted to the blastocoel of another can contribute to all three germ layers (Heasman *et al.*, 1984). This is true even of cells containing germ plasm, so the possession of this plasm does not imply an irreversible determination for the germ-line (Wylie *et al.*, 1985b).

While the cells of the two polar areas exhibit very distinct properties the interesting question is whether cells at intermediate levels are determined as mesoderm. For a long time it was thought that determinants already localised in the grey crescent area of the zygote ensured that this area in the gastrula became an organisation centre for the embryo (see pp. 60 and 135). In that case no determinative effects would be expected in the intermediate stages, and there has been some recent support for such ideas. Imoh (1984, 1985) has claimed that a yolk-poor equatorial 'mesoplasm' is recognisable at all pregastrula stages in *Xenopus* and is segregated to the mesodermal and neural areas, and he suggests that it may contain determinants. On experimental grounds too Gurdon *et al.* (1985*c*) have proposed that the subequatorial region is determined for mesoderm formation even in the zygote (p. 55), and they have provided similar evidence at cleavage and blastula stages (Mohun, Brennan & Gurdon, 1984*b*; Gurdon *et al.*, 1985*b*). The cells concerned, however, are apparently derivatives of the third tier of the 32-cell embryo and this will in fact produce only a part of the mesoderm (Fig. 4.23). For the second tier, Gurdon *et al.* (1985*b*; Gurdon & Fairman, 1986) agree that mesoderm arises inductively (see below).

Other studies of the developmental properties of equatorial regions indicate that mesodermal properties arise there epigenetically over morula and blastula stages. The ability to self-differentiate as mesoderm is apparently acquired at morula stages (Nakamura & Matsuzawa, 1967; Koebke, 1977), and the inductive properties required for later 'organiser' activity only by late blastulae (Nakamura *et al.*, 1971*b*). It could be argued that the early explants contain determinants and only fail to differentiate because of the large size and fragility of the cells, but there are other reasons for believing that equatorial cells become determined as a result of interactions with more polar cells. When the most animal and vegetal cells of morulae and blastulae, including no prospective mesoderm, are combined, mesodermal tissues are formed (Ogi, 1967; Nakamura, Takasaki & Ishihara, 1971*a*). This recalls the production of normal sea urchin larvae from combinations of polar cells (p. 148) or of whole *Euscelis* embryos when posterior poles are grafted onto anterior fragments (p. 173).

These findings suggested to the Japanese workers that a double-gradient system could determine cell fates along the animal–vegetal axis in amphibians as in sea urchins. Nieuwkoop (1969*a*), however, concluded that the interaction is all in one direction – the vegetal cells inducing mesoderm from cells of the 'animal half'. This would produce an equatorial ring of mesoderm in normal development, with the cells of the animal pole being too far away to be induced. In fact Pasteels (1953) had found that the most animal cells can also develop axial structures including dorsal mesoderm if late blastulae are centrifuged so that the blastocoel roof collapses onto the endoderm; and this does seem to be due to induction by the endoderm rather than a direct effect of centrifugation (Karasaki & Yamada, 1955). The yolky vegetal cells can even

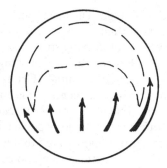

Fig. 6.14. Induction by the endoderm in the urodele blastula. Apparent differences in the strength of the inductive activity are indicated by the size of the arrows. They may explain why different kinds of mesoderm are induced in different parts of the marginal zone with pharyngeal endoderm appearing above the dorsal lip. (After Nieuwkoop, 1977.)

induce some endoderm from cells of the animal half and this may be how pharyngeal endoderm normally develops from cells above the blastopore (Sudarwati & Nieuwkoop, 1971; Nieuwkoop & Ubbels, 1972). This would make the interaction a general vegetalisation with the line between the inducing cells and those which can respond being at the level of the future blastopore (Fig. 6.14). The formation of mesoderm by early subequatorial isolates, as seen by Gurdon's group, may represent an early stage in this vegetalisation.

Nakamura's group still believe that a reciprocal influence from the animal pole also affects determination. Their examples of possible animalising effects changing fates in partial embryos (Nakamura & Takasaki, 1971b), however, remain incompletely authenticated. When cells of the animal half are combined with vegetal yolky cells, almost all of the resulting mesoderm arises from the animal cells (Sudarwati & Nieuwkoop, 1971; Nieuwkoop & Ubbels, 1972) but a little may arise from the vegetal ones (Dale, Smith & Slack, 1985). Although the fate of individual vegetal cells can be changed simply by transplantation, Nieuwkoop (1977) was unable to change the fate of this area by any experimental treatment. However, it must be remembered that changes in the fate of the most vegetal cells have only rarely been obtained in the more extensive experimentation with sea urchin eggs (see p. 150). If an animal gradient exists, it might, by analogy with the sea urchin work, be detected as a variable resistance to vegetalisation (see p. 150). This is effectively seen in the different reactions to vegetalisation of 'animal half' cells according to their position of origin along the animal–vegetal axis (Sutasurja & Nieuwkoop, 1974; Nieuwkoop, 1977).

The similarity with sea urchin embryos is supported by the vegetalising effect of lithium ions. This is particularly clear after a three-hour treatment beginning at the 16-cell stage. The blastopore lips develop at a level further from the vegetal pole than usual (Smith, Osborn & Stanisstreet, 1976), and many embryos become exogastrulae with many more endo-dermal cells than normal (Osborn, 1977). Fairly high concentrations of lithium ions can directly vegetalise animal (Grunz, 1968) or marginal (Koebke, 1977) cells, while lower concentrations enhance the induction of the former by yolky cells (Nieuwkoop, 1970). Nieuwkoop, however, believes that the direct effect is due to non-specific damage. Kao, Masui & Elinson (1986) have recently demonstrated that lithium ions can cause an overdevelopment of dorsal structures (see also Breckenridge, Warren & Warner, 1987) or affect dorso-ventral polarity in ways to be described below. Agents which cause animalisation of sea urchin embryos have less clear-cut effects on amphibians although some studies will be reviewed in the next chapter (p. 227) as the most interesting effects have been seen with tissues from gastrulae.

Nieuwkoop (1969*b*) found that the vegetal cells also decide the dorso-ventrality of the embryos formed when they are combined with 'animal halves'. The orientation of the animal cells had a slight effect on that of the blastopore, but later the dominance of the yolky cells was effectively complete. In fact one to three dorsal vegetal cells can rescue embryos which have lost their own dorso-ventrality following uv irradiation of the zygote (Gimlich & Gerhart, 1984; see Fig. 6.15). The same cells grafted ventrally to normal 64-cell embryos can cause the production of a second complete body axis there. Similar effects have now been achieved by injecting lithium ions into single vegetal cells of normal or uv-radialised embryos (Kao *et al.*, 1986). The dorsal vegetal cells are therefore able to decide the dorso-ventrality of the rest of the embryo. The first stage of their action is probably to ensure that the prospective mesoderm is already dorso-ventrally polarised when it is first determined. Dorsal subequatorial blastomeres of the 32-cell embryo already show some polarising ability which may be mosaically determined, but this increases through blastula stages apparently by the inductive effect of more vegetal cells (Gimlich, 1985, 1986). The ventral marginal cells of blastulae, when explanted, can form only blood islands and mesothelium, while dorsal marginal cells usually form axial mesoderm (Nakamura & Matsuzawa, 1967). That dorsal and ventral yolky cells have different inductive effects on the 'animal half' was directly demonstrated by Boterenbrood & Nieuwkoop, (1973): again only the dorsal cells induced axial mesoderm. The mesoderm will then in turn determine the dorso-ventrality of the ectoderm as described in the following chapter. Dorsal marginal cells probably only acquire the ability to induce a second embryonic axis in a host embryo at blastula stages (Malacinski *et al.*, 1980), and the ability to neuralise ectoderm at late blastula (Nakamura *et al.*, 1971*b*).

It should be noted that this interpretation does not require any special

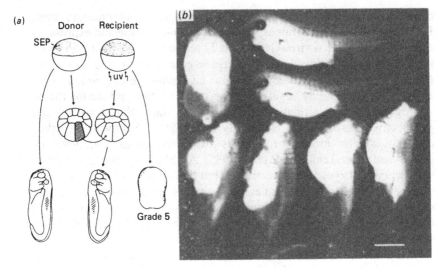

Fig. 6.15. Rescue of uv-radialised *Xenopus* embryos by normal, dorsal, vegetal cells. (*a*) summary of the experiment. Recipient embryos were irradiated as zygotes and if not further treated would form radially symmetrical (Grade 5) embryos. They could be rescued at the 64-cell stage if 1–3 vegetal cells were replaced by dorsal vegetal cells from a normal embryo. SEP, sperm entry point. (*b*) a batch of recipient embryos showing partial or complete rescue except in 1 case (top left). Scale bar = 1 mm. (From Gimlich & Gerhart, 1984.)

properties of the grey crescent area. Its significance has already been challenged for other reasons (see p. 60), and it now seems likely that the essential role of the movements of symmetrisation in the zygote is to polarise the presumptive endoderm, the grey crescent being only a visible expression of these movements. The early assignment of the dorsal and ventral sides is experimentally reversible and probably due to quantitative differences (pp. 63 and 136): indeed this polarity can even be changed in late blastulae according to Landström & Løvtrup (1975). The determination of different mesodermal tissues is probably a second step in the fixation of this polarity as well as its transformation into qualitative differences. Further complexity can arise as these mesodermal tissues induce the ectoderm, so it does not seem unreasonable that an apparent dorsal 'organisation centre' (p. 215) can have arisen epigenetically by the early gastrula stage.

The extent of spatial determination in pregastrula amphibian embryos has recently been re-examined by studying the expression of tissue-specific genes. Even if the cells of the embryo are dissociated throughout this interval, and repeatedly dispersed at stages equivalent to the morula and blastula, they go on to activate genes specific to epidermis and to endoderm (Sargent, Jamrich & Dawid, 1986). These two pathways of development therefore seem to be determined in a cell-autonomous way. Following the same treatment, however, mesoderm-specific expression of the α-actin gene is strongly inhibited, supporting the suggestion that mesoderm is an induced tissue (Sargent *et al.*, 1986). Gurdon *et al.* (1985*a*; Gurdon & Fairman, 1986) also give evidence that animal pole cells can be induced by vegetal ones so that they later transcribe the α-actin gene, but, as already discussed, they believe that a part of the prospective mesoderm is mosaically determined. Synthesis of other chordamesoderm-specific products is also activated in combinates of animal and vegetal cells (Dale *et al.*, 1985). Another interesting result was seen when embryos were dissociated but cultured as a mound of cells within the vitelline membrane. Here it was epidermis-specific gene expression which was later inhibited, supporting the suggestion that all the cells of the animal half can be induced by vegetal cells if they are sufficiently close together (Sargent *et al.*, 1986).

This evidence that induction can occur between cells which are contiguous but have been dissociated suggests that no specialised cell junctions are required for the transfer of mesodermalising stimuli. Indeed, this transfer occurs when gap junctional communication is blocked (Warner & Gurdon, 1987), and even through filters with a pore diameter of $0.4\,\mu m$ where no cell processes can be seen in the pores (Grunz & Tacke, 1986). There is a fall in the degree of cell coupling (Bozhkova *et al.*, 1973) and in intracellular sodium levels (first seen by Kostellow & Morrill, 1968) at blastula stages, but their significance is not clear. Computer modelling suggests that mesodermal induction could be effected by the diffusion of a small inducer molecule (Weyer, Nieuwkoop & Lindenmayer, 1977), but different molecules may be involved in the dorsal and other meridians (Dale *et al.*, 1985). Recently, some candidates for the responsible molecules have been proposed. Smith (1987) reported that a factor, possibly a small protein, secreted by a *Xenopus* cell line is capable of inducing mainly dorsal mesoderm from isolated animal pole regions. Slack *et al.* (1987) found that three purified growth factors induced more ventral mesodermal tissues, although the distinction is not absolute. The growth factors were of mammalian origin but Slack *et al.* (1987) gave some evidence that similar molecules may act in the amphibian blastula, since all of the factors bind to heparin and heparin placed between interacting vegetal and animal cells strongly inhibits mesodermalisation. Grunz (1987) believes that all of these factors can induce any mesodermal or endodermal tissue if concentrations and treatment times are sufficiently varied, which would suggest that they have a general vegetalising effect.

A striking event in the amphibian midblastula is the abrupt activation of nuclear transcription (Bachvarova & Davidson, 1966; Gurdon & Woodland, 1969; Newport & Kirschner 1982). This coincides with the slowing of cleavage cycles and other important changes making up the MBT (see p. 189). The lengthening of the cell cycles is apparently the primary event, allowing time for transcription and possibly releasing the machinery for motility (Kimelman, Kirschner & Scherson, 1987) and its timing seems to be affected by the nucleus to cytoplasm ratio (Newport & Kirschner, 1982). After the MBT, cytoplasmic factors stimulating transcription are detectable according to Crampton & Woodland (1979). There do not seem to be distinct early and late forms of the histones as there are in sea urchins (see Byrd & Kasinsky, 1973; Flynn & Woodland, 1980; van Dongen, Moorman & Destrée, 1983), but an acetylation of the histones accompanies the transcriptional activation (Poupko, Kostellow & Morrill, 1977). Little histone H1 is formed until late blastula stages (Adamson & Woodland, 1974) and it is present at low levels in blastula chromatin (Imoh, 1977).

The spectrum of blastula transcripts is dominated by heterogeneous RNAs (Gurdon & Woodland, 1969), most of which were once thought to be stage specific (Davidson, Crippa & Mirsky, 1968). Later studies have mainly used gastrulae and have detected very few species different from the maternal spectrum, although these few are rapidly produced apparently from the MBT (Dworkin & Dawid, 1980$b$; Sargent & Dawid, 1983). At the same time, syntheses of transfer and 5S RNAs (Newport & Kirschner, 1983), mitochondrial RNAs (Dawid *et al.*, 1985) and possibly ribosomal RNAs (Shiokawa, Misumi & Yamana, 1981$b$) are also activated in *Xenopus*. In the axolotl, rRNAs may be formed even during the synchronous cleavages (Strelkov & Ignat'eva, 1975).

Development is blocked at the midblastula when embryos lack functional nuclei (Briggs, Green & King, 1951) or RNA synthesis is inhibited with $\alpha$-amanitin (Brachet, Hubert & Lievens, 1972). X-irradiation before the midblastula also causes blastula arrest, but later treatment allows gastrulation, which suggests a role for the new transcripts in this process (Ursprung, Leone & Stein, 1968; see also the actinomycin studies of Nakamura & Takasaki, 1971$a$). Work on the offspring of homozygous $o$ mutant female axolotls (see p. 27) is also of interest here. In such embryos cell division rates fall sharply at the mid to late blastula and nuclear transcription is not activated (Carroll, 1974); they die during gastrulation (Briggs & Cassens, 1966). They can be rescued by the injection of material from wild-type oocyte GVs or egg cytoplasm, presumably due to the presence of the $o^+$ substance there (Briggs & Cassens, 1966). Brothers (1976) has studied the mode of action of $o^+$ using nuclear transplantation. She concludes that the $o^+$ substance moves into the nuclei and changes them in midblastulae in a way which is heritable through many mitotic divisions. Nuclei which have not been affected by $o^+$ are actually damaged by the late blastula stage and cannot support development even after transfer to normal eggs. The substance is probably a protein (Briggs & Justus, 1968), and may be one of the factors involved in the change of developmental programme at the MBT (see above). Dreyer, Scholz & Hausen (1982) have described a protein in *Xenopus* which also passes from the GV to the cytoplasm and returns to the nuclei in early blastulae. Such proteins would appear to affect all nuclei, and so not to be involved in spatial determination, although as Brothers (1976) points out the stability of the above effect on nuclei compares well with the stability of determination in other developing systems.

Work on the spatial distribution of the blastula transcripts has produced conflicting results. Early data suggested that rates of synthesis per cell were highest in the prospective endoderm and some mesoderm (Bachvarova & Davidson, 1966; Woodland & Gurdon, 1968; Brachet & Hulin, 1969). Dziadek & Dixon (1977) reported an opposite gradient in late blastulae, with most synthesis in the ectoderm. Newport & Kirschner (1982) reported that all nuclei are involved in the transcriptional activation, and noted no differences according to the embryonic area. Probes for specific transcripts show that several of the qualitatively new ones are localised in late blastulae (Jamrich, Sargent & Dawid, 1987) or gastrulae (see p. 242), but these probably appear too late to have a role in mesoderm determination. Rates of polyadenylation of maternal RNAs also appear to vary in space, being highest in the floor and dorsal wall of the blastocoel where different yolk compartments are mixing (Phillips, 1985).

Protein synthesis patterns show quite extensive changes, of both quantitative and qualitative kinds, at these stages (Malacinski, 1971, 1972; Brock & Reeves, 1978; Bravo & Knowland, 1979). Most are due to the programmed usage of different maternal mRNAs (Crippa & Gross, 1969; Malacinski, 1972; Ballantine, Woodland & Sturgess, 1979), and two of the axolotl maternal mutations (including the *o* mutation) affect the pattern in blastulae (Malacinski, 1971). The first zygotic transcripts are retained for some hours in the nuclei (Woodland & Gurdon, 1969), but by the late blastula many have entered polysomes (Crippa & Gross, 1969) and a few paternally coded proteins are detectable (Woodland & Ballantine, 1980). Animal and vegetal halves of blastulae show different patterns of protein synthesis at least partly due to a faster synthesis of histones in the faster dividing animal half (Lützeler & Malacinski, 1974). No qualitative differences between dorsal and ventral halves were reported before gastrulation (Smith & Knowland, 1984).

The morula and blastula stages have other interesting biochemical properties. The activities of an endonuclease (Ford, Pestell & Benbow, 1975) and some enzymes of polyamine biosynthesis (Russell, 1971) are rising, while levels of a protease inhibitor fall from a peak in the morula (Wittenberg, Kohl & Triplett, 1978). Blastulae seem to use respiratory pathways quite unlike later embryos (Legname, 1968). Many indices of metabolic activity show a gradual fall from the animal to vegetal pole, but at least some of these disappear when calculated on the basis of non-yolk volumes (Melton & Smorul, 1974). Endoderm and mesoderm show high levels of free radical processes at the time of their inductive interaction (Melekhova, 1976). Cells of the three future germ layers already have different electrophoretic mobilities, and the changes for ectodermal cells do not require induction (Nakamura, Aochi & Shiomi, 1970).

## Ascidians and Amphioxus

As shown in Chapter 4 these groups provide perhaps the best examples of mosaic development, with determinants localised in the zygote being passed to specific cells at cleavage and so determining their fate from the outset. We also saw, however, that not all cells are so determined: in particular the formation of neural and sensory structures seems to require influences from other embryonic regions. According to Reverberi & Minganti (1947*b*,*c*) the influences affecting neural development in ascidians are complex involving mutual induction between the derivatives of the anterior animal and anterior vegetal cells, and an inhibition exercised by the posterior vegetal cells unless they are balanced by the presence of the posterior animal cells! If this is correct, then all parts of the embryo will affect the chances that neural structures can develop, and this could imply the existence of embryo-wide gradients much like those proposed for other animal groups. In later work, however, only the inductive interactions between cells of the anterior half have been considered, and, since this probably occurs mainly during gastrulation, it is dealt with in Chapter 8.

Some evidence that determination is affected by gradient-distributed factors comes from the effects of those artificial treatments which also change cell fates in sea urchins and other embryos. Culturing ascidian embryos in lithium ions causes exogastrulation, poor differentiation of notochord (Ranzi & Ferreri, 1944; Nieuwkoop, 1953) and a failure to form brain and sense organs (Farinella-Ferruzza, 1955). However, only the latter effect seems to be a clear change in determination patterns and there are conflicting reports on the effects of thiocyanate ions (Ranzi & Ferreri, 1945; Ortolani, 1969), an animaliser in the sea urchin system. Clearer effects are obtained when the treatments are applied to just the animal quartet (prospective ectoderm) of the eight-cell embryo. Ferruzza & Farinella (1981) found that, on their own, these cells form only a blastula, but after a three-hour lithium treatment endoderm, neural tissues and melanin granules also form. They suggested that the effect may be a vegetalisation, as in sea urchins, and that the endoderm then induces neural and sensory structures from the ectoderm. Trypsin, an animaliser of sea urchin embryos, produces neural and sensory structures

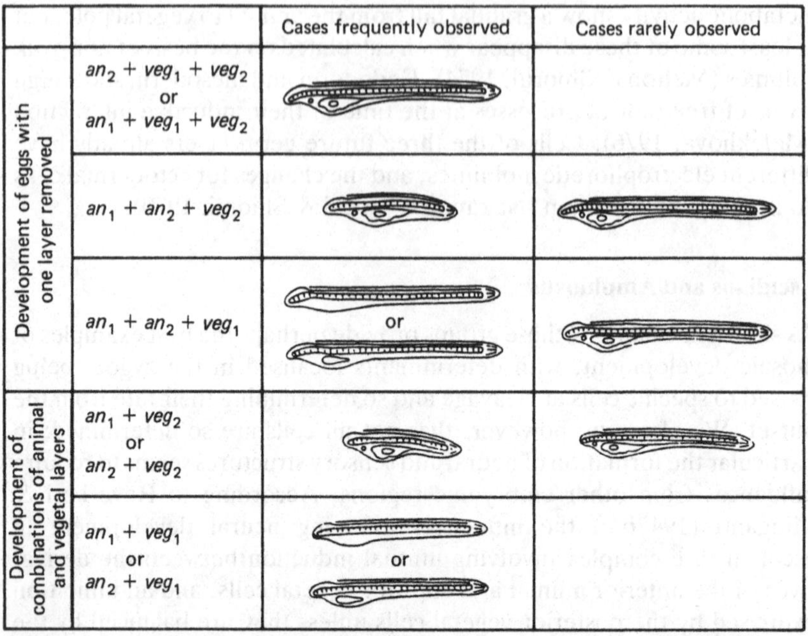

| | | Cases frequently observed | Cases rarely observed |
|---|---|---|---|
| **Development of eggs with one layer removed** | $an_2 + veg_1 + veg_2$ or $an_1 + veg_1 + veg_2$ | | |
| | $an_1 + an_2 + veg_2$ | | |
| | $an_1 + an_2 + veg_1$ | or | |
| **Development of combinations of animal and vegetal layers** | $an_1 + veg_2$ or $an_2 + veg_2$ | | |
| | $an_1 + veg_1$ or $an_2 + veg_1$ | or | |

Fig. 6.16. The development of partial embryos of amphioxus recombined from two or three of the four cell layers in the 32-cell embryo. (From Tung *et al.*, 1960*a*.)

apparently directly from the animal quartet (Ortolani, Patricolo & Mansueto, 1979; see also p. 263).

Tung (1934) felt that some of his data from operative work on ascidians suggested possible animal–vegetal interactions, and his group later produced far clearer evidence of this using amphioxus (Tung *et al.*, 1960*a*). They designated the tiers of cells in the 32-cell embryo $an_1$, $an_2$, $veg_1$ and $veg_2$, and, in work comparable with that of Hörstadius on sea urchins, removed one or two of these layers leaving the others in an abnormal combination (Fig. 6.16). When $veg_2$ was lacking little or no endoderm formed, but otherwise normal larvae could form from any three cell layers. In combinations of only $an_1$ and $veg_2$, muscle, notochord and nervous system all develop well, yet these layers contain little or none of the prospective areas for these tissues (see Fig. 4.7) and neither layer in isolation normally produces them. Several parallels with more regulative embryos can be drawn from these data. The vegetal endoderm is most fixed in its determination (while its differentiation may require a stimulation from more animal cells). The animal half is more labile and changes its fate in response to inductions from more vegetal cells. Determination of the intermediate muscle and notochord areas also seems quite labile (see too Tung, Wu & Tung, 1961, 1962*b*), and, in view of their formation

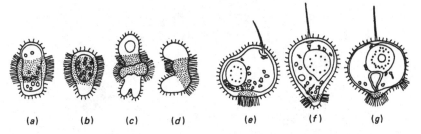

Fig. 6.17. The development of partial embryos of *Cerebratulus* following operations at the 16-cell stage. (*a*)–(*d*) larvae from the middle two layers, $an_2+veg_1$. (*e*)–(*g*) larvae from the most animal and vegetal cell layers, $an_1+veg_2$. (From Hörstadius, 1937*b*.)

by $an_1$ : $veg_2$ combinations, these tissues may even arise as a result of interactions between the poles. It would be most interesting to perform comparable experiments with ascidian embryos where there is evidence that histospecific determinants are localised in the zygote and something is even known of their mode of action.

## Spirally cleaving embryos

Like the ascidians and amphioxus spiralian embryos segregate developmentally important materials into several different lineages at the early cleavages (see Chapter 3). The blastula stage is in any case short and the cell number still low when gastrulation begins, but there are still several clear cases of interactions occurring among the cells.

A good example of mosaic determination, with fates being unaffected by interactions among the cells, is provided by Hörstadius' (1937*b*) work with the nemertine *Cerebratulus*. Abnormal combinations made from various fragments of the 16-cell embryo seemed always to form just those structures expected from the fate map. Many of the combinations tried were closely comparable with his own sea urchin work (pp. 148 *et seq.*) and the contrast with the regulation seen there was very marked (see Fig. 6.17). For example, combinations of just the four most animal ($an_1$) and four most vegetal ($veg_2$) cells apparently formed no structures normally derived from the intermediate ($an_2$ and $veg_1$) layers. This result contrasts not only with sea urchin data but with those from other animal groups cited earlier in this chapter and including the generally 'mosaic' amphioxus. However, it is of interest that the $veg_2$ cells can differentiate here when they cannot in isolation, since this may indicate an activating influence from more animal cells as has been proposed for other embryos.

Other work already described in Chapter 3 has, however, shown that isolated cells of many spiralian embryos do not produce the structures expected from their prospective fates. In particular, cells of the D

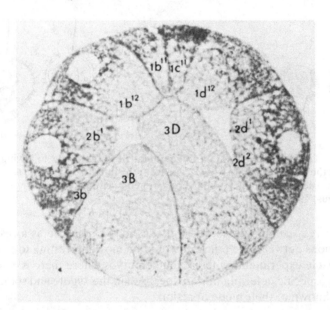

Fig. 6.18. An almost medial section through a *Patella* embryo 70 min after the start of the resting phase (at 32-cells in *Patella*). 3D is the only macromere that borders on the animal micromeres. (From van den Biggelaar, 1977*a*.)

quadrant commonly form too many structures and cells of the other quadrants too few. Further interactions must be required to ensure that each quadrant produces those structures appropriate to it; and in the case of polar lobe-bearing species such as *Ilyanassa* and *Dentalium* it has been shown that these interactions include inductive and inhibitory influences from the D quadrant (see pp. 88–9). Without them the A and even C quadrants seem to develop as ventral, B, quadrants, so, in effect, the A and C quadrants are being accorded intermediate status along the dorso-ventral axis. In other groups considered earlier in this and the previous chapter interactions have been mainly shown to specify intermediate regions along the animal–vegetal axis, so it may be worth recalling here that the D quadrant owes its special status to the possession of the vegetal pole plasm.

In those spiralian embryos where the first two cleavages are equal (e.g. *Lymnaea*, *Patella* and *Physa*) the interaction of the D quadrant with neighbouring cells has received particular study. It occurs during a pause in cleavage (often at the 24-cell stage) when the cells increase their contacts, gradually obliterating the blastocoel. During this process, one macromere extends further internally than the others and becomes the only one to make contact with the first quartet micromeres, $1a^1$–$1d^1$ (Raven, 1974; van den Biggelaar, 1976*a*, 1977*a*: see Fig. 6.18): this macromere then takes the role of 3D. It is almost certainly the interaction

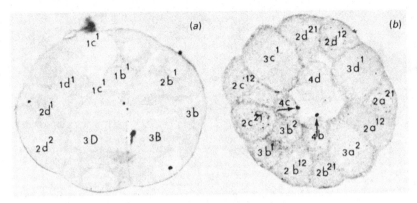

Fig. 6.19. The fate of the ectosomes seen in methyl-green pyronin-stained *Lymnaea* embryos. (*a*) section almost through the bilateral plane 170 min after the onset of the 24-cell stage: the ectosomes in 3D are scattered along the common borders with $1b^1$ and $2b^1$, while in 3B they are concentrated into a dense body; (*b*) horizontal section through a 48-cell embryo: ectosomes are being extruded into the blastocoel from 4b and 4c (arrows). (From van den Biggelaar, 1976*b*.)

with the micromeres which finally decides which quadrant will become D, since the choice is strongly affected by the deletion of specific micromeres (Arnolds *et al.*, 1983: see p. 91). If the interaction is prevented by keeping 24-cell embryos in cytochalasin B (where the blastocoel is not obliterated), no quadrant takes the role of D and there is a tendency for all to form ventral structures (Martindale, Doe & Morrill, 1985; see too van den Biggelaar & Guerrier, 1979). (There is no equivalent to this role of the micromeres in the polar lobe-bearing *Bithynia*, where headless but dorso-ventrally organised larvae can develop even when all the first quartet micromeres are deleted: van Dam & Verdonk, 1982.) In equally cleaving embryos the D quadrant is thus a target of intercellular signals selecting it as D; but once selected it is also (as in other spiralians) a source of signals affecting pattern throughout the embryo including its dorso-ventrality (see p. 92). Computer simulations can best account for the dorso-ventral organisation of further development if an 'induction centre' is postulated at the contact points of the first quartet micromeres with 3D (Bezem & Raven, 1975).

The interaction of the first quartet cells with 3D soon has demonstrable effects on both participants (see Fig. 6.19). In the first quartet cells, lipid globules and mitochondria accumulate near the contact points with 3D (Raven, 1974). In the macromeres, RNA-rich 'ectosomes' have begun moving inwards from the vegetal pole earlier in the 24-cell stage (see p. 91). The ectosomes in 3D disperse very close to the contacts with the apical micromeres, while in the other macromeres they remain compact

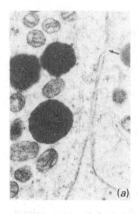 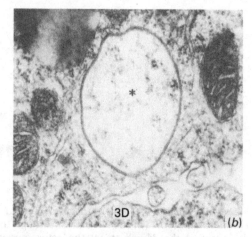

Fig. 6.20. Electron micrographs of 3D: micromere contact zones in *Lymnaea*. (*a*) a possible cytoplasmic connection (arrow) between 3D (left) and a micromere (right). × 10,800; (*b*) a gap junction-bounded vesicle inside a micromere which touches 3D: it was not clear whether this protruded from 3D. × 34,500. ((*a*) from Luchtel, 1976; (*b*) from Dohmen & van de Mast, 1978.)

and may be cleaved to the 4a to 4c cells or extruded to the cleavage cavity (van den Biggelaar, 1976*b*). The fate of the RNA in 3D has not been clearly established: Wierzejski (1905) and Minganti (1950) thought it was transferred through delicate cytoplasmic bridges to the micromeres, but it is also possible that it remains in 3D, its dispersal being a result of an activation from the micromeres (van den Biggelaar, 1976*b*). Also, over the period of the 24-cell stage, the first quartet micromeres become measurably smaller and 3D probably larger (there is a problem in identifying 3D in early 24-cell stages), which may indicate that any transfer of materials occurs mainly from the former to the latter (Bezem, Wagemaker & van den Biggelaar, 1981). Studies of a *Lymnaea* maternal-effect mutation suggest that the ectosomes play their major role in the vegetal half (Arnolds, 1982*a,b*). Affected eggs apparently lack ectosomes and the vegetal halves never develop dorso-ventral polarity, but head structures of dorsal, lateral and ventral quadrants can all still form and it is only their arrangements which are abnormal.

There have been several studies of the contact points between the micromeres and 3D (Fig. 6.20). The suggestion of cytoplasmic bridges received some support from Luchtel's (1976) electron microscopy work with *Lymnaea*, where one possible direct cytoplasmic connection between the cells was seen. Dohmen & van de Mast (1978) saw no such open connections but described an annular gap junction surrounding a vesicle or protrusion possibly extending from 3D which could have a role in the transfer of small molecules. Normal gap junctions are present in *Lymnaea* from earlier stages and on most cell contacts (Dohmen & van de Mast, 1978; Dorresteijn *et al.*, 1981). In *Patella* they seem to be specifically absent from the 3D: first quartet contacts (Dorresteijn *et al.*, 1982), and injected Lucifer yellow is not transferred between these cells at the time that they are interacting, although the dye is passed from 3D to other neighbouring cells (Dorresteijn *et al.*, 1983). An extracellular matrix, possibly of proteoglycans, is also seen where 3D contacts the micromeres, and is probably secreted by the latter (Kühtreiber *et al.*, 1986).

In leeches, more details of spatial pattern can be shown to be mosaically determined than in any other spiralian embryos (p. 76). Yet even here there is some indeterminacy. In particular, the O and P bandlets which normally contribute quite different neurons to each body segment will exchange fates if their positions are exchanged by crossing over (Weisblat & Blair, 1984). This and other results show that the difference between these two bandlets is determined by position (and therefore interactions with neighbouring cells) rather than by differences in the teloplasm they receive from their respective teloblasts. The stages by which they are later determined have been investigated by Shankland & Weisblat (1984; see too Ho & Weisblat, 1987). There is also some plasticity in the epidermal contribution which each ectoteloblast can make following experimental interference (Blair & Weisblat, 1984), but the neural contributions of N, O/P and Q do seem to be mosaically determined. Late interactions between mesoderm and ectoderm are required before either can show full segmentation (Blair, 1982).

Spiralian embryos, like many others considered earlier, show gross developmental abnormalities following treatment with lithium ions. Some of these, such as exogastrulation in *Lymnaea* (Raven, 1948), overdevelopment of endoderm in *Eisenia* (Devriès, 1976) and the lack of an apical tuft in *Nereis* (Henley, 1946) could be interpreted as vegetalisation, but Raven (1952) found no evidence that extra cells become endodermal in *Lymnaea* exogastrulae. In other treated *Lymnaea* embryos dorso-ventral organisation is suppressed and cyclocephalic larvae result (Raven, 1948), but here too a relatively trivial explanation is available as no one macromere achieves preferential contacts with the micromeres (van den Biggelaar, 1977*b*). Many other treatments can mimic at least some of the effects of lithium ions (see p. 91), which again suggests a relatively unspecific effect. Various ions can cause exogastrulation with differing effectiveness (Raven, 1956), and it is interesting that Raven suggested from this that they act by changing colloid properties in ways similar to those proposed for sea urchins by Ranzi (see p. 160). Elbers (1983) has, however, found that little lithium enters *Lymnaea* cells, and suggests an action at the plasmalemma.

Many of the biochemical changes occurring at blastula stages of spiralian development have already been described in Chapter 3, but it is rarely known whether these changes are dependent upon interactions between cells. One point to add here is that new subtypes of histone appear in molluscs (Gabrielli & Baglioni, 1975; Mackay & Newrock, 1982) and the echiuroid *Urechis* (Das, Micou-Eastwood & Alfert, 1982) as they do in sea urchins (p. 165). Since mechanisms of spatial determination are generally so different in these groups, this suggests that histone transitions have some other role in early development.

### Hydrozoans

The previous two sections have dealt with embryos where determinants seem segregated to specific cells at early cleavage: hydrozoans provide a good contrast since even one of the first 16 cells can form a whole larva (p. 113). Teissier (1931) and Freeman (1981*b*) have also isolated fragments from morulae and blastulae, including anterior and posterior halves, and shown that they all regulate completely. In those species where the prospective ectodermal and endodermal areas are always concentrically arranged, the regulative ability of fragments might have little significance. However, in other species all the endoderm arises from the posterior half, and then the regulation of fragments means that even the choice between ectoderm and endoderm has not been made irrevocably at these stages. How later embryos deal with this problem is considered on p. 265.

As in early cleavage (p. 113) all fragments of morulae and blastulae retain their polarity when they regulate to form a larva (Teissier, 1931; Freeman, 1981*b*). The importance of polarity is demonstrated even more clearly when different fragments from blastulae are combined (Freeman, 1981*b*): if two posterior poles are present each will begin gastrulation independently (though later regulation may produce a single compromise axis), and, if two anterior poles are present, gastrulation begins midway between them. Even dissociated and reaggregated cells can regulate and form a larva, and Freeman (1981*b*) found that a small piece of undissociated blastula wall can impose its polarity on a much larger reaggregate, i.e. this piece will form the area expected from its fate and the reaggregate will form the rest in a harmonious whole. Freeman (1981*b*) supposes that a reaggregate on its own attains a single polarity in a similar way, starting from an area where by chance several cells have similar polarities.

A few biochemical data on *Hydractinia* embryos were provided by Edwards (1975): development seems to be maternally programmed even through gastrulation, and indeed zygotic RNA synthesis (RNase-sensitive uridine incorporation) is not detectable in blastulae. From patterns of sensitivity to toxic treatments, Teissier (1931) suggested that a physiological gradient reverses in blastulae, the posterior pole being most sensitive before then and the anterior pole thereafter.

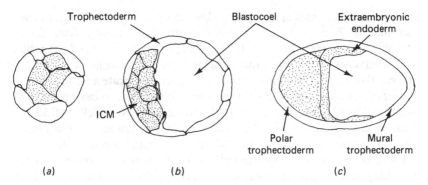

Fig. 6.21. The diversification of lineages in the mouse morula (*a*) and blastocyst (*b*),(*c*). ICM lineages are dotted. ((*a*),(*b*) after Ducibella, 1977; (*c*) after Copp, 1978.)

## Mammals

In Chapter 4, mammalian development was described to the compacted eight-cell stage. All eight cells still seem to be roughly equivalent, but the inner and outer areas of each cell differ in many ways including apparently their fates. This polarisation arises as a reaction to contacts with other cells, and becomes far more marked as this contact increases at compaction (see pp. 139–41). The next cleavage cycle produces a 16-cell morula (Fig. 6.21(*a*)) with a small but variable number of inner cells completely enclosed by outer ones (Handyside, 1981). After fifth cleavage a higher proportion of cells is enclosed (Handyside, 1978; Pedersen, Wu & Bałakier, 1986; Fleming, 1987) and cavitation occurs, a small cleavage cavity appearing among the cells (Smith & McLaren, 1977). Its enlargement produces a blastocyst (Fig. 6.21(*b*)), the outer cells of the morula becoming a single-layered trophectoderm, while the inner cells are left as an inner cell mass (ICM) contacting the trophectoderm at one pole of the blastocyst. Further differentiation then occurs within each of these tissue types (Fig. 6.21(*c*)). The polar trophectoderm over the ICM continues division, while in the mural trophectoderm elsewhere chromosomes are endoreplicated and the cells transform to giant cells. Within the ICM a surface layer delaminates as extra-embryonic endoderm, so that all of the embryo must arise from the remaining ICM cells.

Many properties of morulae and blastocysts can be traced back to the polar organisation of the early blastomeres and the effects of compaction. The outer cells inherit the close junctional contacts formed at compaction (p. 140) and further develop these into zonular tight junctions capable of excluding lanthanum and antibodies (Ducibella *et al.*, 1975; Magnuson, Demsey & Stackpole, 1977; Magnuson, Jacobson & Stackpole, 1978).

Still later desmosomes appear there (Ducibella *et al.*, 1975). Outer cells also take up fluid from the external medium and release it from their internal surfaces (see Borland, 1977). This probably happens during the early cleavages, explaining the vesicles seen in the cells (Calarco & Brown, 1969), but, once the zonular tight junctions create a permeability barrier, this fluid is retained between the cells resulting in cavitation and blastocyst formation. The general adhesiveness of inner cells will keep them together in the ICM, where there are gap junctions but few more specialised junctions (Ducibella *et al.*, 1975; Magnuson *et al.*, 1977).

There is no evidence that the polarity of the blastocyst (with ICM at one pole) bears any relationship to the earlier polarity of the egg. In any case the ICM can apparently later migrate to that side of the trophoblastic vesicle which has attached to the uterine walls at implantation (Kirby, Potts & Wilson, 1967). Thereafter blastocysts show a fairly constant spatial relationship with the axis of the oviduct and environmental cues may provide positional information for further development (Smith, 1980, 1985). Another controlling factor seems required for the timing of developmental events such as cavitation. It is apparently not based upon time since ovulation, cell number or number of cleavages (Smith & McLaren, 1977) or on the number of DNA replication cycles (see Johnson, McConnell & van Blerkom, 1984), leaving nucleocytoplasmic ratio as the most likely index to be used.

The divergence of the first two cell types, trophectoderm and ICM, like other early events, seems to originate in the intracellular differences which appear in the early cleavages and intensify at compaction. In fact the blastomeres of the 16-cell embryo already show properties typical of these two cell types, with the majority being larger, less adhesive and polar, and a minority smaller, more adhesive and apolar (Handyside, 1980; Johnson & Ziomek, 1981*a*). Some of the first eight cells have apparently cleaved tangentially, separating the inner and outer poles with their different properties, and more of the outer cells do so at fifth cleavage, building up the population of inner cells (Pedersen *et al.*, 1986; Fleming, 1987). Cell allocation to the two layers may also be affected by division order (Kelly *et al.*, 1978; Surani & Barton, 1984) and the active movement of daughter cells (Sołtyńska, 1982). When a single polar and a single apolar cell are placed together in culture the former is able to encircle the latter (Ziomek & Johnson, 1981). Even if it is not required for the establishment of two layers, this behavioural difference probably helps to maintain their relative positions. Position then apparently acts back upon cell properties: inner cells of older embryos being increasingly easy to encircle, and outer cells increasingly difficult to encircle, when paired with eight-cell blastomeres (Kimber, Surani & Barton, 1982).

The determination of the inner and outer cells for their different fates requires a similarly extended period. A 16-cell embryo experimentally produced from all polar, or all apolar, cells can form a blastocyst, so cells which are diverging morphologically and behaviourally must still retain totipotency (Ziomek, Johnson & Handyside, 1982). In this case, polar cells probably produce new apolar ones by a division across the cell axis, while apolar ones probably respond to an outside position on the aggregate by polarising. Only in quite late blastocyst stages do outer and inner cells seem to lose this regulative ability (Rossant, 1975; Rossant & Vijh, 1980) and other experimental techniques suggest that it can be called forth even later (Wiley, 1978). The differences inherited by outer and inner cells may predispose them to develop along different pathways, but this predisposition can be overridden by new positional information. In normal development the two cell types may act largely as separate lineages, but their regulative ability will allow them to correct any disturbance in the ratio of the two types.

Compaction supplies at least some of the cues used in this first developmental choice. If it is prevented until the 32-cell stage normal blastocysts still form, but as compaction is delayed still longer more abnormal blastocysts with little or no ICM result (Johnson *et al.*, 1979; see also Shirayoshi, Okada & Takeichi, 1983). The increased junction formation and isolation from the external environment may be the important effects of compaction here, but the diffusible components of the blastocoel fluid cannot commit cells to the ICM pathway (Pedersen & Spindle, 1980). Contact on all sides is almost certainly important in keeping inner cells apolar, and it is interesting that even in blastocysts processes from the trophectoderm cells cover most of the surface of the ICM which faces towards the blastocoel (Fleming *et al.*, 1984; see Fig. 6.21.(*b*)). This may not be a sufficient condition however, as inner cells seem to be surrounded in the uncompacted embryos of Johnson *et al.* (1979).

The next stages of cell diversification in the blastocyst also involve recognition of position and interactions between cells. When peripheral ICM cells form extraembryonic endoderm they are almost certainly reacting to their position (Rossant, 1975). This is, of course, the same signal which causes younger ICM cells to transform into trophectoderm, and the different responses seem entirely due to the changed competence of the ICM cells with age. The different properties of mural and polar trophectoderm seem determined by effects of the ICM which underlies the latter. ICM stimulates the division of trophectoderm cells, but as they move further from the ICM this influence decreases until the most distant trophectoderm cells replicate their chromosomes without cleavage, transforming to giant cells (Copp, 1978, 1979). According to Copp (1979) this interaction can also explain how the blastocyst transforms to an egg cylinder, the next stage in development.

Biochemical studies of the early cleavages in mammals showed an unusually great range of syntheses and indicated that even some differentiation may be beginning (see p. 141). By blastocyst stages the genome itself may be changed (see Kiessling & Weitlauf, 1981; Manes & Menzel, 1982; Yamagishi *et al.*, 1983) or parts of it quite permanently inactivated (Johnson *et al.*, 1984). Development is largely under zygotic control, which is quite immediate for early events like cavitation (Braude, 1979) but seems to use longer-lived transcripts from hatching (Schindler & Sherman, 1981; see too Kidder & Pedersen, 1982). The competence change by ICM which changes its reaction to isolation requires both new transcription (Johnson, 1979) and translation (McCue & Sherman, 1982). Levels of protein synthesis are rising rapidly (Monesi & Salfi, 1967), but there are few qualitative changes in the pattern (van Blerkom & Manes, 1974; van Blerkom & Brockway, 1975) until late blastocyst stages (Failly-Crépin & Martin, 1979). Among the earlier new species there are however several which are specific either to the ICM or to the trophectoderm (van Blerkom, Barton & Johnson, 1976; Handyside & Johnson, 1978).

Proteins of the cell surface (Gooding, Hsu & Edidin, 1976; McCormick & Babiarz, 1984), cytoskeleton (Jackson *et al.*, 1980; Sobel, 1983) and extracellular matrix (Leivo *et al.*, 1980; Sherman *et al.*, 1980) are among those segregated to specific early lineages, and all may be involved in the morphogenetic events of these stages. The changing cell surface properties may also depend on other membrane constituents, including lipids (Pratt, 1982). One glycoprotein which promotes both compaction and endoderm formation is known as a cell adhesion molecule in other epithelial cell types (Richa *et al.*, 1985; and see p. 141), again indicating that the attainment of epithelial properties is an important part of both these events. Antibodies to this molecule decompact early embryos and prevent the formation of an ICM (Shirayoshi *et al.*, 1983). Some enzymes of respiratory metabolism show dramatic increases in activity in morulae (e.g. Wudl & Chapman, 1976), presumably as a result of the metabolic reorganisation undertaken during cleavage (p. 141). Glycogen concentrations then rise rapidly in the peripheral and later the trophectodermal cells (Edirisinghe, Wales & Pike, 1984).

Lithium ions which affect developmental gradients and choices in other embryos, reduce the number of ICM cells in mouse embryos, but this seems to be a simple result of slower cleavages forming less inner cells by the time of determination (Izquierdo & Becker, 1982). A possible true change of fate from embryonic to extraembryonic structures is seen when periimplantation embryos are treated with hyaluronic acid (Hamasima, 1982). Mammals are clearly unusual in the way the first developmental decisions are taken. There is apparently no determination centre instructing other regions, nor a stable polarity, but intercellular signalling is involved since it is position relative to other cells which determines fate. The same could be said of the two-cell stage in any regulative embryo:

one blastomere will form half an embryo *in situ* but a whole one in isolation. Evolution has interpolated extra choices into mammalian development before the usual ones over germ layers or organ primordia, and there is no good reason to expect that the same biochemical mechanisms will be involved.

## Conclusions

The work reviewed in this chapter shows that interactions between cells have a role in determination in very many, perhaps all, animal embryos. The situation seems most like the sea urchin in insects, amphibians and, most surprisingly, in amphioxus. In all of these, there is strong evidence for interactions along the animal–vegetal axis (antero-posterior in insects), and their nature shows many parallels with the sea urchin. Animal halves in isolation can form only a few of the most polar structures expected there from fate maps, the formation of their other derivatives depending on a 'vegetalising' influence from the vegetal half. Vegetal cells often seem determined for their fates but may fail to differentiate unless activated by more animal cells, providing a clearer illustration of an animal effect than is known in sea urchins. In some insect embryos influences from anterior regions also affect posterior determination patterns. By blastula (or blastoderm) stages these polar influences have determined intermediate regions as mesoderm in amphibians and thoracic and other segments in many insects, but we do not know when equatorial regions are finally determined in amphioxus. Data on dorso-ventral determination are compatible with the existence of a single dominant zone which may be ventral in insects as in sea urchins, and dorsal in amphibians; but in amphioxus there are probably qualitative differences in the dorso-ventral plane (see Chapter 4).

The data available for other arthropod (crustacean and chelicerate) and for starfish embryos imply that cell fates are determined by interactions perhaps occurring at blastoderm or blastula stages. Cell fates in hydrozoan embryos are not determined at the end of the blastula stages, but cells communicate such information as their polarity. Some determination has occurred by blastocyst stages of mammalian development, and it requires interactions between cells, but in ways quite unlike those in sea urchins, and lacking any apparent determination centre.

*Cerebratulus* seems to provide a good example of mosaic development even when embryonic fragments are brought together in abnormal combinations. However, in many of the spiralian embryos where the D quadrant plays a dominant role in development, D blastomeres exert inductive and inhibitory influences on other quadrants, and in leeches the fate of a few lineages are affected by position, which presumably means effects of neighbouring cells. Even in ascidians, abnormal combinations of blastomeres reveal interactions particularly affecting the determination of neural tissue.

As in sea urchins, lithium ions can change spatial patterns in favour of vegetal tissues at least in starfish and amphibians, and can apparently directly vegetalise animal halves in ascidians and amphibians. Their effects on spiralian and mammalian embryos should however warn us of the danger of non-specific actions.

Work with *Drosophila* mutations has provided us with considerable information on the genetic control of spatial determination. It seems that maternal transcripts gradually establish quantitative gradients perhaps of gene regulators and probably in the periplasm. Blastoderm nuclei then respond to local levels of these materials along the anterior–posterior axis by transcribing specific combinations of genes which finally determine its subdivision into a linear array of compartments and the nature of the individual compartments (a process which is not completed until postblastoderm stages). There are possible parallels in amphibian embryos, where interactions along the animal–vegetal axis may have determined the mesoderm while development is still maternally controlled, and transcription is then abruptly activated and new developmental programmes apparently initiated. In *Lymnaea* and some other spiralian embryos maternal RNAs from the vegetal pole move up to the point where the 3D macromere is interacting with micromeres, but their fate and developmental role are uncertain. In mammals RNA synthesis is required at least for changes in cell determination; and in hydrozoan blastulae, where no cells are determined, transcription may still not have begun.

# 7

## Interactions between moving cells: the case of amphibian gastrulae

The last two chapters have demonstrated that the extent of spatial determination is often greatly increased at morula and blastula stages by interactions involving large areas of the embryo. Even in late blastulae, however, this determination is commonly incomplete, with little more than the three germ layers being defined in amphibians.

The next stage of development, gastrulation, is one of morphogenetic movements which broadly speaking bring the germ layers to their correct positions in a multilayered embryo. Inner layers may be directly separated by delamination, but more commonly some cells move in actively by an individual ingression or as a coordinated invagination or involution of whole sheets of cells (see Trinkaus, 1984, p. 11, for definitions of these terms). Wherever cells move and make new contacts, new opportunities for interactions will arise. This chapter will concern the amphibian gastrula where such interactions have received particular study, and have been shown to greatly increase the complexity of spatial determination.

### Morphogenetic movements

The movements taking place in amphibian gastrulae are an intensification of those already described in blastulae (p. 189). There is a spreading movement of cells away from the animal pole called epiboly, and an inward movement of cells in the vegetal area. During gastrulation, both of these activities occur much the fastest on the dorsal side. Figure 7.1(a)–(e) shows some stages of gastrulation in sagittal view in the simpler case of the urodeles. Externally, the first visible sign of gastrulation is an accumulation of pigment in the dorsal vegetal area of the embryo. The pigment is in the narrow necks of 'bottle cells' which elongate towards the centre of the embryo forming a surface depression (Rhumbler, 1902; Holtfreter, 1944). A little later a distinct transverse furrow is seen in the same position, and a whole sheet of cells is found to be involuting around its upper edge – the much-studied dorsal lip of the early gastrula. The furrow extends into more lateral and finally ventral regions, while at the

211

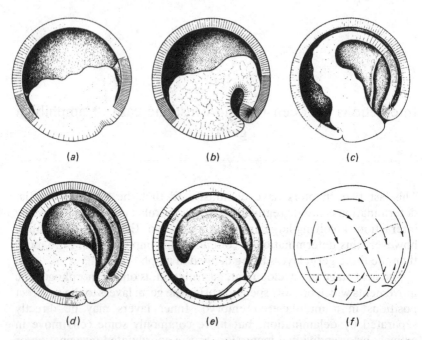

Fig. 7.1. Gastrulation in urodeles. (*a*)–(*e*) diagrammatic sagittal views. (*f*) lateral view showing movements in the outer cell layer (solid arrows) and after involution (broken arrows): the dotted lines separate the prospective areas of the three germ layers. ((*a*)–(*e*) from Spemann, 1938; (*f*) from Waddington, 1956.)

same time the invagination becomes increasingly obvious. Most of the involuting cells are provided by the continuing epiboly of cells from around the animal pole, and they 'fold' inwards around the upper edge of the furrow. This means that the lips contain a transient population of cells which have flowed in an animal–vegetal direction and are turning to return in the opposite direction as an inner layer. (This is what distinguishes involution from the simple inpocketing of cells termed invagination.) The edges of the furrow soon meet on the ventral side to form a complete ring, the blastopore, around a vegetal yolk plug.

This description should make clear that the gastrulation movements start in different meridians in a temporal sequence which progresses from the dorsal to the ventral side. The differences are then further enhanced because the cells do not move in a strictly animal–vegetal direction: both before and after invaginating they also tend to move towards the dorsal side (Fig. 7.1(*f*)). This explains the particularly rapid advance of the invaginating cells on the dorsal side, and means that the newly forming cavity, the archenteron, extends from the dorsal side. As it does so, it obliterates the blastocoel cavity and changes the centre of gravity of the embryo so that it floats with its dorsal side uppermost. The roof of the archenteron is prospective notochord and its floor prospective endoderm.

In anuran gastrulation, far fewer cells involute from the original outer cell layer. Dettlaff (see 1983) has long claimed that the outer cells contribute only a little prospective notochord to the chordamesoderm, and this has recently been confirmed by Smith & Malacinski (1983). The remainder of the chordamesoderm originates from the inner cells of the blastocoel walls (these being many cells thick in anurans and only one-cell thick in urodeles). According to Nieuwkoop & Florschütz (1950; and see Keller, 1976) they involute around an inner lip quite independently of the outer cells and starting before gastrulation is externally recognisable. These differences should be remembered when comparing data obtained from early gastrulae of the two groups.

Amphibian gastrulation has received much recent study at the level of individual cell behaviour. Epiboly involves the stretching of the cells in the animal half as well as rearrangement of the cells in the deep layers (Keller, 1978, 1980). The initial formation of a blastoporal groove is probably the main role of the bottle cells (Keller, 1981). Thereafter a variety of kinds of pseudopodia develop from the involuting cells and those of the blastocoel walls, and the former migrate actively over the latter (Nakatsuji, 1974, 1975, 1976) by a microfilament-dependent process (Nakatsuji, 1979). The prospective endoderm of the archenteron floor seems to be passively pulled inwards. The migrating cells attach to, and may be guided by, a network of extracellular fibrils beneath the blastocoel roof (Nakatsuji, Gould & Johnson, 1982; Nakatsuji, 1984). Fibronectin is a constituent of the fibrils, and its destruction prevents mesodermal attachment and migration and even the epibolic spreading of animal pole cells (Boucaut *et al.*, 1984). Another cell adhesive glycoprotein, laminin, is also present (Nakatsuji, Hashimoto & Hayashi, 1985). Cells which have rounded the gastrular lips show many behavioural changes (see LeBlanc & Brick, 1981). The convergent extension of the circumblastoporal region involves interdigitation of cells in the marginal zone, which may be active in the involuting cells and passive in those remaining outside the blastopore (Keller *et al.*, 1985). The resulting closure of the blastopore is another activity in which microfilaments seem to be involved (Moran, 1985). The gastrulation movements may soon be explicable in terms of such cellular activities, and current evidence weakens the case for other factors such as inside–outside gradients or attraction points for migrating cells (Kubota & Durston, 1978).

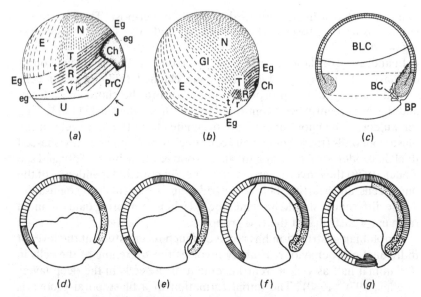

Fig. 7.2. Fate maps through gastrulation in amphibians. Lateral views of *Cynops pyrrhogaster* before (*a*) and at the end of gastrulation (*b*). Ch, notochord; E, epidermis; eg, Eg, boundary of involution at the end of gastrulation (eg) and at closure of the blastopore (Eg); Gl, neural crest; J, position where involution will begin; N, neural tissue; PrC, prechordal plate; R, trunk somites; r, trunk lateral plate mesoderm; T, tail somites; t, tail lateral plate; U, position of later gastrular lip and upper boundary of the prospective endoderm; V, pronephros. (*c*) sagittal view of *Xenopus* early gastrula, showing that all prospective mesoderm (dotted) is already internal and is involuting around an inner lip. Prospective ectoderm is radially hatched and includes deep and superficial layers; prospective endoderm is unhatched except for the prospective pharyngeal endoderm (cross-hatched). BC, bottle cells; BLC, blastocoel; BP, blastopore. (*d*)–(*g*) sagittal views of *Cynops pyrrhogaster* through gastrulation. Prospective fates are indicated as follows: epidermis, spaced radial hatching; neural tissue, close radial hatching; notochord, coarse stippling; prechordal mesoderm, fine stippling; ventral mesoderm, tangential hatching; endoderm, unhatched. ((*a*),(*b*) from Nakamura, 1942; (*c*) after Keller, 1976; (*d*)–(*g*) original using sources cited in the text: the boundaries shown should be considered as approximations.)

## Fate maps through gastrulation

Many later sections of this chapter will consider what small groups of cells from gastrulae can form after explantation, implantation to other embryos or other experimental treatments. Obviously, the significance of such data must be judged in comparison with the fates of these same cells in a normal embryo. Vogt (1929) checked these fates by marking cells with vital dyes and his studies have been extended for urodeles (Nakamura, 1942; Okada & Hama, 1945*b*) and anurans (see Keller, 1975, 1976). The major findings are presented in Figure 7.2.

Urodeles provide the simpler case with cells of the future ectoderm, mesoderm and endoderm in an animal–vegetal sequence in the early gastrula. Within each germ layer, future tissues are arranged primarily in a dorso-ventral sequence. Prospective chordamesoderm is widest and attains its most animal extent on the dorsal side because epiboly and involution are most active there. There is also a little prospective endoderm above the dorsal lip. In the superficial cell layer of anuran early gastrulae there is very little prospective chordamesoderm (see earlier) so that prospective ectoderm and endoderm are in direct contact (see Fig. 7.2(*c*)), but the dorso-ventral organisation of the deep layer is very like that in urodele prospective chordamesoderm.

Prospective endoderm is the first tissue involuted in gastrulation, and it will always form the lining of the archenteron in anurans. In urodeles, prospective mesodermal cells are rapidly removed from the archenteron lining after their involution, leaving endoderm in direct contact with the lateral edges of the notochord (Lundmark, 1986). Of the chordamesodermal tissues, prospective head mesoderm is the first to involute at the dorsal side in both amphibian groups, and it is followed by prospective notochord. The convergence and extension movements convert the latter from a broad crescent before involution to a narrow strip within the dorsal midline of the archenteron roof. Prospective somite, pronephros and lateral plate mesoderm involute around successively more ventral parts of the blastopore lips, but then show a decreasing tendency to elongate. Most of the uninvaginated cells which overlie the axial mesoderm in late gastrulae will become neural tissue, but just outside the blastopore some prospective tail mesoderm is still uninvaginated (Bijtel, 1931).

### Spemann's organiser

Spemann & Mangold (1924) removed the dorsal lip from one quite early gastrula and grafted it into the ventral side of another one. The latter went on to form two dorsal axes: the host's own complete one, and a second axis usually lacking some structures but still showing very extensive organisation (see Fig. 7.3). Two different *Triturus* species were used to provide the graft tissue and the host, so the origin of the cells remained recognisable from their pigment structure when the resulting 'double embryo' was examined histologically (Fig. 7.3(*b*)). In the secondary axis, the grafted cells had formed most of the notochord and some somites as well as sometimes making smaller contributions to the nervous system and gut roof, i.e. they had differentiated much as would be expected from their original prospective fate, but with a tendency to form more of the dorsal structures than would be expected from such a small piece of tissue. The most interesting change of fate was shown by the neighbour-

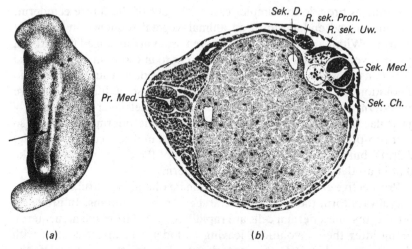

Fig. 7.3. The induction of a second embryonic axis in a newt embryo when a second dorsal lip was grafted into the ventral ectoderm. (*a*) external view: the induced axis on the host's left side (arrow) ends at the level of the ear vesicles. (*b*) cross-section: graft cells (paler coloured) contribute all the notochord of the secondary axis (Sek. Ch.), and part of the somites (R. sek. Uw.) and neural tube (Sek. Med.). Pr. Med., primary neural tube; R. sek. Pron., secondary pronephric duct; Sek. D., secondary gut. (From Spemann, 1938, after Spemann & Mangold.)

ing host cells which should have formed epidermis and lateral plate mesoderm, and, instead, formed dorsal structures such as neural tube, somite and even some notochord. These results have been confirmed many times, and in more recent years by the use of more reliable cell markers, although the extent of regulation by the grafted lip cells is less in *Xenopus* according to Smith & Slack (1983) and Recanzone & Harris (1985).

Spemann (1918) had previously shown that the fates of prospective epidermal and neural tissue, like those of ventral marginal tissue, are labile in early gastrulae. They stood in strong contrast to the dorsal lip which 'asserts itself in its new environment and makes use of this for its own purposes, namely, for the formation of a new embryonic anlage' (Spemann, 1925). Spemann thus proposed that the dorsal lip is an organisation centre for the embryo. He always stressed the harmony of the induced axes, with graft and host cells cooperating at a very intimate level, and with bilaterally symmetrical axes forming even if the graft was taken from one side of the dorsal lip. Nevertheless he also began the analysis of the organising effect. In particular he showed (Spemann, 1931) that the secondary axes induced by dorsal lips from early gastrulae contained mainly head structures, while those induced by late gastrula dorsal lips were of trunk and tail structures (Fig. 7.4).

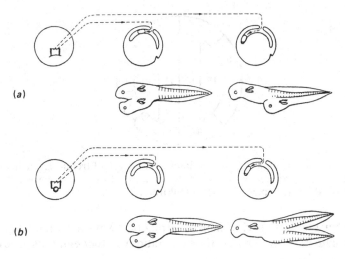

Fig. 7.4. Spemann's evidence for head and tail organisers in the newt. (*a*) dorsal lip tissue from an early gastrula, placed in the blastocoel of another embryo, always induces a head whatever its position in the host. (*b*) dorsal lip tissue from a late gastrula induces a head if it ends up in the head region but a tail elsewhere. (From Waddington, 1956.)

Spemann (see 1938, pp. 56 *et seq.*) had also studied the induction of the lens by the optic cup in older amphibian embryos, and he envisaged developmental decisions being made by a hierarchial system of organising influences spreading through the embryo. In this system the dorsal lip would be the primary organiser. Its effect has become known as primary embryonic induction (see Saxén & Toivonen, 1962), although of course earlier inductive interactions are now known (p. 190). The term is not well defined, being used by some authors to refer to neural induction distinct from the other dorsalising effects shown by the lip.

More recently, Cooke has re-examined the effects of a grafted organiser mainly using *Xenopus*. His work confirms both the autonomy of organiser development and its widespread effect on spatial determination in the host (Cooke, 1972*a,b*). Smith & Slack (1983) have checked that there is no reciprocal influence of a grafted ventral marginal zone on adjacent organiser fates. Two organiser regions in one embryo seem to compete in recruiting the remaining cells to two different embryonic fields (Cooke, 1973*b*). In terms of positional information theories, such properties suggest that organisers are signalling regions controlling cell fates in a dorso-ventral field. Simple models would, however, predict that the presence of two signalling (dorsal) regions would lead to the suppression of the opposite (ventral) cell types (compare the model of *bicaudal* development in *Drosophila* in Fig. 1.14), and this does not apparently happen in amphibian gastrulae. According to Cooke (1981)

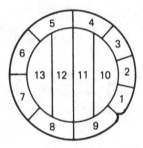

Fig. 7.5. Some examples of the isolates made by Holtfreter. Even his data showed that the frequency of axial differentiations from piece 2 was substantially lower than from piece 1. (After Holtfreter, 1938*a*.)

cells enter the various chordamesodermal pathways from notochord to lateral plate in approximately the same ratios whether embryos have one or two dorsal axes.

### The development of explants and implants from the early gastrula

If we are to understand fully what happens during 'primary induction' it is obviously important to know the state of determination of cells from various embryonic regions at the beginning of gastrulation. Holtfreter provided such data from extensive studies of the crescent-shaped lip stage of urodele (1938*a*) and anuran (1938*b*) gastrulation (see also Holtfreter & Hamburger, 1955). He had developed simple saline media in which small pieces of amphibian embryos could develop after their explantation from the rest of the embryo. As an example Figure 7.5 shows some of the areas which Holtfreter (1938*a*) explanted. From the tissues produced by the various isolates he constructed the map shown as Figure 7.6(*b*). For comparison, Figure 7.6(*a*) shows the normal fate map for this stage, redrawn from Vogt's (1929) data. According to Holtfreter, the isolates develop in one of three main ways which map closely in the areas of prospective ectoderm, chordamesoderm and endoderm. Prospective ectoderm formed atypical epidermis but no neural plate; and later workers have shown that the epidermis is structurally normal except in cell alignment (Grunz, 1973) and forms normal molecular markers (Slack, 1984*b*). Prospective chordamesoderm was highly regulative, being capable of forming all the tissues of the embryo, but its properties were also graded in the dorso-ventral direction with ventral mesoderm often self-differentiating as blood islands, etc. and dorsal areas showing the greatest regulative ability. Prospective endoderm never formed any non-endodermal structures, but it appeared to require the presence of mesoderm before it could self-differentiate.

Holtfreter (1936, 1938*a*,*b*) used these same data to map the head and

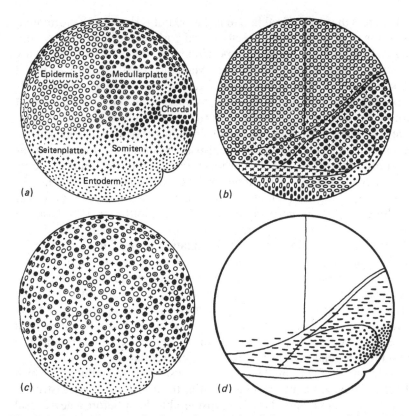

Fig. 7.6. Spatial patterns in the early gastrula of the newt according to Holt-freter. (*a*) Vogt's fate map: the tissue symbols are also used in (*b*) and (*c*). (*b*) differentiation capacity shown by isolates with extra symbols: (x), lateral plate mesoderm; ( • ), head endoderm; (○), oesophagus and stomach; (0), intestine; (◐), nutritive yolk. (*c*) prospective potency, or the kinds of tissue which can be experimentally produced. (*d*) the distribution of head (.) and trunk–tail (-) inducing ability. ((*a*),(*c*) from Holtfreter, 1936; (*b*),(*d*) from Holtfreter, 1938*a*.)

trunk–tail organisers in the crescent-lip gastrula (Fig. 7.6(*d*)). This was done by scoring all the neural structures produced by isolates according to whether they were parts of the brain or spinal cord. This was a very different test from that originally used by Spemann (1931) to separate these organisers, but did appear to show that regional differences in inductive power were already present in quite early gastrulae. This is further discussed on p. 223.

Finally, in his surveys of the early amphibian gastrula, Holtfreter (1936) mapped the 'prospective potency' of the various embryonic regions, i.e. what they could be induced to form following such operations as implantation to another embryonic area (Fig. 7.6(*c*)). This provided most extra information about prospective ectoderm, which formed less than its normal fate in isolation but is shown to be totipotent following implantation. Small ectodermal grafts conform to the development of the surrounding area, appearing to be 'infected' with the ability to induce if grafted into the lip area (Mangold, 1923). As already noted for blastulae (p. 190), 'animal halves' can be vegetalised by vegetal cells to produce all other tissues: even primordial germ cells can be induced although it is difficult to obtain them from those cells nearest the animal pole (Sutasurja & Nieuwkoop, 1974). Implants of prospective mesoderm again showed regulative ability but also a considerable autonomy in self-differentiating even in abnormal locations. Endodermal implants only developed endodermal derivatives (and see Heasman *et al.*, 1984).

By bringing together all these data we can summarise the properties of the three prospective germ layers according to Holtfreter. Ectoderm is a tissue showing very little determination but a very high potentiality. Chordamesoderm shows some determination even of its dorso-ventral subdivisions but it also retains the ability to regulate and produce ectoderm and endoderm. Endoderm seems completely determined, but requires some influence from the mesoderm to allow its differentiation. Greatest determination in the vegetal area has been a property of many other embryonic types and stages discussed in this book, and in several cases discussed in Chapter 6 this determined area requires an influence from more animal regions before it can differentiate.

Paterson (1957) has made an interesting study of the influences which affect the development of the equatorial third of *Rana* early gastrulae. When it develops alone, there is a considerable delay before a new dorsal lip forms, but the isolate can then go on to form a tadpole. In combination with the vegetal third (the site of the original dorsal lip) development is hardly delayed and again extensive. The animal two-thirds of the gastrula, however, differentiates only to folded and pitted tissue, which suggested to Paterson that excess animal influence was preventing the equatorial third from differentiating. There is much other evidence that mesoderm arises as a result of animal–vegetal interactions (see pp. 190–4), and this work implies that it is still not fully determined in the early gastrula.

Holtfreter's data have been refined in even more important ways by Japanese studies with the newt *Cynops pyrrhogaster* (Fig. 7.7(*a*)). They used earlier stages where the future dorsal lip was only recognisable as a pigment concentration. This is important because the animal–vegetal positions of the cells will then be minimally affected by involution. At this stage they found that only a small area above the future lip showed any determination as chordamesoderm, while above this the remaining prospective notochord, like prospective ectoderm, formed atypical epidermis (Takaya, 1953*a*; Kato, 1958; Kanéda & Hama, 1979). This suggests that endoderm only induced a narrow ring of mesoderm from the 'animal half' at blastula stages (p. 190), and that the remaining parts of the future chordamesoderm still remain to be induced. At the upper edge of the induced area the structures which differentiate at lower frequencies include posterior neural ones according to Kato (1959) and ventral mesoderm according to Kanéda & Hama (1979). The tissue within the induced band on the dorsal side usually differentiated to notochord and somites (Takaya, 1953a; Kato, 1958, 1959; Kanéda & Hama, 1979), and if it was grafted into a host embryo (Okada & Takaya, 1942) or explanted with early gastrula ectoderm (Okada & Hama, 1943; Kato, 1958) it induced posterior neural structures (spinal cord). These are surprising properties since the area is presumptive head mesoderm, and since there is no trace of the head organiser reported by Spemann (see Fig. 7.4) and Holtfreter (see Fig. 7.6(d)). There are, of course, differences in the species and techniques used, but other studies by the Japanese suggest an explanation for all the data and provide a very different and more dynamic view of cell interactions during gastrulation than the earlier German interpretation. These studies will now be considered.

### Changes in spatial properties during gastrulation

The Japanese workers extended their studies of *Cynops* to investigate the state of determination and the inductive power of all the tissue which passes through the dorsal lip during gastrulation. Their findings are summarised in Table 7.1 and Figure 7.7, and the latter should be compared with the fate maps through gastrulation given as Figure 7.2. One important feature of this work was that the outer and involuted cells at the dorsal lip were investigated separately: the lips studied by earlier workers included both layers.

The very first cells in the outer lip position are prospective head endoderm and they never differentiate into mesodermal structures (Suzuki, Mifune & Kanéda, 1984). Thereafter, however, all cells which are just about to involute can differentiate into notochord and somites and induce trunk and tail derivatives if implanted to host embryos or explanted with early gastrula ectoderm. Sometimes more anterior structures are produced, particularly when lip tissue is explanted alone, but Hama (1949) believes that this is due to changes occurring after explantation. At later gastrula stages the abilities to differentiate and to induce posterior tissues appear in uninvaginated cells at successively greater distances from the dorsal lip. (This is one explanation of the differences from Holtfreter, 1938a, as he used a rather later stage.) As with early gastrulae there may first be a transient stage in which the cells show the properties of posterior neural tissue (Kato, 1959) or ventral mesoderm (Kanéda & Hama, 1979) before they acquire those of posterior dorsal mesoderm. It seems that new cells are continually being mesodermalised

Table 7.1. *Changes in the self-differentiation and the inductive ability of Cynops pyrrhogaster mesoderm during gastrulation.*

| Prospective fate | Position when tested | | | | | | | |
|---|---|---|---|---|---|---|---|---|
| | Well before invagination | | Just before invagination | | Soon after invagination | | In final position | |
| | Forms | Induces | Forms | Induces | Forms | Induces | Forms | Induces |
| Prechordal plate | — | — | Notochord (Takaya, 1953a) | Trunk and tail (Okada & Takaya, 1942) | Mesenchyme (Takaya, 1953a) | Head (Okada & Takaya, 1942) | Prechordal plate | Forebrain (Okada & Hama, 1945a, Masui, 1960a) |
| Anterior chorda | Epidermis (Kato, 1958); some neural (Kato, 1959) or ventral mesoderm (Kaneda & Hama, 1979) | Mostly none (Kato, 1958) | Notochord (Takaya 1953a) | Trunk and tail (Okada & Takaya, 1942) | Mesenchyme (Takaya, 1953a), notochord (Kato, 1958) | Head (Okada & Takaya, 1942) hindbrain, or trunk (Kato, 1958) | Notochord | Hindbrain (Okada & Hama, 1945a) |
| Posterior chorda | Notochord (Kato, 1958) some neural (Kato, 1959) | Trunk and tail (Kato, 1958) | Notochord (Takaya, 1953a) | Trunk and tail (Okada & Takaya, 1942) | Notochord (Takaya, 1953a) | Trunk and tail (Okada & Takaya, 1942) | Notochord | Trunk structures (Okada & Hama, 1945a) |

(After Wall, 1973: the data have been confirmed by later studies cited in the text.)

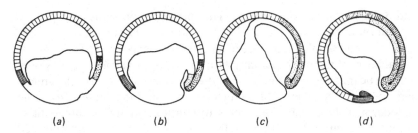

(a)                (b)                (c)                (d)

Fig. 7.7. Changes in developmental potency during gastrulation in *Cynops*. The figures summarise the development shown by small pieces when they are isolated from the four stages shown. Compare with Fig. 7.2 (*d*)–(*g*): the same symbols are used except that areas reported to develop either as ventral mesoderm or neural tissue are cross-hatched. (The boundaries should be considered as approximations.)

as they approach the dorsal lip, and the same change can be induced in prospective ectoderm which is grafted into a position somewhat above the early lip (Kanéda, 1981; Kanéda & Suzuki, 1983).

Cells which have involuted around the dorsal lip show very different changes (Table 7.1 and Fig. 7.7). The first cells to round the lip show a rapid change in properties so that they differentiate to mesenchyme and induce head structures from ectoderm. In both features this is an anteriorisation compared with the properties of the same cells just before involution. As they then pass forward in the archenteron roof their inductive power weakens considerably. Cells which have just rounded the lip in more advanced gastrulae retain their posterior properties and are anteriorised more gradually as they move forwards in the archenteron roof. The determination of the trunk organiser for notochord seems actually to be more firmly fixed following its involution (Ohara & Hama, 1979*a,b*). The last cells involuted have mesodermal inducing activity.

It is the properties of the involuted cells which probably explain the different conclusions of Spemann (1931) and Holtfreter (1938*a*). They presumably provide the head organising abilities of the early lip, while both components of later lips will have tail organising ability. The importance of the distinction can be seen when it is realised that the head organiser has only arisen as cells rounded the lip: it did not exist as a dorsal specialisation of the pregastrular mesoderm as might easily be imagined from Figure 7.6(*d*).

By the end of gastrulation the major fates of all areas are sufficiently determined to allow self-differentiation in isolation, as can be seen by comparing Figures 7.2 and 7.7. Anterior pieces from the archenteron roof will form endoderm and mesenchyme and posterior pieces notochord and somites. Pieces from the overlying tissue form, in antero-posterior sequence, structures of the forebrain, hindbrain, spinal cord and finally somitic mesoderm. The presence of some prospective mesoderm in posterior neural plate was noted earlier (p. 215).

How is this complex sequence of changes to be explained? The simplest explanation of the mesodermalisation of cells approaching the lip would seem to be that it is a continuation of the mesodermalisation of the animal half seen in blastulae. This induction spreads tangentially from the endoderm in blastulae, and seems to continue to spread tangentially during gastrulation from those cells which are already mesodermalised (Takaya, 1957; Kanéda, 1981; Kanéda & Suzuki, 1983; Hama *et al.*, 1985). Explants from some distance above the early lip do not acquire mesodermal properties during ageing *in vitro* so it is not an autonomous change (Kato, 1959), nor are they mesodermalised by the underlying involuted cells (Kanéda, 1980). Mesodermalisation can also spread tangentially through a piece of prospective ectoderm which has been artificially mesodermalised (see p. 228) at one end (Kurihara, 1981). In older gastrulae the posterior archenteron roof has mesodermalising ability (see below) so vertical inductions may also contribute to the extension of chordamesodermal properties. The fact that it is posterior mesodermal structures (and perhaps posterior neural ones) which are induced in the uninvaginated cells is most surprising at early stages when the tissue concerned is the prospective head organiser. However, Pasteels (1953, 1954) also found in other species that posterior axes were induced in the blastocoel roof when it collapsed onto the endoderm in centrifuged late blastulae and early gastrulae.

The anteriorisation of cell properties as they round the early dorsal lip will occur if uninvaginated cells are aged *in vitro* and so is an autonomous property (Okada & Takaya, 1942; Takaya, 1953*b*; Suzuki, 1968; Hama *et al.*, 1985). However, some of these authors find that the change takes longer *in vitro* than *in vivo*, and it will also occur if cells from outside the *late* dorsal lip are aged *in vitro* and this will not happen *in vivo* (Miyagawa & Suzuki, 1969). This suggests that interactions with neighbouring cells modulate the change *in vivo* and such interactions could most easily occur with the overlying cells of the outer lip layer (Wall, 1973; Nieuwkoop & Weijer, 1978). The cells which are being folded together at the lip were previously separated along the dorsal meridian where only short distances separate cells with very different developmental properties. The first mesoderm to involute will thus be brought into contact with more animal cells which may well be responsible for the anteriorisation of its properties, although a direct demonstration of this is still lacking. It is possible to see this as an 'animalisation' comparable with some effects of animal cells in, for example, the sea urchin, (see also p. 228). The mesodermalisation of suprablastoporal endoderm when it is combined with early gastrula ectoderm (Kato & Okada, 1956) can be explained in the same way. The changes in motility patterns seen as cells round the lip

(p. 213) may be another effect of the interaction, and could in turn affect later aspects of primary induction as well as morphogenesis (see Takaya, 1978).

The most widely studied interaction in the gastrula is of course the reciprocal effect of the archenteron roof on the overlying tissue. Having been anteriorised as they rounded the early lip, the first archenteron roof cells tend to induce forebrain structures. However, the overlying cells are also receiving mesodermalising stimuli tangentially, and the major effect of the vertical induction may be to dorsalise the kinds of mesoderm produced (Kanéda, 1980), although some vertical neuralisation may also occur (see too Kato, 1959). Cells rounding the lip at late stages would retain posterior-inducing properties for some time and change more gradually towards the induction of forebrain. The anterior neural plate will thus receive only forebrain-inducing stimuli, while more posterior neural areas will have received forebrain-inducing stimuli from the first cells to pass beneath them followed by inducing stimuli of more posterior kinds. Cells of the posterior neural plate will presumably have received a succession of inductive influences of increasingly posterior nature, and are finally underlain by cells with an almost pure mesodermalising activity (Okada & Hama, 1945a). This is the sequence of inductions described as neural activation and transformation by Nieuwkoop (see p. 230), but this need not imply that posterior induction is always a two-stage process. The dramatic change in properties as cells round the early lip was attributed above to the bringing together of cells with very different properties. By the end of gastrulation the two cell layers at the dorsal lip have almost equivalent properties and both will contribute chiefly mesodermal tissue to the tail. This convergence of properties of the two lip layers can probably be explained in principle by a two-way theory of interactions. As gastrulation progresses, the cells in the outer lip position will already have been underlain by an increasing number of involuted cells, and this may well have moderated their developmental properties. This could render them less capable of anteriorising the inner lip cells, which would in turn have a posterior inducing effect on the next cells to reach the outer lip position, etc. Figure 7.8 summarises the interactions which appear to be occurring in the gastrula. Suzuki *et al.* (1984) have provided evidence that the differentiation of the late archenteron roof can still be affected by the kind of ectoderm with which it is combined.

By the end of gastrulation both the outer and involuted cell layers will vary in their properties along the anterior–posterior axis. This alone could determine that different embryonic organs will develop in an anterior–posterior sequence in each layer. The two layers are still in contact, however, and their further interactions appear also to have important developmental roles which will be discussed on p. 246.

It is of course very important that conclusions derived from work with

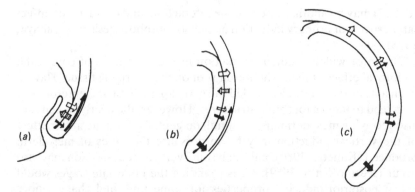

Fig. 7.8. Summary of the inductive interactions occurring in the amphibian gastrula. Black arrows, mesodermalisation; white arrows, neuralisation; hatched arrows, anteriorising effects. The strength of the induction is indicated by the width of the arrow.

one species, *Cynops pyrrhogaster*, should be tested with others, and this has often still not been done. Gallera (1959, 1960) has, however, shown with *Triturus alpestris* that specific areas of the prospective chordamesoderm acquire more anterior inductive properties during gastrulation, while Spofford (1948) showed that the posterior archenteron roof of late *Ambystoma* gastrulae is a mesodermal inductor. The concept of two-way interactions is supported by studies of *Triturus alpestris* where the differentiation of archenteron roof is affected by the ectoderm with which it is combined (Niewkoop & Weijer, 1978), and of axolotls, where the age and mass of the interacting tissues was varied (Leyhausen, 1987). Of work with anurans, Holtfreter (cited by Deuchar, 1972) found that early *Xenopus* dorsal lips induced posterior structures, but such cells may already have anterior properties before involution in *Rana* (Okada & Takaya, 1942; Srivastava & Srivastava, 1969).

### Inductive effects of simple culture media

Studies of the interactions occurring in amphibian gastrulae soon revealed that abnormal conditions, sometimes of a totally artificial kind, could mimic many of the inductive effects. Waddington, Needham & Brachet (1936) stained gastrula ectoderm with methylene blue, and showed that it was itself neuralised and could neuralise further ectoderm in a host embryo. Such inductions are, however, very different from the secondary axes which Spemann & Mangold (1924) obtained with a grafted dorsal lip. The neural tissue is usually of a poor level of organisation and often cannot be attributed to any particular part of the nervous system. Neural tissues which can be recognised are usually derivatives of the forebrain (archencephalic inductions). Needham, Waddington & Needham (1934) suggested that this 'evocation' of neural tissue is just one component of the organiser's activity, and that its organisation into the major components of the central nervous system required a second step of 'individuation'.

The ease with which amphibian gastrula ectoderm is evocated to neural tissue varies according to the species. Barth (1941) claimed that even control ectoderm of *Ambystoma punctatum* neuralised during *in vitro* culture, but Holtfreter (1945) showed that this could be prevented by more careful control of the pH of the culture medium. If the pH was raised or lowered sufficiently, or if calcium ions were omitted from the medium, even *Triturus* ectoderm neuralised. Holtfreter (1947) and Karasaki (1957) suggested that the first stage of evocation is a non-specific 'sublethal cytolysis' seen as a solation of endoplasm and formation of pseudopodia. Much of Barth's later work has been with small explants (about 125 cells) of *Rana pipiens* ectoderm, which can be induced to neural and pigment cell types by many ions and non-ionic solutes. This supports the idea that only an unspecific releasing stimulus is required, but Barth's group found that the induced tissues only differentiated in media with high sodium ion concentrations (Barth & Barth, 1969) or high osmotic pressures (Barth & Barth, 1974a). They suggested that a release of ions into the blastocoel is actually required to *prevent* neural differentiation, so that a trapping of ions by the presence of the archenteron roof is enough to ensure the formation of the neural plate (Barth & Barth, 1974a). Barth & Barth (1969), however, accept that larger ectodermal explants can be induced even in media with low sodium ion concentrations; and such explants cannot be induced merely by changing ion concentrations (Siegel et al., 1985). Derivatives of cyclic AMP can also induce nerve and pigment cell types from small, but not from large, explants of early gastrula ectoderm (Wahn et al., 1975; Wahn, Taylor & Tchen, 1976). The very high sensitivity of small explants (see too Løvtrup & Perris, 1983) seems to make them an unsuitable test system for inductive activities.

The precytolytic effects of neuralising treatments are reminiscent of the denaturing effects claimed for animalising treatments in sea urchin embryos by Ranzi (see p. 160). Ranzi (1962) has, in fact, confirmed that treatments which animalise sea urchins have the same denaturing effects on amphibian embryos, although their effects on development are not a straightforward animalisation. One such treatment, thiocyanate ions, evocates archencephalic structures from large explants of *Triturus* ectoderm (Ogi, 1958a). This again suggests a parallel between archencephalic induction and animalisation.

In contrast, the addition of lithium ions can mesodermalise or even endodermalise *Cynops* ectoderm (Masui, 1960b, 1961). This result has been confirmed for other urodeles by other authors, and for small ectodermal explants from *Rana* by Barth & Barth (1974*b*), but Slack (pers. commun.) could demonstrate no inductive effects on *Xenopus* ectoderm using lithium or other salts (and see p. 245). Ogi (1961) has described lithium's effect as a vegetalisation as it is on the sea urchin embryo. It was, of course, already known that early gastrula ectoderm is competent to form derivatives of the other germ layers but this remains apparently the only inorganic treatment capable of inducing the change.

Another developmental pathway into which early gastrula ectoderm can be directed is the formation of cement gland cells. Media containing ammonium chloride induce this change in *Xenopus* ectoderm (Picard, 1975), while high sodium ion concentrations induce it in *Rana japonica* (Yoshizaki, 1981). That the same tissue can be directed along so many different developmental pathways by different artificial treatments at least ensures that the effects cannot all be dismissed as non-specific damage.

Treatments known to animalise or vegetalise sea urchin embryos also have opposite effects on the mesodermal tissues in amphibians. Ventral mesoderm from gastrulae can be dorsalised by thiocyanate ions, zinc ions or urea so that it forms notochord and somites (Ogi, 1958*a,b*). In contrast, when whole gastrulae are treated with lithium ions, mesoderm may be ventralised causing local absence of the notochord and dorsally fused somites (Lehmann, 1937). Lithium ions also have interesting effects on prospective prechordal plate (Masui, 1960*a*), a tissue which shows posterior mesodermal properties before involution and anterior ones thereafter or following culture *in vitro*. They completely prevent the anteriorising effects of *in vitro* culture, and can even reverse these changes somewhat in cells which have involuted (Masui, 1960*a*). It was suggested earlier that the anteriorisation of properties occurring at the lip is a kind of animalisation, and the fact that lithium ions oppose the changes would support this interpretation. The cyclopic and related abnormalities which were some of the first known effects of lithium (or magnesium) ions on vertebrate embryos are probably consequences of changed properties in the prechordal mesoderm (Adelmann, 1934).

**Heterogenous inductors**

It is as easy to cause inductions in amphibian gastrulae or their ectoderm using foreign tissues as it is by changing the culture medium. By 1934*b* Holtfreter had produced a very long list of inductively active tissues taken from a wide variety of animals. As in the previous section, some order can be introduced into this wealth of data by considering the kinds of structures induced by different treatments. Again some treatments produced archencephalic and others mesodermal inductions, but still others

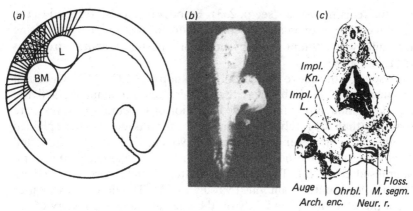

Fig. 7.9. The induction of a complete secondary axis when archencephalic and mesodermal inducers are implanted close together. (*a*) the operation and its interpretation: the areas of influence (straight lines) from liver (L) and bone marrow (BM) overlap. (*b*),(*c*) the resulting induction seen in external view and in section: Arch. enc., archencephalon; Auge, eye; Floss., fin; Impl. Kn., implanted bone marrow; Impl. L., implanted liver; M. segm., segmented muscle; Neur. r., spinal cord; Ohrbl., ear vesicle. (From Toivonen & Saxén, 1955*a*, (*a*) redrawn.)

produced derivatives of the hindbrain region (deuterencephalic) or tails containing spinal cords (spino-caudal). Pretreatment of the inductor tissue also changed the balance of the induced structures, most often leading to an increase in archencephalic inductions and a decrease of more posterior types. This led Chuang (1939) to propose that there must be at least two different inductive principles which differ in their lability.

A great advance was made when alcohol-fixed guinea pig liver (archencephalic) and bone marrow (mesodermalising) were implanted or explanted close to each other beneath *Triturus* early gastrula ectoderm (Toivonen & Saxén, 1955*a,b*). In this case well-integrated axial systems were induced containing neural and sensory structures from forebrain to tail level as well as mesodermal tissues. This provided good evidence that two factors from heterogenous tissues could induce axes with much of the harmony of Spemann's organiser as long as they were allowed to interact in distinct but overlapping concentration gradients (Fig. 7.9). Other work confirmed that deuterencephalic and spino-caudal inducing fractions could always be further separated to yield pure archencephalic and mesodermalising ones (Tiedemann, Becker & Tiedemann, 1963), and that these 'pure' fractions could be recombined in varying ratios to produce neural structures of different antero-posterior level (Tiedemann & Tiedemann, 1964). Toivonen & Saxén (1955*a,b*; Saxén & Toivonen, 1962) have proposed that normal primary induction also requires two inductive agents acting in similar spatial patterns, although the final result will also be affected by changes occurring in time both in the inductive and the reacting tissues (see p. 233). Ectoderm can also respond to the combined effects of an archencephalic inductor tissue and a vegetalising medium (with lithium ions) by producing posterior neural structures (Masui, 1960*b*).

Despite all the above evidence, Nieuwkoop (see 1973, 1977; Gebhardt & Nieuwkoop 1964) believes that a third distinct inductive activity, transformation, is required for the formation of all neural structures more posterior than the forebrain. The evidence for this idea can be explained in other ways, however (see Saxén & Toivonen, 1962, pp. 193–204), and treatments which mesodermalise early gastrula ectoderm also transform neuralised ectoderm. These include posterior archenteron roof (Eyal-Giladi, 1954) and lithium ions (Masui, 1960*b*). The different responses seem therefore to depend only on the age and inductive history of the reacting tissue, and even so the final result of transformation in late gastrulae would be the induction of mesoderm in the posterior neural plate (Nieuwkoop, 1973). The case for distinct transforming inductions therefore remains unproven, but the role of the reacting tissue is further considered on p. 233, and the evidence for postgastrula transforming effects on p. 250.

Mesodermalising fractions can also induce endoderm and even primordial germ cells from urodele gastrula ectoderm (Takata & Yamada, 1960; Wang, Mo & Shen, 1963; Kocher-Becker & Tiedemann, 1971). Apart from demonstrating that primordial germ cells are inducible from inappropriate cells in urodeles (in anurans they may be determined by localised materials: see pp. 28–9), this suggests that the induction is a general vegetalisation. One vegetalising fraction, obtained from older chick embryos, has been subjected to extensive further purification without ever separating its mesodermalising and endodermalising effects (Geithe *et al.*, 1975). This would mean that the factors required for primary embryonic induction are an archencephalically inducing neuraliser and a vegetaliser. An alternative is that the latter factor is an endodermaliser, and that mesoderm arises secondarily when this endoderm acts upon cells which were previously uninduced (Minuth & Grunz, 1980). Secondary interactions may also affect the dorso-ventrality of the induced mesoderm (Asahi *et al.*, 1979; Grunz, 1983).

A special case of an abnormal inductor tissue is the neural plate from a neurula stage embryo. It has itself only just been induced by the archenteron roof, but if implanted beneath the ectoderm of an early gastrula it can induce a further neural plate (Mangold & Spemann, 1927). The phenomenon was called homoiogenetic or assimilatory induction, and suggests that the neural plate acquires at least some of the properties of the chordamesoderm during primary induction. Mangold (1932) showed that anterior neural plates induce brains and posterior neural plates induce tails, but also noted an interesting difference in their mode

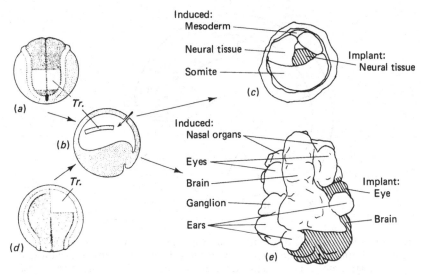

Fig. 7.10. Complementary (*a*)–(*c*) and autonomous (*d*),(*b*),(*e*) induction in newt embryos. Different parts of the neural plate (*a*),(*d*) are transferred to the blasto-coel of an early gastrula (*b*). (*c*) section through a tail induced by the posterior neural plate, in which graft and host cells co-operate. (*e*) reconstruction of head tissue induced by one half of the anterior neural plate: the graft has self-differentiated to a lateral half head and induced a separate bilaterally symmetrical head from the host. (After Mangold, 1932.)

of action (Fig. 7.10). In inductions by the posterior neural plate, graft and host cooperate intimately in the production of neural tube and somites within a single harmonious tail. The effect is much like that of an earlier tail organiser except that no notochord is formed. In inductions by the anterior neural plate, graft and host form largely separate structures where even the plane of symmetry can be quite different. Mangold described the latter effect as autonomous induction, and a trend towards the simple evocation of archencephalic structures (as seen elsewhere in this and the previous section) is clearly visible here.

## The competence of the ectoderm

Early gastrula ectoderm is totipotent (p. 220) and this means that it has the competence to produce all other tissues. Clearly the ectodermal cells must make a choice between these pathways, and we have so far considered how external conditions acting upon them affect that choice. This section considers the effects of conditions within the ectoderm itself.

The role of the reacting tissue is highlighted by the ease with which it can be induced, particularly in the evocation of archencephalic tissue. This suggests that the ectoderm contains almost all the factors required

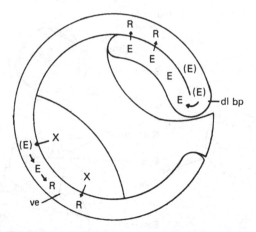

Fig. 7.11. Some possible routes for neuralisation of amphibian ectoderm. Ectoderm may normally contain evocator molecules (E) in a masked form (indicated by brackets); these could be activated at the dorsal lip (dl bp) and then release the response (R) of neuralisation in overlying cells. Artificial inducing agents (X) could activate the masked evocator in ventral ectoderm (ve) or release the neuralisation response directly. (After Needham, 1942.)

for neural differentiation, and requires only a releasing cue of low specificity. Several authors have suggested that materials required for development as neural tissue are already present in ectoderm but complexed with inhibitors (e.g. Needham, 1942, pp. 176–188). Inducer molecules would destroy the complex, releasing the components and thus initiating development along a neural pathway. Artificial neuralising treatments could also break the complex or they could act at another level, e.g. to take the place of the ectoderm's own neuralising factors (see Fig. 7.11). If they act in the latter way then studies of artificial neuralisation can tell us nothing about the normal induction process, but only about reactions normally occurring within the ectoderm.

Although early gastrula ectoderm is not so easily vegetalised, this can be achieved by at least one totally non-physiological mechanism: the use of lithium ions. It is therefore possible to argue that factors for vegetalised development must also pre-exist in the ectoderm perhaps again complexed with inhibitors. Tiedemann's group have in fact found that chick tissues which vegetalise amphibian ectoderm also contain an inhibitor of vegetalisation, and that amphibian gastrulae may also contain an inhibitor (Born, Tiedemann & Tiedemann, 1972; see also p. 239). The same arguments could be applied to ectodermal competence for cement gland formation, which is again inducible by artificial conditions. Arguments of this sort suggest that early gastrula ectoderm is already 'primed' for development along each of several alternative pathways. Yet if we overemphasise the effects of such 'priming' we may obscure the fact that it is external factors which decide which pathway is followed. In the absence of all such factors most authors would agree that the ectoderm forms epidermis (although see Løvtrup & Perris, 1983). All other differentiations must therefore require an inductive action and, even if it is a fairly simple releaser, it must supply sufficient information to specify one of the major alternatives: archencephalic, vegetalised and perhaps cement gland. Moreover, vegetalisation includes the induction of a wide variety of tissues, both mesodermal and endodermal, and the choice amongst these is also decided by external factors (Grunz, 1983), whether they are quantitative variations in vegetaliser activity or second inductive actions (see p. 230). Yet another group of developmental pathways is followed by ectoderm induced by both the archencephalic and the vegetalising factors, their ratio deciding whether deuterencephalic or spino-caudal structures will form. It seems unlikely that ectoderm is prepared in any specific way to develop into all of these very different types of tissue and organ.

One major way in which competence can affect the choices available to ectodermal cells is that competence patterns change with developmental age. These changes have been estimated from the response to media with changed ionic compositions (e.g. Chuang, 1955; Gebhardt & Nieuwkoop, 1964), to heterogenous inducing tissues (e.g. Leikola, 1963, 1965; Sasaki *et al.*, 1976) and to the natural inducing factors in the embryo (e.g. Nieuwkoop, 1958; Ohara & Hama, 1979*a*,*b*; Suzuki & Ikeda, 1979; Ohara, 1980, 1981). There is a broad measure of agreement using these varied methods that competence for vegetalisation (usually seen as mesodermalisation) shows the first peak, between midblastula and early gastrula stages, while competence for archencephalic induction (often described as neural activation) is maximal in early gastrulae (see Fig. 7.12). The period of competence for the normal neural inducer may be narrower than the figure suggests, as Ohara (1980, 1981) believes that it does not develop until after gastrulation has started. It may end as cells leave the fifteenth cleavage cycle (Suzuki & Ikeda, 1979). Tissue which has already been neuralised retains the competence for transformation to more posterior neural structures for many more hours into neurula stages (Nieuwkoop, 1958). Ectoderm which has not been subjected to any previous inductive influences, however, loses its competence for posterior neural structures rather earlier than its competence for brain induction (Ohara & Hama, 1979*a*,*b*). Nieuwkoop believes that transformation is a third distinct inducing activity (see also pp. 230 and 250), but it may rather be that competence for vegetalising effects is retained longer by cells which have been neuralised, where it results in the production of more posterior neural structures. Already-induced ectoderm certainly shows an extended competence for the effects of the

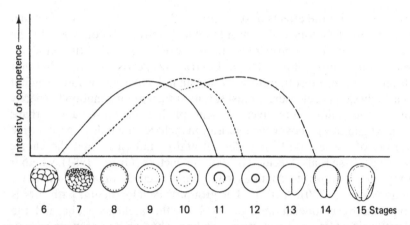

Fig. 7.12. Changes of competence with time in the animal half of urodele embryos. Mesodermal competence (———); neural competence (- - - - -); competence for transformation (— —). (From Nieuwkoop, 1973.)

neural inducer which then results in the production of more anterior neural structures (Nieuwkoop & van der Grinten, 1961).

The changing pattern of ectodermal competence with time almost certainly affects the final arrangement of the induced tissues, but it is probably not the major influence. Put another way the result of an induction usually depends more upon the nature of the inducer than upon the age of the ectoderm. Position of origin within the ectoderm also seems to play a relatively minor role: all early ectodermal areas are totipotent and in particular the choice between epidermis and neural tissue is primarily decided by external factors (Spemann, 1918; Holtfreter, 1947; Gimlich & Cooke, 1983). This must limit any influence of 'mosaic' factors, such as the inheritance of 'mesoplasm' by the prospective neural area (Imoh, 1985, and see p. 190), but the induction of neural-specific gene expression does seem to occur more easily in dorsal, equatorial regions of the ectoderm than elsewhere (Sharpe *et al.*, 1987). There are probably also differences in competence between the inner and outer cell layers of anuran ectoderm although their extent is not yet clear (Asashima & Grunz, 1983; Dettlaff, 1983).

## The transmission of inductive stimuli

There are potentially many ways in which inductive stimuli could be transmitted between cells in amphibian gastrulae. Small inducer molecules may diffuse between cells while larger ones would probably require more specialised transfer mechanisms. Steric interactions between surface-bound molecules could alter surface properties and so change the developmental pathway of the whole cell (Weiss, 1958), or communication may be mediated by extracellular materials. Descriptive and experimental work can help to discriminate amongst such possibilities, and it should be remembered that several different kinds of interaction appear to be occurring and each could use a different kind of mechanism.

Early suggestions of intercellular bridges between interacting cells (Eakin & Lehmann, 1957; Karasaki cited by Yamada, 1961) have not been supported by more recent studies with improved techniques (Kelley, 1969; Grunz & Staubach, 1979*a*), and would appear unlikely since the cells move quite rapidly over each other. Grunz & Staubach (1979*a*) did, however, see close contacts between the two cell layers, which are also coupled in gastrulae although the degree of coupling then decreases (Ito & Ikematsu, 1980). It is now possible to test the significance of gap junctions in inductions by injecting antibodies to the major gap junction protein (see pp. 138 and 194), but this has not yet been done specifically for the interactions occurring in gastrulae. Following the injection of one dorsal, animal cell of the eight-cell embryo, the eye may fail to develop and the brain be reduced on the injected side (Warner, Guthrie & Gilula, 1984), but an inhibition of 'primary' induction is only one of the possible explanations for these results. Yamada (1962) suggested that inducers could be taken into ectodermal cells by pinocytosis, and vesicles have been seen near the surfaces of the interacting cells and apparently opening into the narrow space between them (Fig. 7.13) by Kelley (1969) and Grunz & Staubach (1979*a*). Kelley (1969) showed that ferritin particles introduced between the two cell layers are taken up by the ectodermal cells. He also described granules and fibrils between the two cell layers and gave some evidence that these too could be transferred to the ectoderm, but Tarin (1973) has shown that development is little affected when these extracellular structures are destroyed enzymatically. Immunofluorescence studies indicate that antigenic molecules (i.e. macromolecules) are taken up into the cytoplasm of ectoderm which has been induced by gastrular chordamesoderm (Rounds & Flickinger, 1958; Flickinger, Hatton & Rounds, 1959) or by heterogenous inductors (Vainio *et al.*, 1962; Yamada, 1962).

Another way of studying the transmission problem has been to separate the interacting cells by porous layers of various kinds. Most early studies failed to demonstrate the transmission of inductive information (see Brahma, 1958). However, Saxén (1961) reported that *Triturus* dorsal lips could induce forebrain and a few hindbrain structures from ectoderm through Millipore filters with an average pore size of 0.8 µm. Cytoplasmic processes could be seen in the pores but no contacts between the layers were observed (Nyholm *et al.*, 1962), so at least the archencephalic inducer seemed to be diffusible. This was confirmed by Toivonen *et al.* (1975) using Nucleopore filters (where the

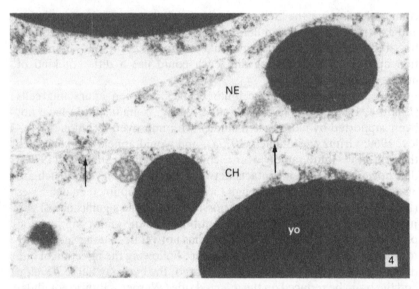

Fig. 7.13. Membrane contacts between a projection from a presumptive neural cell (NE) and a presumptive chordamesoderm cell (CH) in a *Triturus* mid-gastrula. Pits and vesicles (arrows) are seen in the chordamesoderm cell near these contacts. yo, yolk. × 16,300. (From Grunz & Staubach, 1979a.)

pores are straight channels of closely controlled diameter). After long culture times ectodermal cell processes do penetrate these filters and contact the mesoderm, but even in such cases all true inductions seem to be archencephalic (Toivonen & Wartiovaara, 1976). Minuth (1978) has obtained some posterior neural and mesodermal inductions after long culture times (chiefly with bone marrow, a heterogenous meso-dermaliser), but Toivonen (1979) suggests that direct cell contacts have been established by this time. Whether posterior inductions resulted could depend upon whether activated ectoderm had lost its competence before contacts were established. In any case such work makes it unlikely that vegetalising and transforming inducers (which may be the same thing) are freely diffusible, since even materials as large as the polio virus should pass through such filters fairly rapidly (Wartiovaara *et al.*, 1974). Such inductions are thus more likely to require direct cell-to-cell contact or contact mediated by matrix materials, unless the size of the total pore area is too low to allow successful induction by a more diffusible inducer (Toivonen & Wartiovaara, 1976). This is surprising since mesodermal inducers in blastulae easily cross filters (see p. 194). We may note that the vegetalising activity of some heterogenous inductors also crosses filters but seems to be due to matrix materials (see Kawakami, 1976).

Work with other techniques indicates that direct cell contact is not obligatory for the transfer of 'primary' inducing stimuli. Cell-free media which have been 'conditioned' by the presence of dorsal lips (Niu & Twitty, 1953; Niu, 1956) or heterogenous inductors (Becker, Tiedemann & Tiedemann, 1959) can induce ectoderm to form various neural and mesodermal tissues. Materials with neuralising activity can also be recovered from the space between the archenteron roof and the neurectoderm (John *et al.*, 1983); but their direct effect on the ectoderm may be limited to the cell membrane as neuralising factors are still active when coupled to large particles to prevent their entry into ectodermal cells (Tiedemann & Born, 1978; Born *et al.*, 1986). In contrast, the vegetalising factor apparently has to enter cells to exert its inductive effect (see below).

### Attempts to identify inducer molecules

Following Spemann & Mangold's (1924) discovery of the 'organiser', teams of biochemists and embryologists began a race to identify the single chemical which many thought must be the responsible agent. Amphibian gastrulae or heterogenous inductor tissues were biochemically fractionated in various ways and the fractions tested for inductive activity (review; Needham, 1942). The overall result was in fact greatly increased confusion: each team found the greatest activity in a different type of molecule, and totally unnatural conditions produced inductions.

The early studies had however begun the analysis of primary induction to the two constituents which have been characterised in previous sections as archencephalic and vegetalising. Those sections also showed that it is the archencephalic inductions which are most readily 'evocated' by artificial conditions. A bewildering range of treatments induce forebrain or non-specific neural differentiation and activity commonly appears in tissues following boiling as it does in all tissues of the gastrula (Holtfreter, 1934*a*). This has commonly been seen as further evidence that the ectoderm already contains factors for neural development in a masked form (see also p. 231): in fact it could equally well mean that virtually any boiled tissue evocates neural tissue but in ways which bear no relation to normal neural induction. Even the dorsal lip itself may induce ectoderm by an entirely different mechanism once it has been killed (Okazaki, 1955). Anything we learn about artificial neuralising treatments may therefore be irrelevant to the normal process of neuralisation.

A more direct way of identifying possible neuralising factors is to analyse the chemicals released by living chordamesoderm cells. Niu (1956) suggested that the materials released by dorsal lips and posterior neural plate cells are nucleoproteins, while Sánchez, Cabada & Barbieri (1983) also gave evidence that RNA, including mRNA, is released from gastrula cells. Mucopolysaccharides are also secreted at the gastrular lips

(Moran & Mouradian, 1975), and one such material, heparan sulphate, has been suggested as the neural evocator (Landström & Løvtrup, 1977*a*) and dorsalises ventral half embryos (Flickinger, 1980). However, it does not induce large ectodermal explants (Born, Davids & Tiedemann, 1987; and see Løvtrup & Perris, 1983). To study neuralising activity relatively free of other inducing factors, it is better to use involuted chordamesoderm, and Kaska & Triplett (1980) report that such cells release one or perhaps two related glycoproteins apparently without any RNA. When material probably released from this same tissue is fractionated using phenol, the neuralising activity is found in the protein fraction (John *et al.*, 1983). Despite the problems in interpreting neuralisations due to fractionated tissues, it is interesting to note that such activities in amphibian gastrulae (Janeczek *et al.*, 1984*a,b*) and in heterogenous inductor tissues (Tiedemann, 1968) have been found in ribonucleoprotein fractions where they are attributable to the protein moiety.

It has been noted earlier (p. 237) that neuralisers do not need to enter ectoderm cells to exert their effect, and this fact allows some further study of the nature of neuralisation. Some lectins are known to have strong neuralising activity (Takata, Yamamoto & Ozawa, 1981), and these molecules are known to react with cell surface glycoproteins. The lectin concanavalin A can also neuralise ectoderm when it is prevented from entering the cells, and its activity appears to depend upon the oligosaccharide residues of cell surface glycoproteins (Takata *et al.*, 1984). This raises the possibility that the normal neuralising factor may act as a lectin, reacting with a glycoprotein receptor on the surface of the ectoderm cells. At present, however, some features of the ectodermal response to concanavalin A seem quite different from the response to chordamesoderm (Grunz, 1985), and the mechanism of the former's activity clearly needs further study (see also Gualandris, Duprat & Rougé, 1987).

As shown in earlier sections, fewer artificial treatments vegetalise ectoderm. Moreover the vegetalising effect of the dorsal lip and of most heterogenous inductors is thermolabile (Chuang, 1939). Most tissues with pure vegetalising effects then lose all activity (Chuang, 1963; but see Kawakami *et al.*, 1977), while those normally producing posterior neural structures become purely archencephalic (see Saxén & Toivonen, 1962, pp. 147–149), as expected if posterior inductions are the product of two inducers acting together. When tissues with vegetalising activity are fractionated by differential centrifugation, the microsomal fraction usually shows the greatest activity (see Yamada, 1958; Kawakami, Iyeiri & Sasaki, 1960), but pure ribosomes induce archencephalically (Kawakami *et al.*, 1960) so the vegetalising component seems to be another microsomal constituent. Biochemical fractionations produce somewhat parallel results: activity is usually in nucleoprotein fractions, but when these are further separated to nucleic acids and proteins only the latter are active (see Saxén & Toivonen, 1962). The mesodermalising activity of bone marrow is unusual in being concentrated in a protein-rich, post-microsomal fraction (Yamada, 1958), but a very similar technique was used by Gianni *et al.* (1972) to obtain reticulocyte mRNPs.

Tiedemann's group have achieved the most extensive purification of any vegetalising fraction, starting from older chick embryos. The activity is found in a protein with a molecular weight of about 30,000, and only 100 pg of it are required to vegetalise competent ectoderm (Tiedemann, 1982). Thioglycolic acid or 2-mercaptoethanol destroy the protein's activity which thus probably depends upon disulphide bonds (Tiedemann *et al.*, 1969). Prolonged denaturing treatments split the protein to two peptides which retain at least some activity (Geithe *et al.*, 1981) but this is lost with any further breakdown (Born *et al.*, 1985). The protein must enter ectoderm cells to exert its inductive effect (Tiedemann & Born, 1978; Born *et al.*, 1980). It is basic with an isoelectric point of about 8 (Geithe, Tiedemann & Tiedemann, 1970), and is known to have an affinity for polyanionic molecules. Among these is heparin (Born *et al.*, 1987), which also inhibits mesodermalisation by growth factors in blastulae (p. 194), suggesting that these inductive materials may operate in similar ways. A proteoglycan fraction (Neufang *et al.*, 1978) and DNA (Tiedemann, Born & Tiedemann, 1972) from chick embryos (the source material for the vegetaliser) can combine with the vegetaliser and reduce its inductive activity. Proteoglycans and nuclear RNAs from *Xenopus* gastrulae can apparently have similar effects (Tiedemann, 1978). A possible interpretation of this is that the vegetaliser acts directly on gene expression, and that the proteoglycan regulates the level of active inducer. If ectoderm always contains inducers complexed with an inhibitor then the latter could be a proteoglycan.

It has not as yet proved possible to isolate purely vegetalising fractions from amphibian gastrulae themselves, let alone to further purify the responsible factors. Faulhaber (1972) found spino-caudal inducing activity in *Xenopus* nuclei (but see Bretzel & Tiedemann, 1986) and deuterencephalic activity in the superficial layer of yolk platelets (see too Faulhaber & Lyra, 1974). Wall & Faulhaber (1976) found spino-caudal activity in membranes separated from the microsomes and in small polysomes. The suggestion that mRNPs may be transferred during vegetalising inductions (Wall, 1973) has still not been adequately tested due to technical difficulties in purifying mRNPs (Wall, 1979), and these two sources should be consulted for the relevant evidence.

## Biochemical events in the gastrula

The identification of inductively active molecules is not the only possible approach to a biochemical understanding of induction. It is also possible to study the activities of the inducing or reacting cells and the effects of experimental interference with these, and so to build up a picture of what is going on in the gastrula. Such an approach is aided by the fact that many changes in enzymatic and other activities, perhaps associated with differentiation, do not become appreciable until postgastrula stages (Løvtrup, 1955; Wallace, 1961).

In continuation of trends established in blastulae cell division rates are quite low in gastrulae (Graham & Morgan, 1966), and slowest of all in prospective endoderm (Asao, 1968; Woodland & Gurdon, 1968). Cell cycles are speeded when gastrular explants are cultured with neuralising concentrations of sodium bicarbonate and slowed when cultured with lithium ions (Flickinger, Lauth & Stambrook, 1970c). The latter observation would agree with the vegetalising effect of lithium ions, but another artificial neuralising treatment also seems to prolong cell cycles (Suzuki & Kuwabara, 1974). Events occurring at the normal dorsal lip may have quite complex effects on mitotic cycles, with delays as the cells approach the lip (Desnitskii, 1978) and then a possibly 'quantal' mitosis as they are underlain by the archenteron roof (Maleyvar & Lowery, 1973). Cell divisions are not, however, required for the determinative events in gastrulae, as mitotically blocked embryos can develop from early gastrula to tail bud stages, and even undergo the respecification of cell fates following the implantation of a second organiser (Cooke, 1973a). The possibility of a link between cell cycle rate and competence was, however, suggested by McDonald (1973). Certain genes may be locally amplified during gastrulation (Lohmann, 1972), and the spectrum of chromosomal proteins changes considerably (Holoubek & Tiedemann, 1978), which may have important implications for gene expression.

All classes of nuclear RNA are being synthesised by amphibian gastrulae, as are mitochondrial rRNAs (Chase & Dawid, 1972). The most prominent poly(A)$^+$RNAs are also mitochondrial (Dworkin *et al.*, 1981), although their synthesis rates are unknown. Gastrulation and neurulation are periods of extensive qualitative change in the poly(A)$^+$RNA population, mainly due to the switching on of new genes (Dworkin & Dawid, 1980a,b; Knöchel & Bladauski, 1980, 1981). Most of these gene activations occur in late gastrulae and early neurulae, probably in association with the differentiation of tissues (Dworkin *et al.*, 1984; Dworkin-Rastl, Kelley & Dworkin, 1986), in which case they are probably responses to spatial determination rather than components of the determination system. Such transcripts code muscle myosin (de Bernardi, 1982) and actin (Mohun *et al.*, 1984*a*), tyrosinase for nerve and pigment cells (Benson & Triplett, 1974), neural cell adhesion molecule (Kintner & Melton, 1987) and products required for the differentiation of ciliated and hatching gland cells (Yoshizaki, 1976). Some activated genes contain sequences with high homology to the homoeo box sequences of *Drosophila* homoeotic genes (see p. 181), one of them having been active also in the oocyte (Carrasco *et al.*, 1984; Müller, Carrasco & de Robertis, 1984; Harvey, Tabin & Melton, 1986; Sharpe *et al.*, 1987). Perhaps the most interesting transcripts are those absent or rare before the MBT and accumulating thereafter (see p. 195). They usually become rare again by larval stages and Sargent & Dawid (1983) call them the DG (differentially expressed in gastrula) sequences.

Ribosomal RNA synthesis has been detected considerably earlier in animal halves than in vegetal ones (Woodland & Gurdon, 1968; Flickinger, 1969; and see Misumi, Kurata & Yamana, 1980), and in dorsal halves before ventral ones (Shiokawa & Yamana, 1979). Overall transcription levels are, however, highest per cell in the endoderm (Flickinger *et al.*, 1967, 1970*c*; Woodland & Gurdon, 1968), perhaps because there is a longer interphase time. Once again, a correlation can be made with the effects of lithium ions which increase transcription rates per cell, while bicarbonate ions decrease them (Flickinger *et al.*, 1970*c*). The response to lithium ions is seen even in cells vegetalised by a three-hour treatment at morula stages and then allowed to develop in normal media (Osborn, Wall & Stanisstreet, 1979). It could be argued that these differences in replication and transcription levels are the fundamental differences along the animal–vegetal axis, and that lithium ions (and other treatments affecting determination) act by disturbing these processes. However, metabolic effects very like those described above for lithium ions can be produced simply by rearing amphibian embryos at low ambient temperatures (Krugelis, Nicholas & Vosgian, 1952; Flickinger, Daniel & Greene, 1970*a*), and such embryos are apparently morphologically normal.

Although changes in the relative rates of DNA and RNA synthesis seem unlikely to affect animal–vegetal determination, they could have a role in the dorso-ventral plane where dorsal cells are smaller from second cleavage (see p. 133). Løvtrup, Landström & Løvtrup-Rein (1978) suggest that invaginating cells differentiate first dorsally because the smaller cells are first to exhaust their endogenous nucleotide pool and switch to slower divisions with more RNA synthesis per cell. By supplying extra nucleotides they could delay this differentiation, while inhibiting DNA synthesis allowed it to occur two cell cycles earlier than usual (Landström, Løvtrup-Rein & Løvtrup, 1975; Landström & Løvtrup, 1977*b*). The invaginating cells of course form the gastrular lips, and the timing with which the lips appear is a good predictor of development in the dorso-ventral plane: 'hyperdorsal' embryos form the whole lip as controls form dorsal lips, and uv-radialised ones as controls form ventral lips (Gerhart *et al.*, 1984*b*) while embryos with partial axial deficiencies result from lesser delays (Cooke, 1985; Cooke & Webber, 1985*a*). Perhaps these timing effects will one day be correlated with effects on synthetic rates for DNA and RNA.

Flickinger *et al.* (1970*b*) reported that lithium and bicarbonate treatments also affected the spectrum of reiterated genes being transcribed. Cold-treatment apparently acts in an entirely quantitative manner (Flickinger *et al.*, 1970*a*), so these effects of the ions could be of more significance for determination. It should also be recalled that the vegetalising protein extracted from chick embryos binds to DNA and may act as a gene-regulatory agent, and that some tissue-specific transcripts are apparently formed in late gastrulae. Transcripts for muscle-specific actins seem to be confined to the somites, and perhaps the more ventral mesoderm, of neurulae (Mohun *et al.*, 1984*a*), and those for neural cell adhesion molecule to the neural plate (Kintner & Melton, 1987). Some homoeo box-containing genes are locally transcribed in late gastrulae, around the blastopore lips and probably especially in prospective posterior neural tissue (Carrasco & Malacinski, 1987; Sharpe *et al.*, 1987). Of the DG genes some, encoding cytokeratins, are expressed in the gastrula ectoderm and later epidermis, while others of an unknown nature are expressed in the endoderm (Dawid & Sargent, 1986). These mRNAs are present sufficiently early to have roles in determination and induction, but epidermis does not require induction in the gastrula and the switching off of the cytokeratin gene in prospective neurectoderm is apparently a response to induction (Jamrich, Sargent & Dawid, 1987). The cytokeratins may be involved in the motility and cell shape changes at ectodermal epiboly and epidermal spreading (compare p. 258).

There are in fact several indications that the information for gastrulation, and perhaps for inductive activity, is transcribed at earlier stages, although this is always a difficult thing to prove (see p. 161). Some information is in fact maternally programmed, as anucleate and uncleaved eggs undergo pseudogastrulation (p. 65; see also Kobayakawa & Kubota, 1981). There are still large amounts of maternal poly(A)$^+$RNA in gastrula polysomes (Shiokawa, Misumi & Yamana, 1981*a*). Further transcription in midblastulae appears to be crucial for development through gastrulation (see p. 195). In 1968, Davidson considered this as the genomic readout on which gastrulation depends, although the requirement may be only for an increase in the same kinds of information already maternally supplied, and it is this aspect which Davidson (1986) has more recently stressed.

To determine whether inducing abilities have become independent of transcription by the beginning of gastrulation it is necessary to discriminate between inhibitory effects on the inducing and reacting tissues. It is impossible to confine the effects of drugs affecting transcription to one of two interacting tissues, but researchers using 5-fluorouracil (Toivonen *et al.*, 1961) and actinomycin (Denis, 1964*a*; Tiedemann, Born & Tiedemann, 1967) have all noted stronger developmental effects on competent ectoderm then on inducing dorsal lips. Reyss-Brion (1964) also showed that local X-irradiation of dorsal lips could not prevent them from inducing neural axes, while the ectoderm shows far higher sensitivity. Moreover, two of these treatments act differentially on the induced tissues, suppressing the archencephalon most readily while spino-caudal differentiation is most resistant (Toivonen *et al.*, 1961; Reyss-Brion, 1964). Archencephalic differentiation thus seems most dependent on transcription in the induced tissue, while some information required in more posterior neural structures may have been transcribed before gastrulation. Unless the prospective posterior neural area of the early gastrula is 'prepared' in some specific way, any extra information would have to be supplied by the inducing mesoderm as RNA or protein.

According to Grunz & Hildegard Tiedemann (cited by Tiedemann, 1968) another event showing little dependency on gastrular transcription is the change from posterior to anterior inducing and differentiative properties seen as cells round the dorsal lip. They found only slight effects of actinomycin on this change while cycloheximide delayed it markedly so that the change may involve the translation of pre-existing RNAs. Suzuki (1968) has, however, reported almost totally opposite results: the prevention of 'anteriorisations' by actinomycin but not by puromycin. Ectodermal competence can be prolonged by culture in cycloheximide according to Grunz (1970).

If preformed mRNAs have roles in gastrulae then controls must exist at the levels of processing, transport or translation or in post-translational events. In some cases where gastrulation is blocked or becomes abnormal RNA seems to be formed at quite normal rates but little is released to the cytoplasm: such cases include some interspecific hybrids (Brachet, 1960, pp. 234–5) and embryos vegetalised by lithium treatment as morulae (Osborn *et al.*, 1979). Flickinger (1970) considered that transport and cytoplasmic conservation of mRNAs become more efficient in post-gastrula development and may be important control levels for spatial determination (see also Shepherd & Flickinger, 1979). Kinetic studies did not reveal any mRNA species which are stored in mRNPs before being translated (Dworkin & Dawid, 1980*b*), but this seems to be precisely what happens to the mRNAs for most ribosomal proteins (Pierandrei-Amaldi *et al.*, 1982). Different mRNAs enter polysomes with different efficiencies, suggesting competition for a limiting translational machinery in the gastrula, while in neurulae all species seem efficiently translated (Dworkin & Hershey, 1981). At least most DG genes seem to be already quite efficiently translated in gastrulae (Sargent & Dawid, 1983). The vegetal tissues, which are forming more RNA per cell than other gastrula areas (Bachvarova & Davidson, 1966; Woodland & Gurdon, 1968), have particularly low levels of this new RNA in the microsomes and much of this appears not to be polysomal (Wall, 1979). Post-transcriptional controls may thus be a feature of normal development, and may be operated differentially within the embryo. It is interesting that, as in sea urchin embryos, actinomycin-treatment can actually stimulate protein synthesis (Tiedemann *et al.*, 1967), suggesting that new transcripts modulate translation of preformed ones: a reciprocal effect has also been indicated (Rollins & Flickinger, 1973).

Actual protein synthesis rates are fairly constant through gastrulation and the following stages (see Woodland, 1974) and are lowest in the endoderm (Denis, 1964b). The species of proteins being formed change quite extensively in gastrulae (Brock & Reeves, 1978; Ballantine *et al.*, 1979; Meuler & Malacinski, 1985), and synthesis may begin for collagen (Green *et al.*, 1968) and α-actin (Ballantine *et al.*, 1979) although the latter appears much later in axolotls (Mohun *et al.*, 1980). In dissected embryos some newly formed protein species are found to be restricted to 'ectodermal' or 'endodermal' fragments, the former usually acquiring such characteristic proteins before the latter (Ballantine *et al.*, 1979; Smith, 1986). According to Smith and Knowland (1984) dorsal and ventral halves also acquire unique proteins in gastrulae: their halves in fact differed also in their animal–vegetal constitution, but the specificity of the dorsal markers is supported by their absence from uv-radialised (see p. 61) embryos. A concentration of some antigens in dorsal (induced) ectoderm is also detectable by late gastrula stages (Stanisstreet & Deuchar, 1972).

As with sea urchin blastulae, there are several indications that amphibian determination is strongly influenced by ill-defined conditions of the bulk cytoplasmic proteins. Agents known as animalisers and vegetalisers primarily in sea urchins have closely comparable effects on the proteins in the two groups (Ranzi, 1957, 1962; see p. 160). Their effects on spatial patterns are also comparable, except that only isolated amphibian tissues react clearly to animalisers while effects on whole embryos are difficult to demonstrate (see p. 192). Since animalisation seems associated with protein denaturation it is noteworthy that protease activities are reported to be highest in dorsal-animal cells (d'Amelio & Céas, 1957) and that animal cells show the first signs of yolk solubilisation (Robertson, 1978).

Cell surface properties appear to change in several ways in gastrulae, although this may be of greater significance for the morphogenetic movements than for determination. The changes affect the adhesiveness and electrophoretic mobility of cells (Macmurdo-Harris & Zalik, 1971;

Schaeffer, Schaeffer & Brick, 1973), and the lateral mobility of membrane components (Johnson & Smith, 1976; Gadenne *et al.*, 1984). They may be associated with the formation of microvilli and secretion of mucopolysaccharides at the gastrular lips (Kosher & Searls, 1973; Moran & Mouradian, 1975; but see Ubbels & Hengst, 1978). Surface properties also change as a result of normal (Barbieri, Sánchez & del Pino, 1980; Suzuki, Ueno & Matsusaka, 1986) and artificial inductions (Grunz & Staubach, 1979*b*; Ito *et al.*, 1984), and the latter can have gross effects on the morphogenetic movements (Kocher-Becker, Tiedemann & Tiedemann, 1965). A loss of particular electrophoretic bands from the cells of induced tissues led Ave, Kawakami & Sameshina (1968) to suggest that induction involves selective cell death, (see too Imoh, 1986) but this seems unlikely (see Dasgupta & Kung-Ho, 1971).

By neurula stages, several areas, including prospective notochord, mesoderm, epidermis and neural tissue, are producing characteristic glycoproteins and other glycoconjugates, most if not all of which are cell surface components (Slack, 1984*a*; Jones, 1985; Smith & Watt, 1985; Jacobson & Rutishauser, 1986). In neural areas the product is neural cell adhesion molecule (Jacobson & Rutishauser, 1986) which has also received study in the development of birds and mammals (e.g. see p. 298). Such molecules have been used as markers to confirm many classical conclusions about determination and induction. Thus explants of early gastrula ectoderm form epidermal markers and marginal explants those of notochord and mesoderm (Cleine & Slack, 1985). When the ectoderm is treated with Tiedemann's vegetalising factor mesodermal markers are formed (Cleine & Slack, 1985) and when it is neuralised neural markers appear (Jacobson & Rutishauser, 1986) and the formation of epidermal markers is inhibited (Slack, 1985; Akers, Phillips & Wessells, 1986). Treatments with lithium ions or cyclic AMP, however, failed to change the pattern of epidermal markers (Cleine & Slack, 1985).

Pathways of respiratory metabolism also change during and after gastrulation (Legname, 1968), perhaps as a result of changes in the mitochondria (Løvtrup-Rein & Nelson, 1982). Rates of oxygen consumption decrease strongly from the animal to the vegetal pole and weakly from the dorsal to the ventral side (Barth, 1942; Boell, 1942; see Fig. 7.14). Weeks & Melton (1987*a*) suggest that mitochondria at the animal pole may be activated by acquiring a mitochondrial ATPase encoded by maternal transcripts which are localised there (see p. 11). Respiratory rates are strongly decreased by lithium-induced vegetalisation (Lallier, 1954). In these terms the properties of the dorsal lip are not remarkable, but cytochemical methods show a marked loss of glycogen in cells rounding the lip (Woerdeman, 1933). The glycogen may simply be secreted from the cells at the lip (e.g. see Ubbels & Hengst, 1978), but

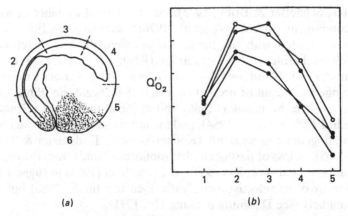

Fig. 7.14. Spatial variations in respiration rate in axolotl gastrulae. (*a*) the positions of the six regions tested. (*b*) the relative rates found for these areas in four different experiments. (From Waddington, 1956, after Boell.)

measurements of respiratory quotient also suggest increased reliance on carbohydrate metabolism in the dorsal lip (Boell, Koch & Needham, 1939). The glycogen loss occurs also, however, in ventral lips, suggesting a role in the provision of energy for involution rather than in primary induction (Jaeger, 1945). Nevertheless, respiratory poisons do block the anteriorisation of developmental properties at the lip (Suzuki, 1968). Lactate inhibits the differentiation of the embryo's cells including that of bottle cells (Landstrom, Løvtrup-Rein & Løvtrup, 1976), but this effect may not be of normal physiological significance (Løvtrup et al., 1978). Wilde (1955) has proposed that another small metabolite, phenylalanine, is a neural crest inducer supplied by the mesoderm, but Løvtrup, Rehnholm & Perris (1984) could not repeat this work.

**What is determined at the end of gastrulation?**

At the beginning of this chapter it was suggested that patterns of embryonic determination might be affected by interactions between the different cell types as they come into contact during gastrulation. Ample evidence of such effects has now been given. Studies of the developmental potential of embryonic isolates show that the spatial complexity of the embryo increases greatly over this period (Fig. 7.7). This final section will consider the nature and extent of determination in the various embryonic areas at the end of gastrulation.

Evidence has been presented in the previous section to show that genes coding tissue-specific products are activated soon after the inductive interactions in gastrulae. It could be argued from this that an inducer has directly selected for activation that battery of genes required for a specific pathway of differentiation. Cells induced solely in this way would never pass through a stage which could be truly described as determined but undifferentiated; and indeed determination might result from following a differentiated pathway to a point where it cannot be reversed. Differentiation precedes determination in this way in mammalian blastomeres (see p. 142), and could do so in some amphibian tissues (see Forman & Slack, 1980).

There are some difficulties in applying this model generally to the state of determination in amphibian late gastrulae. For one thing, it appears to be organ primordia rather than tissue types which become determined (see Fig. 7.7). Many of these primordia will later develop several tissue types, and each tissue will appear in several of these primordia, e.g. neural tissue in forebrain, hindbrain and spinal cord. A system based on spatial rather than histological units recalls compartment-specification in insects (pp. 177–184). If this parallel applies, determination would involve the activation of genes defining a spatial compartment, and only in response to this spatial information would further genes be activated, this time coding tissue-specific products. The *eyeless* locus of the axolotl acts in a domain which is spatially rather than histologically defined: the gene is required only in a small area of the neural plate, but its mutation affects both the eye and hypothalamus (van Deusen, 1973). This could therefore be a selector gene for a spatial unit, although such a small area would probably be determined at a later stage than the gastrula.

Another characteristic of the compartment specification system in insects is that gradients in the level of a few morphogens seem to decide the combination of selector genes activated. Small changes in morphogen levels would mean that a particular area develops the structures of a different but contiguous compartment, and as the change gets greater successively more distant compartments would be specified. When amphibian gastrula cells fail to show their expected determination the kinds of change seen suggest similar relationships among the various embryonic areas. This would not be expected if the fate of each area was specified by the activation of a unique histospecific battery of genes.

Such shifts in determination, suggesting a relatively slight spatial reprogramming, are most obvious among the various parts of the future mesoderm. If a sufficiently large piece of prospective somite is explanted from an early neurula it is sufficiently determined to self-differentiate as muscle (Muchmore, 1957). Smaller mesodermal explants, however, form more ventral structures than would be expected from fate maps. Prospective somite forms pronephric tubules, and prospective pronephros forms blood islands (Yamada, 1940).

Yamada also found that the same explants made in combination with pieces of prospective notochord (a mid-dorsal derivative) are able to develop in accordance with their prospective fate or to produce structures

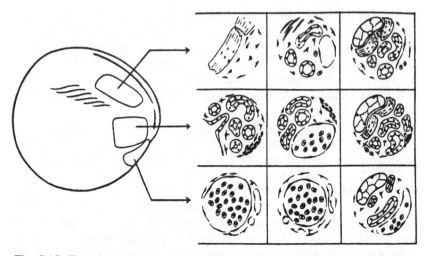

Fig. 7.15. Experiments on determination in the trunk mesoderm of the early neurula in the newt. The positions of the three pieces studied is shown on the left, their prospective fates (left column) being somite, pronephros and blood islands. The effects of isolation and combination with notochord are illustrated (middle and right columns respectively) and described in the text. (From Yamada, 1940.)

of still more dorsal types (Fig. 7.15). Conversely, grafting prospective somite tissue ventrally in another neurula completely ventralises it to form blood islands (Forman & Slack, 1980). Such results suggest that the determination of the mesodermal areas may be controlled by the differing levels of one or a few morphogens. For instance, a morphogen with dorsalising effects could be produced by prospective notochord and exist in a decreasing dorso-ventral gradient. Leakage of this morphogen could explain the ventralisation of small explants. Cooke (1983) has proposed a rather more complex theory in which each mesodermal domain, once determined, inhibits other undetermined areas from following the same developmental pathway.

Similar relationships probably exist among the differently determined areas of the early neural plate. Toivonen's group have used cells determined as archencephalic, spino-caudal or trunk mesodermal, taken from early neurulae (Toivonen & Saxén, 1966) or from artificially induced ectoderm (Saxén, Toivonen & Vainio, 1964). They differentiate as expected even if the cells are first dissociated and allowed to reaggregate. If however the dissociated cells of such areas are mixed, intermediate structures including deuterencephalon can also form. Toivonen & Saxén (1968) were able to produce successively more posterior neural regions simply by increasing the ratio of mesodermal cells in the reaggregate (Fig. 7.16). Like the dorso-ventral sequence of mesodermal areas,

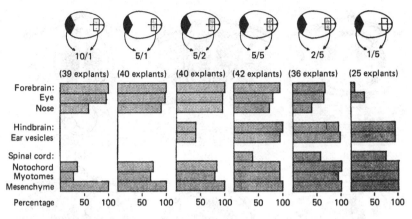

Fig. 7.16. The development of mixed cell reaggregates from presumptive fore-brain (black) and axial mesoderm (stippled) in *Triturus* neurulae. The two cell types were mixed in different ratios, and the distribution (%) of various tissues formed after 14 days culture is shown in the histograms. (From Toivonen & Saxén, 1968. © 1968 AAAS)

the antero–posterior sequence of neural primordia may thus depend on morphogen gradients. It is also possible that each of these primordia has already activated the appropriate combination of selector genes: this does not prevent the cells of insect imaginal discs from changing their compartmental determination following the same extreme challenge of mixing with the cells of another compartment (Haynie & Bryant, 1976).

There are thus indications that the initial specification of mesodermal and neurectodermal areas is still reversible if local conditions change. These local conditions are not restricted to a single germ layer. In some conditions spinal cord is more efficient than notochord in dorsalising mesoderm (Holtzer, Lash & Holtzer, 1956), while uninduced older ectoderm may have rather the opposite effect (Ohara & Hama, 1979*b*; Suzuki *et al.*, 1984). A reciprocal effect of the mesoderm on neural tissue is also known, but it is still difficult to assess its significance for spatial determination. Those parts of the neural tube in contact with notochord or mesenchyme divide and use their yolk reserves only slowly, while parts in contact with somites proliferate and so increase in thickness (Leh-mann, 1926, 1928; Holtfreter & Hamburger, 1955; Takaya, 1955, 1956*a,b*; Takaya & Watanabe, 1961). Different arrangements of the mesodermal tissues could thus explain the different morphologies of archencephalon, deuterencephalon and spinal cord (see Fig. 7.17). Explanted neural plates, lacking these cues, form only simple tubular structures in the region of the hindbrain and anterior spinal cord (Takaya, 1959). The most posterior neural plate area has of course been induced to mesoderm (see p. 225), and this may explain how the tail spinal cord

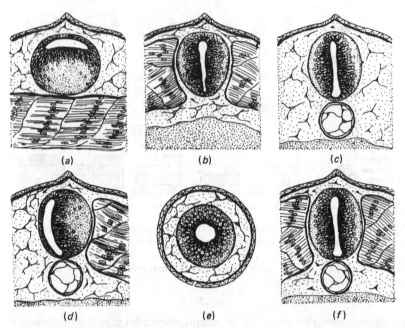

Fig. 7.17. Semi-diagrammatic illustration of the formative effects of somites, notochord and mesenchyme upon the shaping of the neural tube. (From Holt-freter & Hamburger, 1955.)

develops with normal morphology. Clearly chordamesoderm continues to influence the morphogenesis of neural tissues long after gastrulation, but this is not evidence that it can switch determination between the various subdivisions of the central nervous system.

Nieuwkoop (e.g. 1973) does in fact believe that invaginated mesoderm continues to act inductively on neural plate at least in early neurula stages. This effect he calls transformation because it would act to transform neurally activated cells to produce all the neural structures more posterior than forebrain (see p. 230). Because of their work with mixed reaggregates, showing great changes in the spatial specification of early neurula cells (see above) but much lesser ones with midneurula cells, Toivonen & Saxén (1966) also concluded that regionalisation in the neural areas occurs relatively late. Toivonen (1978) has thus accepted the concept of transformation. In the judgment of the present author, however, several difficulties still remain. First, it needs to be established that posterior neural plate explants from the early neurula develop structures more anterior than is their prospective fate. Takaya (1955) thought that they did, but as noted above he later reported that they failed to form any local structures (Takaya, 1959). Toivonen's group, as also noted above, claimed that even dissociated cells retained their regionality unless the different cell types were mixed. A second problem would arise even if it was established that explants form structures which are 'too anterior': the posterior area might still have attained its correct regional specification *in vivo*, but require influences from other local tissues to maintain this, as suggested elsewhere in this section for other primordia. If the state of determination is indeed changing in the neural plate *in vivo* a final problem would remain: is this a distinct third inductive principle (transformation) or a continuation of vegetalising activity? (see p. 233).

In the third germ layer of the gastrula, endoderm, there is also evidence for graded differences in developmental potential along the antero-posterior axis and for mesodermal effects on developmental fates. Holtfreter (1938*a,b*) recognised that endodermal differentiation required the presence of mesoderm but considered this to be a non-specific effect. Okada (1955*a,b*, 1957, 1960) and Takata (1960), however, showed that the kind of mesoderm added affected which endodermal structures developed: with mesenchyme anterior endodermal structures formed, and with lateral plate mesoderm posterior ones. Intrinsic differences within the endoderm also seemed to influence developmental choices and times of determination in their work (and see Albert, 1972), and there may be a further inductive effect from neural crest derivatives (Gipouloux & Girard, 1986).

Other interactions occurring in postgastrula development of course increase the diversity of cell types in the hierarchical way exemplified by eye and lens induction (see p. 217) and suggested by Spemann (1938). The difficulty with many of the interactions described in this section is in deciding whether they too change the developmental potential of cells or only maintain and doubly assure a previously allocated status. In any case interactions continue after gastrulation, just as they began before it. At the end of gastrulation the differences between the areas seem often to be of a graded, quantitative and reversible nature, and it is interesting that cells from these areas also show differences in *in vitro* behaviour which seem programmed by the end of gastrulation and to be of a quantitative nature (Jones & Elsdale, 1963).

## Conclusions

When amphibian embryos begin gastrulation three areas, arranged along the animal–vegetal axis, display distinct developmental properties. To a first approximation these are the prospective areas of the three germ layers and as they are rearranged into these layers during gastrulation interactions occur among them.

One component of these interactions is apparently a continuation of the vegetalisation of the cells of the animal half. The induction of all the future chordamesoderm is completed in this way, and the same factor may act jointly with a neuraliser to form the posterior neural structures. Neuralisation itself is a completely distinct inductive activity exerted by involuted cells on the overlying ones, and causing production of forebrain structures when it acts alone. There is also evidence for reciprocal effects – of the outer cells on the involuted ones – and this may explain the dramatic changes occurring as cells round the dorsal lip in early gastrulae. The effects of rounding the lip decrease later in gastrulation, and by its close cells in both lip layers have been induced to form mesoderm.

The effects of these interactions can often be mimicked by artificial conditions, lithium ions having vegetalising effects here as in sea urchin morulae, while some animalising treatments neuralise ectoderm and dorsalise mesoderm. From the ease with which ectoderm can be neuralised it has been proposed that factors for neural development are already present in the ectoderm complexed with inhibitors, and require only a releaser of low specificity. The same argument has been applied to vegetalisation, but in the final analysis it has to be the external (inducing) factors which select the developmental pathway followed, even if factors such as the age of the reacting tissue also affect the result.

Current evidence indicates that the neuralising factor may be a protein which can diffuse through filters and acts on the surface membrane of ectodermal cells. The vegetalising factor is probably also a protein, but cell fractionation studies and its poor transmission through filters indicate that it is normally present in a large complex, and other studies suggest that it must enter cells to exert its effect.

Induction appears relatively independent of contemporary transcription, while the changing patterns of inductive activity and competence are affected by inhibition of translation or respiratory metabolism. The genes activated in midblastulae (p. 195) may be very important for determination and differentiation in the gastrula, and are certainly being transcribed in spatially specific patterns. At about the end of gastrulation many more genes coding tissue-specific products are activated, programming the differentiation of larval tissues. In general, however, it seems that organ primordia rather than tissues are determined, that the early differences between them are of a graded quantitative nature, and that further interactions among these primordia normally ensure that each develops along the appropriate pathway.

# 8

## Spatial determination in the gastrulae of other groups

Separate germ layers form in the development of all diploblastic and triploblastic animals. In a few cases the layers form simply by tangential divisions of the cells in an earlier single layer (delamination). In all other cases some cells move from an outer to an inner position by some form of active cell migration and/or inward folding of a cell sheet. All such rearrangements offer opportunities for new interactions among the cells.

This chapter is concerned with any such interactions, in non-amphibian holoblastic embryos, which affect the determination of areas within the embryo. Some of these embryos have, however, completed the determination of their major areas at earlier stages, and only refine it in minor ways during gastrulation. They may give us some clues as to which processes are required for the morphogenetic movements and which for determination, but our treatment of them can be brief. Other cases, where there is positive evidence for inductive interactions, are also reviewed here, while interactions in meroblastically cleaving embryos are dealt with in Chapter 9.

### Sea urchins

The prospective areas of the major sea urchin larval organs are arranged along the animal–vegetal axis of the early embryo, and are largely determined at morula and blastula stages (Chapter 5). During gastrulation cells at different animal–vegetal levels show different behaviour patterns, but their interactions seem to have little effect on determination in this axis. Dorso-ventrality is also fixed before gastrulation, but the final extent of each primordium in this plane is probably still labile (see below).

The ingression of cells from the surface of the sea urchin blastula begins in a stage often separated as the mesenchyme blastula. The migrating cells are called the primary mesenchyme and come from a ring around the vegetal pole (Katow & Solursh, 1980). They are probably daughters of the large micromeres in the 32-cell embryo, and leave eight cells, which

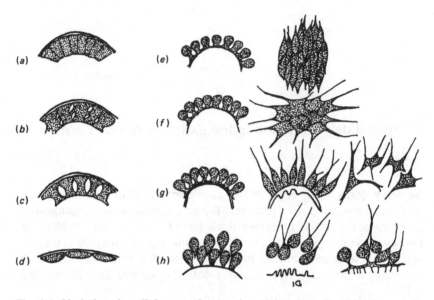

Fig. 8.1. Variations in cellular morphogenetic activity along the animal–vegetal axis of the sea urchin embryo. (*a*)–(*d*) different ectodermal regions with decreasing contact between cells. (*e*) gut and (*f*) coelom regions, where cells show pulsatile and then pseudopodial activities mostly within a sheet. (*g*) secondary mesenchyme where these stages are followed by dissociation of the cells. (*h*) primary mesenchyme where the cells form pseudopodia and separate at an early stage. (From Gustafson & Wolpert, 1967.)

would be the daughters of the small micromeres, at the vegetal pole. Later the cells of a large vegetal area undergo a coordinated invagination led by secondary mesenchyme cells which are derivatives of the macromeres. The tube of invaginated cells, the archenteron, eventually makes contact at its tip with the ectoderm of the oral field on the ventral side of the embryo.

Gustafson & Wolpert (1967) suggested that many features of gastrulation could be traced to changes in a few simple cell properties, such as adhesiveness and pseudopodial activity, occurring at different times and in a different sequence according to the animal–vegetal position of the motile cells (see Fig. 8.1). The pattern of these events was probably determined in the morula and blastula (see Chapter 5), and, in fact, patterns of cell adhesiveness and mesenchyme distribution change when the animal–vegetal balance is disturbed by lithium or zinc ions (Okazaki, Fukushi & Dan, 1962; see too the isozyme studies of O'Melia, 1972). Later work has stressed different aspects of local cell behaviour (Ettensohn, 1984, 1985; Hardin & Cheng, 1986), but their pattern is presumably still controlled by an earlier established spatial organisation. For instance, the filopodia of secondary mesenchyme may act mainly as sensory organelles, but they would then have to recognise patterns beneath the ectodermal layer such as those seen from hatching by Galileo & Morrill (1985).

Cell-binding assays show that mesenchyme cells lose their affinity for hyalin and for other cells, and show increased affinity for basal lamina constituents, at the stage when they would normally exhibit motility (Fink & McClay, 1985). Materials in the basal lamina or blastocoel cavity which could be important here include sulphated mucopolysaccharides (Sugiyama, 1972; Kinoshita & Saiga, 1979), glycoproteins (Schneider, Nguyen & Lennarz, 1978) and collagen (Mizoguchi & Yasumasu, 1982). Collagen synthesis is increased by vegetalising treatments and decreased by zinc ions (Mizoguchi & Yasumasu, 1983a,b): this could explain how zinc ions are able to affect determination even at late stages (see p. 156), but in general such effects may be consequences, not causes, of the changed determination.

There is very little evidence that patterns of cell determination are changed as new cell contacts form in gastrulae. The extent of each primordium along the animal–vegetal axis seems to have been quite irreversibly determined at earlier stages, and there is no longer any dorso-ventral regulation when early gastrulae are halved vertically (see Chapter 5). However, the addition of sugars to sea-water only affects spatial pattern after hatching, causing an enlargement of the oral field (Hörstadius, 1959; see Fig. 8.2). Some determinative events may therefore occur in gastrulae, and Czihak (1962) has proposed that influences from the oral field determine the positions of the ciliary band and the triradiate spicules of the skeleton (Fig. 8.3). Interactions between the ectoderm and skeletal rods are involved in the formation of the larval arms (see Hörstadius, 1973a, pp. 23–24), but arms do not grow out if the rods contact inappropriate parts of the ectoderm (von Ubisch, 1932). The state of determination within the archenteron is reminiscent of that in amphibian endoderm (p. 251): some regulation is possible among its different parts, but differences in developmental potential are already detectable along its length (Jenkinson, 1911a).

Development through and beyond gastrulation requires neither DNA synthesis nor mitosis (Stephens *et al.*, 1986). The mRNA species formed by gastrulae are almost all a subset of the maternal ones which are thought to code for the 'morphogenesis proteins' required to construct an embryo (Galau *et al.*, 1976; Hough-Evans *et al.*, 1977). They are often present at lower levels in gastrulae than at earlier or later stages, presumably because most maternal copies have been lost and embryonic transcription has not yet fully replaced them (Flytzanis *et al.*, 1982). Some preferential decay may also occur (Shepherd & Nemer, 1980).

Expression does begin for some genes not represented in the maternal pool, including one with high homology to *Drosophila* homoeo boxes

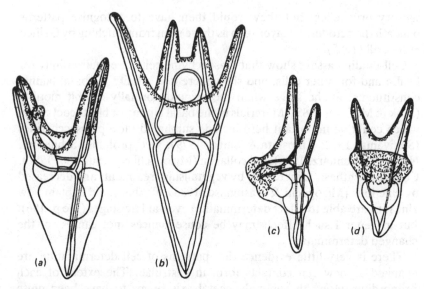

Fig. 8.2. The effects of sugar solutions on sea urchin development. (*a*),(*b*) control plutei. (*c*),(*d*) plutei with enlarged ciliated fields following development with galactose (*c*) or glucose (*d*). Isotonic sugar solutions were diluted 1:3 with sea water. (From Hörstadius, 1959.)

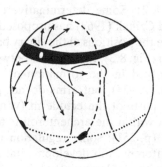

Fig. 8.3. Czihak's concept of an oral field gradient. The field is shown black with a white centre from which determination spreads. The ciliary band develops on its border (dashed line) and the triradiate spicules where this line crosses the circle of primary mesenchyme cells (dotted). (From Hörstadius, 1973*a*, after Czihak.)

(Dolecki *et al.*, 1986). Others are expressed in restricted areas of the embryo (Fig. 8.4). They include mRNAs coding a family of calcium-binding proteins (Carpenter *et al.*, 1984) and two forms of cytoskeletal actin (Shott *et al.*, 1984) all of which seem formed exclusively by dorsal ectodermal cells (Bruskin *et al.*, 1981; Lynn *et al.*, 1983; Angerer & Davidson, 1984; see Fig. 8.4, *a,d*). Of the mRNAs coding other

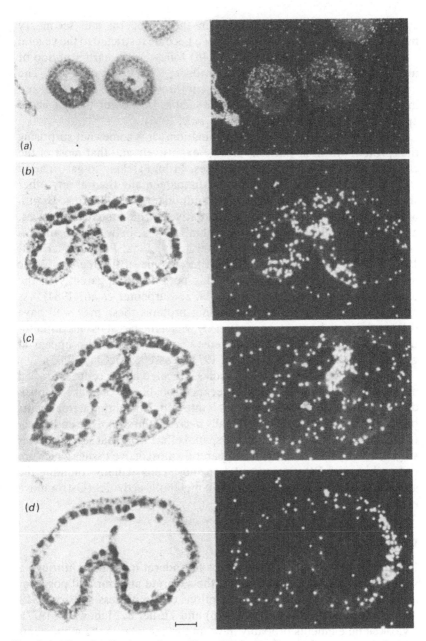

Fig. 8.4. The distribution of specific mRNAs in *Strongylocentrotus* embryos checked by *in situ* hybridisation and viewed under phase contrast (left) or darkfield (right) illumination. (*a*) Spec1 mRNA (coding calcium-binding proteins) in a mesenchyme blastula (× 230). (*b*)–(*d*) mRNAs for the cytoskeletal actins CyIIb (*b*), CyIIa (*c*) and CyIIIa (*d*) in 35 h gastrulae. Scale bars = 10 μm. ((*a*) courtesy of Professor R. C. Angerer; (*b*)–(*d*) from Cox *et al.*, 1986.)

cytoskeletal actins, one is formed in the endoderm and secondary mesenchyme (Fig. 8.4c) and another two become restricted to the vegetal tissues and ventral ectoderm (Fig. 8.4b) following an earlier period of ubiquitous expression (Angerer & Davidson, 1984; Cox *et al.*, 1986). The mRNA coding metallothionein is present throughout the ectoderm, and in zinc-animalised embryos is also inducible in the gut though never apparently in mesenchyme (Angerer *et al.*, 1986).

In view of the data on mRNA populations, it is somewhat surprising that protein synthesis patterns change extensively and that most of the changes are new or increased syntheses. From hatching to gastrulation these patterns are transformed more fundamentally than at any other stage from the egg to the pluteus (Brandhorst, 1976; Bédard & Brandhorst, 1983). There are extensive quantitative and qualitative differences, affecting particularly the major protein species, between the mesenchymal and other cells (Harkey & Whiteley, 1982; Pittman & Ernst, 1984). The translation products of at least some of the region-specific mRNAs (see above) appear over this period and apparently in the expected locations (Bruskin *et al.*, 1982; Carpenter *et al.*, 1984). As cytoskeletal actins and calcium-binding proteins these may well have roles in the morphogenetic processes by which the larval tissues differentiate. Cell-surface antigens restricted to specific tissues also appear at gastrulation (McClay & Chambers, 1978; Wessel & McClay, 1985).

Child's (1936a) dye reduction studies indicated a new centre of raised metabolic activity at the vegetal pole, and probably also a weaker ventrodorsal gradient. In this connection it is interesting that lithium treatments begun just before gastrulation actually *decrease* the size of the endoderm (Child, 1936b). This is the opposite spatial effect from that seen at earlier stages, but in both cases it is the apparently most active tissues which are reduced in size. Later in gastrulation those cells actually rounding the gastrular lips seem to show the highest metabolic activities (Ostroumova *et al.*, 1977).

### Insects

In the gastrulation of insect embryos mesoderm invaginates through a ventral furrow and endoderm from the separate anterior and posterior areas marked on Figure 6.2. Descriptions of the process in *Drosophila* have been provided by Rickoll (1976) and Turner & Mahowald (1977). Cytoskeletal elements probably have important roles in the movements with one effect being the production of wedge-shaped cells which may assist the folding of the cell sheet. The cells still retain open connections with the yolk syncytium, which may play an active role in morphogenesis (see Rickoll, 1976). Gastrulation can at least begin where local cellularisation has failed to occur, and this has led Swanson & Poodry (1981) to speculate that determinants for this were localised in the egg (compare amphibian pseudogastrulation: p. 65). During and after gastrulation the germ band extends, folding around the posterior pole of the egg (see Turner & Mahowald, 1977; Rickoll & Counce, 1980). In many other insect species it undergoes a far more complex series of movements, which are particularly marked in short-germ species.

In many insects determination along at least a part of the antero-posterior axis is complete to the level of segments or subsegmental compartments in cellular blastoderms, but in others the ligation of an egg with a distinct germ anlage still results in a gap several segments long (see Table 6.1, p. 171). The new spatial relationships established during germ-band extension may supply extra information for determination, particularly of any segments which arise late from a posterior proliferation centre, which means most of the embryo in short-germ species.

A feature of insect development which has received considerable attention is the possession of a differentiation centre, which is usually in a thoracic segment. Processes such as gastrulation and segmentation begin in this area and seem to spread to other parts of the embryo (Seidel, 1934, 1961; Haget, 1953). Egg regions separated from the differentiation centre by even a loose ligature fail to develop any embryonic structures (Seidel, 1934; see Fig. 8.5). This treatment is unlikely to prevent the movement of a determinant, and may instead inhibit the transmission of physical stimuli, such as the contraction of the syncytial yolk system. The differentiation centre can thus be visualised as a 'centre of morphodynamic movements' (Bodenstein, 1953) rather than a determination centre. Suggestions that it also plays an instructive role (Schnetter, 1934; Haget, 1953; Seidel, 1961) have been opposed by Sander (1976).

Gastrulation appears to have more important effects on determination patterns in the dorso-ventral plane. Indeed, Naidet *et al.* (1987) have recently reported that, when gastrulation is prevented, the whole surface secretes cuticle of a dorsal type. Following the sagittal halving of a *Tachycines* gastrula no regulation is seen in the mesoderm and the missing half is replaced from neighbouring extraembryonic material (Krause, 1952: see Fig. 8.6). In this situation at least it therefore seems that inductive influences can spread tangentially through the tissue layer from the mesoderm, a situation reminiscent of that in the outer layer of the amphibian dorsal lip (p. 224). The results of local cauterisations suggest, however, that mesoderm is usually the recipient of inductive signals from the ectoderm (Seidel, Bock & Krause, 1940; Haget, 1953). In such cases ectoderm differentiates fairly normally in the absence of mesoderm (although morphogenesis is somewhat disturbed), but mesoderm fails to differentiate in the absence of ectoderm (see Fig. 8.7)

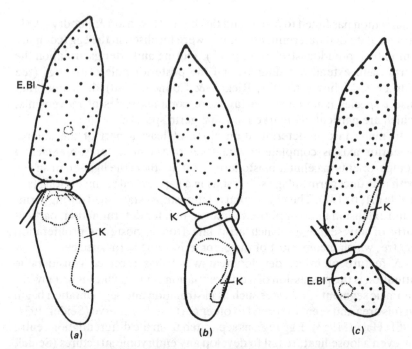

Fig. 8.5. Some of Seidel's evidence for a differentiation centre in *Platycnemis*. The egg was loosely constricted at the blastoderm stage (*a*) anteriorly to, (*b*) in or (*c*) posteriorly to the proposed centre. An embryonic anlage (K) formed only in those parts containing a part of the centre; elsewhere only extraembryonic blastoderm cells (E. Bl) were seen. (From Seidel, 1934.)

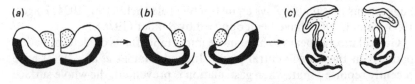

Fig. 8.6. Development of the cricket *Tachycines* following sagittal halving of the gastrula: (*a*) the separated halves in transverse section: mesoderm dotted, ectoderm white and amnion black; (*b*) following healing new embryonic tissues are induced from the amnion (arrows); (*c*) two bilaterally symmetrical germ bands result. (From Sander, 1976, after Krause.)

and its regional specification may be affected by factors from the ectoderm. Work on cell commitment within the various tissues of *Drosophila* gastrulae has been reviewed by Technau (1987).

Dorso-ventral determination within the prospective ectodermal areas of *Drosophila* has received particular study. The most ventral cells seem already to be determined as neural tissue in young gastrulae, and commitment seems to spread to more dorsal (epidermal) areas (Technau

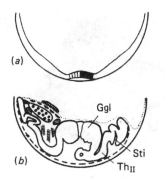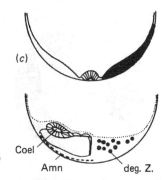

Fig. 8.7. The relationship between ectodermal and mesodermal development in *Chrysopa perla* (Neuroptera). When one lateral half of the mesoderm is removed from the early gastrula (*a*) the ectoderm of that side still undergoes morphogenesis and differentiation (*b*). When ectoderm is removed (*c*) the mesoderm of that side fails to differentiate (*d*). Amn, amnion; Coel, coelom; deg. Z., degenerated cells; Ggl, ganglion; Sti, spiracle; th$_{II}$, second thoracic segment. (From Seidel *et al.*, 1940.)

& Campos-Ortega, 1986). Some genes which appear to be concerned in neural and sensory determination are transcribed in the neurogenic region during or even before gastrulation (Cabrera, Martinez-Arias & Bate, 1987; Romani, Campuzano & Modolell, 1987). Other loci may be concerned in the choice between neural and epidermal development as mutants show a hypertrophy of the nervous system with the sensitive phase extending from blastoderm to segmentation stages (Lehmann *et al.*, 1983). The patterns in which these latter genes are expressed probably depend on spatial information produced by maternal *dorsal* and *Toll* (see p. 23) activity (Campos-Ortega, 1983).

Some genes are newly activated in *Drosophila* gastrulae (Zusman & Wieschaus, 1985) and there are great changes in protein synthesis patterns (Trumbly & Jarry, 1983; Summers *et al.*, 1986; see too Wilcox & Leptin, 1985). Raised alkaline phosphatase activity has been associated with the differentiation centre (Yao, 1950).

It is worth noting here that in spider development a region known as the cumulus seems to have an important inducing role from gastrular stages (Holm, 1952). It arises from the endodermal area and seems to determine the dorsal side (see too Ehn, 1963*b*).

### Spiralian embryos

In these groups gastrulation starts when cell number is still relatively low, but involves cell behaviour patterns already seen in other embryos (Verdonk & van den Biggelaar, 1983; see too Hess, 1971). The induction of the shell gland in some molluscs is the only case where the new

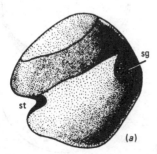 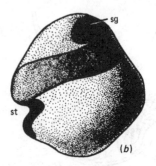

Fig. 8.8. Development of the shell gland in lithium-treated *Lymnaea* embryos. These embryos have a reduced pretrochal area (top) and the shell gland (sg) may form post-trochally (*a*) or pretrochally (*b*). In both cases it forms in the ectoderm with which the endoderm makes contact. st, stomodaeum. (From Raven, 1952.)

positional relationships established at gastrulation clearly affect determination patterns. Following lithium treatment of the pulmonate *Lymnaea* no shell gland developed in complete exogastrulae, but where a partial endodermal invagination occurred the gland developed in that ectoderm which was underlain by endoderm (Raven, 1952). Normally it would develop from post-trochal ectoderm, but if contact was established with the prospective head ectoderm of the pretrochal region the gland appeared there (Fig. 8.8). Hess (1956*a,b*) confirmed a similar inductive relationship in the prosobranch *Bithynia*, where all of the endoderm seems to have inductive ability. Other inductive interactions in spiralians seem to occur mainly at the 24-cell stage (see Chapters 3 and 6), but may continue into gastrulation which is only a few cleavage cycles later. In *Eisenia* endoderm fails to differentiate in the absence of mesoderm (Devriès, 1970).

Although lobeless *Ilyanassa* embryos fail to develop most organs they still show qualitatively normal protein synthesis patterns, even during gastrulation and organogenesis (Collier, 1983*a*, 1984). Some proteins are however produced at higher or lower rates than normal, and this seems to apply mainly to the major protein species. Collier (1983*a*) compares this with the sea urchin embryo where the cell-type-specific proteins are among the most prominent species (p. 258). Emanuelsson & Heby (1978) suggest that polyamine metabolism is important for gastrulation in the polychaete *Ophryotrocha*, and indeed in other embryos.

### Ascidians and Amphioxus

Gastrulation occurs early in these groups too (descriptions: Conklin, 1905*a*, 1932), and involves many cell behaviour patterns familiar from other embryos (Satoh, 1978). In both groups the strongest evidence for inductive interactions concerns the determination of the nervous system (although see von Ubisch, 1963). In ascidians, neural development requires mutual interactions between the anterior-animal (a4.2) and anterior-vegetal (A4.1) lineages (p. 100), but Ortolani (1959) has shown that the a4.2 lineage is still unable to form a brain if the derivatives of the A4.1 lineage are destroyed in the last cell cycle before gastrulation. It seems therefore that neural induction occurs during gastrulation as it does in amphibians. The inductor in this case would be chordaendoderm (not chordamesoderm as in amphibians), and any single cell from these parts of the A4.1 lineage seems able to induce neural structures providing good contacts are established with the reacting cells (see Reverberi, Ortolani & Farinella-Ferruzza, 1960). Another difference from the amphibian is that not all the prospective ectoderm is equally competent for neural induction: the competence of the posterior–animal (b4.2) lineage is low (Rose, 1939) or absent (Reverberi & Minganti, 1947c). Neural differentiation by the a4.2 lineage seems thus to require both intrinsic and extrinsic factors, and the latter may be of low specificity since they can be mimicked by trypsin treatment (Ortolani *et al.*, 1979). This recalls amphibian autoneuralisations and neural evocations, and suggests an initial reaction at the level of the cell membrane.

In amphioxus Tung *et al.* (1962b) used grafts to the blastocoel of a host to show that pieces of the blastoporal lips self-differentiating as notochord can induce neural tissue and promote the differentiation of somites – both effects familiar from amphibians. Pieces self-differentiating as endoderm appeared to have no inductive activity despite the activity of the $veg_2$ layer in morulae (see p. 198). Since eight-cell embryos could develop normally after the animal blastomeres had been rotated through 90 or 180° on the vegetal ones (Tung, Wu & Tung, 1960b), all ectoderm appears competent for neural induction, again as in amphibians but apparently unlike ascidians.

## Some lower vertebrates

Since there are so many similarities between the neural induction systems of amphioxus and amphibians, it is interesting to look briefly at the embryos of the lower vertebrates which occupy phylogenetically intermediate positions. Of course these groups have continued to evolve since their original separation, and in the case of the teleost and cartilaginous fishes the modifications have been so profound that they will be dealt with in a separate chapter. The evidence available for an agnathan, the lamprey, and for non-teleost bony fishes does, however, suggest the operation of common basic mechanisms in early chordate and vertebrate development.

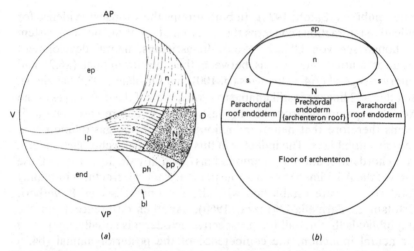

Fig. 8.9. Fate maps: (*a*) for the lamprey, *Petromyzon*, early gastrula viewed from the left, and (*b*) for the sturgeon, *Acipenser*, viewed dorsally before involution starts at the equator. bl, blastoporal groove; end., endoderm; ep, epidermis; lp, lateral plate; n, neural plate; N, notochord; ph, pharynx endoderm; pp, prechordal plate; s, somites. In (*b*) few presumptive somite or neural cells are in the most superficial layer. ((*a*) from Nieuwkoop & Sutasurya, 1979, after Weissenberg and Dalcq; (*b*) after Ballard & Ginsburg, 1980.)

A similarity with amphibians is suggested by the fate maps of lamprey (Weissenberg, 1934) and sturgeon (Ballard & Ginsburg, 1980) early gastrulae (Fig. 8.9), the former recalling urodele and the latter anuran fate maps. Many aspects of the gastrular movements in sturgeons again recall those in anurans (Ballard & Ginsburg, 1980). The inductive ability of the dorsal lip has been confirmed experimentally in these lower vertebrate groups. In lampreys, the implantation of a lip to the blastocoel results in the formation of a secondary embryo in which the graft forms chorda, somites and pharyngeal endoderm and the nervous system is induced from the host (Bytinski-Salz, 1937; Yamada, 1938). The comparison between the sturgeon and amphibian systems has been taken to quite a detailed level by Ignatieva (1960, 1961). At the beginning of gastrulation cells outside the future lip induce posterior neural structures, and prospective notochord has usually not acquired the ability to self-differentiate. The first cells to round the dorsal lip acquire head-inducing abilities as they do so, while the cells which involute later retain their posterior properties. Studies of the holostean fish *Amia* (Ballard, 1986) suggest that the transition to the very different pattern in teleosts began with the evolution of this group.

A few other data are worthy of a brief mention here. Lithium treatment of lamprey embryos causes such familiar effects as exogastrulation and microcephalia (Ranzi & Janeselli, 1941; Ranzi, 1957). There

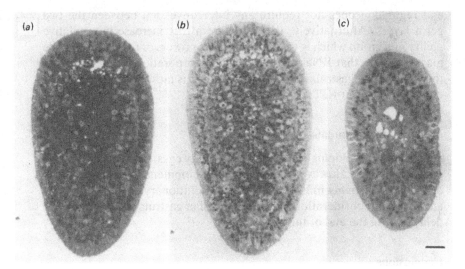

Fig. 8.10. The development of half embryos separated at midgastrula in *Phialidium*. Planula larvae formed by a lateral half (*a*) a posterior half (*b*) and an anterior half (*c*). Note the absence of endoderm and basement membrane in (*c*). Bar length = 20 μm. (From Freeman, 1981*b*.)

has been considerable study of the transition to cleavage asynchrony which occurs in sturgeon blastulae (see Dettlaff, 1964; Chulitskaia, 1970): the phenomenon seems similar to the MBT of amphibians (p. 189) and to analogous phenomena in teleosts (p. 281). If this comparison is valid the transcription required for gastrulation may begin at or near this stage.

## Hydrozoans

In all of the groups so far considered, at least some of the major embryonic primordia had been determined before gastrulation: but this is apparently not so for the hydrozoans. Gastrulation mechanisms are unusually varied in this group, but most often involve some form of ingression or the direct delamination of an inner layer (see Mergner, 1971). In *Phialidium*, where gastrulation is by unipolar ingression with endoderm originating at the posterior pole, Freeman (1981*b*) found that anterior and posterior halves of the early gastrula could still regulate to form normal larvae, but by the midgastrula there was some evidence of determination. Anterior halves then formed little if any endoderm and posterior halves developed with a relatively thin ectodermal layer (Fig. 8.10). By postgastrula stages however regulative ability has returned to such transversely separated halves (Freeman, 1981*b*). Perhaps by this time gastrulation has brought endodermal cells into the anterior halves and regulation does not require any interconversion between the two germ layers. Alternatively, there may be a true increase in regulative ability in a group which is well known for its powers of regeneration. We may note here that RNase-labile uridine incorporation is detectable for the first time in gastrulae of *Hydractinia*, and is most pronounced in the endoderm (Edwards, 1975).

### A note on the mammals

As embryos developing from small, non-yolky eggs, mammals should be considered here. However, mammalian development beyond the blastocyst stage is strongly influenced by their evolutionary origin from lower amniotes. A consideration of the mammalian gastrula will therefore be delayed until the end of the next chapter.

### Conclusions

A few inductive events can be clearly associated with the gastrulation movements of the embryos considered in this chapter. These include the induction of the shell gland by endoderm in some molluscs and induction of neural tissue by chordamesoderm or chordaendoderm in several protochordate and lower chordate groups. In sea urchin and many insect embryos animal–vegetal (or antero-posterior) determination is already quite precise before gastrulation but it seems to be refined in the ventrodorsal plane. This is probably most important in the insects where mesodermal invagination occurs ventrally and mesodermal, neural and epidermal prospective areas are (in broad terms) segregated ventrodorsally. The first determinative event in the development of hydrozoans may also be associated with gastrulation.

It is often difficult to separate events important for determination from those involved in morphogenetic movements and/or overt differentiation. The differentiation centre of insect embryos now appears to take a leading role in morphogenesis and differentiation without issuing determinative instructions to other regions, while both roles may be taken to a relatively minor extent by the sea urchin oral field and more spectacularly by the chordate dorsal lip. Many features common to a variety of animal gastrulae seem to be associated with morphogenesis (changing cell surface affinities, mucopolysaccharide secretion) or differentiation (tissue-specific RNA and protein syntheses) rather than with determination.

# 9

## Determination in embryos showing partial cleavage

Many animals lay large, very yolky eggs which develop directly to a small adult: presumably they can omit larval feeding stages because of their large food store. Often such eggs cleave partially, or 'meroblastically', to form the embryo in a restricted area, and then the uncleaved yolk is gradually absorbed into this embryo. Such adaptations have evolved independently in a large number of animal groups, but in this chapter we shall consider only those groups where it is the normal mode of development. In at least these cases it is the area around the animal pole which forms the embryo and the more vegetal parts which are nutritive (indeed this is the origin of the terms 'animal' and 'vegetal': see p. 10). Earlier chapters have shown for other groups that spatial differences are most pronounced along the egg's animal–vegetal axis and are very important for embryonic determination. How then does determination operate in cases where a large part of the animal–vegetal axis has an extraembryonic fate?

### Cephalopods

The cephalopods are a molluscan group but show no signs of the spiral cleavage patterns so characteristic of other groups in this phylum. As noted in the first two chapters of this book the axes of bilateral symmetry are often already visible in the egg (p. 10) and fertilisation is followed by cytoplasmic streaming to form a blastodisc at the animal pole (p. 32). The account of development given here is based primarily on *Loligo pealii* which has been described by Watasé (1891) and Arnold (1965a, 1971; Arnold & Williams-Arnold, 1976). External views of its development are given as Figure 9.1. The first two cleavages cut vertically into the blastodisc with the positions of both furrows being slightly displaced from its apex. First cleavage separates the future left and right sides, second cleavage dorsal and ventral areas, and the bilateral organisation is particularly clear after third cleavage. Subsequent cleavages are asynchronous and unequal. From about the 16-'cell' stage, horizontal furrows also cut between the nucleated area and the yolk beneath, forming the

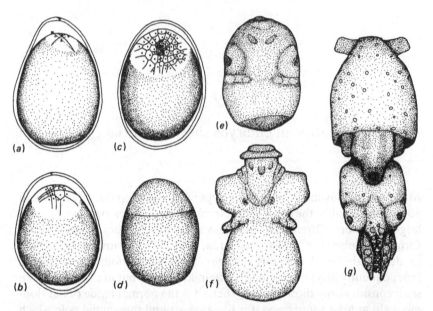

Fig. 9.1. Stages in development of the squid *Loligo pealii*. (*a*)–(*c*) show second, fourth and sixth cleavages. In (*d*) the blastoderm covers about one-third of the egg, and in (*e*) it covers almost the whole egg and is thickened to form the major organ primordia. (*f*) shows a further stage in organogenesis, and (*g*) the form at hatching. (*e*)–(*g*) are ventral views and all are about × 28. (From Arnold, 1965*a*.)

first true cells in the central area. Further divisions produce a nearly circular area of cells, the blastoderm, which is at first one cell thick. At its periphery there remain incomplete 'cells' which thus retain syncytial connections with each other and with the yolk. These are more clearly seen in *Sepia* (see Fig. 9.2) where Vialleton (1888) called them blastocones. They continue to add to the expanding blastoderm by divisions from their upper margins.

The processes described above produce a single-layered blastoderm of 60 to 100 cells in *Loligo pealii* (stage 9 of Arnold, 1965*a*). Thereafter, separate germ layers are formed. An inner layer of cells is first seen at the periphery of the blastoderm and gradually comes to underlie the more central areas too. Beneath both cellular layers, nuclei also spread from peripheral regions towards the animal pole within the thin cytoplasmic layer which still surrounds the yolk, so forming a syncytial yolk epithelium. Arnold (1971) has suggested that the nuclei of this syncytial layer originate from divisions of the blastocone nuclei. The origin of the inner cellular layer is still the subject of controversy, although the application of modern tracer and other techniques might well settle the issue. Its appearance at, and extension from, the peripheral regions led to early suggestions of an invagination or involution of cells there. An

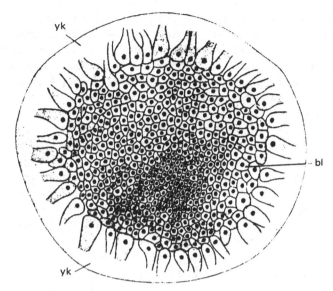

Fig. 9.2. A late cleavage stage in *Sepia* showing the incompletely cleaved blasto-cones at the margin of the blastoderm (bl). yk, yolk. (After Vialleton, 1888.)

alternative of delamination from the original single cell layer was sup-ported by Arnold in 1971. Later, Arnold & Williams-Arnold (1976) cited unpublished work by Singley suggesting that cells three rows from the blastodermal margin first expand with a fibroblast-like motility over the more peripheral cells, and that the latter then increase their mitotic rates and generate the inner cell layer. They also claimed that further expan-sion of the blastoderm is due to continued epiboly by the cells, leaving the original egg surface almost undisturbed beneath them. This conflicts with Arnold's own earlier data (1961) showing that the egg surface itself is cellularised as the blastoderm expands. Since Arnold believes that the cortex bears a map of determinants (see below) it is important that this issue be resolved.

The blastoderm expands as a three-layered structure (of two cellular and one syncytial layers) throughout the prospective embryonic area. This is the animal half of the egg in *Loligo*, but can be a much larger or smaller proportion in other species. As expansion continues over the more vegetal area it involves only a single cell layer and the yolk epithelium: this region will form the external yolk sac, an extraembryonic structure.

Before the blastoderm reaches the vegetal pole, organogenesis has begun in the animal half as a thickening and folding of the blastoderm. The primordium of the mantle with a central shell gland forms first at the animal pole, then those of the eyes laterally, funnel folds ventrally, arms in an equatorial ring, etc. (see Fig. 9.1). The external yolk sac becomes increasingly constricted from the embryonic area, and undergoes contractions which apparently circulate the products of yolk digestion: the yolk epithelium of both the embryonic and extraembryonic areas is responsible for this digestion. Mantle folds extend from the animal pole to enclose many of the organs in the mantle cavity. By hatching the animal is recognisable as a squid.

The fates of the early layers, like their origin, are very uncertain and warrant re-examination by modern techniques. The most favoured interpretation is that the outer layer is entirely ectodermal and the syncytial epithelium entirely extraembryonic. The inner cellular layer must then be mesendodermal, and it seems that even some of the nervous system originates there (Arnold & Williams-Arnold, 1976).

Arnold (1965*b*) developed techniques for culturing dechorionated embryos and tissue explanted from them. At stages 16 and 17 (when organ primordia are just becoming detectable) he found that the explanted cells of a future primordium could only differentiate if accompanied by the underlying yolk epithelium. He considered that the role of the yolk epithelium might be inductive rather than nutritive, and confirmed that differentiation occurred when denuded yolk epithelium became covered again by cell migration or by reaggregated cells taken from another area, but failed if the yolk epithelium was damaged. This suggests a special kind of mosaic determination in which the developmental map is restricted to the yolk epithelium, while the cells themselves are undetermined until they receive inductive signals from the map at or after stage 16. In 1968 Arnold used three further techniques to test this hypothesis. The strongest supporting evidence came from studies with a uv microbeam applied tangentially to the prospective areas of specific organs at early cleavage (long before nuclei enter these areas). The irradiated areas cellularised normally but then often failed to produce the appropriate organ (Fig. 9.3), although it must be said that in many other cases local organ formation did still occur. The second technique was the separation of an area of the surface at stage 10 using a ligature and it resulted in a failure to develop organs appropriate to this area. The significance of this is not clear, however, as the ligature usually removed cells as well as yolk epithelium and where it did not only the size of the external yolk sac, and in one case the arms, were affected. Thirdly, Arnold (1968) centrifuged the egg, displacing the cytoplasm; when organs formed their sizes depended on the amount of cytoplasm available, but their positions were unchanged.

Boletzky (1970) has since extended the ligation evidence for determinants to the mid- and hind-gut, but it is not clear whether cells were included in the ligated areas. Arnold believes that all such data strengthen the case for a surface-associated mosaic of determinants which

(a)　　　　　　(b)

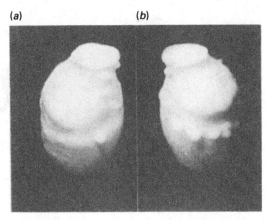

Fig. 9.3. Two views of a squid embryo irradiated on the right side in the arm region during early cleavage. Only the second arm has formed on the side shown in (a). × 25. (From Arnold, 1968.)

are uv labile but not displaceable by centrifugation. He suggests that they were formed and localised during oogenesis by interactions with the follicle cells. Since 1968 Arnold has described the location of these determinants as cortical, although the subsurface layers usually considered as cortical must enter the cellular layer at least around the animal pole and according to some theories over the rest of the surface too. The idea of a cortical map of course recalls Raven's concept of determination in a very different mollusc, *Lymnaea*, where subcortical patches reflect the pattern of follicle cells and probably provide spatial information used in embryonic development (see pp. 16 and 91). However, it was concluded in Chapter 3 that spatial information in *Lymnaea* is used not as a static point-by-point map but to establish the axes and polarity of the embryo. In cephalopods too there are other data which oppose the idea of a detailed map.

Okada (1927) first showed that cephalopod organs can develop in abnormal positions when he ligated squid eggs transversely and obtained dwarf embryos from the animal fragments, with all primordia, including the external yolk sac, in approximately normal proportions (see too Marthy, 1975). As with many other embryos, the positions of organs can also be changed in opposite ways by rearing in the presence of lithium (Ranzi, 1928, 1957) or thiocyanate (Ranzi, 1944) ions (Fig. 9.4). Because of the reversed orientation of the cephalopod embryo compared with vertebrates, the lithium-induced reduction of organs begins 'behind' the eyes in the mantle region, but this is still a vegetalisation and can result in cyclopia or cyclocephalic formations recalling work with *Lymnaea* (p. 203). All such evidence suggests the operation of gradient-distributed factors rather than a fixed map.

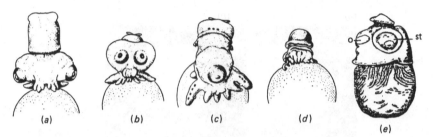

Fig. 9.4. Ionic effects on developmental patterns in the squid: (*a*) control embryo. (*b*)–(*d*) effects of Li$^+$: (*b*) convergent eyes; (*c*) cyclopy; (*d*) anophthalmy. (*e*) effects of SCN$^-$: the stomodaeum (st) is greatly enlarged. o, eye. ((*a*)–(*d*) from Ranzi, 1957; (*e*) from Ranzi, 1944.)

Since 1970 Marthy has assembled evidence that the early blastoderm already contains sufficient developmental information to differentiate, and he suggests that the role of the yolk epithelium is essentially nutritive. In 1972 he reported the production of complete dwarf embryos from isolated blastoderms: apparently the yolk epithelium and some yolk remained adherent to them but they had not yet spread over some of the area which Arnold claims contains organ determinants. In direct contradiction of Arnold (1965*b*), Marthy (1970) claimed that cells transplanted onto the yolk epithelium of another region differentiated according to their original fate not their new position, but this differentiation does seem to be far from complete (Marthy, 1973). Marthy (1970, 1973) did find that an eye developed when the prospective region was denuded of cells, but suggested (with some evidence) that the potential to produce an eye may be distributed in a field around the normal site and so be present in the cells migrating from around the denuded area. Eyes also formed successfully following destruction of the yolk epithelium beneath the rudiment (Marthy, 1973). Most recently, Marthy (1978) has begun to assemble evidence for an alternative mechanism of organ determination, involving induction of the outer cell layer by the inner cells. A few inner cells were removed when this layer was still only a ring, and this resulted in the deletion of a complex of organs at the position which these inner cells should have come to occupy. The outer layer thus seems completely undetermined but even the inner layer may not have a complex point-by-point map of determinants at this stage.

It is possible that elements of the theories of Arnold and of Marthy are both applicable to the determination system used by cephalopods. For instance, determinants acquired during oogenesis could first occupy a ring outside the immediate animal pole area and then be segregated to the cells which will form the inner blastodermal layer.

The biochemical nature of these proposed determinants remains unknown, although their apparent sensitivity to uv irradiation has led

Arnold since 1968 to suggest that they may be masked maternal messenger RNAs. The inhibition of cytoplasmic streaming in the fertilised egg (using cytochalasin B) has gross effects on later development, which suggests that streaming helps to localise or fix the pattern of determinants in some unknown way (Arnold & Williams-Arnold, 1974). At later stages the same drug applied locally causes organ deletion, supporting the concept of surface-associated determinants (Arnold & Williams-Arnold, 1976). The biochemical nature of the inductive process has not been studied, but there are some data relevant to the problem of intercellular communication. Arnold reported in 1974 that open cytoplasmic bridges are retained between the cells of the blastoderm, and seemed to include bridges between the two cell layers. In the further study of Cartwright & Arnold (1980), however, no bridges were seen between the layers, but they were shown to link large areas of the external layer. It seemed possible that the boundaries of these areas (at stage 16) were coincident with those of the organ primordia, so that their significance could be in allowing a uniform response to inductive signals. In considering the transfer of signals between layers it may be more relevant to note that all embryonic tissues tested by Potter, Furshpan & Lennox (1966) were electrically coupled to the yolk, even across many cell boundaries, and that a dye of about 1000 molecular weight could sometimes pass between cells. The implication that there is a well developed system of gap junctional communication has now been confirmed by Ginzberg *et al.* (1985).

## Teleosts

Meroblastic cleavage is characteristic of both the teleosts and the cartilaginous fish. This section deals almost exclusively with the former, which have been studied quite extensively, but a few relevant data for the cartilaginous fish are collected at the end.

The eggs of teleosts vary considerably in size, some being smaller than amphibian eggs. The yolk may be in platelets or in a structure like a single vacuole surrounded by a yolk membrane of high resistivity (Bennett & Trinkaus, 1970). Around it is a cytoplasmic layer which is thickest at the animal pole particularly following the ooplasmic flows in the newly fertilised egg (pp. 32–3). The early cleavages are restricted to this area and form a blastoderm. The first furrows cut vertically to form incomplete cells but are followed by tangential cleavages which separate the cells from the yolk which retains a thin cytoplasmic covering. A detailed study of the cleavage stages in the zebrafish (see Fig. 9.5) has been provided by Kimmel & Law (1985*a,b,c*). There are 11 or 12 synchronous cleavages in this species, but embryonic symmetry is not consistently related with the early cleavage planes and later mixing of the cells means that fates are

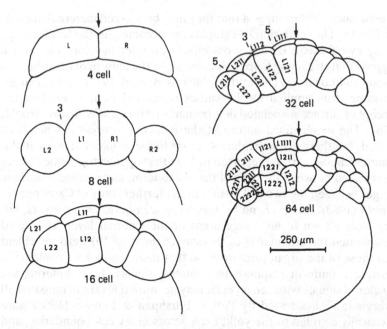

Fig. 9.5. Cleavage stages in the zebrafish embryo. Second cleavage is in the plane of the page and only the cells in the near half of the embryo are shown. By the 64-cell stage some cells are not visible at the surface. (From Kimmel & Law, 1985*a*.)

completely unrelated with early cell positions. From fifth cleavage central cells are completely separated from the underlying yolk, but the marginal blastomeres retain open connections with it. These marginal cells then collapse into the yolk cell usually at tenth cleavage, supplying nuclei to form a yolk syncytial layer or periblast. Tangential cleavages within the blastoderm also separate loosely arranged deep cells from an enveloping layer with marked epithelial properties. There is considerable evidence that cytoplasmic materials continue to flow from the vegetal pole to beneath the blastoderm throughout early development (e.g. see Oppenheimer, 1936*a*; Svetlov, Bystrov & Korsakova, 1962); and according to Thomas (1968) further cells are added by divisions from the periblast.

Two major events occur during the following stages of teleost development. One is the spread of the periblast and blastoderm to envelop the rest of the egg in a yolk sac (see Fig. 9.6(*a*)). The other is the formation of the germ layers, the equivalent of gastrulation in other embryos, as an embryonic primordium forms within the blastoderm. The motive force for the epibolic spreading seems to be largely if not entirely provided by the periblast, perhaps by microfilament action (Betchaku & Trinkaus, 1978; Trinkaus, 1984). At the margins of the blastoderm the cells of the

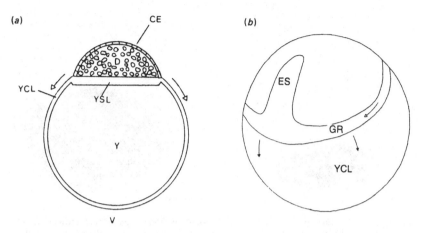

Fig. 9.6. Morphogenetic movements in early teleost development. (*a*) diagrammatic vertical section to illustrate the parts of the egg and the way in which the blastoderm spreads (arrows). CE, cellular envelope; D, deep cells; V, vegetal pole of egg; Y, yolk; YCL, yolk cytoplasmic layer; YSL, yolk syncytial layer or periblast. (*b*) later stage when the blastoderm has spread (closed arrows) to cover almost half the yolk sphere. Deep cells have accumulated near its rim to form a germ ring (GR) and then converged within the ring (open arrow) to form an embryonic shield (ES). (From Long, 1984.)

enveloping layer form tight junctions with the periblast, and these authors conclude that the former is passively stretched as the latter spreads. Extra material for periblast extension is probably provided by the flow of materials from within the egg (Long, 1980*a*). As the periblast extends vegetally its cytoplasm and possibly the nuclei also concentrate towards one side of the egg, usually called the posterior side, where the embryo will later form (Long, 1980*a,b*).

Teleost embryos were for a long time considered to gastrulate by involution of cells in a way quite closely comparable with amphibians (e.g. see Oppenheimer, 1947). However, Ballard (1966*a,b*) showed that chalk particles injected among the cells of the enveloping layer, or vital stains allowed only a shallow penetration of the surface, were later restricted to epidermal areas. It is now generally accepted that the enveloping layer takes no part in gastrulation or neurulation: it is shed at hatching, so that even the larval epidermis originates from the deep cells (Wourms, 1972; Bouvet, 1976). Ballard (1966*c*, 1968, 1973*a*; Ballard & Dodes, 1968) studied the movements of the deep cells chiefly by implanting chalk particles among them. They too spread away from the animal pole and converge towards the posterior side. This produces a concentration of deep cells just within the edges of the enveloping layer as a germ ring, and by stage 7 of *Salmo* a particularly thick 'nubbin' of cells on the posterior side. Further deep cells move around the germ ring to extend

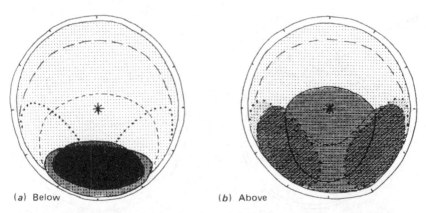

(a) Below (b) Above

Fig. 9.7. Fate map of the stage 7 blastodisc of *Salmo gairdneri* viewed from below (a) or above (b) with the posterior side at the bottom. Starting from the innermost cells, layers can be recognised as presumptive endoderm (heavy stipple), notochord (vertical lines), mesoderm (light stipple, with anterior somites behind the line of large dots and further stippled in (b)) and nervous system (horizontal lines). The enveloping layer is only shown at the rim. (From Ballard, 1973b, where further details can be found.)

this nubbin into an embryonic shield as the blastoderm expands (Fig. 9.6(b)). According to Ballard cells of the different germ layers reach their final positions within the shield without any involution movement. A compromise view has been proposed by de Vos (1981) who accepts that there is no active invagination of cells but believes that a passive inrolling may occur as the deepest cells are anchored internally and more superficial ones are swept along beneath the enveloping layer.

Ballard (1973b) has provided a three-dimensional map of stage seven (Fig. 9.7), according to which the cells of the prospective germ layers are already in approximately their correct relative positions on the posterior side. On the anterior side, almost all deep cells are prospective mesoderm. These fates are not determined by lineage, as the (marked) progeny of early blastomeres mix extensively with unlabelled cells and contribute to a wide variety of embryonic tissues (Kimmel & Law, 1985c). Cell fates may therefore be decided by later positional cues from their surroundings. The progeny of single cells marked at 'gastrula' stages do seem to populate single tissue types (Kimmel & Warga, 1986), but this need not imply that their fates are determined by this stage. There are some indications from Kimmel & Warga's (1986) work that division patterns in gastrulae ensure a periodic and bilaterally symmetrical distribution of clonally related cells in some axial organs.

Most experimental studies of teleost determination have concerned the role of the yolk cell and periblast. Oppenheimer (1936a) isolated *Fundulus* blastoderms at various stages: up to the 16-cell stage they formed hyperblastulae which rarely differentiated any recognisable tissues, but older blastoderms could 'gastrulate' and form quite complete axes at least at the head level. Similar effects were later seen in other fish species although the stage of blastodermal autonomy varies. Several possible explanations for the change can be suggested (e.g. see Waddington, 1956, p. 231). Some involve relatively unspecific factors such as an increasing 'toughness' of the older cells or the supply of a simple nutrient from the yolk. These ideas have been tested by improving culture conditions, but it has usually proved impossible to bring forward the stage of independence from the yolk. Kostomarova (1969) found this for the loach *Misgurnus*, where independence is achieved at midblastula stage. Rott *et al.* (1978) succeeded in bringing forward the stage of independence by several cell cycle times using media with high levels of potassium ions, but still the youngest blastoderms failed to differentiate.

Positive evidence favouring the supply of a complex factor from the yolky egg regions has come particularly from Tung and his coworkers. In latitudinal separations of goldfish embryos they found that more than half the yolk was needed for development from the one- or two-cell stage, a smaller amount at four-cells and none at eight-cells (Tung *et al.*, 1945). This suggested a factor which moved up from more vegetal regions of the egg over this period. Later they used centrifugation to increase the size of the animal cap of cytoplasm by 2.5 times, when even the cap alone can sometimes form an embryo (Tung, Wu & Tung, 1955*b*). This suggests that centrifugation has moved a determinative substance towards the animal pole. Oppenheimer (1936*a*) pointed out that the effect was exerted at the periphery of the blastoderm where mesoderm will arise. At the time of this work determinative effects from the yolky area seemed to be an unusual phenomenon and very different from the amphibian system with its chordamesodermal 'organiser'. Now of course it is clear that the ring of mesoderm in amphibians is also a product of inductive influences from the vegetal pole (see pp. 190–1). In teleosts, however, the significance of nutritive effects from the yolk cell still needs to be fully assessed, and in species where blastodermal autonomy is acquired relatively late the role of late cell cleavage from the periblast (see Thomas, 1968) also needs to be checked.

The comparison with amphibian development can be taken further when we consider determination in the antero-posterior direction. When blastomeres are extirpated from the animal pole any one of the first two, or two of the first four, remaining blastomeres could often give rise to a complete embryo (see Nicholas & Oppenheimer, 1942). The data indicated that there is no determination centre like the dorsal one in amphibians. However, when the yolk as well as the blastoderm was bisected, the results were much more like those in amphibians. In some cases both halves formed embryos, and in others one did so while the other formed only a hyperblastula (Tung & Tung, 1944; Tung, 1955; Tung, Lee & Tung, 1955a). Tung's group interpreted the former as right and left pairs and the latter as posterior and anterior ones. Of course hyperblastulae may result from unspecific damage and in some cases both halves developed in this way, but improvements in the bisection technique reduced the number of such cases and gave a better fit with the posterior determinant theory. Since the blastomeres themselves do not seem determined in the antero-posterior direction Tung *et al.* (1945) concluded that the determinative materials supplied by the yolk cell also fix the plane of bilateral symmetry. Long (1983) has since transplanted young radially symmetrical blastoderms onto the yolk and periblast from stages when this plane is recognisable. He confirmed that it is the latter which usually decide the antero-posterior axis of the former. All this of course recalls the amphibian situation where the same inductive interaction seems to vegetalise and polarise 'animal halves' (see pp. 192–3).

An implication of this work is that the whole egg is bilaterally organised by very early cleavage stages. Tung & Tung (1944) gave further evidence of this by fusing two such embryos and obtaining twinned axes in most cases with a few large single axes presumably where the two antero-posterior orientations were compatible. The question of when the egg is symmetrised has however caused some controversy. Vakaet (1955) suggested for *Lebistes* that bilateral organisation is already recognisable in the distribution of plasms within the oocyte. Using the same species, which is viviparous, Clavert & Filogamo (1966) noted that the blastodermal edge which is lowest at cleavage stages becomes posterior, and claimed to be able to reverse this symmetry by inverting a female containing embryos at this stage. They concluded that symmetrisation is mediated by gravity, as it can be in amphibians, but rather later in development. Long & Speck (1984), however, suggest that the embryo (marking the posterior side of the egg) is only seen on the low side of the blastoderm because it is heavy and causes a rotation within the chorion: in dechorionated *Oryzias* eggs it forms randomly relative to gravity. Moreover, Cherdantseva & Cherdantsev (1985) claim to be able to recognise the future anterior and posterior sides of the zebrafish blastodisc before first cleavage by slight differences in shape. However, such data do not preclude an effect of gravity on the newly fertilised egg, as can occur in amphibians, and the problem certainly warrants a careful re-examination.

We do not know whether the inductive signals passed from the yolk cell to the anterior and posterior margins of the blastoderm differ qualitatively or only quantitatively. One possibility is that the signals are received rather earlier on the future posterior side, since anterior half blastoderms still require contact with the yolk cell when posterior halves have become independent of it (Richard, Devillers & Colas, 1956). Even when the normal symmetry of the blastoderm has been decided the differences between the parts of the germ ring are reversible. Luther (1936*a*) divided blastoderms to quadrants at a stage when the posterior side is already recognisable from the greater thickness of the germ ring. Each quadrant was grafted into the yolk sac epithelium of an older host embryo, and all were found capable of developing the main embryonic organs though in a somewhat haphazard arrangement. A similar experiment performed on rather older embryos showed that the ability to form embryonic organs is becoming increasingly restricted to the posterior side, with other sectors failing in particular to develop notochord and anterior somites (see too Luther, 1937). In part, these results presumably reflect the fact that cells destined for the embryo originate in all quadrants and converge towards the posterior side during epiboly, but it seems likely that true regulation is also occurring. Other techniques indicate that the posterior side already has some dominance by the time that it is first recognisable (Luther, 1937), but this may be reversed by increasing the thickness of the originally anterior side (Cherdantseva & Cherdantsev, 1985).

In exercising its dominance through further development the posterior side of the teleost germ ring apparently acts in a similar way to the amphibian dorsal lip. When the edge of the embryonic shield is grafted into the anterior germ ring a fairly normal secondary embryo is induced (Oppenheimer, 1934, 1936*b*; Luther, 1935). Luther (1935) also obtained secondary embryos by grafting anteriorly only those inner cells which were at the time considered to be invaginating posteriorly to form an archenteron roof (Fig. 9.8), and Eakin (1939) claimed that different regions of this 'roof' already show the properties of prechordal plate, head organiser and tail organiser. Luther (1936*b*) claimed that prospective neural plate and epidermis could be exchanged without affecting development, indicating that the choice between these pathways is decided inductively. In the light of our modern knowledge of teleost fate maps and gastrulation movements, however, all of these claims should be re-examined. We need to be sure that host cell fates are really changed when a secondary embryo is formed anteriorly. The properties of the inner cell layers should be checked, since these cells are now believed to reach their final positions directly and not via involution, and since the properties of the equivalent cells in amphibians are so strongly influenced by their passage through the dorsal lip. Did Luther (1936*b*) really exchange prospective ectodermal areas or only two areas of the enveloping layer?

The very different organisation of the amphibian and teleost embryos is likely to have great effects on their inductive interactions, but the results of Oppenheimer, Luther and Eakin surely indicate a basic similarity. Moreover, secondary embryos can be induced by older brain

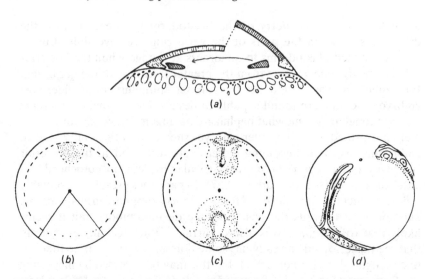

Fig. 9.8. Induction of a second embryo in the trout following the transfer of early embryonic shield material from the posterior to the anterior margin of the germ ring: (*a*) the operation scheme; (*b*) the implant has healed into place; (*c*) two medullary plates form, and (*d*) the induced embryo has somites and ear vesicles and the donor side an apparently complete head. (From Luther, 1935.)

tissue or newt liver implanted anteriorly (Luther, 1935), suggesting the existence of a non-specific evocating component as in amphibians. Lithium ions have the familiar effect of producing cyclopic embryos (McClendon, 1912). This disturbance of the antero-posterior gradient was first reported using magnesium ions (Stockard, 1907), a less active vegetaliser in some other embryos, and can be obtained in teleosts using a variety of other treatments (Laale & Lerner, 1981).

There have been several studies of the junctional connections between the embryonic cells (e.g. Trinkaus & Lentz, 1967; Hogan & Trinkaus, 1977), which may be useful in considering how inductive information is transferred. Fluorescent dyes pass between blastomeres (Bennett, Spira & Spray, 1978), but measurements of electrical coupling indicate decreasing communication during later cleavage (Bozhkova *et al.*, 1971). Longer distance communication through the extracellular space can also occur because the whole embryonic compartment is separated both from the external medium and from the yolk vacuole by high resistance barriers (Bennett & Trinkaus, 1970). Thomas (1968) reported that in some places a row of vesicles, rather than continuous membranes, separates the blastomeres from the periblast.

Physiological and biochemical studies, mainly of the loach *Misgurnus*, have provided several possible explanations of the blastoderm's dependence on the yolk cell. Materials which are apparently transferred

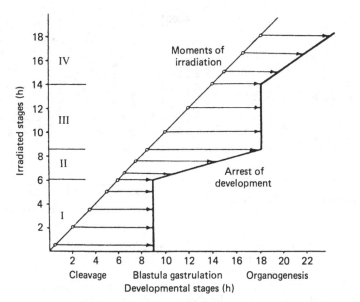

Fig. 9.9. The amount of development possible in *Misgurnus* after 40 kr X-irradiation at various stages. Open circles indicate times of irradiation, arrows to their right further development and the thick line the stage of arrest. From the data four stages can be recognised and are marked (I–IV) on the ordinate. (From Neyfakh, 1964.)

from the latter to the former include potassium ions (Beritashvili, Kvavilashvili & Kafiani, 1969; Beritashvili *et al.*, 1970), massive amounts of ribosomes (Aitkhozhin, Belitsina & Spirin, 1964; see also the cytochemical studies of Hagenmaier, 1969; Kostomarova & Nechaeva, 1970; Dasgupta & Singh, 1981) and some enzymes (Klyachko & Neifach, 1980; Klyachko *et al.*, 1982). Thomas' (1968) work, noted earlier, provides two possible explanations of these massive transfers: the late cleavage of extra cells from the periblast, and the existence of syncytial connections between periblast and cells. Other ways in which the yolk may affect the metabolism of the blastoderm have been suggested by Mel'nikova *et al.* (1972) and Timofeeva & Solovjeva (1973).

Biochemical studies of the loach embryo have now provided a considerable body of information, and some plausible connections with developmental phenomena. Most RNA synthesis appears to be cytoplasmic until the sixth hour midblastula, when nuclear transcription of heterogeneous RNAs is activated, and even in the 'midgastrula' ribosomal RNA synthesis was not detected (Kafiani *et al.*, 1969). The transcripts produced between 6 and 8½ hours appear to be required for development through 'gastrulation' as X-irradiation (Fig. 9.9) or actinomycin treatment before this stage prevent 'gastrulation' while

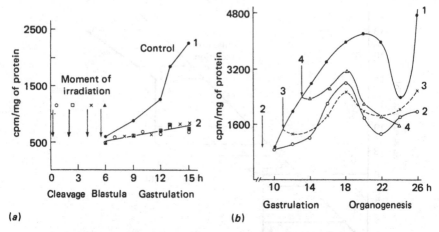

Fig. 9.10. Patterns of protein synthesis following X-irradiation (20 kr) of loach embryos: (*a*) treatments between 15 min and 5.5 h after fertilisation block the rise normally seen during 'gastrulation' (compare curves 1 and 2); (*b*) treatments at 9–13 h allow a later rise and fall in synthesis (curves 2–4) reminiscent of that in controls (curve 1). (From Krigsgaber & Neyfakh, 1972.)

treatments after 8½ hours allow it (Neyfakh, 1959, 1964, 1971; see also Hagenmaier, 1969, on *Salmo* species). The same treatments started at times from 8½–14 hours (stage III of Fig. 9.9) always cause arrest at the same stage, as though the transcripts produced over this period have other, possibly 'house-keeping', functions. Of course, such data are never conclusive (see p. 161) but here the effects on protein synthesis patterns (Krigsgaber & Neyfakh, 1972) do seem to support the interpretation that transcripts produced at 6–8½ hours are translated some hours later during 'gastrulation' (Fig. 9.10). Both the sixth hour transcriptional activation (Dontsova & Neyfakh, 1969) and the later translational activation (Krigsgaber, Kostomarova & Burakova, 1975) fail in blastoderms isolated from the yolk at early stages, so this may be the cause of their failure to 'gastrulate'. Both activations, moreover, begin in those deep cells which are nearest to the yolk cell and spread progressively to the upper cell layers (Neyfakh, 1971; Burakova & Kostomarova, 1975).

If midblastula transcripts are only required later in development, they could be retained for some hours in the nuclei (see Neyfakh, Kostomarova & Burakova, 1972) or masked in small mRNPs or 'informosomes', which were in fact first described in loach embryos (Spirin *et al.*, 1964). There is however no kinetic evidence that zygotic mRNAs can be chased from mRNPs to polysomes (Neyfakh, 1971; and see pp. 163 and 243). Overall protein synthesis levels may be determined by properties of the translation machinery with the various mRNAs competing for its components (Neyfakh, 1971; see also pp. 5–6, 164 and 243). Some new protein species are synthesised from the late blastula, and many more in 'gastrulation' (Ermolaev, Korzh & Neyfakh, 1984), but it is not clear whether these were coded by transcripts of the 6–8½ hour period.

In considering the significance of the loach data we should remember that in other teleost species nuclear transcription, nuclear 'morphogenetic function' and blastodermal autonomy from the yolk do not all begin at the same stage (e.g. in the trout: Mel'nikova *et al.*, 1972; Ignat'eva & Dunaeva, 1973). In other aspects of metabolism and biochemistry, teleosts show similarities even with invertebrates, e.g. there may be a histone transition (Vorobyev, Gineitis & Vinogradova, 1969). Although 'gastrulation' is morphologically so unlike that of amphibians, there are many similarities at the biochemical level, including increased glycogen breakdown and a rise in respiratory quotient (Hishida & Nakano, 1954; Boulekbache, 1981). The mitochondria also differentiate in some undefined way (Abramova & Vasil'eva, 1973).

There have been few studies of early development in cartilaginous fish. As in teleosts cleavage is partial, a yolk syncytium is formed beneath the blastomeres, and the embryonic shield develops on the posterior side. According to Vandebroek (1936) the shield is formed by true gastrulation movements, particularly a posterior involution of cells, but it would be valuable to have these claims checked. Wintrebert (1922) claimed that he could decide the bilateral plane experimentally by tilting cleavage-stage blastoderms, when the uppermost edge became posterior. However, he also suggested that the posterior side has the lowest density because of the eccentrically placed blastocoel, so it would seem wise to check that the selected side does really remain uppermost in tilted blastoderms. The regulative ability of the dogfish blastoderm has been confirmed by splitting it medially and showing that each half can form an embryo (Vivien & Hay, 1954).

## Reptiles

Reptiles, like birds, lay very large eggs in which the egg cell itself is often bathed in albuminous secretions and always surrounded by a shell. Most of the volume of the egg cell is occupied by the yolk, stored in large yolk platelets and sufficient to support development to hatching. Cleavage is restricted to a small germinal disc at the animal pole. It gives rise to a blastoderm of a single epithelial layer over a few dispersed deep cells.

Gastrulation movements have been studied by the classical methods, but confirmation is really required using modern methods. According to Pasteels (1957*a,b*) there is some variation in the mechanisms used. In chelonians the first inner cells arise by ingression near the future posterior edge of the blastoderm and extend anteriorly as an endoblast beneath the

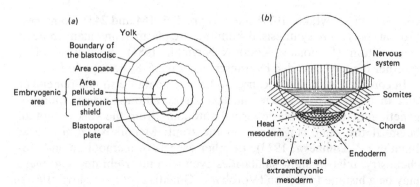

Fig. 9.11. Early development in the turtle *Clemmys*: (*a*) general view of the egg at the blastoderm stage; (*b*) fate map of the embryogenic area. (From Pasteels, 1937, (*a*) redrawn.)

central (future embryonic) zone. In many other reptiles the blastoderm apparently becomes two layered by delamination, the inner layer being called an endophyll, but its cells still move anteriorly. In either case, a blastopore then appears posteriorly and a whole tube of cells is soon involuted forwards between the previous two layers. Prospective notochord involutes over the upper lip (the equivalent of the amphibian dorsal lip) and mesoderm in its sides. In lacertilians the floor of the tube probably provides all of the embryonic endoderm (Pasteels, 1957*a*) with the endophyll becoming extraembryonic (Hubert, 1962), but according to Pasteels (1937) no endoderm involutes at the blastopore in the turtle.

A fate map of the early turtle blastoderm (Fig. 9.11) confirms that there are similarities with the arrangement in amphibians (Fig. 7.2) and birds (Fig. 9.14). Indeed, the brief account given above may help as an introduction to the rather complicated development of the chick described later.

These organisational similarities presumably reflect common developmental mechanisms at the cellular, and perhaps even the molecular, level. In particular, there is evidence that the plane of bilateral symmetry is fixed in a way closely comparable to that used by birds. As will be described more fully for birds (see p. 290), the orientation of most embryos follows the rule of von Baer, and seems likely to be controlled by rotation of the egg in the uterus (see Clavert, 1962). It has been confirmed that the *Anguis* egg is rotated, and that most embryos then appear with the head pointing in the direction of the rotation (Raynaud, 1961). If the blastoderm remains on the upper side of the egg, by egg rotation within the egg investments as occurs in birds, then it would be the upper blastodermal edge which is determined as posterior (as it is in birds). Raynaud (1961), however, believes that the *Anguis* egg cell rotates together with its investments.

# Birds

## (a) Descriptive embryology

The egg cells of birds (the 'yolk' of the egg) include the largest known cells and have massive supplies of yolk. All of the embryo, however, develops from a very small area of clear cytoplasm with the nucleus at the animal pole. The earliest stages of development occur inside the mother and while the non-living coats of albumen, membranes and shell are being secreted around the egg cell. This has hindered observation, but recent studies have greatly increased our knowledge of the stages up to and rather beyond laying (Eyal-Giladi & Kochav, 1976; Kochav, Ginsburg & Eyal-Giladi, 1980; see Fig. 9.12). The pattern of the early cleavages does not seem very precise and soon becomes asynchronous. The furrows seem impeded from entering the yolk, and curving furrow planes soon separate complete cells from the yolk. It is still not clear whether there are nuclei in the uncleaved yolk (see Bellairs, Lorenz & Dunlap, 1978). Cleavage produces a plate of cells called the blastoderm, and by the time of laying two regions can be recognised within it. An outer ring of cells is more yolky and opaque (area opaca) while the central cells (area pellucida) have less yolk platelets and are separated from the main yolk by a fluid-filled subgerminal cavity. The formation of the area pellucida involves the shedding of all subepithelial cells in the central area and has been further studied by Fabian & Eyal-Giladi (1981). It begins on the side which will form the posterior end of the embryo and progresses anteriorly (Fig. 9.12($b$)), the detached cells also accumulating beneath the future anterior side of the blastoderm.

Before describing further events within its central area it is important to note that, soon after laying, the whole blastoderm begins a rapid expansion to cover an increasing proportion of the uncleaved yolk. New (1959) showed that the edge cells adhere to and move across the vitelline membrane, setting up a tension in the blastoderm, and that cell proliferation occurs particularly just behind the blastodermal margin. The blastoderm appears not to require yolk from the uncleaved area for 48 hours after laying (Bellairs, 1963$a$).

From about the time of laying the bird blastoderm becomes two layered. A lower layer or hypoblast is first seen near the posterior margin of the area pellucida and gradually extends anteriorly beneath the upper layer or epiblast (Fig. 9.12($c$)). The first hypoblast cells are in loosely arranged clumps which suggests an ingression or polyinvagination of cells from the original single cell layer followed by coalescence of the cells into a layer (Eyal-Giladi & Kochav, 1976; Weinberger & Brick, 1982). However, it is clear that other hypoblast cells are contributed from the marginal zone at the edges of the area pellucida and particularly from its

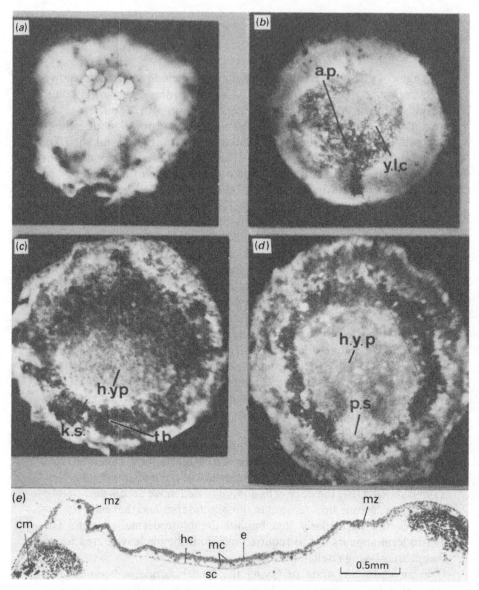

Fig. 9.12. Early stages in chick development: (*a*) upper view of the germ during cleavage (stage II); (*b*)–(*d*) lower views of blastoderms during the formation of the area pellucida (ap) at st. VIII (*b*), the hypoblast (hyp) at st. XII (*c*) and primitive streak (ps) at Hamburger & Hamilton st. 2 (*d*). All these events begin on the posterior side. ks, Koller's sickle; tb, transparent belt separating hypoblast and area opaca; ylc, yolk laden cells; (*e*) section through a st. XIII blastoderm when the central area has complete layers of epiblast (e) and primary hypoblast cells (hc) as well as some middle cells (mc). cm, carbon mark showing the posterior side; mz, marginal zone; sc, sub-blastodermic cavity. ((*a*)–(*d*) from Eyal-Giladi & Kochav, 1976; (*e*) from Kochav *et al.*, 1980.)

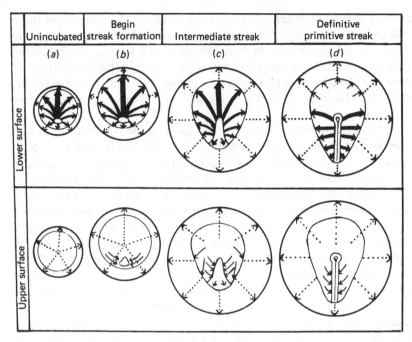

Fig. 9.13. Diagram to illustrate the major morphogenetic movements on the lower and upper surfaces of the early chick blastoderm, as well as its general growth expansion. (From Spratt & Haas, 1960*a*.)

posterior edge (Spratt & Haas, 1960*a*; Vakaet, 1962). Spratt & Haas (1960*a*) described a fountain-like flow of cells anteriorly from the posterior marginal zone (Fig. 9.13). They suggested that these cells were formed by the mitotic activity of a ring-shaped growth zone which is most active posteriorly, and thought it unlikely that any were contributed by ingression from the epiblast (Spratt & Haas, 1962*b*). Weinberger & Brick (1982), however, described a depression in the upper epiblast surface near its posterior margin.

The next event of bird development, primitive streak formation, also begins just inside the posterior edge of the area pellucida (Fig. 9.12(*d*)). As with the hypoblast, the streak extends anteriorly but as a far narrower structure, and the length of the streak is also increased because the blastodermal area is increasing rapidly by this time with the greatest extension of the area pellucida being seen posteriorly (Fig. 9.13). All cell layers of the blastoderm are united at the streak, with the epiblast being raised and gradually developing a median groove. A variety of data indicated a large-scale ingression of cells through the streak into both an intermediate mesodermal layer and the lower endodermal layer (Hunt 1937*a*,*b*; Fraser, 1954). This was firmly established by grafting [³H]-thymidine labelled cells (see especially Rosenquist, 1966) or cells of another species (Vakaet, 1973) into the epiblast. The progeny of these cells are later recovered from both inner layers and from both sides of the streak, which demonstrates that extensive cell mixing occurs during ingression. Cells from the streak resemble amphibian bottle cells in their form and ultrastructure (Balinsky & Walther, 1961; Granholm & Baker, 1970), and Balinsky & Walther note that some birds do form a shallow archenteron.

Cell marking experiments have also allowed the construction of fate maps for streak stage chick embryos. All the cells of the future embryo apparently originate in the epiblast. Some primordial germ cells leave this layer singly before streak formation (Ginsburg & Eyal-Giladi, 1986), but all the prospective somatic tissues are still in the epiblast at early streak stages. The most central cells, particularly around the anterior end of the streak, are predominantly future embryonic endoderm, although because of cell mixing the boundaries around areas are less definite than Figure 9.14(*a*) indicates. Outside this central area is a ring of predominantly future mesoderm, and outside this is future ectoderm which will remain in the epiblast when ingression at the streak is completed. Different kinds of endoderm and mesoderm immigrate at different antero-posterior levels of the streak (see also Nicolet, 1971). The endoderm originating from the streak becomes inserted into the centre of the previous hypoblast which is pushed out towards the edges of the area pellucida and becomes extraembryonic endoderm. The arrangement of the prospective areas in the three cell layers after ingression is shown in Figure 9.14(*b*)–(*d*).

By the definitive streak stage, prospective chorda is entirely confined to Hensen's node, a small but marked thickening at the anterior end of the streak. The central groove of the streak is also deeper here forming the primitive pit through which prechordal mesoderm and then some prospective chorda cells immigrate anteriorly. This rod of axial mesoderm ahead of the node is visible through the epiblast and is called the head process. The further extension of the notochord is not achieved by anteriorly directed immigration but rather by posterior movement of the node. The node is, in fact, seen to move posteriorly along the streak so that the streak shortens or regresses from its anterior end. (At the same time, however, the streak is extending posteriorly as the blastoderm continues its expansion.) As the node moves backwards it contributes more cells to the mesodermal layer to extend the notochord. Cells destined for other axial tissues are associated transitorily with the node as it regresses, but only those for notochord seem to move back with it acting like a stem-cell population.

A typical vertebrate dorsal axis soon becomes recognisable in the anterior regions through which the node has travelled. The segmentation of presomitic mesoderm in fact begins somewhat ahead of the node

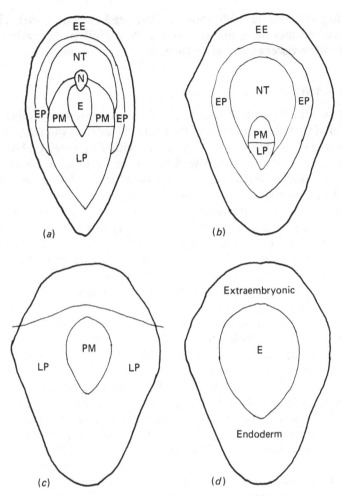

Fig. 9.14. Fate maps of the chick blastoderm before and after immigration of cells through the streak: (*a*) the epiblast at streak stages; (*b*)–(*d*) the ectodermal (*b*), mesodermal (*c*) and endodermal (*d*) layers at head process stage. E, gut endoderm; EE, extraembryonic ectoderm; EP, epidermis; LP, lateral plate mesoderm; N, notochord; NT, neural tube; PM, paraxial mesoderm. (After Rosenquist, 1966.)

(Meier, 1979; Triplett & Meier, 1982). Other features of organogenesis, particularly in the mesodermal layer, were described by Trelstad, Hay & Revel (1967). A neural plate appears in the ectodermal layer and rolls into a neural tube. This formation of axial organs progresses posteriorly behind the regressing node so that transverse sections reveal organised axes anteriorly when there is still only a simple streak posteriorly. A left–right asymmetry is also seen in bird development, beginning at neurulation when the left side shows a slight lead (Lepori, 1966). This asymmetry becomes more marked with further development, with the heart and other viscera forming on the right side of the body axis.

### (b) Causal analysis

The earliest developmental decision which has proved accessible to experimental study is the fixation of the plane of bilateral symmetry. Von Baer long ago showed that most eggs opened with the pointed end to the right contained embryos with the head away from the observer. Vintemberger and Clavert carried out an extensive series of investigations (reviewed by Clavert, 1962) which showed that most eggs pass through the uterus with the pointed end towards the cloaca, and are rotated there at ten to fifteen revolutions per hour. A minority pass through with the blunt end towards the cloaca, and were shown to have a reversed direction of coiling of the chalazae in the albumen (the system suspending the 'yolk') and a reversed orientation of the embryo. The exact angle of the embryonic axis was also affected by the orientation of the uterus relative to gravity. The critical period for the determination of bilateral symmetry was found to be eight to six hours before the time of laying, which is as the area pellucida appears, off centre, within the blastoderm, and Clavert (1960) could show that area pellucida formation began at the future posterior edge of this area.

Even rotation apparently acts via its effects on orientation relative to gravity. Kochav & Eyal-Giladi (1971) showed that blastoderms stay on the upper side of the 'yolk' in rotated eggs but with the margin in the direction of the rotation pointing downwards. The future posterior side is towards the uppermost edge of the blastoderm (Fig. 9.15(a)). Eggs taken from the uterus before the critical stage and kept stationary with the pointed end upwards later formed embryos with the posterior end towards the egg's pointed end (Fig. 9.15(b) and (c)). Area pellucida formation was also shown to proceed from the upwardly directed side. Embryo orientation can no longer be changed when more than half of the area pellucida has been formed by polarised cell shedding (Eyal-Giladi & Fabian, 1980). No embryo forms at all in quail blastoderms isolated before area pellucida formation unless they are maintained obliquely or vertically (Olszańska, Szołajska & Lassota, 1984). Many of these findings recall the situation in the fertilised amphibian egg (pp. 58 *et seq.*), where the plane of bilateral symmetry is normally fixed by a rotation or contraction, fails to develop in the absence of these reactions and can be changed by manipulations relative to gravity. There are indications that similar factors may operate in various fish (e.g. pp. 278 and 283) and reptile (p. 284) embryos, although there are also some obvious differences, e.g. in the time of axis fixation.

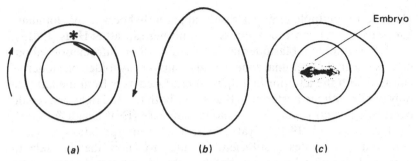

Embryo

(a)        (b)        (c)

Fig. 9.15. The effects of gravity on the orientation of the chick embryo: (a) the position of the blastoderm in a rotating uterine egg: the arrows mark the direction of rotation and the asterisk the future posterior side; (b),(c) experimental determination of embryo orientation: if a uterine egg is maintained for 10 h with its pointed end upwards (b) the embryo later forms with its posterior end towards the pointed end (c). ((a) original; (b),(c) from Kochav & Eyal-Giladi, 1971. © 1971 AAAS)

The epigenetic determination of orientation appears to rule out the theory that the posterior centre owes its special properties to an ooplasm surviving from oogenesis (see Spratt & Haas, 1961a). Even after laying more drastic interference is able to produce embryonic axes with abnormal orientations, e.g. Lutz (1949) bisected unincubated duck blastoderms and showed that both parts could form complete embryos. Those formed from the normal posterior area retained their polarity, but polarity was labile for those formed elsewhere. Spratt & Haas (1960b) found that any part of the marginal zone between the opaque and pellucid areas (see Fig. 9.12(e)) is capable, after isolation, of initiating an axis directed across the pellucid area. This arrangement recalls the teleost germ ring, a structure which seems to be induced by the yolk syncytium (see pp. 277–9). There is no evidence for such an inductive action by the uncleaved area in birds, but we know too little about the developmental potential of the early blastoderm to be able to exclude it. However, the ability to initiate an axis seems to be present even in central areas isolated from the newly laid egg (Eyal-Giladi & Spratt, 1965), perhaps because cutting stimulates local metabolism. The ability is then progressively localised first to the marginal zone and then to its posterior area.

Although so many areas could form an axis, only one axis normally forms, initiating in the posterior marginal zone. Even when blastodermal halves are combined in abnormal arrangements one axis normally forms: from the older of two posterior halves or the younger of two anterior halves (Spratt & Haas, 1962a). In such cases one area seems to act as a dominant zone, suppressing axis formation elsewhere in the marginal ring. A single embryo can even develop when three chick blastoderms are fused together before incubation; but with any more than three blastoderms multiple embryos result, apparently because the dominant zones are too far apart to suppress one another (Spratt & Haas, 1961*b*). Work with folded blastoderms revealed further information about dominance in the marginal zone. Young transversely folded blastoderms cultured with the anterior half upward produced axes from the left side only: while with the posterior half upwards both left and right sides could overcome posterior dominance and initiate axes (Eyal-Giladi & Spratt, 1964; Eyal-Giladi, 1969). Eyal-Giladi's (1970) interpretation of this is that folding provides a sufficient stimulus to allow the left side to overcome posterior dominance, but the right side can only do so when the posterior centre has the further disadvantage of being away from the medium. She also found that local cutting can further increase the axis-forming potential of the right side. Dominance has thus been shown to operate in the left–right direction as well as antero-posteriorly even at this early stage. (The folding work also revealed another factor which may influence normal developmental patterns: the gut always opened on the upper side of the blastoderm even when this location was inappropriate.)

The distribution of axis-forming potential suggests a link with hypoblast formation, which can occur from the whole marginal ring but is soon largely restricted to its posterior area. As long ago as 1933 Waddington rotated the hypoblast 180° relative to the epiblast of streak stage embryos, and found that a second axis sometimes developed with its polarity decided by the hypoblast. With more modern culture techniques only a single axis develops, but its orientation is still decided by the hypoblast if the rotation is made sufficiently early (Azar & Eyal-Giladi, 1981). Any deviations in the pathway followed by the hypoblast as it grows forwards are later reflected in the form of the primitive streak (Spratt & Haas, 1960*b*, 1961*a*). The extent of hypoblast development affects both streak polarity and later left–right asymmetry in fragmented blastoderms. Anterior fragments form streaks with normal polarity if they contain hypoblast (Lutz, 1953); and the right fragment of a longitudinally split blastoderm forms an embryo with reversed asymmetry if cut when the hypoblast has reached half way across the area pellucida (Lutz & Lutz-Ostertag, 1957).

The first hypoblast cells, derived directly from the epiblast by polyinvagination, appear to have no axis-inducing potential, which supports the link between induction and the marginal zone (Azar & Eyal-Giladi, 1979). The marginal zone itself, however, loses its ability to initiate an axis once the hypoblast has formed (Khaner, Mitrani & Eyal-Giladi, 1985; see also Khaner & Eyal-Giladi, 1986). The hypoblast's inductive power is greatest posteriorly and reduces towards the anterior side, although this may again reflect the relative contributions of polyinvaginated and marginally derived hypoblast cells (Mitrani, Shimoni & Eyal-Giladi, 1983). Hypoblast action has apparently polarised the epiblast before a primitive streak appears. If such an epiblast is combined with a disaggregated and reconstituted hypoblast it is the former which decides the orientation of the streak, but the hypoblast is still required for its induction (Mitrani & Eyal-Giladi, 1981). It is interesting to compare these effects with mesodermal induction by the endoderm in amphibian blastulae (p. 190). In both cases the endoderm appears to be imposing its polarity on the other germ layers. The effect looks very different at first sight, but we should remember that when the amphibian blastocoel roof is collapsed onto the endoderm (a situation much more like the bird embryo) axial structures seem to be directly induced from the roof cells (Pasteels, 1953, 1954).

The primitive streak and its node appear to be the avian equivalent of the amphibian blastoporal lips, and were soon recognised as the potential site of the 'organiser'. Waddington (1932) inserted anterior and middle streak fragments beneath the epiblast and confirmed that neural grooves were induced. Mulherkar (1958) also found inductive activity at lower levels in the areas immediately around the node and on the left side of the anterior streak, i.e. the node is the centre of a gradient field in which even left dominance is again detectable. Nodes, like amphibian dorsal lips, also show a great ability to self-differentiate even in abnormal locations, producing chordamesoderm, neural tissues and gut endoderm (Gallera & Ivanov, 1964; McCallion & Shinde, 1973).

It is now clear that the node and streak can have two quite different inductive effects upon competent epiblast: the induction of neural tissue or of a further streak (Gallera & Ivanov, 1964). The choice between these outcomes is affected by both the age and the position of origin of both the inducing and reacting tissues. The ability of one streak to induce another is high when the streak is still lengthening and has already fallen greatly by the definitive streak stage, at which time neural inducing ability is strong (Eyal-Giladi & Wolk, 1970). Epiblast competence for streak formation is lost by the midstreak stage (Azar & Eyal-Giladi, 1981), again well before that for neural structures (Gallera & Nicolet, 1969; Eyal-Giladi & Wolk, 1970) and even neural competence becomes increasingly limited to brain formation (Gallera & Ivanov, 1964). Epiblasts from which streaks have been induced retain their neural competence longer than usual (Gallera, 1968). Position along the streak affects inductive activity, with anterior fragments inducing neurally and more posterior ones inducing streaks (Vakaet, 1964, 1965; Sanders & Prasad, 1986). Competence also varies according to position within the epiblast, that from the area pellucida responding to an implanted node after a shorter contact time than that from the area opaca (Gallera, 1970b; and see below). Even within the area pellucida competence for somite induction appears to decrease with increasing distance from the streak (Hornbruch, Summerbell & Wolpert, 1979).

The effects of age and position on the inductive activity of streak fragments are probably in part a reflection of the fact that different tissues move through the streak in different temporal and spatial patterns. In general, endoderm ingresses before mesoderm, and prospective chordamesoderm and pharyngeal endoderm ingress more anteriorly than other mesodermal and endodermal tissues. As might therefore be expected, streak pieces which self-differentiate predominantly as chordamesoderm tend to induce neural structures, while differentiation as endoderm can be accompanied by the induction of either neural tissue or streaks (Gallera & Nicolet, 1969). Hara (1961) succeeded in isolating just the newly immigrated mesoderm and confirmed that it is a neural inductor predominantly of brain structures.

There are many obvious similarities between these interactions and those in the amphibian gastrula. The induction of streaks by the hypoblast has already been compared with mesodermal induction in amphibian blastulae, and the induction of one streak by another would seem to be a continuation of this process, just as mesodermalising (vegetalising) inductions continue in amphibian gastrulae. The induction of brains can obviously be compared with neuralising inductions in amphibians, although in birds they are not restricted to forebrain structures. One possible explanation for this is that much of the epiblast has already received some vegetalising influence from the hypoblast: in amphibians this combination would specify more posterior neural structures (p. 229). (The earlier influence of the hypoblast could also explain the differences in competence between area pellucida and area opaca epiblast.) Temporal patterns of competence are also similar in the two groups, with competence for vegetalisation or streak formation generally preceding that for neuralisation. Even the fact that endoderm can show both kinds of inductive activity has a parallel in the amphibians, where endoderm is generally vegetalising but invaginated pharyngeal endoderm is a forebrain inductor.

At later stages, as the node regresses posteriorly along the streak, it is almost certainly the source of inductive signals. For instance axial mesoderm forms wherever the node regresses, even if this is in an abnormal location (Spratt, 1957) or the postnodal streak has been removed (Ooi, Sanders & Bellairs, 1986). However, it has proved impossible to find any single tissue which is responsible for this effect (Fraser, 1960; Bellairs, 1963b; Nicolet, 1970). Up to the definitive streak stage the whole node can be removed, but axes still form lacking notochord but with somites, fused medially, and a spinal cord (Grabowski, 1956). It has been noted earlier that the node is the centre of an inductive field, so presumably tissues from the field periphery have taken over the inductive role in this case. During regression the node is apparently dorsalising mesoderm and neuralising ectoderm, which by analogy with work on amphibians suggests that it acts purely as a neuraliser. As noted in the previous paragraph, the neuralisation of already vegetalised tissues would be expected to produce posterior neural structures, so there should be no need for a subsequent transforming induction (compare p. 230). Rao (1968) claimed, however, to have distinguished activation and transformation stages, at least in the induction of those brain structures posterior to the procephalon but produced ahead of the node.

Several other interactions between the tissues have been reported for poststreak stages of bird development. The arrangement of the chordamesodermal tissues affects the morphology of the neural tube in ways closely comparable with those in amphibians (Watterson, Goodheart & Lindberg, 1955). No reciprocal interaction has apparently been claimed but epiblast which has been neuralised can homoiogenetically induce another neural plate in competent ectoderm again as in amphibians (Rasilo & Leikola, 1976). The endoderm apparently continues to have an inductive role, since in its absence mesoderm fails to form the heart (Orts-Llorca, 1963) and blood islands (Wilt, 1965a). Reciprocal effects of the mesoderm on endodermal derivatives have also been proposed (Bellairs, 1957; Bennett, 1973), although the later differentiation of the various gut epithelia can occur in the absence of mesoderm (Sumiya, 1976).

## (c) Biochemical aspects

Avian development illustrates once again (usually for the chick) many of the features noted elsewhere for the early development of other embryos. Cleavage asynchrony increases, and mitotic rates fall, from about the 30-cell stage (Emanuelsson, 1965). Syntheses of messenger and transfer RNAs are detectable from a similar stage in the quail (Olszańska et al., 1984), but rRNA synthesis probably begins only as the area pellucida is formed by cell shedding (Raveh, Friedländer & Eyal-Giladi, 1976; see too Wylie, 1972). A massive breakdown of glycogen also occurs at this time and may provide the energy for the process.

Areas with the capacity to form hypoblast are characterised by high mitotic rates. They are highest in the posterior marginal zone, but the rest of the marginal ring (especially the left side) also acts as a growth centre (Spratt, 1966). In their combinations of anterior and posterior half blastoderms of different ages Spratt & Haas (1962a) in fact found that axis initiation always occurred from the part with the greatest growth rate. Cell division is not required, however, for the formation of the hypoblast or the acquisition of its streak-inducing ability (Weinberger & Brick, 1982; Mitrani, 1984). The resistance of these processes to bromodeoxyuridine (Zagris & Eyal-Giladi, 1982) and of streak formation to (partial) transcriptional inhibition (Olszańska & Kludkiewicz, 1983) suggests that they may be controlled by long-lived mRNAs.

Eyal-Giladi *et al.* (1975) have investigated the rates and patterns of protein synthesis in epiblast and hypoblast over the period when the latter induces the primitive streak. Rates are always higher in the epiblast (particularly in the marginal zone: Zagris & Matthopoulos, 1985), but they double in the hypoblast at the time of its inductive activity and show a peak in the epiblast at the short streak stage. In association with these rate rises two apparently new protein species are detected first in the hypoblast and then in the epiblast. Eyal-Giladi *et al.* (1975) therefore suggested that induction involves either the transfer of one or both of these proteins or the stimulation of their synthesis in the epiblast. Wolk & Eyal-Giladi (1977), in a study of the antigenic differences between these layers, also gave evidence of antigen transfer from the hypoblast to the epiblast. Epiblast also synthesises one protein not formed by hypoblast (Zagris & Matthopoulos, 1985), but still produces a narrower range of proteins than hypoblast when streak formation is complete (Lovell-Badge, Evans & Bellairs, 1985). Some streak induction can apparently occur through filters with an average pore diameter of 0.45 μm, but these streaks never develop notochord or somites (Eyal-Giladi & Wolk, 1970).

Once a primitive streak appears it, like the marginal ring, is characterised by a high mitotic index, and far more of the spindles are vertically oriented than in other surface areas (Spratt, 1966; Efremov, Morozova & Sergovskaya, 1980). There are several other indications that the streak and particularly the node are areas of high metabolic activity, e.g. in dye reduction rates and susceptibility to poisons (Spratt, 1955, 1958). Neural induction is not prevented by actinomycin treatment of the node or of the reacting area opaca epiblast (Gallera, 1970a); a reaction is possible even if the node is removed some hours previously and actinomycin is present throughout the latent period (Gallera, 1969). These results suggest that neural inductivity and competence have become largely independent of transcription by the definitive streak stage or even earlier, but the extent of synthetic inhibition in Gallera's work was not checked and there are many other uncertainties associated with such studies (see p. 161). The neuralising stimulus can cross filters but its effect is then attenuated or at least delayed (Gallera, Nicolet & Baumann, 1968).

In amphibians many treatments which neuralise ectoderm also dorsalise mesoderm (see Chapter 7). Effects of this kind can be assayed in bird embryos using postnodal pieces, i.e. the posterior fragments of blastoderms transected behind the node. Without special treatment these rarely form neural structures or axial mesoderm, but, as in amphibians, these inductions are all too easily released. Successful treatments include the use of DNA fractions, even after DNase (Butros, 1960), follicle-stimulating hormone (Sherbet & Mulherkar, 1963), cyclic AMP (Deshpande & Siddiqui, 1976) and small, sulphydryl-containing compounds (Waheed & Mulherkar, 1967; Chauhan & Rao, 1970). The relevance of these effects for normal inductive mechanisms remains highly questionable. Some reagents affecting sulphydryl groups do indeed inhibit inductions and axial differentiation (Lakshmi, 1962a,b; Joshi, 1968), but the effects of 2-mercaptoethanol seem much less specific (see Pohl & Brachet, 1962). Nerve growth factor induces neural tissue without dorsalising mesoderm (Lee, Nagele & Roisen, 1985). RNA and protein-rich extracts from specific chick tissues have very different effects on postnodal pieces or whole blastoderms: the former promote formation of the homologous tissue (Butros, 1965; Arnold, Innis & Siddiqui, 1978) while the latter inhibit it (Lenicque, 1959).

There are other indications that neural induction, as in other embryos, requires only a relatively simple evocating stimulus. Neural structures can be induced from area pellucida epiblast using dead nodes (Waddington, 1933b; Leikola & McCallion, 1968) and heterogenous tissues (Pasternak & McCallion, 1962) although these may not be active on area opaca epiblast (work of Gallera cited by Gallera et al., 1968). In the chick it is possible that the induction of mesoderm and endoderm is also relatively unspecific, since these too differentiate following prolonged culture of the induced tissue in the coelom of older embryos (Viswanath, Leikola & Rostedt, 1968). However, this work lacked adequate checks for the effects of the coelomic environment, and indeed for the effects of the hypoblast on the epiblast before the experiment began (see p. 294). Certainly the alcohol-killed chick node used in the work only induced archencephalic structures from newt ectoderm (Viswanath, Leikola & Toivonen, 1969).

It has proved difficult to artificially vegetalise the epiblast of streak-stage embryos. Deuchar (1969) obtained only slight reactions to the factor isolated from later chick embryos which is so active in vegetalising amphibian embryos. Rostedt (1971) found no difference in the response to bone marrow (a vegetaliser in the amphibian system) and to heterogenous neuralisers, and attributed this to the low competence for streak induction at the definitive streak stage. Surprisingly, lithium ions are among the substances which confer nodal properties on postnodal pieces (Waheed & McCallion, 1969): perhaps this should be compared with the dorsalisation of amphibian embryos following early lithium treatment (p. 192). When whole chick embryos are cultured with lithium ions the cyclopic defects familiar from other vertebrate classes are seen (Rogers, 1963).

Some biochemical properties of the streak may relate with the ingression movement rather than with inductive activities. This seems to be true of the loss of glycogen at the amphibian blastopore (p. 246), so may well be for the equivalent change in birds (noted by Jacobson, 1938).

The separation of the cell layers may depend upon a switch in the adhesion molecules expressed on the cell surfaces (Thiery *et al.*, 1984; Hatta & Takeichi, 1986). Ingressing cells also show raised acetyl-cholinesterase activities (Drews, Kussäther & Neuman, 1985) and reduced acid phosphatase activities and yolk content (Laasberg & Kyarner, 1979), but more complex spatial patterns are established in these cases. Mesodermal migration can be delayed by inhibiting polyamine synthesis, (Löwkvist, Emanuelson & Heby, 1983; Löwkvist *et al.*, 1986). Levels of cyclic AMP vary at the gross and ultrastructural levels (Reporter & Rosenquist, 1972; Neuman, Laasberg & Kärner, 1983), and there is evidence that an extracellular cyclic AMP signal can control the direction of streak extension (see Gingle & Robertson, 1979). There is a wave-like component to the movement (Stern & Goodwin, 1977; Robertson, 1979) which is consistent with such an extracellular signal. Electrical currents have also been detected leaving the upper surface of the streak and returning elsewhere through the epiblast (Jaffe & Stern, 1979), but the resulting fields probably do not guide cells migrating from the streak (Stern, 1981). It is more likely that they result from a system of sodium pumps in the epiblast used to transport fluid into the embryo (Stern, Manning & Gillespie, 1985) much as occurs in the cavitating mammalian blastocyst (p. 206).

As in many other groups extracellular materials apparently play an important part in the gastrulation movements. The migrating cells are probably guided by oriented fibrils of fibronectin beneath the epiblast (Critchley *et al.*, 1979), and mucopolysaccharides may mediate this interaction as well as maintaining the extracellular spaces (Fisher & Solursh, 1977; Harrisson, van Hoof & Foidart, 1984). Enzymes (Shur, 1977) and lectins (Zalik, Milos & Ledsham, 1983) recognising the surface residues of glycoproteins may also play a part in this system, and it appears that mesoblast cells remodel the overlying basement membranes during their migration (van Hoof & Harrisson, 1986). It is possible that the extracellular materials or surface-associated activities also have inductive effects (Vanroelen & Vakaet, 1981; Shur, 1977), or that the former, acting simply as a barrier, affect the outcome of inductive interactions (Duband & Thiery, 1982; see too Sanders & Prasad, 1986).

Of the late inductive interactions, that in which endoderm induces blood islands from mesoderm has received some further study. The activity can cross Millipore filters and is destroyed by lethal heating of the endoderm (Miura & Wilt, 1969). The synthesis of haemoglobin in the blood islands seems to be translationally controlled following an earlier transcriptional event (Wilt, 1965*b*; Miura & Wilt, 1971), but the relation between the inductive interaction and this gene expression pattern is unknown. By organogenesis stages many of the early primordia are synthesising distinctive marker proteins although this appears still not to be true of the ectoderm (Lovell-Badge *et al.*, 1985).

## A note on the mammals

Mammals of course develop by the complete cleavage of small non-yolky eggs, but their method of gastrulation clearly betrays their evolutionary relationships with the reptiles and birds, and means that they can be most easily discussed here. In Chapter 6 development was described to advanced blastocyst stages, where there is a thin trophectodermal layer enclosing an inner cell mass (ICM) and a large cavity, and where the ICM cells facing the cavity already form a recognisably distinct layer. All of the trophectoderm and some of the ICM will be devoted to the formation of extraembryonic structures, but these processes will receive almost no consideration here. This can be justified not only because extraembryonic structures are a peripheral interest for this book but also because their development is both complex and highly variable among different mammalian species (see Balinsky, 1975, pp. 270–283).

It is the further development of the two-layered ICM which recalls so strongly development from the two-layered stages of reptile, and particularly bird, embryos. In fact, the two layers are frequently described as epiblast and hypoblast, and the whole ICM as a blastodisc. A streak appears in the posterior part of the blastodisc with a node at its anterior end. Through these structures cells immigrate very much as they do in birds, ultrastructural and cinemicrographic studies confirming the close similarity of the streak in the two groups (Solursh & Revel, 1978; Nakatsuji, Snow & Wylie, 1986). In some mammals, including man, a distinct canal equivalent to an archenteron is formed within the notochord as cells move in at the node. The ingressing cells apparently give rise to all the mesoderm including extra-embryonic mesoderm and to endodermal cells which invade the hypoblast (Heuser & Streeter, 1941). It is likely that, as in birds, all the embryonic endoderm arises in this way (Levak-Švajger & Švajger, 1974; Beddington, 1983).

Švajger and his co-workers have developed methods for the isolation of small tissue areas from prestreak to head-fold stage rat embryos, and their culture in extrauterine sites in adults (Švajger & Levak-Švajger, 1975; Švajger, Levak-Švajger & Škreb, 1986). The pieces develop into teratomas with many recognisable tissues. In the absence of fate maps it is, however, difficult to know whether these isolates are self-differentiating or regulating, and it cannot be assumed that culture conditions in homografts are inductively neutral. The data suggest that streak epiblast is totipotent (Švajger *et al.*, 1981) and that even the later ectoderm in regions where a neural plate is forming can regulate to produce epidermal and mesodermal derivatives (Svajger & Levak-Švajger, 1976). In contrast, the differentiation of isolates cultured *in vitro* suggested extensive mosaicism with the possibility of clonal mechanisms of determination (Snow, 1981), but in this work cells of all germ layers were present in the isolates. Beddington has provided information on both normal fates and regulative abilities by injecting or grafting labelled epiblast or streak cells into same-age hosts. Normal fates were checked by transferring cells to the same position in recipient embryos. This confirmed that prospective mesoderm and endoderm are still present in the epiblast at streak levels while more anterior epiblast will contribute mainly to epidermis and neurectoderm (Beddington, 1981), and showed that prospective mesodermal cells are in the same sequence along the streak as in the chick (Tam & Beddington, 1987). From such work it should soon be possible to construct a fate map. When cells were injected to different positions they developed primarily according to their new position, demonstrating their regulative ability, although this ability seemed already to be decreasing in anterior ectoderm at late streak stages (Beddington, 1982).

The organisation of regulative development in other vertebrates, however, requires an organiser region which can self-differentiate even in an abnormal location, and this was not obvious in Beddington's work. In particular, cells which were apparently prospective notochord failed to self-differentiate or affect the development of recipient embryos when injected to other sites. Waddington (1937) who grafted rabbit streak fragments beneath lateral epiblast also failed to get any clear inductions. Rabbit streaks did, however, show inductive activity when tested on chick epiblast, and rabbit epiblast was competent to react to chick streaks (Waddington, 1934). The importance of the mouse node for the formation of the axial tissues is also suggested by the strong inhibitory effects which follow its irradiation with X-rays (Benoit, 1969).

Another approach to the problems of early mammalian development has been made by studies of mutational effects, of which the most interesting concern loci in a restricted region of chromosome 17 in the mouse. Early work (reviewed by Gluecksohn-Waelsch, 1953) suggested that the complex *T* locus might be required for the formation and normal functioning of chordamesoderm, while neighbouring loci had roles in organising phenomena like those known in other vertebrate classes. Later work suggests, however, that the *T* locus may be more concerned with morphogenetic movements than with determination. Thus, homozygotes for one *t* allele form an abnormally large streak at which mesodermal ingression is much delayed (Spiegelman & Bennett, 1974), and a mutant antigen is expressed on the mesodermal cells (Ben-Shaul, Artzt & Bennett, 1983). Moreover, in this case changed mitotic patterns have been detected even in prestreak embryos (Snow & Bennett, 1978).

The streak stages of the mouse seem characterised by rapidly increasing rates of metabolism (Clough & Whittingham, 1983) and of mitosis (Snow, 1977). Mitosis would be particularly active in a region near the node according to Snow (1977), while Poelmann (1980) noted more dividing cells in prospective epidermis than in prospective neurectoderm. The similarity between the streaks of birds and mammals extends to the biochemical level, with extracellular fibronectin (Zetter & Martin, 1978; Wartiovaara, Leivo & Vaheri, 1979) and mucopolysaccharides especially hyaluronate (Solursh & Morriss, 1977) having been detected there, and a cell adhesion molecule characteristic of epithelial cells being lost as mesodermal cells ingress (Vestweber & Kemler, 1984). Cytoskeletal changes in the mesoderm, including a switch to vimentin formation, may also be important for motility (Franke *et al.*, 1982). Biochemical differences between the tissues are noted at particularly early stages in mammalian development (see pp. 140 and 208), and cell-surface associated carbohydrates show complex spatial patterns by streak stages (Fenderson, Hahnel & Eddy, 1983). Genes containing homoeoboxes are also being transcribed (Jackson, Schofield & Hogan, 1985), at least one of them in a spatially restricted way (Gaunt *et al.*, 1986) which is suggestive of a selector gene role like that in *Drosophila*. Even before this, different marker proteins are being formed by the epiblast (Martin, Smith & Epstein, 1978), the endoderm in and around the embryo (Dziadek & Adamson, 1978; Adamson, 1982) and the endoderm lining the trophectoderm (Strickland, Reich & Sherman, 1976). As in the chick, endodermal tissues apparently synthesise a broader range of proteins than the epiblast (Evans *et al.*, 1979). Interactions with other cells may affect some of these syntheses (Adamson, 1982), but we still have no evidence on the nature of the inductive signals in early mammalian development.

## Conclusions

Even in the development of large partially cleaved eggs many of the principles noted earlier for totally cleaving species are recognisable. For instance, the cells nearest to the animal pole seem to contain little developmental information and to require induction from more vegetal areas. In the cephalopods Arnold believes that cells cleaved animally are induced as they spread over the uncleaved area. There is evidence both for and against his idea that determinants are arranged in a fixed cortical map, but in Marthy's interpretation it is still cells moving from more vegetal levels which pass inductive information to the overlying cell layer.

In the vertebrate classes showing partial cleavage, cells at the animal pole seem to contain only the information appropriate for differentiation as epidermis, and more peripheral ones again play dominant developmental roles. In fact in teleosts it now seems that the whole enveloping cell layer produces no more than a temporary epidermis, while the axis is initiated amongst deep cells in a germ ring at the edge of the embryonic region. In contrast, the outer cell layer of bird embryos gives rise to all the embryonic tissues, but it would probably produce only epidermis if it were not for a sequence of inductive interactions again originating in a germ ring. The teleost germ ring apparently acquires its developmental properties inductively from the uncleaved yolky part of the egg, but this part of the avian egg is not known to have any informational role in development.

Although teleost and avian axes originate from a germ ring they show bilateral, rather than radial, symmetry from an early stage. This is because one side of the ring, at the future posterior end of the embryo, acts as a dominant centre and suppresses axis formation elsewhere. In birds the future posterior side is epigenetically determined as that side of the embryonic area which is uppermost at a critical stage of intrauterine development. This role of gravity recalls amphibian egg symmetrisation, and similar systems may operate in reptiles and bony and cartilaginous fish, but the evidence for the bony fish can be interpreted in other ways. In any case teleost bilaterality is at first a property of the uncleaved area which polarises the blastoderm as it induces the embryonic tissues there. This recalls the combined inducing and symmetrising effects of the vegetal endoderm on the animal half of amphibian blastulae, and in birds too the endoderm induces and symmetrises the epiblast during streak induction. Moreover, this is a vegetalisation in which endoderm and mesoderm are induced from the cells at the animal pole.

The induction of neural tissue occurs after the major phase of vegetalisation in amphibians and is a property of cells involuted at gastrulation. Classical evidence for a similar system in teleosts requires re-evaluation, since it is now usually thought that no involution occurs. In birds neuralising and dorsalising inductions are again late steps in the determination sequence, exercised mainly by the node, but its posterior regression along the streak is more difficult to relate with amphibian gastrulation.

The evidence of homology in the causal embryology of different vertebrate classes, when their visible processes of development are so different, is one of the most interesting findings of this chapter. It suggests that the exact spatial parameters within which the inductive interactions operate have very profound effects on development, and the development of some egg-brooding hylid frogs (*Gastrotheca* species) supports this conclusion. These eggs cleave completely but are large (3 to 4 mm diameter) and archenteron extension is delayed. An axis forms in the disc of cells which overlies the archenteron, and looks superficially like that of a bird (see Fig. 9.16) although its mode of formation is different (del Pino & Elinson, 1983).

Biochemical studies of meroblastically cleaving embryos have pro-

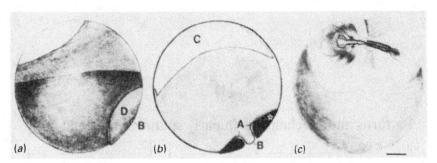

Fig. 9.16. Development from a large egg (3 mm in diameter) in *Gastrotheca riobambae*. (*a*),(*b*) external (*a*) and sagittal (*b*) views of a late gastrula: A, small archenteron; B, blastopore; C, large blastocoel cavity; D, embryonic disc of yolk-poor cells, coloured black in (*b*) with the future head position marked by an asterisk. (*c*) a later stage in which most of the embryo's body has developed from the expanded disc. (From del Pino & Elinson, 1983.)

vided some interesting information. It is claimed that such complex materials as enzymes and ribosomes are transferred between the teleost yolk cell and blastoderm, and information on the transfer mechanisms would be very valuable. Transcripts required for gastrulation are apparently produced at blastula stages in the loach (and perhaps the chick), and many processes occurring at this stage seem to require the presence of the yolk. In studies of streak induction by the hypoblast of the chick there is some evidence for protein inducing factors. Neuralising and dorsalising effects are released by dead or heterogenous tissues in teleosts and birds as they are in amphibians. Indeed, work with cephalopods, where spatial pattern is disturbed by lithium ions and RNA-containing determinants have been postulated, suggests that some spatial determination mechanisms may be common to a far wider range of animals.

# 10

## Patterns and mechanisms in early spatial determination

The preceding chapters have reviewed the available data on spatial determination in the early development of a great variety of animals. It is hoped that this will provide a useful database for researchers active in or entering the field, both circumscribing the possibilities and suggesting new avenues of approach for the model builders and the molecular embryologists. It is particularly encouraging when the data from different animal groups indicate common phenomena and possibly common underlying mechanisms. In this final chapter we will seek to identify such basic features and consider how they may have been modified in the development of different types of embryo.

### Patterns

It is clear that oocytes possess little preformed spatial pattern. In some cases not even the primary polarity is determined, e.g. in ctenophores where it is apparently fixed at fertilisation. In other cases it is probable that this axis is fixed but few developmentally important materials have been localised along it, since fragments cut from oocytes and eggs so often succeed in developing as miniature larvae. Determinants of various larval structures could, however, be present even in concentrically organised oocytes, as is shown by ascidian development. There three plasms can be recognised in an approximately concentric arrangement within the unfertilised egg, and following fertilisation they flow to new positions defining a bilaterally symmetrical pattern. This pattern is of undoubted importance for further development, although whether it represents a mosaic of tissue-specific determinants remains to be discussed later.

The determination of the bilateral plane in ascidian and spiralian embryos, like that of polarity in ctenophores, seems to depend on the sperm entry point. This and gravity are two of the cues commonly used to fix bilateral symmetry in vertebrates. In all these cases, however, it is likely that three-dimensional patterns are first established using maternal components, and it is only their final arrangement which is environmentally decided. Cleaving mammalian embryos rearrange maternal com-

304

ponents according to the environmental cue of inside–outside position. Even after these rearrangements, however, little embryonic organisation has been determined. In the sea urchin, classical studies suggested maternal control of polarity and bilaterality while almost all other features of development depended upon the embryo's own genome. In mammals even the axes remain to be fixed later in development.

Another method of assessing the maternal role in embryonic determination is to observe the development of eggs laid by mutant females. If maternal genes coded tissue-specific determinants their mutation should lead to the development of offspring lacking particular tissues. We now know several maternal mutations which cause a deletion of the germ cells in *Drosophila* but none which delete specific somatic tissues or organs in this or any other species. (It would be interesting to know whether such deletions are obtainable in ascidians.) Other *Drosophila* maternal mutations do however affect somatic patterns – displacing larval primordia, reversing polarity over large areas, deleting terminal regions or affecting segmentation patterns. This supports the conclusion that the maternal contribution to somatic determination is to define a few simple parameters such as polarity and dorso-ventrality. At present there is conflicting evidence over whether the germ-line is determined together with other derivatives of the posterior pole or by a separate system of tissue-specific determinants.

According to both genetics and experimental embryology, therefore, the details of local determination must be fixed epigenetically. Since the theoretical contributions of Wolpert (1969) and Goodwin & Cohen (1969) there has been considerable interest in how cells may be determined by recognising their distance from a few reference points. These modelling studies have been discussed by Meinhardt (1982) and Slack (1983), and here we will concentrate rather on the empirical evidence.

The different parts of animal eggs and early embryos are not all equally important for development. In many cases the vegetal pole plays a particularly important part. Vegetal cells are often the first to be determined for a somatic fate and the germ-line may originate there. Even the development of some animal pole structures like the spiralian apical tuft is at first dependent on factors localised vegetally, and in most groups vegetal halves regulate to produce structures appropriate to the animal half as well as their own derivatives. In contrast isolated animal halves almost always produce less than they should, commonly ceasing development as hyperblastulae in spiralians, sea urchins, ascidians, amphioxus, teleosts and amphibians. Such blocked embryos exhibit epithelial properties and may develop structures (e.g. apical tuft or test material) or molecular markers of the epidermis. Cells of the animal half seem therefore to have the potential for autonomous epidermal differentiation (and perhaps morphogenesis, e.g. in epibolic spreading) but to require signals from the vegetal half before they can differentiate (or show the behaviour patterns) of other cell types.

We can explain the role of the vegetal pole in such cases in at least three ways: 1) that it is a determination centre, receiving determinants for all embryonic structures except the epidermal ones, 2) that it is a reference point, or 3) that it is a precociously determined area which then affects the developmental choices of other areas. Most researchers studying 'mosaic' embryos talk in terms of organ-specific determinants, e.g. proposing that apical tuft determinants are localised in the polar lobe of many spiralian embryos. However, since the polar lobe also has profound effects on the animal–vegetal and dorso-ventral organisation of the whole embryo, and since organ-specific inhibitors must also be invoked in some cases, it may be more appropriate to think of the vegetal pole as a reference point. In its absence positional information throughout the embryo would be grossly disturbed. Data from 'regulative' embryos are more habitually explained in such terms, e.g. when an isolated animal half from a sea urchin embryo fails to form ciliated band and stomodaeum we propose that the vegetal half is needed to supply not specific determin- ants but positional information. More specifically, in such embryos the vegetal area changes the fates of more animal cells in a vegetal direction, i.e. it vegetalises them. In spiralian embryos too the vegetal area promotes the development of the post-trochal region, but its role in apical tuft determination clearly cannot be seen as a vegetalisation.

Cells from the animal half of most embryos can be shown to 'animalise' the fates of more vegetal cells only in rather special circumstances, e.g. when the animal and vegetal poles are brought together during the closure of a sea urchin meridional half embryo. An analogous explana- tion has been offered in Chapter 7 for events occurring at the amphibian dorsal lip, where cells from different animal–vegetal levels are brought into contact by the gastrulation movements and the more vegetal cells change their properties in ways which can be interpreted as an animalisa- tion. The induction of spinal cord in ascidians may be an animalisation of the vegetal half, but we need to know what fate change is occurring here. Determining effects by the animal pole plasm in spiralians could explain why a D cell isolate apparently regulates better in *Tubifex* than in *Dentalium*: only the former has the normal balance between the two polar plasms. Many authors have interpreted the need for an animal– vegetal balance as evidence for antagonism between two polar influences, and sometimes an animal gradient can be recognised by differences in resistance to vegetalisation. Cells from animal areas are not just a neutral substrate for induction by the vegetal pole. Moreover, in many cases the vegetal area, though determined, seems unable to differentiate unless activated by cells from more animal levels.

Another point of general importance has emerged from studies of animal–vegetal interactions in sea urchin, amphibian and *Euscelis* embryos. This is that no part of the axis is indispensable for embryo formation. Equatorial areas, and combinations of the two polar areas, can both regulate successfully, presumably because both have a balance between animal and vegetal influences. Even fragments which lack such a balance can regulate successfully following artificial treatments, e.g. a sea urchin animal half regulates following treatment with lithium ions. This suggests that the polar influences are not dependent on complex, pre-formed materials but are inducible in any cell by relatively unspecific treatments. We may in fact be dealing with animal and vegetal metabolic types, as discussed in the other main section of this chapter. Even in the ascidian embryo, where the evidence for organ-specific determinants is probably strongest, lithium ions can induce endoderm from animal halves.

We have dealt here at some length with those embryos where the vegetal pole plays a major role in spatial determination, but this is not the only system to be used. Some insect species show a comparable reliance on the posterior (approximately vegetal) pole, while in others there is a trend to increase the importance of the anterior pole, culminating in systems where the two poles have approximately equal significance as in chironomids. In the ctenophore zygote if there is a determination centre it is at the fertilisation site which is usually at the anterior pole. Fragments from early hydrozoan or starfish embryos can regulate even when they lack an animal–vegetal balance, but a recognition of polarity does seem to be important, and it is possible that it is used later to define a reference point vegetally. In mammalian embryos it is a recognition of concentric organisation which is required for the first developmental choice, and there may be no polar organisation once the polar bodies have been emitted.

The early embryos of most animal groups show not only polar but also bilateral organisation. The second axis is described in different ways in different groups of embryos according to its orientation in later develop-ment. It is the dorso-ventral axis in spiralians, sea urchins, insects and amphibians, and the antero-posterior axis in ascidians, amphioxus and the meroblastic vertebrate classes. In ctenophores there is the tentacular plane, although both 'ends' of this plane develop in equivalent ways rather than being different extremes of an apparent single gradient. No second axis is obvious in the early development of hydrozoans, a group noted for radial symmetry throughout the life cycle. In mammals all axes arise late, and it appears that stages with concentric organisation have been interpolated before those recognisably derived from the vertebrate tradition.

Determination in this second plane seems to be different in principle from that along the polar axis, as first became clear in work with sea urchins. When the influence of the vegetal pole was attenuated in some way animal pole derivates became overdeveloped, spreading to more vegetal levels than usual. When the influence of the ventral side was attenuated the dorsal side developed as a ventral one. In contrast to the antagonism between the animal and vegetal poles, determination in the dorso-ventral plane seems to be controlled by a single dominant zone on the ventral side. There appears to be a similar dominant side dorsally in amphibians and most spiralians, posteriorly in the meroblastic vertebrates and perhaps ventrally in insects. Structures usually character-istic of the dominant side can develop in other meridians if they are separated from the dominant side (teleosts, birds), placed at the 'favour-able' end of an environmental gradient (amphibians) or stimulated in some other way (birds). In many spiralians the dominant side is mosaically decided at the first two cleavages, but if these are experi-mentally equalised it seems that once again all quadrants have the potential to assume dominance. To change the animal–vegetal polar-ity of these same embryos may be completely impossible or require a major redistribution of the ooplasms or the local destruction or addition of materials.

Despite these differences between the polar and bilateral determin-ation systems there is much evidence that the two are linked. Where the vegetal area acts as a determination centre or reference point it seems also to be the first area to be bilaterally symmetrised. The more animal cells show complete radial symmetry (or slight differences, as in the competence of ascidian cells) until symmetrised by interactions with the vegetal cells. The vegetal area is therefore inducing new tissues along the animal–vegetal axis and, apparently at the same time, symmetrising these tissues, e.g. amphibian endoderm induces mesodermal tissues in a bilaterally symmetrical pattern, and the dorsal mesoderm then induces neural tissues from yet more animal cells. The link is most obvious in those spiralian embryos where the dorsal side is determined by the unequal segregation of the vegetal pole plasm. If this plasm is removed, a radially symmetrical larva develops with all four quadrants forming ventral derivatives. Vegetal uv irradiation of the amphibian zygote produces a similar radially symmetrical ventral ground-state. In this case, moreover, we have good evidence that there are no specific dorsal determinants, since all that is needed to re-establish the bilateral organis-ation is a rotation of the egg. Lithium ions, known usually for their vegetalising effects, can also promote the development of dorsal structures.

The simplest explanation of these data is that embryos acquire bilateral symmetry by moving the vegetal reference point slightly to one side of the pole. Explanations of this sort, involving displaced polar centres or

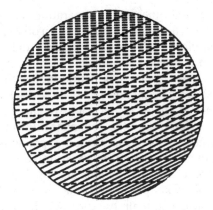

Fig. 10.1. Two polar factors can specify bilateral patterns, rather than radial ones, if the centre producing one factor is displaced to one side of the pole.

different transmission rates for polar influences, were suggested long ago for sea urchin embryos. In equally cleaving spiralian embryos too the dominant quadrant is that in which the animal and vegetal plasms interact. Effectively, this means that bilaterality arises as a distortion of the animal–vegetal determination system (see Fig. 10.1), with the different meridians at first differing only quantitatively. Such ideas will probably prove most difficult to apply to the ascidians, where three different tissues already seem determined along the antero-posterior axis in the vegetal half of the zygote. Even in this case, however, it is at least possible that the three zygotic areas differ quantitatively, the different levels providing positional information which indirectly determines that they will have different fates.

A system which can specify position in two planes is formally capable of explaining determination in bilaterally symmetrical embryos. That it is position which is first specified is supported by what we know of compartment specification in insects, and by the apparently graded nature of the differences between determined areas in amphibians even at the end of gastrulation. In both cases differentiation (the synthesis of tissue-specific macromolecules) soon follows, and it is not difficult to imagine ways in which this is possible in a cell which 'recognises' its position. Simple systems such as those already discussed in this chapter are probably capable of establishing quite complex spatial patterns, particularly as morphogenetic movements are often changing the spatial relationships between differently determined cells. There is a reciprocal effect here too, since induction frequently changes the motile properties of cells.

Time is another factor which should be taken into account in considering early embryonic determination. Events such as tissue-specific syntheses in ascidians and micromere separation in sea urchins are programmed to occur at a certain time even when other processes such as cleavage have been blocked. Time is particularly important for spatial determination when it affects the inductive activity or competence of cells. Such changes may explain why neural induction is delayed until gastrula stages in ascidians despite the fact that the A4.1 and a4.2 lineages are always in contact. It could be precocious timing rather than greater signal strength which determines the dominant side in bilateral systems. Competence patterns also change as a result of previous inductions, the general finding in amphibians being that exposure to one inductive signal prolongs competence for reaction to other signals. It has been suggested in this book that Nieuwkoop's 'transformation' may be a prolonged response to vegetalising stimuli by cells which have already been neuralised.

The extent to which common spatial patterns are recognisable in different animal groups is both surprising (since it seems to have been so little acknowledged in the past) and encouraging (for the future quest for a 'basic' system and its mechanism). The use of the vegetal pole as a determination centre or reference point may have evolved because some animal pole cytoplasm is lost in the polar bodies at maturation. The ctenophores would offer an exception, as 'determinants' apparently accumulate in a position where polar bodies are usually emitted, but most accumulation probably occurs after maturation in this case.

Even a basically homologous system for spatial determination will of course have been subject to modification according to the requirements of development in different groups. It is noteworthy that many embryos with mosaic determination develop with unusual rapidity: some ascidian and ctenophore species can produce a motile larva within eight hours of fertilisation. Such groups may have adopted mosaicism because there is little time for interaction among the cells. Interaction between animal and vegetal determinants or metabolic types would still be possible so long as the responsible materials were correctly partitioned at the early cleavages. Indeed the possibility that axial determination depends upon two polar gradients was probably first suggested for the mosaic nematode embryo in these now familiar terms: 'the given ratio of two substances *An* and *Ve* in a blastomere decides whether the cell adopts the quality AB or $P_1$' (Boveri, 1910, p. 205, see also his p. 199). The inflexibility of mosaicism does however have its costs: cell deaths may cause the deletion of whole organs, and small deviations in the morphogenetic movements would cause a misalignment of the organs. It may be that all embryo groups retain some regulative ability to ensure the harmony of their development. In particular the inductive origin of neural tissue in ascidians, amphioxus and probably all vertebrates almost certainly reflects the importance of the correct alignment of chorda and neural tube within the embryonic axis.

A related factor which probably also affects determination is the complexity of the organism to be constructed. The ascidian larva has comparatively few organs many of which are primarily composed of a single cell type. According to Conklin enough plasms are segregated in the zygote to foreshadow much of the larval organisation. This situation is suitable not only for mosaicism but also for direct determination of tissues, rather than determination of position which is then interpreted by specific differentiation. The amphibian tadpole, despite its similar overall organisation, displays far greater histological complexity. In ascidians 'muscle determination' effectively means 'tail determination', but in amphibians the tail contains many tissues and muscle makes a major contribution to many organs. It would be difficult to specify such a complex pattern mosaically, although we should note that such controls are largely responsible for the formation of miniature adults in leeches and oligochaetes. This is apparently possible because of the extreme precision of the cleavage pattern, probably linked with the repeated nature of annelid organisation.

In those cases where determination requires interactions among the cells, the size of the embryo must also have important effects. Egg diameters vary greatly, being of the order of $100\,\mu$m in many marine invertebrates, 1 mm or more in amphibians and even many centimetres in some birds. On the assumption that morphogens are small diffusible molecules (with molecular weights of 300–500) Crick (1970) estimated that fairly linear concentration gradients could be established in three hours over distances up to 1 mm. With macromolecular morphogens this distance would have to be reduced, although the existence of syncytial connections and other specialised mechanisms would facilitate transfer. A limit of this order is consistent with the establishment of gradients along the whole axial length of the sea urchin embryo, while only an equatorial ring of cells is affected by interactions in the amphibian blastula. Further spread of the inductive influences in the latter case can apparently only occur when morphogenetic movements have brought cells from the two polar areas closer together. In meroblastically cleaving eggs the problems of large size are solved by restricting inductive interactions to a part of the egg volume and to particularly short linear distances, e.g. between two cell layers or the uncleaved cytoplasm and the nearest cells. Perhaps the feasibility of direct interactions explains the absence of normal gastrulation movements in teleosts and the differences in competence between avian area pellucida and area opaca. The relative movements of the layers, however, probably still affect the patterns of induction.

**Mechanisms**

In seeking the underlying mechanisms of determination, we have to remember both the findings of experimental embryology, discussed above, and those of the molecular embryologists. The former make some possible mechanisms seem very unlikely indeed but also suggest several clues to help in the search. For instance, we could concentrate our effort on places acting as determination centres and on stages when determination is progressing. Unfortunately, these clues have frequently been ignored in the past and quite a lot of our knowledge of molecular embryology is of little help for the determination question.

Many researchers have tried to find 'determinants' in the ooplasms which so often seem important for the patterns of development. In most cases centrifugal stratification of the larger organelles has little effect on the patterns, making it unlikely that such organelles bear determinants. The most surprising exception is the reversal of polarity when the amphibian egg is inverted in unit gravity. This could indicate that very large organelles carry important spatial information. If we propose instead that smaller components are responsible but move with the mass flow following inversion, we must explain why this does not happen during the centrifugation of other eggs. It could be, for instance, that some movement is permissible in a relatively large egg which will show regulative development, but must be prevented by anchoring the 'determinants' when eggs are small and/or require accurate localisations for mosaic development.

Where patterns are little affected by moderate centrifugations, 'determinants' could be 1) present in components too small to be displaced by such forces, 2) held in place by the cytoskeleton, or 3) held in place beneath the egg surface. Many authors have supported ideas of cortical maps but there is little direct evidence for them. In fact, development seems little affected when quite large areas of the surface are withdrawn experimentally in ascidians or nematodes or lost in the enlarged polar bodies of other eggs. The success of ooplasmic transplants in ascidians and *Drosophila* supports an internal 'map', but a similar transplantation of polar lobe plasm to non-D quadrants failed to affect development in *Dentalium*. The inversion work shows that amphibian egg polarity does not depend upon surface-associated materials, although these may have a role in the determination of bilateral symmetry.

Even if the cortex contains no determinants itself it could play an indirect role by affecting the localisation or segregation of internal components. Ooplasmic segregation commonly involves a contraction of cortical elements and an accumulation of inner materials subcortically in specific areas (often vegetally, though as several patches in *Lymnaea*). Properties of the vegetal cortex are important for the unequal cleavage which forms the sea urchin micromeres and which may have far-reaching consequences for development. More generally, an interplay between cortical and inner components controls cleavage pattern and so the segregation if not the localisation of all components. There is now considerable interest in the relationships between the cytoskeleton and the cell surface, and beween different cytoskeletal elements, and this may in the near future greatly increase our understanding of the cortical role in determination.

In a few cases the existence of a determination centre or reference point has allowed investigators to focus their studies on this area and identify some likely 'determinants'. Granular bodies can be associated with the germ-line in many groups: in *Drosophila* they are almost certainly involved in the segregation if not the determination of the germ-line and possibly in the determination of posterior somatic structures too. In *Bithynia* the 'vegetal body' is apparently responsible for much of the polar lobe's significance, and in equally cleaving spiralians the fate of the 'ectosomes' suggests a role in spatial determination. These are all relatively large components, but the vegetal body at least is not easily moved by centrifugation. They all also contain RNA, which apparently plays an indispensable role at least in pole cell determination in *Drosophila*. The evidence is less complete in spiralian embryos but, even in *Ilyanassa* where we know less about the responsible organelles, maternal RNAs segregated via the polar lobe and translated over the period of embryonic determination could provide the basis of the D quadrant's special role (see Chapter 3).

In most animal eggs it is impossible to find organelles which are specifically localised within restricted areas. Less direct evidence suggests, however, that RNA-containing determinants exist in many of these cases. In the chironomid *Smittia* they seem certain to be involved in the determination of the anterior pole and likely to be so at the posterior pole. Maternal RNAs (some but not all of them localised) provide at least some of the information required for spatial determination in *Drosophila*, as has been shown by RNA injection experiments. There are at least quantitative differences in the RNA populations of the various ascidian plasms, and transcripts for tissue-specific proteins are present although we do not know whether they are localised or act as determinants. Several maternal mRNAs are unevenly distributed in the amphibian egg, and the protein sequences encoded by two of these suggest causal links with the particular properties of the two polar areas (see further below). The effects of a uv microbeam have led Arnold to suggest that cephalopod determinants contain RNA, but other lines of evidence are really needed. In the amphibian zygote, for instance, uv effects which were interpreted as a destruction of determinants now seem likely to be due to effects on the cytoskeleton and symmetrisation movements. Of course, effects on the cytoskeleton could indirectly affect the distribution of RNAs if they are bound to it, as they are in some eggs.

In the first section of this chapter we saw that few areas of animal eggs show special developmental properties and that little more than polarity and bilaterality is maternally determined. From the evidence of the last paragraph localised RNAs, presumably mRNAs, could provide an adequate explanation for patterns of this order. Further details of embryonic organisation would have to be epigenetically determined, and we can most easily visualise how this occurs in *Drosophila*. There, products of the maternally active genes affecting patterns are probably distributed in quite simple gradients, but this information is apparently used to ensure the local activation of embryonic genes in complex patterns. Only when positions are defined in this more precise way are genes coding tissue-specific products activated. We do not yet know how general such mechanisms will prove to be. DNA sequences with homology to the homoeo box and the DNA-binding region of the *Krüppel* gene are present in a wide variety of animals (Holland & Hogan, 1986; Schuh *et al.*, 1986), but their function there is unknown. Homoeo box-containing genes in *Xenopus* and the mouse are expressed in spatially restricted patterns from late gastrulation, which supports a role in spatial determination.

Even if common molecular mechanisms operate in many groups, however, the timing may vary considerably. Many of the classically defined mosaic embryos can develop to relatively advanced stages when zygotic transcription is strongly inhibited, suggesting that more precise spatial and even tissue-specific information may be provided maternally. On the other hand some regulative embryos may inherit very little maternal spatial information – in mammals perhaps none. An interesting case is provided by the sea urchins, which have typical regulative embryos with a vegetal determination centre or reference point. Like many more mosaic embryos they undergo ooplasmic flows which establish a special vegetal domain. There is no evidence that any maternal RNAs are localised here (some, in fact, seem to be specifically excluded), but the existence of the vegetal domain and a clock mechanism normally ensures that small cells form here at fourth cleavage. These micromeres form a different spectrum of RNAs from the other blastomeres and because of their small size and long interphase time zygotic transcripts accumulate to higher levels than elsewhere. It is possible that these early localisations and gradients are the functional equivalent of those established maternally in other groups, and that similar gradients of zygotic transcripts are established in the loach, as the blastoderm acquires independence of the yolk, and even at the MBT in amphibians. However, these are speculative suggestions, and we must note that there is still doubt about the importance of the unequal cleavage, and even of the vegetal domain, in sea urchins.

If transcripts localised at one or a few points are to affect developmental patterns throughout the embryo then they or their products must spread either by intracellular transport before cleavage separates the areas or by transmission between cells. A direct transfer of informational molecules, whether of mRNA or proteins, would require specialised transfer mechanisms. These are certainly available in the sea urchin where micromeres form syncytial connections with neighbouring cells; and similar connections have also been claimed between the micromeres and the 3D macromere of equally cleaving spiralians and the yolk syncytium and deep cells of teleosts. In the amphibian gastrula, though not the blastula, vegetalising inductions usually require direct cell contact and so fail to cross filters. This suggests a macromolecular inducer which now seems most likely to be a protein. Since work on the other chapters of this book was completed it has been shown that one of the maternal mRNAs localised at the *Xenopus* vegetal pole encodes a protein which is very closely related to the growth factors known to mesodermalise *Xenopus* blastula ectoderm (Weeks & Melton, 1987*b*; and see p. 194). It is extremely encouraging to be able to relate heterogenous inducing factors with a localised endogenous molecule, and suggests that the latter is involved in mesodermal induction. Alternatively, as a vegetaliser, it could act directly in the vegetal cells, explaining their early specification as endoderm. In either case the mechanism by which such a protein would affect developmental properties is as yet entirely unknown.

We know very little about the biochemical properties of areas which do not act as reference points, e.g. the animal halves of most eggs and early embryos. Perhaps we should not expect informational molecules to be localised there, since these areas have so little developmental potential. At a less specific level, however, the animal pole does seem to be characterised by a particular type of metabolism. This is where cleavages usually begin and the potential for metabolic activity seems to be greatest. The discovery that one mRNA localised at the animal pole of *Xenopus* eggs encodes a mitochondrial ATPase led Weeks & Melton (1987*a*) to speculate that choices between developmental pathways could be affected by molecules which change the physiological state of cells as well as by those which directly change gene expression patterns. We should recall here that embryonic cells are very easily animalised, or neuralised, by artificial treatments of low specificity, and that these treatments have broad and quite consistent effects on colloid properties (see Chapter 5). Neuralising stimuli, from chordamesoderm or heterogenous sources, cross filters far more rapidly than most vegetalising ones, again suggesting the involvement of a small molecule probably with low information content.

The interaction occurring along the primary axis of most animal embryos seems therefore to be between a determined but metabolically less active vegetal pole and an active animal pole where any spatial cues are of low specificity. The former could supply informational molecules to the latter or cause them to synthesise their own determinants. Perhaps the animal 'half' activates more vegetal areas, which do indeed often fail to differentiate in isolation despite being determined. We would, of course, like to know which metabolic pathway requires activation vegetally. In many oocytes and embryos the rate of protein synthesis seems to be limited by factors other than the supply of mRNA. I have suggested (Wall, 1973) that the key effect of animalisation may be to activate translation, and that this may operate antagonistically to a transfer of mRNPs in vegetalisation. Lithium ions are known to inhibit translation at the initiation step (Lubin, 1963) and many other treatments which vegetalise sea urchin embryos also inhibit protein synthesis there. These suggestions have still not been tested in any satisfactory way, because of technical difficulties in the amphibian system (see Wall, 1979), but more generally because there is little interest in ideas which seem vague and unsophisticated to molecular biologists. It is hoped that this book will encourage a change in these attitudes.

We concluded in the first section of this chapter that determination in the bilateral plane is effectively only a distortion of that in the primary axis. This is supported by what little we know of the mechanisms of bilateral determination in embryos with a vegetal reference point. The dominant side may receive more than its share of vegetal plasm apparently with determinants, as in many spiralians, or it may show the same signs of raised metabolic activity as the animal pole. Even in *Drosophila*, where several maternally acting genes are concerned with ventro-dorsal determination, the first products of gene action appear to be equally distributed in all meridians, and an ill-defined activation of the $Toll^+$ product may be the first ventro-dorsally polarised activity. Evidence that the $snake^+$ product is a serine protease bearing a calcium-binding site (DeLotto & Spierer, 1986) may be our first clue in tying 'activation' down to a precise molecular mechanism.

These facts and hypotheses do not seem to fit well into current concepts of the roles of gene expression in early development. Davidson (1986), who has reviewed this subject in detail, suggests that the construction of an embryo is largely the responsibility of a very large maternal gene set. These are of course the genes expressed in the oocyte (and/or the nurse cells) but they also form the bulk of the transcripts in the early embryo although many of them are not expressed in adult cells. Davidson's group suggests that this set codes for morphogenesis proteins, i.e. those required for the production of three-dimensional structures involved in cellular morphogenesis (Hough-Evans *et al.*, 1977; see Davidson, 1986, pp. 172–173). Cells in different parts of an embryo obviously show different morphogenetic behaviour, but Davidson appears to believe that the products of the maternal gene set are present in all cells but differentially assembled according to other spatial cues. The formation of tissue-specific products would be the responsibility of a far smaller set of embryonic late genes which are not significantly expressed in oogenesis but are activated in appropriate areas of the early embryo.

Many of the RNAs considered in this section were produced maternally but are localised, often at one pole of the egg, and seem to have more specific effects on development than would be expected if they coded morphogenesis proteins. Yet neither they nor the zygotic transcripts considered seem to code tissue-specific products, opposing their inclusion in the embryonic late genes. Their developmental effects appear to be less direct ones: perhaps providing the spatial information used to ensure that tissue-specific gene activations occur in the right cells.

Davidson has of course been interested for a long time in the gene-regulatory systems which control the expression of producer genes. He points out that the activation of particular genes usually requires an interaction with specific regulatory molecules, and proposes that in early embryos these might be maternal determinants localised to particular regions, inducers passed from neighbouring cells or the products of earlier zygotic activity by a few master genes (see Davidson, 1986, p. 43). The *Drosophila* segmentation and homoeotic genes would apparently be such master genes, producing proteins which collect in the nuclei and affect gene expression patterns, although how directly they control tissue-specific gene expression is still unknown. In 1971 Davidson & Britten showed how systems of the first two kinds, using maternal determinants and inducer molecules, might operate in mosaic and regulative development respectively. In mosaic determination, different activator RNAs would be localised in different areas of the oocyte: when cleavage brought embryonic nuclei to these areas, each activator RNA would recognise a different receptor sequence in the genome and so activate a different battery of producer genes (see Fig. 10.2(*a*)). In regulative determination, the materials localised in the oocyte would be sensor proteins (or information required for their synthesis): in the embryo these would recognise regulatory gene sequences, but a signal from neighbouring cells (an inducer) would also be required for the activation of the regulatory, and thus indirectly the producer, genes (see Fig. 10.2(*b*)).

The system which appears to be best suited to explanations of the Davidson/Britten type is the ascidian egg. Here the differentiation of many lineages correlates quite closely with the kind of ooplasm they receive (although perhaps not absolutely, e.g. we do not know whether the A4.1 and b4.2 cells receive a little of the myoplasm). Moreover, specific producer genes are activated in the cells which receive particular ooplasms. We can postulate that myoplasm contains an activator RNA

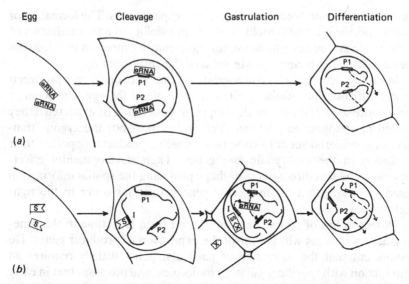

Fig. 10.2. Diagrams illustrating mechanisms for 'mosaic' and 'regulative' determination proposed by Davidson & Britten (1971): (*a*) in mosaic determination different activator RNAs (aRNA) are localised in different parts of the egg; during cleavage each interacts with specific sites in the genome to activate a particular spectrum of producer genes (P1, P2) coding for tissue-specific products. (*b*) in regulative determination sensor proteins (S) could be localised in the egg and interact during cleavage with integrator genes (I). Later, if the cell received an inducer molecule (X), this would complex S and allow the transcription of aRNAs and so of producer gene sets.

which activates a battery of genes including those coding myosin and acetylcholinesterase. Neuroplasm could contain a sensor protein which, in the presence of an inducer, indirectly activates a battery of genes including that coding tyrosinase. The data available for these lineages are therefore compatible with the Davidson/Britten models, but we need more information to demonstrate that these do indeed operate. In any case, it would seem that the producer genes for some other tissue-specific proteins were already activated in the oocyte.

However, several features of the Davidson/Britten models seem to contradict conclusions drawn in the first section of this chapter. Most animal embryos seem, like *Drosophila*, to determine cells first according to position rather than tissue type, so the ascidian case may well be exceptional. No known maternal mutation causes the deletion of specific somatic tissues as would be expected if the regulatory genes were affected. Studies with egg fragments usually either reveal no spatial differences in developmental potential or find all apparent 'determinants' in one area (often the vegetal pole). It could be argued that areas forming only an epithelium in isolation do contain sensor proteins for other tissues, but require induction before these can act. However, any part of the animal 'half' in the amphibian early gastrula is probably competent to produce any embryonic tissue: does this mean that each area has sensor proteins appropriate for differentiation as any neural, mesodermal or endodermal cell type?

It is also difficult to apply models of the Davidson/Britten type to species where RNA localisations are detectable by microscopy (in ecto- somes, a vegetal body or polar granules) or cytochemistry (in sea urchin micromeres). It is very unlikely that gene-regulatory agents would be produced in such large amounts, and sometimes there is positive evidence that these are transcripts from producer genes. In *Ilyanassa* the products of the polar lobe RNAs are sufficiently prominent to alter the profile of gel-separated polypeptides. In *Drosophila*, translation of the polar granule RNAs seems to have a direct effect on development in the precocious separation of the pole cells.

These last two examples illustrate some other important points. Translation of the polar granule RNAs seems to have a straightforward morphogenetic effect, but if this fails a specific cell type, the germ cells, never forms. This suggests that morphogenesis and differentiation can be far more intimately connected than Davidson (1986) envisages. The allocation of mammalian blastomeres to ICM and trophectoderm pro- vides another example: positional cues are amplified by morphogenetic activities such as adhesive changes and intracellular polarisation until two microenvironments are created. In other embryos morphogenetic move- ments such as gastrulation place cells in different environments, which can potentially provide them with positional information. Indeed, the cellular activities required for the execution of these movements could themselves provide information for determination. The *Ilyanassa* polar lobe RNAs illustrate another, but possibly related, point. Their products affect protein synthesis patterns quantitatively but not apparently qualitatively, implying that the lobe contains the same major species of RNA as the rest of the egg but in different ratios. In many other embryos the differences between cells in the early stages of specification appear to be quantitative and reversible ones, possibly based on poorly defined metabolic differences rather than the presence or absence of 'determinants'. Rather than dismissing such differences as less interest- ing, perhaps we should conclude that determination may not always require high-grade information such as localised gene-regulatory agents in the ooplasm.

# References

Abbate, C. & Ortolani, G. (1961). The development of *Ciona* eggs after partial removal of cortex or ooplasm. *Acta Embryol. Morph. exp.* **4**, 56–61.

Abramczuk, J. & Sawicki, W. (1974). Variation in dry mass and volume of nonfertilized oocytes and blastomeres of 1-, 2- and 4-celled mouse embryos. *J. exp. Zool.* **188**, 25–33.

Abramova, N. B. & Vasil'eva, M. N. (1973). Some properties of embryonic mitochondria of the loach. *Ontogenez* **4**, 288–293 (in translation 265–270).

Abreu, S. L. & Brinster, R. L. (1978). Synthesis of tubulin and actin during the preimplantation development of the mouse. *Expl Cell Res.* **114**, 135–141.

Adamson, E. D. (1982). The location and synthesis of transferrin in mouse embryos and teratocarcinoma cells. *Devl Biol.* **91**, 227–234.

Adamson, E. D. & Woodland, H. R. (1974). Histone synthesis in early amphibian development: histone and DNA synthesis are not co-ordinated. *J. molec. Biol.* **88**, 263–285.

Adelmann, H. B. (1934). A study of cyclopia in *Amblystoma punctatum*, with special reference to the mesoderm. *J. exp. Zool.* **67**, 217–281.

Agrell, I. (1958). A cytoplasmic production of ribonucleic acid during the cell cycle of the micromeres in the sea urchin embryo. *Ark. Zool.* **11**, 435–441.

Agrell, I. (1964). Natural division synchrony and mitotic gradients in metazoan tissues. In *Synchrony in Cell Division and Growth*, ed. E. Zeuthen, pp. 39–67. New York: Wiley.

Aït-Ahmed, O., Thomas-Cavallin, M. & Rosset, R. (1987). Isolation and characterization of a region of the *Drosophila* genome which contains a cluster of differentially expressed maternal genes (*yema* gene region). *Devl Biol.* **122**, 153–162.

Aitkhozhin, M. A., Belitsina, N. V. & Spirin, A. S. (1964). Nucleic acids during early development of fish embryos (*Misgurnus fossilis*). *Biokhimiya* **29**, 169–175 (in translation 145–152).

Akam, M. E. & Martinez-Arias, A. (1985). The distribution of *Ultrabithorax* transcripts in *Drosophila* embryos. *EMBO J.* **4**, 1689–1700.

Akers, R. M., Phillips, C. R. & Wessells, N. K. (1986). Expression of an epidermal antigen used to study tissue induction in the early *Xenopus laevis* embryo. *Science, N.Y.* **231**, 613–616.

Albert, J. (1972). Analyse expérimentale des interactions endomésodermiques chez l'embryon de *Rana dalmatina* (Amphibien anoure). *Bull. Soc. zool. Fr.* **97**, 461–469.

Albertson, D. G. (1984). Formation of the first cleavage spindle in nematode embryos. *Devl Biol.* **101**, 61–72.

Alexandre, H., de Petrocellis, B. & Brachet, J. (1982). Studies on differentiation without cleavage in *Chaetopterus*. Requirement for a definite number of DNA replication cycles shown by aphidicolin pulses. *Differentiation* **22**, 132–135.

Al-Mukhtar, K. A. K. & Webb, A. C. (1971). An ultrastructural study of primordial germ cells, oogonia and early oocytes in *Xenopus laevis*. *J. Embryol. exp. Morph.* **26**, 195–217.

Ancel, P. & Vintemberger, P. (1948). Recherches sur le déterminisme de la symétrie bilatérale dans l'oeuf des amphibiens. *Bull. biol. Fr. Belg. Suppl.* **31**, 1–182.

Anderson, D. T. (1966). The comparative embryology of the Polychaeta. *Acta zool., Stockh.* **47**, 1–42.

Anderson, D. T. (1973). *Embryology and Phylogeny in Annelids and Arthropods*. Oxford: Pergamon Press.

Anderson, E. (1964). Oocyte differentiation and vitellogenesis in the roach *Periplaneta americana*. *J. Cell Biol.* **20**, 131–155.

Anderson, K. V., Bokla, L. & Nüsslein-Volhard, C. (1985*a*). Establishment of dorso-ventral polarity in the *Drosophila* embryo: the induction of polarity by the *Toll* gene product. *Cell* **42**, 791–798.

Anderson, K. V., Jürgens, G. & Nüsslein-Volhard, C. (1985*b*). Establishment of dorso-ventral polarity in the *Drosophila* embryo: genetic studies on the role of the *Toll* gene product. *Cell* **42**, 779–789.

Anderson, K. V. & Lengyel, J. A. (1980). Changing rates of histone mRNA synthesis and turnover in *Drosophila* embryos. *Cell* **21**, 717–727.

Anderson, K. V. & Nüsslein-Volhard, C. (1984*a*). Information for the dorso-ventral pattern of the *Drosophila* embryo is stored as maternal mRNA. *Nature, Lond.* **311**, 223–227.

Anderson, K. V. & Nüsslein-Volhard, C. (1984*b*). Genetic analysis of dorso-ventral embryonic pattern in *Drosophila*. In *Pattern Formation: A Primer in Developmental Biology*, ed. G. M. Malacinski & S. V. Bryant, pp. 269–289. New York: Macmillan.

Andreuccetti, P., Taddei, C. & Filosa, S. (1978). Intercellular bridges between follicle cells and oocyte during the differentiation of follicular epithelium in *Lacerta sicula* Raf. *J. Cell Sci.* **33**, 341–350.

Angerer, L. M., Kawczynski, G., Wilkinson, D. G., Nemer, M. & Angerer, R. C. (1986). Spatial patterns of metallothionein mRNA expression in the sea urchin embryo. *Devl Biol.* **116**, 543–547.

Angerer, R. C. & Davidson, E. H. (1984). Molecular indices of cell lineage specification in sea urchin embryos. *Science, N.Y.* **226**, 1153–1160.

Arceci, R. J. & Gross, P. R. (1980*a*). Histone gene expression: progeny of isolated early blastomeres in culture make the same change as in the embryo. *Science, N.Y.* **209**, 607–609.

Arceci, R. J. & Gross, P. R. (1980*b*). Histone variants and chromatin structure during sea urchin development. *Devl Biol.* **80**, 186–209.

Arezzo, F. & Giudice, G. (1983). The lack of cell contact causes genomic activation in early sea urchin embryos. *Cell Biol. int. Rep.* **7**, 5–10.

Arnold, H.-H., Innis, M. A. & Siddiqui, M. A. Q. (1978). Control of embryonic development: effect of an embryonic inducer RNA on *in vitro* translation of mRNA. *Biochemistry, N.Y.* **17**, 2050–2054.

Arnold, J. M. (1961). Observations on the mechanism of cellulation of the egg of *Loligo pealii*. *Biol. Bull. mar. biol. Lab., Woods Hole* **121**, 380–381.

Arnold, J. M. (1965*a*). Normal embryonic stages of the squid, *Loligo pealii* (Lesueur). *Biol. Bull. mar. biol. Lab., Woods Hole* **128**, 24–32.

Arnold, J. M. (1965*b*). The inductive role of the yolk epithelium in the development of the squid, *Loligo pealii* (Lesueur). *Biol. Bull. mar. biol. Lab., Woods Hole* **129**, 72–78.

Arnold, J. M. (1968). The role of the egg cortex in cephalopod development. *Devl Biol.* **18**, 180–197.

Arnold, J. M. (1971). Cephalopods. In *Experimental Embryology of Marine and Fresh-Water Invertebrates*, ed. G. Reverberi, pp. 265–311. Amsterdam: North-Holland.

Arnold, J. M. (1974). Intercellular bridges in somatic cells: cytoplasmic continuity of blastoderm cells of *Loligo pealii*. *Differentiation* **2**, 335–341.

Arnold, J. M. & Williams-Arnold, L. D. (1974). Cortical–nuclear interactions in cephalopod development: cytochalasin B effects on the informational pattern in the cell surface. *J. Embryol. exp. Morph.* **31**, 1–25.

Arnold, J. M. & Williams-Arnold, L. D. (1976). The egg cortex problem as seen through the squid eye. *Am. Zool.* **16**, 421–446.

Arnolds, W. J. A. (1982a). A mutation with a maternal effect in *Lymnaea stagnalis* L., affecting development of bilateral symmetry and dorsoventrality and preventing gastrulation. *Proc. K. ned. Akad. Wet. C* **85**, 1–20.

Arnolds, W. J. A. (1982b). Early development of maternally determined mutant dauerblastulae in *Lymnaea stagnalis* L.: cellular interactions in the dorsalization and lateralization of head quadrants. *Proc. K. ned. Akad. Wet. C* **85**, 635–662.

Arnolds, W. J. A., van den Biggelaar, J. A. M. & Verdonk, N. H. (1983). Spatial aspects of cell interactions involved in the determination of dorsoventral polarity in equally cleaving gastropods and regulative abilities of their embryos, as studied by micromere deletions in *Lymnaea* and *Patella*. *Wilhelm Roux Arch. dev. Biol.* **192**, 75–85.

Artzt, K., Dubois, P., Bennett, D., Condamine, H., Babinet, C. & Jacob, F. (1973). Surface antigens common to mouse cleavage embryos and primitive teratocarcinoma cells in culture. *Proc. natn. Acad. Sci. U.S.A.* **70**, 2988–2992.

Asahi, K.-I., Born, J., Tiedemann, H. & Tiedemann, H. (1979). Formation of mesodermal pattern by secondary inducing interactions. *Wilhelm Roux Arch. dev. Biol.* **187**, 231–244.

Asao, T. (1968). The variation of cell number during development of *Triturus pyrrhogaster* embryo. *J. Fac. Sci. Univ. Tokyo IV* **11**, 453–458.

Asashima, M. & Grunz, H. (1983). Effects of inducers on inner and outer gastrula ectoderm layers of *Xenopus laevis*. *Differentiation* **23**, 206–212.

Atkinson, J. N. (1971). Organogenesis in normal and lobeless embryos of the marine prosobranch gastropod *Ilyanassa obsoleta*. *J. Morph.* **133**, 339–352.

Audet, R. G., Goodchild, J. & Richter, J. D. (1987). Eukaryotic initiation factor 4A stimulates translation in microinjected *Xenopus* oocytes. *Devl Biol.* **121**, 58–68.

Ave, K., Kawakami, I. & Sameshina, M. (1968). Studies on the heterogeneity of cell populations in amphibian presumptive epidermis, with reference to primary induction. *Devl Biol.* **17**, 617–626.

Azar, Y. & Eyal-Giladi, H. (1979). Marginal zone cells – the primitive streak-inducing component of the primary hypoblast in the chick. *J. Embryol. exp. Morph.* **52**, 79–88.

Azar, Y. & Eyal-Giladi, H. (1981). Interaction of epiblast and hypoblast in the formation of the primitive streak and the embryonic axis in chick, as revealed by hypoblast-rotation experiments. *J. Embryol. exp. Morph.* **61**, 133–144.

Bachvarova, R. & Davidson, E. H. (1966). Nuclear activation at the onset of amphibian gastrulation. *J. exp. Zool.* **163**, 285–296.

Bachvarova, R. & De Leon, V. (1980). Polyadenylated RNA of mouse ova and loss of maternal RNA in early development. *Devl Biol.* **74**, 1–8.

Bäckström, S. (1953). Studies on the animalizing action of iodosobenzoic acid in the sea urchin development. *Ark. Zool.* **4**, 485–491.

Bäckström, S. (1959). Activity of glucose-6-phosphate dehydrogenase in sea urchin embryos of different developmental trends. *Expl Cell Res.* **18**, 347–356.

Bäckström, S. & Gustafson, T. (1953). Lithium sensitivity in the sea urchin in relation to the stage of development. *Ark. Zool.* **6**, 185–188.

Baker, E. J. & Infante, A. A. (1982). Nonrandom distribution of histone mRNAs into polysomes and nonpolysomal ribonucleoprotein particles in sea urchin embryos. *Proc. natn. Acad. Sci. U.S.A.* **79**, 2455–2459.

Bakken, A. H. (1973). A cytological and genetic study of oogenesis in *Drosophila melanogaster*. *Devl Biol.* **33**, 100–122.

Bakulina, E. D., Pankova, N. V. & Mitrofanov, V. G. (1984). Phenogenetics of the mutation *rough* with maternal effect in *Drosophila virilis*. *Ontogenez* **15**, 283–289 (in translation 187–192).

Bałakier, H. & Pedersen, R. A. (1982). Allocation of cells to inner cell mass and trophectoderm lineages in preimplantation mouse embryos. *Devl Biol.* **90**, 352–362.

Balinsky, B. I. (1975). *An Introduction to Embryology*, 4th edition. Philadelphia: W.B. Saunders Co.

Balinsky, B. I. & Walther, H. (1961). The immigration of presumptive mesoblast from the primitive streak in the chick as studied with the electron microscope. *Acta Embryol. Morph. exp.* **4**, 261–283.

Ballantine, J. E. M., Woodland, H. R. & Sturgess, E. A. (1979). Changes in protein synthesis during the development of *Xenopus laevis*. *J. Embryol. exp. Morph.* **51**, 137–153.

Ballard, W. W. (1966a). The role of the cellular envelope in the morphogenetic movements of teleost embryos. *J. exp. Zool.* **161**, 193–200.

Ballard, W. W. (1966b). Origin of the hypoblast in *Salmo*. I. Does the blastodisc edge turn inward? *J. exp. Zool.* **161**, 201–210.

Ballard, W. W. (1966c). Origin of the hypoblast in *Salmo*. II. Outward movement of deep central cells. *J. exp. Zool.* **161**, 211–220.

Ballard, W. W. (1968). History of the hypoblast in *Salmo*. *J. exp. Zool.* **168**, 257–272.

Ballard, W. W. (1973a). Morphogenetic movements in *Salmo gairdneri* Richardson. *J. exp. Zool.* **184**, 27–48.

Ballard, W. W. (1973b). A new fate map for *Salmo gairdneri*. *J. exp. Zool.* **184**, 49–73.

Ballard, W. W. (1986). Morphogenetic movements and a provisional fate map of development in the holostean fish *Amia calva*. *J. exp. Zool.* **238**, 355–372.

Ballard, W. W. & Dodes, L. M. (1968). The morphogenetic movements at the lower surface of the blastodisc in salmonid embryos. *J. exp. Zool.* **168**, 67–84.

Ballard, W. W. & Ginsburg, A. S. (1980). Morphogenetic movements in acipenserid embryos. *J. exp. Zool.* **213**, 69–103.

Baltus, E., Brachet, J., Hanocq-Quertier, J. & Hubert, E. (1973). Cytochemical and biochemical studies on progesterone-induced maturation in amphibian oocytes. I. Ribonucleic acid and protein synthesis (effects of inhibitors and a 'maturation promoting factor'). *Differentiation* **1**, 127–143.

Bánki, O. (1929). Die Entstehung der äusseren Zeichen der bilateralen Symmetrie am Axolotlei, nach Versuchen mit örtlicher Vitalfärbung. *Proc. 10th int. Congr. Zool. (Budapest, 1928).* **1**, 377–385.

Barbieri, F. D., Sánchez, S. S. & del Pino, E. J. (1980). Changes in lectin-mediated agglutinability during primary embryonic induction in the amphibian embryo. *J. Embryol. exp. Morph.* **57**, 95–106.

Barros, C., Hand, G. S. & Monroy, A. (1966). Control of gastrulation in the starfish, *Asterias forbesii*. *Expl Cell Res.* **43**, 167–182.

Barth, L. G. (1941). Neural differentiation without organizer. *J. exp. Zool.* **87**, 371–383.

Barth, L. G. (1942). Regional differences in oxygen consumption of the amphibian gastrula. *Physiol. Zoöl.* **15**, 30–46.

Barth, L. G. & Barth, L. J. (1969). The sodium dependence of embryonic induction. *Devl Biol.* **20**, 236–262.

Barth, L. G. & Barth, L. J. (1974a). Ionic regulation of embryonic induction and cell differentiation in *Rana pipiens*. *Devl Biol.* **39**, 1–22.

Barth, L. J. & Barth, L. G. (1974b). Effect of the potassium ion on induction of notochord from gastrula ectoderm of *Rana pipiens*. *Biol. Bull. mar. biol. Lab., Woods Hole* **146**, 313–325.

Bates, W. R. & Jeffery, W. R. (1987). Alkaline phosphatase expression in ascidian egg fragments and andromerogons. *Devl Biol.* **119**, 382–389.

Beachy, P. A., Helfand, S. L. & Hogness, D. S. (1985). Segmental distribution of bithorax complex proteins during *Drosophila* development. *Nature, Lond.* **313**, 545–551.

Beams, H. W. & Kessel, R. G. (1974). The problem of germ cell determinants. *Int. Rev. Cytol.* **39**, 413–479.

Becker, U., Tiedemann, H. & Tiedemann, H. (1959). Versuche zur Determination von embryonalem Amphibiengewebe durch Induktionsstoffe in Lösung. *Z. Naturf.* **14b**, 608–609.

Bédard, P.-A. & Brandhorst, B. P. (1983). Patterns of protein synthesis and metabolism during sea urchin embryogenesis. *Devl Biol.* **96**, 74–83.

Bédard, P.-A. & Brandhorst, B. P. (1986a). Cytoplasmic distributions of translatable messenger RNA species and the regulation of patterns of protein synthesis during sea urchin embryogenesis. *Devl Biol.* **115**, 261–274.

Bédard, P.-A. & Brandhorst, B. P. (1986b). Translational activation of maternal mRNA encoding the heat-shock protein hsp90 during sea urchin embryo-genesis. *Devl Biol.* **117**, 286–293.

Beddington, R. S. P. (1981). An autoradiographic analysis of the potency of embryonic ectoderm in the 8th day postimplantation mouse embryo. *J. Embryol. exp. Morph.* **64**, 87–104.

Beddington, R. S. P. (1982). An autoradiographic analysis of tissue potency in different regions of the embryonic ectoderm during gastrulation in the mouse. *J. Embryol. exp. Morph.* **69**, 265–285.

Beddington, R. S. P. (1983). The origin of the foetal tissues during gastrulation in the rodent. In *Development in Mammals*, vol. 5, ed. M. H. Johnson, pp. 1–32. Amsterdam: Elsevier Science Publishers.

Beeman, R. W. (1987). A homoeotic gene cluster in the red flour beetle. *Nature, Lond.* **327**, 247–249.

Beetschen, J.-C. (1979). Recherches expérimentales sur la symétrisation de l'oocyte et de l'oeuf d'Axolotl: facteurs conditionnant l'apparition précoce du croissant gris à la suite d'un choc thermique. *C. r. hebd. Séanc. Acad. Sci., Paris D* **288**, 643–646.

Beetschen, J.-C. & Fernandez, M. (1979). Studies on the maternal effect of the semi-lethal factor *ac* in the salamander *Pleurodeles waltli*. In *Maternal Effects in*

*Development*, ed. D. R. Newth & M. Balls, pp. 269–286. Cambridge: The University Press.

Beetschen, J.-C. & Gautier, J. (1987). Heat-shock induced grey crescent formation in axolotl eggs and oocytes: the role of gravity. *Development* **100**, 599–609.

Belanger, A. M. & Rustad, R. C. (1972). Movements of echinochrome granules during the early development of sea urchin eggs. *Nature new Biol.* **239**, 81–83.

Bellairs, R. (1957). Studies on the development of the foregut in the chick embryo. IV. Mesodermal induction and mitosis. *J. Embryol. exp. Morph.* **5**, 340–350.

Bellairs, R. (1963*a*). Differentiation of the yolk sac of the chick studied by electron microscopy. *J. Embryol. exp. Morph.* **11**, 201–225.

Bellairs, R. (1963*b*). The development of somites in the chick embryo. *J. Embryol. exp. Morph.* **11**, 697–714.

Bellairs, R., Lorenz, F. W. & Dunlap, T. (1978). Cleavage in the chick embryo. *J. Embryol. exp. Morph.* **43**, 55–69.

Bender, W., Akam, M., Karch, F., Beachy, P. A., Peifer, M., Spierer, P., Lewis, E. B. & Hogness, D. S. (1983). Molecular genetics of the bithorax complex in *Drosophila melanogaster*. *Science, N.Y.* **221**, 23–29.

Benford, H. H. & Namenwirth, M. (1974). Precocious appearance of the grey crescent in heat-shocked axolotl eggs. *Devl Biol.* **39**, 172–176.

Bennett, K. L. & Ward, S. (1986). Neither a germ line-specific nor several somatically expressed genes are lost or rearranged during embryonic chromatin diminution in the nematode *Ascaris lumbricoides* var *suum*. *Devl Biol.* **118**, 141–147.

Bennett, M. V. L., Spira, M. E. & Spray, D. C. (1978). Permeability of gap junctions between embryonic cells of *Fundulus*: a reevaluation. *Devl Biol.* **65**, 114–125.

Bennett, M. V. L. & Trinkaus, J. P. (1970). Electrical coupling between embryonic cells by way of extracellular space and specialized junctions. *J. Cell Biol.* **44**, 592–610.

Bennett, N. (1973). Study of yolk-sac endoderm organogenesis in the chick using a specific enzyme (cysteine lyase) as a marker of cell differentiation. *J. Embryol. exp. Morph.* **29**, 159–174.

Benoit, J. (1969). Irradiation *in vitro* de la région du noeud de Hensen de la gastrula de souris. *C.r.hebd. Séanc. Acad. Sci., Paris* **269**, 724–727.

Ben-Shaul, Y., Artzt, K. & Bennett, D. (1983). Immunoscanning electron microscopy of antigenic determinants of T/t complex ($t^{w18}$) mouse embryos. *Cell Differ.* **13**, 159–170.

Benson, S. C. & Triplett, E. L. (1974). The synthesis and activity of tyrosinase during development of the frog *Rana pipiens*. *Devl Biol.* **40**, 270–282.

Berg, W. E. (1956). Cytochrome oxidase in anterior and posterior blastomeres of *Ciona intestinalis*. *Biol. Bull. mar. biol. Lab.*, *Woods Hole* **110**, 1–7.

Berg, W. E. (1965). Rates of protein synthesis in whole and half embryos of the sea urchin. *Expl Cell Res.* **40**, 469–489.

Berg, W. E. (1968*a*). Effect of lithium on the rate of protein synthesis in the sea urchin embryo. *Expl Cell Res.* **50**, 133–139.

Berg, W. E. (1968*b*). Rates of protein and nucleic acid synthesis in half embryos of the sea urchin. *Expl Cell Res.* **50**, 679–683.

Berg, W. E. & Humphreys, W. J. (1960). Electron-microscopy of 4-cell stages of the ascidians *Ciona* and *Styela*. *Devl Biol.* **2**, 42–60.

Berg, W. E. & Long, N. D. (1964). Regional differences of mitochondrial size in the sea urchin embryo. *Expl Cell Res.* **33**, 422–437.

Beritashvili, D. R., Kvavilashvili, I. Sh. & Kafiani, C. A. (1969). Redistribution of K⁺ in cleaving eggs of a fish *Misgurnus fossilis*. *Expl Cell Res.* **56**, 113–116.

Beritashvili, D. R., Kvavilashvili, I. Sh., Rott, N. N. & Ignat'eva, G. M. (1970). The accumulation of potassium in the developing blastoderm of the trout. *Ontogenez* **1**, 628–630 (in translation 464–465).

Betchaku, T. & Trinkaus, J. P. (1978). Contact relations, surface activity, and cortical microfilaments of marginal cells of the enveloping layer and of the yolk syncytial and yolk cytoplasmic layers of *Fundulus* before and during epiboly. *J. exp. Zool.* **206**, 381–426.

Bezem, J. J. & Raven, C. P. (1975). Computer simulation of early embryonic development. *J. theor. Biol.* **54**, 47–61.

Bezem, J. J., Wagemaker, H. A. & van den Biggelaar, J. A. M. (1981). Relative cell volumes of the blastomeres in embryos of *Lymnaea stagnalis* in relation to bilateral symmetry and dorsoventral polarity. *Proc. K. ned. Akad. Wet.* C **84**, 9–20.

Bibring, T. & Baxandall, J. (1977). Tubulin synthesis in sea urchin embryos: almost all tubulin of the first cleavage mitotic apparatus derives from the unfertilized egg. *Devl Biol.* **55**, 191–195.

Bier, K. (1967). Oogenese, das Wachstum von Riesenzellen. *Naturwissenschaften* **54**, 189–195.

Bier, K., Kunz, W. & Ribbert, D. (1967). Struktur und Funktion der Oocytenchromosomen und Nukleolen sowie der Extra-DNS, während der Oogenese panoistischer und meroistischer Insekten. *Chromosoma* **23**, 214–254.

Bijtel, J. H. (1931). Über die Entwicklung des Schwanzes bei Amphibien. *Arch. EntwMech. Org.* **125**, 448–486.

Bird, J. M. & Kimber, S. J. (1984). Oligosaccharides containing fucose linked α(1–3) and α(1–4) to N-acetylglucosamine cause decompaction of mouse morulae. *Devl Biol.* **104**, 449–460.

Black, S. D. & Gerhart, J. C. (1986). High-frequency twinning of *Xenopus laevis* embryos from eggs centrifuged before first cleavage. *Devl Biol.* **116**, 228–240.

Blair, S. S. (1982). Interactions between mesoderm and ectoderm in segment formation in the embryo of a glossiphoniid leech. *Devl Biol.* **89**, 389–396.

Blair, S. S. (1983). Blastomere ablation and the developmental origin of identified monoamine-containing neurons in the leech. *Devl Biol.* **95**, 65–72.

Blair, S. S. & Weisblat, D. A. (1984). Cell interactions in the developing epidermis of the leech *Helobdella triserialis*. *Devl Biol.* **101**, 318–325.

Bluemink, J. G. & Hoperskaya, O. A. (1975). Ultrastructural evidence for the absence of premelanosomes in eggs of the albino mutant (aᴾ) of *Xenopus laevis*. *Wilhelm Roux Arch. dev. Biol.* **177**, 75–79.

Bock, S. C., Campo, K. & Goldsmith, M. R. (1986). Specific protein synthesis in cellular differentiation. VI. Temporal expression of chorion gene families in *Bombyx mori* strain C108. *Devl Biol.* **117**, 215–225.

Bodenstein, D. (1953). Embryonic development. In *Insect Physiology*, ed. K. D. Roeder, pp. 780–821. New York: Wiley.

Boell, E. J. (1942). Biochemical and physiological analysis of organizer action. *Growth* **6**, *Suppl.* 37–53.

Boell, E. J., Koch, H. & Needham, J. (1939). Morphogenesis and metabolism: studies with the cartesian diver ultramicromanometer. IV. Respiratory quotient of the regions of the amphibian gastrula. *Proc. R. Soc.* B **127**, 374–387.

Bohrmann, J., Dorn, A., Sander, K. & Gutzeit, H. (1986a). The extracellular electrical current pattern and its variability in vitellogenic *Drosophila* follicles. *J. Cell Sci.* **81**, 189–206.

Bohrmann, J., Huebner, E., Sander, K. & Gutzeit, H. (1986*b*). Intracellular electrical potential measurements in *Drosophila* follicles. *J. Cell Sci.* **81**, 207–221.

Boletzky, S. V. (1970). On the lay-out of the mid-gut rudiment in *Loligo pealei* Le Sueur. *Experientia* **26**, 880–881.

Bonfig, R. (1925). Die Determination der Hauptrichtungen des Embryos von *Ascaris megalocephala*. *Z. wiss. Zool.* **124**, 407–456.

Bonhag, P. F. (1955). Histochemical studies of the ovarian nurse tissues and oocytes of the milkweed bug, *Oncopeltus fasciatus* (Dallas). *J. Morph.* **96**, 381–439.

Borland, R. M. (1977). Transport processes in the mammalian blastocyst. In *Development in Mammals*, vol. 1, ed. M. H. Johnson, pp. 31–67. Amsterdam: Elsevier-North-Holland.

Born, G. (1885). Biologische Untersuchungen. I. Ueber den Einfluss der Schwere auf das Froschei. *Arch. mikr. Anat.* **24**, 475–545.

Born, J., Davids, M. & Tiedemann, H. (1987). Affinity chromatography of embryonic inducing factors on heparin-Sepharose. *Cell Differ.* **21**, 131–136.

Born, J., Grunz, H., Tiedemann, H. & Tiedemann, H. (1980). Biological activity of the vegetalizing factor: decrease after coupling to polysaccharide matrix and enzymatic recovery of active factor. *Wilhelm Roux Arch. dev. Biol.* **189**, 47–56.

Born, J., Hoppe, P., Janeczek, J., Tiedemann, H. & Tiedemann, H. (1986). Covalent coupling of neuralizing factors from *Xenopus* to Sepharose beads: no decrease of inducing activity. *Cell Differ.* **19**, 97–101.

Born, J., Hoppe, P., Schwarz, W., Tiedemann, H., Tiedemann, H. & Wittmann-Liebold, B. (1985). An embryonic inducing factor: isolation by high performance liquid chromatography and chemical properties. *Biol. Chem. Hoppe-Seyler* **366**, 729–735.

Born, J., Tiedemann, H. & Tiedemann, H. (1972). The mechanism of embryonic induction: isolation of an inhibitor for the vegetalizing factor. *Biochim. biophys. Acta* **279**, 175–183.

Boswell, R. E. & Mahowald, A. P. (1985). *tudor*, a gene required for assembly of the germ plasm in *Drosophila melanogaster*. *Cell* **43**, 97–104.

Boterenbrood, E. C. & Nieuwkoop, P. D. (1973). The formation of the mesoderm in urodelean amphibians. V. Its regional induction by the endoderm. *Wilhelm Roux Arch. EntwMech. Org.* **173**, 319–332.

Boucaut, J. C., Darribère, T., Boulekbache, H. & Thiery, J. P. (1984). Prevention of gastrulation but not neurulation by antibodies to fibronectin in amphibian embryos. *Nature, Lond.* **307**, 364–367.

Boulekbache, H. (1981). Energy metabolism in fish development. *Am. Zool.* **21**, 377–389.

Bounoure, L. (1937). Les suites de l'irradiation du déterminant germinal, chez la grenouille rousse par les rayons ultra-violets: résultats histologiques. *C. r. Séanc. Soc. Biol.* **125**, 898–900.

Bouvet, J. (1976). Enveloping layer and periderm of the trout embryo (*Salmo trutta fario* L.). *Cell Tissue Res.* **170**, 367–382.

Boveri, T. (1887). Über Differenzierung der Zellkerne während der Furchung des Eies von *Ascaris megalocephala*. *Anat. Anz.* **2**, 688–693.

Boveri, T. (1901*a*). Die Polarität von Ovocyte, Ei und Larve des *Strongylocentrotus lividus*. *Zool. Jb. Abt. Anat. Ont.* **14**, 630–653.

Boveri, T. (1901*b*). Über die Polarität des Seeigel-Eies. *Verh. phys.-med. Ges. Wurzb.* **34**, 145–176.

Boveri (1903). Über den Einfluss der Samenzelle auf die Larvencharaktere der Echiniden. *Arch. EntwMech. Org.* **16**, 340–363.

Boveri, T. (1910). Die Potenzen der *Ascaris*-Blastomeren bei abgeänderter Furchung. Zugleich ein Beitrag zur Frage qualitativ-ungleicher Chromosomen-Teilung. In *Festschrift zum sechzigsten Geburtstag Richard Hertwigs*, vol. 3, pp. 131–214. Jena: Gustav Fischer.

Bownes, M. (1976). Defective development after puncturing the periplasm of nuclear multiplication stage *Drosophila* embryos. *Devl Biol.* **51**, 146–151.

Bownes, M. & Sander, K. (1976). The development of *Drosophila* embryos after partial u.v. irradiation. *J. Embryol. exp. Morph.* **36**, 394–408.

Bownes, M. & Sang, J. H. (1974). Experimental manipulations of early *Drosophila* embryos. I. Adult and embryonic defects resulting from micro-cautery at nuclear multiplication and blastoderm stages. *J. Embryol. exp. Morph.* **32**, 253–272.

Boyer, B. C. (1971). Regulative development in a spiralian embryo as shown by cell deletion experiments on the acoel, *Childia. J. exp. Zool.* **176**, 97–106.

Bozhkova, V. P. & Isaeva, V. V. (1984). Interference with cellular interactions in sea urchin embryos caused by sodium dodecyl sulphate. *Ontogenez* **15**, 465–470 (in translation 283–287).

Bozhkova, V. P., Kovalev, S. A., Chailakhyan, L. M. & Shilyanskaya, E. N. (1971). Study of electrical communication between loach embryo cells in early stages of development. *Ontogenez* **2**, 512–516 (in translation 411–414).

Bozhkova, V. P., Kvavilashvili, I. Sh., Rott, N. N. & Chailakhyan, L. M. (1973). Measurement of electrical coupling between cells of axolotl embryos during cleavage divisions. *Ontogenez* **4**, 523–526 (in translation 480–482).

Bozhkova, V. P., Nikolaev, P. P., Petryaevskaya, V. B. & Chailakhyan, L. M. (1983). Participation of cytoskeletal structures in the formation of micromeres in sea urchins. *Ontogenez* **14**, 247–254 (in translation 156–161).

Bozhkova, V. P., Nikolaev, P. P., Petryaevskaya, V. B. & Shmukler, Yu. B. (1982). Intercellular interactions in early sea urchin embryos. IV. Orientation of blastomere cleavage planes as a factor of micromere formation. *Ontogenez* **13**, 596–604 (in translation 361–367).

Brachet, A. (1906). Recherches expérimentales sur l'oeuf non-segmenté de *Rana fusca. Arch. EntwMech. Org.* **22**, 325–341.

Brachet, J. (1937). La differenciation sans clivage dans l'oeuf de chétoptère envisagée aux points de vue cytologique et métabolique. *Archs Biol.* **48**, 561–589.

Brachet, J. (1940). Étude histochimique des protéines au cours du développement embryonnaire des poissons, des amphibiens et des oiseaux. *Archs Biol.* **51**, 167–202.

Brachet, J. (1960). *The Biochemistry of Development.* London: Pergamon Press.

Brachet, J. (1965). The role of nucleic acids in morphogenesis. *Prog. Biophys. mol. Biol.* **15**, 97–127.

Brachet, J. (1977). An old enigma: the grey crescent of amphibian eggs. *Curr. Top. dev. Biol.* **11**, 133–186.

Brachet, J. & Donini-Denis, S. (1978). Studies on maturation and differentiation without cleavage in *Chaetopterus variopedatus*. Effects of ions, ionophores, sulfhydryl reagents, colchicine and cytochalasin B. *Differentiation* **11**, 19–37.

Brachet, J., Hanocq, F. & van Gansen, P. (1970). A cytochemical and ultrastructural analysis of *in vitro* maturation in amphibian oocytes. *Devl Biol.* **21**, 157–195.

Brachet, J., Hubert, E. & Lievens, A. (1972). The effects of α-amanitin and rifampicins on amphibian egg development. *Revue suisse Zool.* **79**, 47–63.

Brachet, J. & Hulin, E. (1972). Studies on nucleocytoplasmic interactions during early amphibian development. I. Localized destruction of the egg cortex. *J. Embryol. exp. Morph.* **27**, 121–145.

Brachet, J. & Hulin, N. (1969). Binding of tritiated actinomycin and cell differentiation. *Nature, Lond.* **222**, 481–482.

Brachet, J., Mamont, P., Boloukhère, M., Baltus, E. & Hanocq-Quertier, J. (1978). Effets d'un inhibiteur de la synthèse des polyamines sur la morphogénèse, chez l'Oursin, le Chétoptère et l'Algue *Acetabularia*. *C. r. hebd. Séanc. Acad. Sci., Paris D* **287**, 1289–1292.

Brahma, S. K. (1958). Experiments on the diffusibility of the amphibian evocator. *J. Embryol. exp. Morph.* **6**, 418–423.

Brandhorst, B. P. 1976). Two-dimensional gel patterns of protein synthesis before and after fertilization of sea urchin eggs. *Devl Biol.* **52**, 310–317.

Brandhorst, B. P. & Newrock, K. M. (1981). Post-transcriptional regulaton of protein synthesis in *Ilyanassa* embryos and isolated polar lobes. *Devl Biol.* **83**, 250–254.

Braude, P., Pelham, H., Flach, G. & Lobatto, R. (1979). Post-transcriptional control in the early mouse embryo. *Nature, Lond.* **282**, 102–105.

Braude, P. R. (1979). Time-dependent effects of α-amanitin on blastocyst formation in the mouse. *J. Embryol. exp. Morph.* **52**, 193–202.

Bravo, R. & Knowland, J. (1979). Classes of proteins synthesized in oocytes, eggs, embryos, and differentiated tissues of *Xenopus laevis*. *Differentiation* **13**, 101–108.

Breckenridge, L. J., Warren, R. L. & Warner, A. E. (1987). Lithium inhibits morphogenesis of the nervous system but not neuronal differentiation in *Xenopus laevis*. *Development* **99**, 353–370.

Breen, T. R. & Duncan, I. M. (1986). Maternal expression of genes that regulate the bithorax complex of *Drosophila melanogaster*. *Devl Biol.* **118**, 442–456.

Brennan, M. D., Weiner, A. J., Goralski, T. J. & Mahowald, A. P. (1982). The follicle cells are a major site of vitellogenin synthesis in *Drosophila melanogaster*. *Devl Biol.* **89**, 225–236.

Bretzel, G. & Tiedemann, H. (1986). Neural-inducing activity of nuclei and nuclear fractions from *Xenopus laevis* embryos. *Wilhelm Roux Arch. dev. Biol.* **195**, 123–127.

Briggs, R. & Cassens, G. (1966). Accumulation in the oocyte nucleus of a gene product essential for embryonic development beyond gastrulation. *Proc. natn. Acad. Sci. U.S.A.* **55**, 1103–1109.

Briggs, R., Green, E. U. & King, T. J. (1951). An investigation of the capacity for cleavage and differentiation in *Rana pipiens* eggs lacking 'functional' chromosomes. *J. exp. Zool.* **116**, 455–499.

Briggs, R. & Justus, J. T. (1968). Partial characterization of the component from normal eggs which corrects the maternal effect of gene *o* in the Mexican axolotl (*Ambystoma mexicanum*). *J. exp. Zool.* **167**, 105–116.

Brock, H. W. & Reeves, R. (1978). An investigation of *de novo* protein synthesis in the South African clawed frog, *Xenopus laevis*. *Devl Biol.* **66**, 128–141.

Brothers, A. J. (1976). Stable nuclear activation dependent on a protein synthesised during oogenesis. *Nature, Lond.* **260**, 112–115.

Brower, D. L., Smith, R. J. & Wilcox, M. (1981). Differentiation within the gonads of *Drosophila* revealed by immunofluorescence. *J. Embryol. exp. Morph.* **63**, 233–242.

Brown, D. D. & Dawid, I. B. (1968). Specific gene amplification in oocytes. *Science, N.Y.* **160**, 272–280.

Brown, E. A. & King, R. C. (1964). Studies on the events resulting in the formation of an egg chamber in *Drosophila melanogaster*. *Growth* **28**, 41–81.

Browne, C. L., Wiley, H. S. & Dumont, J. N. (1979). Oocyte-follicle cell gap junctions in *Xenopus laevis* and the effects of gonadotropin on their permeability. *Science, N.Y* **203**, 182–183.

Broyles, R. H. & Strittmatter, C. F. (1971). Hexose monophosphate shunt dehydrogenases during sea urchin development. *Expl Cell Res.* **67**, 471–474.

Bruhns, E. (1974). Analyse der Ooplasmaströmungen und ihrer strukturellen Grundlagen während der Furchung bei *Pimpla turionellae* L. (Hymenoptera). I. Lichtmikroskopisch-anatomische Veränderungen in der Eiarchitektur koinzident mit Zeitrafferfilmbefunden. *Wilhelm Roux Arch. EntwMech. Org.* **174**, 55–89.

Bruskin, A. M., Bédard, P.-A., Tyler, A. L., Showman, R. M., Brandhorst, B. P. & Klein, W. H. (1982). A family of proteins accumulating in ectoderm of sea urchin embryos specified by two related cDNA clones. *Devl Biol.* **91**, 317–324.

Bruskin, A. M., Tyner, A. L., Wells, D. E., Showman, R. M. & Klein, W. H. (1981). Accumulation in embryogenesis of five mRNAs enriched in the ectoderm of the sea urchin pluteus. *Devl Biol.* **87**, 308–318.

Buehr, M. L. & Blackler, A. W. (1970). Sterility and partial sterility in the South African clawed toad following the pricking of the egg. *J. Embryol. exp. Morph.* **23**, 375–384.

Bull, A. L. (1966). *Bicaudal*, a genetic factor which affects the polarity of the embryo in *Drosophila melanogaster*. *J. exp. Zool.* **161**, 221–241.

Burakova, T. A. & Kostomarova, A. A. (1975). Autoradiographic investigation of protein synthesis in the loach blastoderm. *Ontogenez* **6**, 225–233 (in translation 189–196).

Burkholder, G. D., Comings, D. E. & Okada, T. A. (1971). A storage form of ribosomes in mouse oocytes. *Expl Cell Res.* **69**, 361–371.

Butros, J. (1965). Action of heart and liver RNA on the differentiation of segments of chick blastoderms. *J. Embryol. exp. Morph.* **13**, 119–128.

Butros, J. M. (1960). Induction by desoxyribonucleic acids of axial structures in post-nodal fragments of chick blastoderms. *J. exp. Zool.* **143**, 259–282.

Byrd, E. W. & Kasinsky, H. E. (1973). Histone synthesis during early embryogenesis in *Xenopus laevis* (South African clawed toad). *Biochemistry, N.Y.* **12**, 246–253.

Bytinski-Salz, H. (1937). Trapianti di 'organizzatore' nelle uova di Lampreda. *Archo ital. Anat. Embriol.* **39**, 177–228.

Cabrera, C. V., Jacobs, H. T., Posakony, J. W., Grula, J. W., Roberts, J. W., Britten, R. J. & Davidson, E. H. (1983). Transcripts of three mitochondrial genes in the RNA of sea urchin eggs and embryos. *Devl Biol.* **97**, 500–505.

Cabrera, C. V., Martinez-Arias, A. & Bate, M. (1987). The expression of three members of the *achaete-scute* gene complex correlates with neuroblast segregation in *Drosophila*. *Cell* **50**, 425–433.

Calarco, P. G. & Brown, E. H. (1969). An ultrastructural and cytological study of preimplantation development of the mouse. *J. exp. Zool.* **171**, 253–283.

Callan, H. G. (1963). The nature of lampbrush chromosomes. *Int. Rev. Cytol.* **15**, 1–34.

Callan, H. G. (1973). DNA replication in the chromosomes of eukaryotes. *Cold Spring Harb. Symp. quant. Biol.* **38**, 195–203.

Cameron, R. A., Hough-Evans, B. R., Britten, R. J. & Davidson, E. H. (1987). Lineage and fate of each blastomere of the eight-cell sea urchin embryo. *Genes & Dev.* **1**, 75–85.

Camey, T. & Geilenkirchen, W. L. M. (1970). Cleavage delay and abnormal morphogenesis in *Lymnaea* eggs after pulse treatment with azide of successive stages in a cleavage cycle. *J. Embryol. exp. Morph.* **23**, 385–394.

Cammarata, M. (1973). Removal of cortical regions from the ascidian egg. *Acta Embryol. exp.* 115–120.

Campos-Ortega, J. A. (1983). Topological specificity of phenotype expression of neurogenic mutations in *Drosophila. Wilhelm Roux Arch. dev. Biol.* **192**, 317–326.

Capco, D. G. & Jeffery, W. R. (1979). Origin and spatial distribution of maternal messenger RNA during oogenesis of an insect, *Oncopeltus fasciatus. J. Cell Sci.* **39**, 63–76.

Capco, D. G. & Jeffery, W. R. (1981). Regional accumulaton of vegetal pole poly(A)$^+$RNA injected into fertilized *Xenopus* eggs. *Nature, Lond.* **294**, 255–257.

Capco, D. G. & Jeffery, W. R. (1982). Transient localizations of messenger RNA in *Xenopus laevis* oocytes. *Devl Biol.* **89**, 1–12.

Capdevila, M. P. & García-Bellido, A. (1974). Development and genetic analysis of *bithorax* phenocopies in *Drosophila. Nature, Lond.* **250**, 500–502.

Capdevila, M. P. & García-Bellido, A. (1981). Genes involved in the activation of the bithorax complex of *Drosophila. Wilhelm Roux Arch. dev. Biol.* **190**, 339–350.

Carpenter, C. D., Bruskin, A. M., Hardin, P. E., Keast, M. J., Anstrom, J., Tyner, A. L., Brandhorst, B. P. & Klein, W. H. (1984). Novel proteins belonging to the troponin C superfamily are encoded by a set of mRNAs in sea urchin embryos. *Cell* **36**, 663–671.

Carpenter, C. D. & Klein, W. H. (1982). A gradient of poly(A)$^+$RNA sequences in *Xenopus laevis* unfertilised eggs and embryos. *Devl Biol.* **91**, 43–49.

Carrasco, A. E. & Malacinski, G. M. (1987). Localization of *Xenopus* homoeobox gene transcripts during embryogenesis and in the adult nervous system. *Devl Biol.* **121**, 69–81.

Carrasco, A. E., McGinnis, W., Gehring, W. J. & de Robertis, E. M. (1984). Cloning of an *X. laevis* gene expressed during early embryogenesis coding for a peptide region homologous to *Drosophila* homeotic genes. *Cell* **37**, 409–414.

Carré, D. & Sardet, C. (1984). Fertilization and early development in *Beroe ovata. Devl Biol.* **105**, 188–195.

Carroll, A. G., Eckberg, W. R. & Ozaki, H. (1975). A comparison of protein synthetic patterns in normal and animalized sea urchin embryos. *Expl Cell Res.* **90**, 328–332.

Carroll, C. R. (1974). Comparative study of the early embryonic cytology and nucleic acid synthesis of *Ambystoma mexicanum* normal and *o* mutant embryos. *J. exp. Zool.* **187**, 409–422.

Carroll, S. B. & Scott, M. P. (1985). Localization of the *fushi tarazu* protein during *Drosophila* embryogenesis. *Cell* **43**, 47–57.

Cartwright, J. & Arnold, J. M. (1980). Intercellular bridges in the embryo of the Atlantic squid, *Loligo pealei*. I. Cytoplasmic continuity and tissue differentiation. *J. Embryol. exp. Morph.* **57**, 3–24.

332    *References*

Cather, J. N. (1967). Cellular interactions in the development of the shell gland of the gastropod, *Ilyanassa. J. exp. Zool.* **166**, 205–224.

Cather, J. N. (1971). Cellular interactions in the regulation of development in annelids and molluscs. *Adv. Morphogen.* **9**, 67–125.

Cather, J. N. (1973). Regulaton of apical cilia development by the polar lobe of *Ilyanassa* (Gastropoda: Nassariidae). *Malacologia* **12**, 213–223.

Cather, J. N. & Verdonk, N. H. (1974). The development of *Bithynia tentaculata* (Prosobranchai, Gastropoda) after removal of the polar lobe. *J. Embryol. exp. Morph.* **31**, 415–422.

Cather, J. N. & Verdonk, N. H. (1979). Development of *Dentalium* following removal of D quadrant blastomeres at successive cleavage stages. *Wilhelm Roux Arch. dev. Biol.* **187**, 355–366.

Cather, J. N., Verdonk, N. H. & Dohmen, M. R. (1976). Role of the vegetal body in the regulation of development in *Bithynia tentaculata* (Prosobranchia, Gastropoda). *Am. Zool.* **16**, 455–468.

Chabry, L. (1887). Contribution à l'embryologie normale et tératologique des ascidies simples. *J. Anat. Physiol., Paris* **23**, 167–319.

Chadwick, R. & McGinnis, W. (1987). Temporal and spatial distribution of transcripts from the *Deformed* gene of *Drosophila. EMBO J.* **6**, 779–789.

Chase, J. W. & Dawid, I. B. (1972). Biogenesis of mitochondria during *Xenopus laevis* development. *Devl Biol.* **27**, 504–518.

Chauhan, S. P. S. & Rao, K. V. (1970). Chemically stimulated differentiations of post-nodal pieces of chick blastoderms. *J. Embryol. exp. Morph.* **23**, 71–78.

Cheney, C. M., Miller, K. G., Lang, T. J. & Shearn, A. (1984). Specific protein modifications are altered in a temperature-sensitive *Drosophila* developmental mutant. *Proc. natn. Acad. Sci. U.S.A.* **81**, 6422–6426.

Cherdantseva, E. M. & Cherdantsev, V. G. (1985). Determination of dorso-ventral polarity in embryos of *Brachydanio rerio* (Teleostei). *Ontogenez* **16**, 270–280 (in translation 165–173).

Child, C. M. (1916). Axial susceptibility gradients in the early development of the sea urchin. *Biol. Bull. mar. biol. Lab., Woods Hole* **30**, 391–405.

Child, C. M. (1936*a*). Differential reduction of vital dyes in the early development of echinoderms. *Wilhelm Roux Arch. EntwMech. Org.* **135**, 426–456.

Child, C. M. (1936*b*). A contribution to the physiology of exogastrulation in echinoderms. *Wilhelm Roux Arch. EntwMech. Org.* **135**, 457–493.

Child, C. M. (1948). Patterns of indicator oxidation in *Triturus* development. *J. exp. Zool.* **109**, 79–108.

Child, C. M (1951). Oxidation–reduction indicator patterns in the development of *Clavellina huntsmani. Physiol. Zoöl.* **24**, 353–367.

Childs, G., Maxson, R. & Kedes, L. H. (1979). Histone gene expression during sea urchin embryogenesis: isolation and characterization of early and late messenger RNAs of *Strongylocentrotus purpuratus* by gene-specific hybridization and template activity. *Devl Biol.* **73**, 153–173.

Chuang, H.-H. (1939). Induktionsleistungen von frischen und gekochten Organteilen (Niere, Leber) nach ihrer Verplanzung in explantate und verschiedene Wirtsregionen von *Triton*keimen. *Wilhelm Roux Arch. EntwMech. Org.* **139**, 556–638.

Chuang, H.-H. (1955). Untersuchungen über die Reaktionsfähigkeit des Ektoderms mittels sublethaler Cytolyse. *Acta Biol. exp. sin.* **4**, 151–186.

Chuang, H.-H. (1963). Studies on the mesoderm-inducing agent from mammalian liver. II. Effect of alcohol and heat treatment on the inductive ability. *Acta Biol. exp. sin.* **8**, 370–387.

Chulitskaia, E. V. (1970). Desynchronization of cell divisions in the course of egg cleavage and an attempt at experimental shift of its onset. *J. Embryol. exp. Morph.* **23**, 359–374.

Chung, H. M. & Malacinski, G. M. (1982). Pattern formation during early amphibian development: embryogenesis in inverted anuran and urodele eggs. *Devl Biol.* **93**, 444–452.

Clavert, J. (1960). Déterminisme de la symétrie bilatérale chez les oiseaux. IV. Existence d'une phase critique pour la symétrisation de l'oeuf. Son stade. *Archs Anat. microsc. Morph. exp.* **49**, 345–361.

Clavert, J. (1962). Symmetrisation of the egg of vertebrates. *Adv. Morphogen.* **2**, 27–60.

Clavert, J. & Filogamo, G. (1966). Investigations on the determination of bilateral symmetry in the fish, *Lebistes reticulatus. Acta Embryol. Morph. exp.* **9**, 105–117.

Clegg, K. B. & Pikó, L. (1983). Quantitative aspects of RNA synthesis and polyadenylation in 1-cell and 2-cell mouse embryos. *J. Embryol. exp. Morph.* **74**, 169–182.

Cleine, J. H. & Slack, J. M. W. (1985). Normal fates and states of specification of different regions in the axolotl gastrula. *J. Embryol. exp. Morph.* **86**, 247–269.

Clement, A. C. (1956). Experimental studies on germinal localization in *Ilyanassa.* II. The development of isolated blastomeres. *J. exp. Zool.* **132**, 427–445.

Clement, A.C. (1962). Development of *Ilyanassa* following removal of the D macromere at successive cleavage stages. *J. exp. Zool.* **149**, 193–216.

Clement, A. C. (1967). The embryonic value of the micromeres in *Ilyanassa obsoleta*, as determined by deletion experiments. I. The first quartet cells. *J. exp. Zool.* **166**, 77–88.

Clement, A. C. (1968). Development of the vegetal half of the *Ilyanassa* egg after removal of most of the yolk by centrifugal force, compared with the development of animal halves of similar visible composition. *Devl Biol.* **17**, 165–186.

Clement, A. C. (1971). *Ilyanassa.* In *Experimental Embryology of Marine and Fresh-Water Invertebrates*, ed. G. Reverberi, pp. 188–214. Amsterdam: North-Holland.

Clement, A. C. (1976). Cell determination and organogenesis in molluscan development: a reappraisal based on deletion experiments in *Ilyanassa. Am. Zool.* **16**, 447–453.

Clement, A. C. & Lehmann, F. E. (1956). Über das Verteilungsmuster von Mitochondrien und Lipoidtropfen während der Furchung des Eies von *Ilyanassa obsoleta* (Mollusca, Prosobranchia). *Naturwissenschaften* **43**, 478–479.

Clough, J. R. & Whittingham, D. G. (1983). Metabolism of [$^{14}$C] glucose by postimplantation mouse embryos *in vitro. J. Embryol. exp. Morph.* **74**, 133–142.

Coggins, L. W. (1973). An ultrastructural and radioautographic study of early oogenesis in the toad *Xenopus laevis. J. Cell Sci.* **12**, 71–93.

Cognetti, G., Settineri, D. & Spinelli, G. (1972). Developmental changes of chromatin non-histone proteins in sea urchins. *Expl Cell Res.* **71**, 465–468.

Cohen, A. & Berrill, N. J. (1936). The development of isolated blastomeres of the ascidian egg. *J. exp. Zool.* **74**, 91–117.

Cohen, L. H., Newrock, K. M. & Zweidler, A. (1975). Stage-specific switches in histone synthesis during embryogenesis of the sea urchin. *Science, N.Y.* **190**, 994–997.

Collier, J. R. (1975). Nucleic acid synthesis in the normal and lobeless embryo of *Ilyanassa obsoleta. Expl Cell Res.* **95**, 254–262.

Collier, J. R. (1977). Rates of RNA synthesis in the normal and lobeless embryo of *Ilyanassa obsoleta. Expl Cell Res.* **106**, 390–394.

Collier, J. R. (1983*a*). Protein synthesis during *Ilyanassa* organogenesis. *Devl Biol.* **100**, 256–259.

Collier, J. R. (1983*b*). The biochemistry of molluscan development. In *The Mollusca*, vol. 3 *Development*, ed. N. H. Verdonk, J. A. M. van den Biggelaar & A. S. Tompa, pp. 253–297. New York: Academic Press.

Collier, J. R. (1984). Protein synthesis in normal and lobeless gastrulae of *Ilyanassa obsoleta. Biol. Bull. mar. biol. Lab., Woods Hole* **167**, 371–377.

Collier, J. R. & McCarthy, M. E. (1981). Regulation of polypeptide synthesis during early embryogenesis of *Ilyanassa obsoleta. Differentiation* **19**, 31–46.

Colman, A. (1983). Maternal mysteries. *Trends biochem. Sci.* **8**, 37.

Conklin, E. G. (1897). The embryology of *Crepidula*, a contribution to the cell lineage and early development of some marine gasteropods. *J. Morph.* **13**, 1–226.

Conklin, E. G. (1905*a*). The organization and cell-lineage of the ascidian egg. *J. Acad. nat. Sci. Philad.* **13**, 1–119.

Conklin, E. G. (1905*b*). Mosaic development in ascidian eggs *J. exp. Zool.* **2**, 145–223.

Conklin, E. G. (1931). The development of centrifuged eggs of ascidians. *J. exp. Zool.* **60**, 1–119.

Conklin, E. G. (1932). The embryology of Amphioxus. *J. Morph.* **54**, 69–151.

Conklin, E. G. (1933). The development of isolated and partially separated blastomeres of Amphioxus. *J. exp. Zool.* **64**, 303–375.

Conrad, G. W., Williams, D. C., Turner, F. R., Newrock, K. M. & Raff, R. A. (1973). Microfilaments in the polar lobe constriction of fertilized eggs of *Ilyanassa obsoleta. J. Cell Biol.* **59**, 228–233.

Cooke, J. (1972*a*). Properties of the primary organization field in the embryo of *Xenopus laevis*. I. Autonomy of cell behaviour at the site of initial organizer formation. *J. Embryol. exp. Morph.* **28**, 13–26.

Cooke, J. (1972*b*). Properties of the primary organization field in the embryo of *Xenopus laevis*. II. Positional information for axial organization in embryos with two head organizers. *J. Embryol. exp. Morph.* **28**, 27–46.

Cooke, J. (1973*a*). Properties of the primary organization field in the embryo of *Xenopus laevis*. IV. Pattern formation and regulation following early inhibition of mitosis. *J. Embryol. exp. Morph.* **30**, 49–62.

Cooke, J. (1973*b*). Properties of the primary organization field in the embryo of *Xenopus laevis*. V. Regulation after removal of the head organizer, in normal early gastrulae and in those already possessing a second implanted organizer. *J. Embryol. exp. Morph.* **30**, 283–300.

Cooke, J. (1981). Scale of body pattern adjusts to available cell number in amphibian embryos. *Nature, Lond.* **290**, 775–778.

Cooke, J. (1983). Evidence for specific feedback signals underlying pattern control during vertebrate embryogenesis. *J. Embryol. exp. Morph.* **76**, 95–114.

Cooke, J. (1985). Dynamics of the control of body pattern in the development of *Xenopus laevis*. III. Timing and pattern after U.V. irradiation of the egg and after excision of presumptive head endo-mesoderm. *J. Embryol. exp. Morph.* **88**, 135–150.

Cooke, J. (1986). Permanent distortion of positional system of *Xenopus* embryos by early perturbation in gravity. *Nature, Lond.* **319**, 60–63.

Cooke, J. & Webber, J. A. (1985*a*). Dynamics of the control of body pattern in the development of *Xenopus laevis*. I. Timing and pattern in the development of dorsoanterior and posterior blastomere pairs, isolated at the 4-cell stage. *J. Embryol. exp. Morph.* **88**, 85–112.

Cooke, J. & Webber, J. A. (1985*b*). Dynamics of the control of body pattern in the development of *Xenopus laevis*. II. Timing and pattern in the development of single blastomeres (presumptive lateral halves) isolated at the 2-cell stage. *J. Embryol. exp. Morph.* **88**, 113–133.

Copp, J. (1978). Interaction between inner cell mass and trophectoderm of the mouse blastocyst. I. A study of cellular proliferation. *J. Embryol. exp. Morph.* **48**, 109–125.

Copp, J. (1979). Interaction between inner cell mass and trophectoderm of the mouse blastocyst. II. The fate of the polar trophectoderm. *J. Embryol. exp. Morph.* **51**, 109–120.

Costello, D. P. (1945). Experimental studies of germinal localization in *Nereis*. I. The development of isolated blastomeres. *J. exp. Zool.* **100**, 19–66.

Costello, D. P. (1948). Ooplasmic segregation in relation to differentiation. *Ann. N.Y. Acad. Sci.* **49**, 663–683.

Costello, D. P. (1961). On the orientation of centrioles in dividing cells, and its significance: a new contribution to spindle mechanics. *Biol. Bull. mar. biol. Lab., Woods Hole* **120**, 285–312.

Costello, D. P. & Henley, C. (1976). Spiralian development: a perspective. *Am. Zool.* **16**, 277–291.

Counce, S. J. (1973). The causal analysis of insect embryogenesis. In *Developmental Systems: Insects*, vol. 2, ed. S. J. Counce & C. H. Waddington, pp. 1–56. London: Academic Press.

Cowan, A. E. & McIntosh, J. R. (1985). Mapping the distribution of differentiation potential for intestine, muscle, and hypodermis during early development in *Caenorhabditis elegans*. *Cell* **41**, 923–932.

Cowden, R. R. & Lehmann, H. E. (1963). A cytochemical study of differentiation in early echinoid development. *Growth* **27**, 185–197.

Cox, K. H., Angerer, L. M., Lee, J. J., Davidson, E. H. & Angerer, R. C. (1986). Cell lineage-specific programs of expression of multiple actin genes during sea urchin embryogenesis. *J. molec. Biol.* **188**, 159–172.

Crampton, H. E. (1896). Experimental studies on gastropod development. *Arch. EntwMech. Org.* **3**, 1–26.

Crampton, H. E. (1897). The ascidian half-embryo. *Ann. N.Y. Acad. Sci.* **10**, 50–57.

Crampton, J. M. & Woodland, H. R. (1979). A cell-free assay system for the analysis of changes in RNA synthesis during the development of *Xenopus laevis*. *Devl Biol.* **70**, 453–466.

Crick, F. (1970). Diffusion in embryogenesis. *Nature, Lond.* **225**, 420–422.

Crippa, M., Davidson, E. H. & Mirsky, A. E. (1967). Persistence in early amphibian embryos of informational RNA's from the lampbrush chromosome stage of oögenesis. *Proc. natn. Acad. Sci. U.S.A.* **57**, 885–892.

Crippa, M. & Gross, P. R. (1969). Maternal and embryonic contributions to the functional messenger RNA of early development. *Proc. natn. Acad. Sci. U.S.A.* **62**, 120–127.

Critchley, D. R., England, M. A., Wakely, J. & Hynes, R. O. (1979). Distribution of fibronectin in the ectoderm of gastrulating chick embryos. *Nature, Lond.* **280**, 498–500.

Crowell, J. (1964). The fine structure of the polar lobe of *Ilyanassa obsoleta*. *Acta Embryol. Morph. exp.* **7**, 225–234.

Crowther, R. J. & Whittaker, J. R. (1983). Developmental autonomy of muscle fine structure in muscle lineage cells of ascidian embryos. *Devl Biol.* **96**, 1–10.

Crowther, R. J. & Whittaker, J. R. (1984). Differentiation of histospecific ultrastructural features in cells of cleavage-arrested early ascidian embryos. *Wilhelm Roux Arch. dev. Biol.* **194**, 87–98.

Crowther, R. J. & Whittaker, J. R. (1986). Differentiation without cleavage: multiple cytospecific ultrastructural expressions in individual one-celled ascidian embryos. *Devl Biol.* **117**, 114–126.

Curtis, A. S. G. (1960). Cortical grafting in *Xenopus laevis*. *J. Embryol. exp. Morph.* **8**, 163–173.

Curtis, A. S. G. (1962). Morphogenetic interactions before gastrulation in the amphibian, *Xenopus laevis* – the cortical field. *J. Embryol. exp. Morph.* **10**, 410–422.

Czihak, G. (1962). Entwicklungsphysiologische Untersuchungen an Echiniden. (Topochemie der Blastula und Gastrula, Entwicklung der Bilateral- und Radiärsymmetrie und der Coelomdivertikel.) *Wilhelm Roux Arch. Entw-Mech. Org.* **154**, 29–55.

Czihak, G. (1965*a*). Evidences for inductive properties of the micromere-RNA in sea urchin embryos. *Naturwissenschaften* **52**, 141–142.

Czihak, G. (1965*b*). Entwicklungsphysiologische Untersuchungen an Echiniden. Ribonucleinsäure-Synthese in den Mikromeren und Entodermdifferenzierung ein Beitrag zum Problem der Induktion. *Wilhelm Roux Arch. EntwMech. Org.* **156**, 504–524.

Czihak, G. (1977). Kinetics of RNA-synthesis in the 16-cell stage of the sea urchin *Paracentrotus lividus*. *Wilhelm Roux Arch. dev. Biol.* **182**, 59–68.

Czihak, G. (1978). Effect of bromodeoxyuridine on differentiation. II. The effect of early BUdR treatment on gastrulation of sea urchin embryos. *Differentiation.* **11**, 103–107.

Czihak, G. & Hörstadius, S. (1970). Transplantation of RNA-labelled micromeres into animal halves of sea urchin embryos. A contribution to the problem of embryonic induction. *Devl Biol.* **22**, 15–32.

Dalcq, A. (1932). Étude des localisations germinales dans l'oeuf vierge d'ascidie par des expériences de mérogonie. *Archs Anat. microsc.* **28**, 223–333.

Dalcq, A. & Dollander, A. (1948). Sur les phénomènes de régulation chez le Triton après séparation des deux premiers blastomères et sur la disposition de la pellicule (coat) dans l'oeuf fécondé et segmenté. *C. r. Séanc. Soc. Biol.* **142**, 1307–1312.

Dale, L. & Slack, J. M. W. (1987). Fate map for the 32-cell stage of *Xenopus laevis*. *Development* **99**, 527–551.

Dale, L., Smith, J. C. & Slack, J. M. W. (1985). Mesoderm induction in *Xenopus laevis*; a quantitative study using a cell lineage label and tissue-specific antibodies. *J. Embryol. exp. Morph.* **89**, 289–312.

D'Amelio, V. & Céas, M. P. (1957). Distribution of protease activity in the blastula and early gastrula of *Discoglossus pictus*. *Experientia* **13**, 152–153.

Damsky, C. H., Richa, J., Solter, D., Knudsen, K. & Buck, C. A. (1983). Identification and purification of a cell surface glycoprotein mediating intercellular adhesion in embryonic and adult tissue. *Cell* **34**, 455–466.

Dan, K. (1979). Studies on unequal cleavage in sea urchins. I. Migration of the nuclei to the vegetal pole. *Dev. Growth Differ.* **21**, 527–535.

Dan, K. & Dan, J. C. (1942). Behavior of the cell surface during cleavage. IV.

Polar lobe formation and cleavage of the eggs of *Ilyanassa obsoleta* Say. *Cytologia* **12**, 246–261.

Dan, K., Endo, S. & Uemura, I. (1983). Studies on unequal cleavage in sea urchins. II. Surface differentiation and the direction of nuclear migration. *Dev. Growth Differ.* **25**, 227–237.

Dan, K. & Ikeda, M. (1971). On the system controlling the time of micromere formation in sea urchin embryos. *Dev. Growth Differ.* **13**, 285–301.

Dan, K. & Inoué, S. (1987). Studies of unequal cleavage in molluscs. II. Asymmetric nature of the two asters. *Int. J. Invertebr. Reprod. Dev.* **11**, 335–353.

Dan, K. & Ito, S. (1984). Studies on unequal cleavage in molluscs. I. Nuclear behavior and anchorage of a spindle pole to cortex as revealed by isolation technique. *Dev. Growth Differ.* **26**, 249–262.

Dan, K., Tanaka, S., Yamazaki, K. & Kato, Y. (1980). Cell cycle study up to the time of hatching in the embryos of the sea urchin, *Hemicentrotus pulcherrimus*. *Dev. Growth Differ.* **22**, 589–598.

Danilchik, M. V. & Gerhart, J. C. (1987). Differentiation of the animal–vegetal axis in *Xenopus laevis* oocytes. I. Polarized intracellular translocation of platelets establishes the yolk gradient. *Devl Biol.* **122**, 101–112.

Dan-Sohkawa, M. & Fujisawa, H. (1980). Cell dynamics of the blastulation process in the starfish, *Asterina pectinifera*. *Devl Biol.* **77**, 328–339.

Dan-Sohkawa, M. & Satoh, N. (1978). Studies on dwarf larvae developed from isolated blastomeres of the starfish, *Asterina pectinifera*. *J. Embryol. exp. Morph.* **46**, 171–185.

Darnbrough, C. & Ford, P. J. (1976). Cell-free translation of messenger RNA from oocytes of *Xenopus laevis*. *Devl Biol.* **50**, 285–301.

Darribère, T., Boucher, D., Lacroix, J.-C. & Boucaut, J.-C. (1984). Fibronectin synthesis during oogenesis and early development of the amphibian *Pleurodeles waltlii*. *Cell Differ.* **14**, 171–177.

Das, N. K., Micou-Eastwood, J. & Alfert, M. (1982). Histone patterns during early embryogenesis in the echiuroid *Urechis caupo*. *Cell Differ.* **11**, 147–153.

Dasgupta, J. D. & Singh, U. N. (1981). Early differentiation in zebra fish blastula: role of yolk syncytial layer. *Wilhelm Roux Arch. dev. Biol.* **190**, 358–360.

Dasgupta, S. & Kung-Ho, C. (1971). Electrophoretic analysis of cell populations in presumptive epidermis of the frog, *Rana pipiens*. *Expl Cell Res.* **65**, 463–466.

Davidson, E. H. (1968). *Gene Activity in Early Development*, 1st edn. New York: Academic Press.

Davidson, E. H. (1986). *Gene Activity in Early Development*, 3rd edn. Orlando: Academic Press.

Davidson, E. H. & Britten, R. J. (1971). Note on the control of gene expression during development. *J. theor. Biol.* **32**, 123–130.

Davidson, E. H., Crippa, M. & Mirsky, A. E. (1968). Evidence for the appearance of novel gene products during amphibian blastulation. *Proc. natn. Acad. Sci. U.S.A.* **60**, 152–159.

Davidson, E. H., Haslett, G. W., Finney, R. J., Allfrey, V. G. & Mirsky, A. E. (1965). Evidence for prelocalization of cytoplasmic factors affecting gene activation in early embryogenesis. *Proc. natn. Acad. Sci. U.S.A.* **54**, 696–704.

Davis, F. C. (1982). Differential utilization of ribosomes for protein synthesis during oogenesis and early embryogenesis of *Urechis caupo* (Echiura). *Differentiation* **22**, 170–174.

Dawid, I. B., Haynes, S. R., Jamrich, M., Jonas, E., Miyatani, S., Sargent, T.

D. & Winkles, J. A. (1985). Gene expression in *Xenopus* embryogenesis. *J. Embryol. exp. Morph.* **89**, Suppl. 113–124.

Dawid, I. B. & Sargent, T. D. (1986). Molecular embryology in amphibians: new approaches to old questions. *Trends Genet.* **2**, 47–50.

de Bernardi, F. (1982). Polyadenylated mRNAs from various developmental stages of *Xenopus laevis. Expl Cell Biol.* **50**, 281–290.

Dekel, N. & Beers, W. H. (1978). Rat oocyte maturation *in vitro*: relief of cyclic AMP inhibition by gonadotropins. *Proc. natn. Acad. Sci. U.S.A.* **75**, 4369–4373.

DeLotto, R. & Spierer, P. (1986). A gene required for the specification of dorsal-ventral pattern in *Drosophila* appears to encode a serine protease. *Nature, Lond.* **323**, 688–692.

del Pino, E. M. & Elinson, R. P. (1983). A novel development pattern for frogs: gastrulation produces an embryonic disk. *Nature, Lond.* **306**, 589–591.

Denich, K. T. R., Schierenberg, E., Isnenghi, E. & Cassada, R. (1984). Cell-lineage and developmental defects of temperature-sensitive embryonic arrest mutants of the nematode, *Caenorhabditis elegans. Wilhelm Roux Arch. dev. Biol.* **193**, 164–179.

Denis, H. (1964*a*). Effets de l'actinomycine sur le développement embryonnaire. I. Etude morphologique: suppression par l'actinomycine de la compétence de l'ectoderme et du pouvoir inducteur de la lèvre blastoporale. *Devl Biol.* **9**, 435–457.

Denis, H. (1964*b*). Effets de l'actinomycine sur le développement embryonnaire. III. Etude biochimique: influence de l'actinomycine sur la synthèse des protéines. *Devl Biol.* **9**, 473–483.

Denis, H., Picard, B., le Maire, M. & Clerot, J.-C. (1980). Biochemical research on oogenesis. The storage particles of the teleost fish *Tinca tinca. Devl Biol.* **77**, 218–223.

Deno, T., Nishida, H. & Satoh, N. (1985). Histospecific acetylcholinesterase development in quarter ascidian embryos derived from each blastomere pair of the eight-cell stage. *Biol. Bull. mar. biol. Lab., Woods Hole* **168**, 239–248.

Deno, T. & Satoh, N. (1984). Studies on the cytoplasmic determinant for muscle cell differentiation in ascidian embryos: an attempt at transplantation of the myoplasm. *Dev. Growth Differ.* **26**, 43–48.

Deppe, U., Schierenberg, E., Cole, T., Krieg, C., Schmitt, D., Yoder, B. & von Ehrenstein, G. (1978). Cell lineages of the embryo of the nematode *Caenorhabditis elegans. Proc. natn. Acad. Sci. U.S.A.* **75**, 376–380.

Dequin, R., Saumweber, H. & Sedat, J. W. (1984). Proteins shifting from the cytoplasm into the nuclei during early embryogenesis of *Drosophila melanogaster. Devl Biol.* **104**, 37–48.

Deshpande, A. K. & Siddiqui, M. A. Q. (1976). Differentiation induced by cyclic AMP in undifferentiated cells of early chick embryo *in vitro. Nature, Lond.* **263**, 588–591.

Desnitskii, A. G. (1978). Cell proliferation in axolotl embryos at different stages of gastrulation. *Ontogenez* **9**, 197–200 (in translation 163–165).

Desplan, C., Theis, J. & O'Farrell, P. H. (1985). The *Drosophila* developmental gene, *engrailed*, encodes a sequence-specific DNA binding activity. *Nature, Lond.* **318**, 630–635.

Dettlaff, T. A. (1964). Cell divisions, duration of interkinetic states and differentiation in early stages of embryonic development. *Adv. Morphogen.* **3**, 323–362.

Dettlaff, T. A. (1983). A study of the properties, morphogenetic potencies and prospective fate of outer and inner layers of ectodermal and chordamesodermal regions during gastrulation, in various Anuran amphibians. *J. Embryol. exp. Morph.* **75**, 67–86.

Deuchar, E. M. (1969). Effects of a mesoderm-inducing factor on early chick embryos. *J. Embryol. exp. Morph.* **22**, 295–304.

Deuchar, E. M. (1972). *Xenopus laevis* and developmental biology. *Biol. Rev.* **47**, 37–112.

Devillers, C. (1961). Structural and dynamic aspects of the development of the teleostean egg. *Adv. Morphogen.* **1**, 379–428.

de Vincentiis, M., Hörstadius, S. & Runnström, J. (1966). Studies on controlled and released respiration in animal and vegetal halves of the embryo of the sea urchin, *Paracentrotus lividus*. *Expl Cell Res.* **41**, 535–544.

de Vos, L. (1981). Étude au microscope électronique à balayage des stades précoces du développement embryonnaire du gardon (*Rutilus rutilus* L.) *Archs Biol., Bruxelles* **92**, 153–173.

Devriès, J. (1970). Sur le rôle du mésoderme dans la différenciation de l'endoderme chez l'embryon d'*Eisenia foetida* (Lombricien). *Bull. Soc. zool. Fr.* **95**, 169–172.

Devriès, J. (1973). Aspect du déterminisme embryonnaire au cours des premiers stades de la segmentation chez le lombricien *Eisenia foetida* (expérience de compression des oeufs). *Ann. Embryol. Morphogen.* **6**, 95–108.

Devriès, J. (1976). Action de lithium sur l'embryon d'*Eisenia foetida* (Lombricien). *Archs Biol., Bruxelles* **87**, 225–243.

Dictus, W. J. A. G., van Zoelen, E. J. J., Tetteroo, P. A. T., Tertoolen, L. G. J., de Laat, S. W. & Bluemink, J. G. (1984). Lateral mobility of plasma membrane lipids in *Xenopus* eggs: regional differences related to animal/vegetal polarity become extreme upon fertilization. *Devl Biol.* **101**, 201–211.

Digan, M.A., Haynes, S.R., Mozer, B.A., Dawid, I.B., Forquignon, F. & Gans, M. (1986). Genetic and molecular analysis of *fs(1)h*, a maternal effect homeotic gene in *Drosophila*. *Devl Biol.* **114**, 161–169.

DiNardo, S., Kuner, J. M., Theis, J. & O'Farrell, P. H. (1985). Development of embryonic pattern in *D. melanogaster* as revealed by accumulation of the nuclear *engrailed* protein. *Cell* **43**, 59–69.

Dittmann, F., Ehni, R. & Engels, W. (1981). Bioelectric aspects of the hempteran telotrophic ovariole (*Dysdercus intermedius*). *Wilhelm Roux Arch. dev. Biol.* **190**, 221–225.

Dohmen, M. R. & Lok, D. (1975). The ultrastructure of the polar lobe of *Crepidula fornicata* (Gastropoda, Prosobranchia). *J. Embryol. exp. Morph.* **34**, 419–438.

Dohmen, M. R. & van de Mast, J. M. A. (1978). Electron microscopical study of RNA-containing cytoplasmic localizations and intercellular contacts in early cleavage stages of eggs of *Lymnaea stagnalis* (Gastropoda, Pulmonata). *Proc. K. ned. Akad. Wet. C* **81**, 403–414.

Dohmen, M. R. & van der Mey, J. C. A. (1977). Local surface differentiations at the vegetal pole of the eggs of *Nassarius reticulatus, Buccinum undatum,* and *Crepidula fornicata* (Gastropoda, Prosobranchia). *Devl Biol.* **61**, 104–113.

Dohmen, M. R. & Verdonk, N. H. (1974). The structure of a morphogenetic cytoplasm, present in the polar lobe of *Bithynia tentaculata* (Gastropoda, Prosobranchia). *J. Embryol. exp. Morph.* **31**, 423–433.

Dohmen, M. R. & Verdonk, N. H. (1979a). The ultrastructure and role of the polar lobe in development of molluscs. In *Determinants of Spatial Organization,* ed. S. Subtelny & I. R. Konigsberg, pp. 3–27. New York: Academic Press.

Dohmen, M. R. & Verdonk, N. H. (1979*b*). Cytoplasmic localizations in mosaic eggs. In *Maternal Effects in Development*, ed. D. R. Newth & M. Balls, pp. 127–145. Cambridge: The University Press.

Dolecki, G. J. & Smith, L. D. (1979). Poly(A)$^+$RNA metabolism during oogenesis in *Xenopus laevis*. *Devl Biol.* **69**, 217–236.

Dolecki, G. J., Wannakrairoj, S., Lum, R., Wang, G., Riley, H. D., Carlos, R., Wang, A. & Humphreys, T. (1986). Stage-specific expression of a homoeo box-containing gene in the non-segmented sea urchin embryo. *EMBO J.* **5**, 925–930.

Dollander, A. & Melnotte, J. P. (1952). Variation topographique de la colorabilité du cortex de l'oeuf symétrisé de *Triturus alpestris* au bleu de Nil et au rouge neutre. *C. r. Séanc. Soc. Biol.* **146**, 1614–1616.

Donohoo, P. & Kafatos, F. C. (1973). Differences in the proteins synthesized by the progeny of the first two blastomeres of *Ilyanassa*, a 'mosaic' embryo. *Devl Biol.* **32**, 224–229.

Dontsova, G. V. & Neifakh, A. A. (1969). Synthesis of high-polymer RNA in isolated blastoderms and dissociated cells of the embryonic loach. *Dokl. Akad. Nauk SSSR* **184**, 1253–1256.

Dorresteijn, A. W. C., Biliński, S. M., van den Biggelaar, J. A. M. & Bluemink, J. G. (1982). The presence of gap junctions during early *Patella* embryogenesis: an electron microscopical study. *Devl Biol.* **91**, 397–401.

Dorresteijn, A. W. C., Bornewasser, H. & Fischer, A. (1987). A correlative study of experimentally changed first cleavage and Janus development in the trunk of *Platynereis dumerilii* (Annelida, Polychaeta). *Wilhelm Roux Arch. dev. Biol.* **196**, 51–58.

Dorresteijn, A. W. C., van den Biggelaar, J. A. M., Bluemink, J. G. & Hage, W. J. (1981). Electron microscopical investigations of the intercellular contacts during the early cleavage stages of *Lymnaea stagnalis* (Mollusca, Gastropoda). *Wilhelm Roux Arch. dev. Biol.* **190**, 215–220.

Dorresteijn, A. W. C., Wagemaker, H. A., de Laat, S. W. & van den Biggelaar, J. A. M. (1983). Dye-coupling between blastomeres in early embryos of *Patella vulgata* (Mollusca, Gastropoda): its relevance for cell determination. *Wilhelm Roux Arch. dev. Biol.* **192**, 262–269.

Doucet-de-Bruïne, M. H. M. (1973). Blastopore formation in *Ambystoma mexicanum*. *Wilhelm Roux Arch. EntwMech. Org.* **173**, 136–163.

Doyle, H. J., Harding, K., Hoey, T. & Levine, M. (1986). Transcripts encoded by a homoeo box gene are restricted to dorsal tissues of *Drosophila* embryos. *Nature, Lond.* **323**, 76–79.

Drews, U., Kussäther, E. & Usadel, K. H. (1967). Histochemischer Nachweis der Cholinesterase in der Frühentwicklung der Hühnerkeimscheibe. *Histochemie* **8**, 65–89.

Dreyer, C., Scholz, E. & Hausen, P. (1982). The fate of oocyte nuclear proteins during early development of *Xenopus laevis*. *Wilhelm Roux Arch. dev. Biol.* **191**, 228–233.

Driesch, H. (1891). Entwicklungsmechanische Studien. I. Der Werth der beiden ersten Furchungszellen in der Echinodermenentwicklung. Experimentelle Erzeugen von Theil- und Doppelbildung. *Z. wiss. Zool.* **53**, 160–178.

Driesch, H. (1892). Entwicklungsmechanische Studien. III–VI. *Z. wiss. Zool.* **55**, 1–62.

Driesch, H. (1895). Von der Entwickelung einzelner Ascidienblastomeren. *Arch. EntwMech. Org.* **2**, 398–413.

Driesch, H. (1900). Die isolirten Blastomeren des Echinidenkeimes. Eine Nachprüfung und Erweiterung früherer Untersuchungen. *Arch. EntwMech. Org.* **10**, 361–410.

Driesch, H. (1908). *The Science and Philosophy of the Organism*, vol. 1. London: A. & C. Black.

Driesch, H. & Morgan, T. H. (1895). Zur Analysis der ersten Entwickelungsstadien des Ctenophoreneies. I. Von der Entwickelung einzelner Ctenophorenblastomeren. *Arch. EntwMech. Org.* **2**, 204–215.

Duband, J. L. & Thiery, J. P. (1982). Appearance and distribution of fibronectin during chick embryo gastrulation and neurulation. *Devl Biol.* **94**, 337–350.

Ducibella, T. (1977). Surface changes of the developing trophoblast cell. In *Development in Mammals*, vol. 1, ed. M. H. Johnson, pp. 5–30. Amsterdam: North-Holland.

Ducibella, T., Albertini, D. F., Anderson, E. & Biggers, J. D. (1975). The preimplantation mammalian embryo: characterization of intercellular junctions and their appearance during development. *Devl Biol.* **45**, 231–250.

Ducibella, T. & Anderson, E. (1975). Cell shape and membrane changes in the eight-cell mouse embryo: prerequisites for morphogenesis of the blastocyst. *Devl Biol.* **47**, 45–58.

Dumont, J. N. (1972). Oogenesis in *Xenopus laevis* (Daudin). I. Stages of oocyte development in laboratory maintained animals. *J. Morph.* **136**, 153–179.

Durante, M. (1956). Cholinesterase in the development of *Ciona intestinalis* (Acidia). *Experientia* **12**, 307–308.

Dworkin, M. B. & Dawid, I. B. (1980*a*). Construction of a cloned library of expressed embryonic gene sequences from *Xenopus laevis*. *Devl Biol.* **76**, 435–448.

Dworkin, M. B. & Dawid, I. B. (1980*b*). Use of a cloned library for the study of abundant poly(A)$^+$RNA during *Xenopus laevis* development. *Devl Biol.* **76**, 449–464.

Dworkin, M. B. & Hershey, J. W. B. (1981). Cellular titers and subcellular distributions of abundant polyadenylate-containing ribonucleic acid species during early development in the frog *Xenopus laevis*. *Mol. cell. Biol.* **1**, 983–993.

Dworkin, M. B. & Infante, A. A. (1976). Relationship between the mRNA of polysomes and free ribonucleoprotein particles in the early sea urchin embryo. *Devl Biol.* **53**, 73–90.

Dworkin, M. B., Kay, B. K., Hershey, J. W. B. & Dawid, I. B. (1981). Mitochondrial RNAs are abundant in the poly(A)$^+$RNA population of early frog embryos. *Devl Biol.* **86**, 502–504.

Dworkin, M. B., Rudensey, L. M. & Infante, A. A. (1977). Cytoplasmic nonpolysomal ribonucleoprotein particles in sea urchin embryos and their relationship to protein synthesis. *Proc. natn. Acad. Sci. U.S.A.* **74**, 2231–2235.

Dworkin, M. B., Strutkowski, A., Baumgarten, M. & Dworkin-Rastl, E. (1984). The accumulation of prominent tadpole mRNAs occurs at the beginning of neurulation in *Xenopus laevis* embryos. *Devl Biol.* **106**, 289–295.

Dworkin-Rastl, E., Kelley, D. B. & Dworkin, M. B. (1986). Localization of specific mRNA sequences in *Xenopus laevis* embryos by *in situ* hybridization. *J. Embryol. exp. Morph.* **91**, 153–168.

Dziadek, M. & Adamson, E. (1978). Localization and synthesis of alpha-foetoprotein in post-implantation mouse embryos. *J. Embryol. exp. Morph.* **43**, 289–313.

Dziadek, M. & Dixon, K. E. (1977). An autoradiographic analysis of nucleic acid synthesis in the presumptive primordial germ cells of *Xenopus laevis*. *J. Embryol. exp. Morph.* **37**, 13–31.

Eakin, R. M. (1939). Regional determination in the development of the trout. *Wilhelm Roux Arch. EntwMech. Org.* **139**, 274–281.

Eakin, R. M. & Lehmann, F. E. (1957). An electronmicroscopic study of developing amphibian ectoderm. *Wilhelm Roux Arch. EntwMech. Org.* **150**, 177–198.

Eckberg, W. R. (1981). An ultrastructural analysis of cytoplasmic localization in *Chaetopterus pergamentaceus*. *Biol. Bull. mar. biol. Lab., Woods Hole* **160**, 228–239.

Eckberg, W. R. & Kang, Y.-H. (1981). A cytological analysis of differentiation without cleavage in cytochalasin B- and colchicine-treated embryos of *Chaetopterus pergamentaceus*. *Differentiation* **19**, 154–160.

Eddy, E. M. (1975). Germ plasm and the differentiation of the germ cell line. *Int. Rev. Cytol.* **43**, 229–280.

Edgar, B. A. & Schubiger, G. (1986). Parameters controlling transcriptional activation during early *Drosophila* development. *Cell* **44**, 871–877.

Edgar, L. G. & McGhee, J. D. (1986). Embryonic expression of a gut-specific esterase in *Caenorhabditis elegans*. *Devl Biol.* **114**, 109–118.

Edirisinghe, W. R., Wales, R. G. & Pike, I. L. (1984). Studies of the distribution of glycogen between the inner cell mass and trophoblast cells of mouse embryos. *J. Reprod. Fert.* **71**, 533–538.

Edwards, M. K. & Wood, W. B. (1983). Location of specific messenger RNAs in *Caenorhabditis elegans* by cytological hybridization. *Devl Biol.* **97**, 375–390.

Edwards, N. C. (1975). Patterns of macromolecular synthesis in a developing hydroid. *Acta Embryol. exp.* 177–200.

Efremov, V. I., Morozova, I. N. & Sergovskaya, T. V. (1980). Proliferative activity and kinetics of cell populations of chick embryo blastoderm during gastrulation and early organogenesis. II. Mitotic index, index of DNA synthesis phase, and composition of cell populations. *Ontogenez* **11**, 234–245 (in translation 144–154).

Ehn, A. (1963a). Morphological and histological effects of lithium on the embryonic development of *Agelena labyrinthica* Cl. *Zool. Bidr. Upps.* **36**, 1–26.

Ehn, A. (1963b). Effects of sulphydryl-blocking substances on development of spider embryos. *Zool. Bidr. Upps.* **36**, 49–72.

Elbers, P. F. (1983). The site of action of lithium ions in morphogenesis of *Lymnaea stagnalis* analysed by secondary ion mass spectroscopy. *Differentiation* **24**, 220–225.

Elinson, R. P. (1975). Site of sperm entry and a cortical contraction associated with egg activation in the frog *Rana pipiens*. *Devl Biol.* **47**, 257–268.

Elinson, R. P. (1983). Cytoplasmic phases in the first cell cycle of the activated frog egg. *Devl Biol.* **100**, 440–451.

Elinson, R. P. (1985). Changes in levels of polymeric tubulin associated with activation and dorsoventral polarization of the frog egg. *Devl Biol.* **109**, 224–233.

Elinson, R. P. & Manes, M. E. (1978). Morphology of the site of sperm entry on the frog egg. *Devl Biol.* **63**, 67–75.

Emanuelsson, H. (1965). Cell multiplication in the chick blastoderm up to the time of laying. *Expl Cell Res.* **39**, 386–399.

Emanuelsson, H. & Anehus, S. (1985). Development *in vitro* of the female germ cells of the polychaete *Ophryotrocha labronica*. *J. Embryol. exp. Morph.* **85**, 151–161.

Emanuelsson, H. & Heby, O. (1978). Inhibition of putrescine synthesis blocks development of the polychaete *Ophryotrocha labronica* at gastrulation. *Proc. natn. Acad. Sci. U.S.A.* **75**, 1039–1042.

Engstrom, L., Caulton, J. H., Underwood, E. M. & Mahowald, A. P. (1982). Developmental lesions in the *agametic* mutant of *Drosophila melanogaster*. *Devl Biol.* **91**, 163–170.

Eppig, J. J. (1977). Mouse oocyte development *in vitro* with various culture systems. *Devl Biol.* **60**, 371–388.

Epstein, C. J. (1969). Mammalian oocytes: X chromosome activity. *Science, N.Y.* **163**, 1078–1079.

Ermolaev, S. V., Korzh, V. P. & Neifakh, A. A. (1984). Investigation of the proteins synthesized during embryonic development of the loach by the method of two-dimensional electrophoresis. *Ontogenez* **15**, 493–504 (in translation 306–316).

Ernst, S. G., Hough-Evans, B. R., Britten, R. J. & Davidson, E. H. (1980). Limited complexity of the RNA in micromeres of sixteen-cell sea urchin embryos. *Devl Biol.* **79**, 119–127.

Ettensohn, C. A. (1984). Primary invagination of the vegetal plate during sea urchin gastrulation. *Am. Zool.* **24**, 571–588.

Ettensohn, C. A. (1985). Gastrulation in the sea urchin embryo is accompanied by the rearrangement of the invaginating epithelial cells. *Devl Biol.* **112**, 383–390.

Evans, M. J., Lovell-Badge, R. H., Stern, P. L. & Stinnakre, M. G. (1979). Cell lineages of the mouse embryo and embryonal carcinoma cells; Forsmann antigen distribution and patterns of protein synthesis. In *Cell Lineage, Stem Cells and Cell Determination*, ed. N. le Douarin, pp. 115–129. Amsterdam: Elsevier North-Holland.

Evans, T., Rosenthal, E. T., Youngblom, J., Distel, D. & Hunt, T. (1983). Cyclin: a protein specified by maternal mRNA in sea urchin eggs that is destroyed at each cleavage division. *Cell* **33**, 389–396.

Ewest, A. (1937). Struktur und erste Differenzierung im Ei des Mehlkäfers *Tenebrio molitor*. *Wilhelm Roux Arch. EntwMech. Org.* **135**, 689–752.

Eyal-Giladi, H. (1954). Dynamic aspects of neural induction in Amphibia. *Archs Biol., Liege* **65**, 179–259.

Eyal-Giladi, H. (1969). Differentiation potencies of the young chick blastoderm as revealed by different manipulations. I. Folding experiments and position effects of the culture medium. *J. Embryol. exp. Morph.* **21**, 177–192.

Eyal-Giladi, H. (1970). Differentation potencies of the young chick blastoderm as revealed by different manipulations. II. Localised damage and hypoblast removal experiments. *J. Embryol. exp. Morph.* **23**, 739–749.

Eyal-Giladi, H. & Fabian, B. C. (1980). Axis determination in uterine chick blastodiscs under changing spatial positions during the sensitive period for polarity. *Devl Biol.* **77**, 228–232.

Eyal-Giladi, H., Farbiasz, I., Ostrovsky, D. & Hochman, J. (1975). Protein synthesis in epiblast versus hypoblast during the critical stages of induction and growth of the primitive streak in the chick embryo. *Devl Biol.* **45**, 358–365.

Eyal-Giladi, H. & Kochav, S. (1976). From cleavage to primitive streak formation: a complementary normal table and a new look at the first stages of the development of the chick. I. General morphology. *Devl Biol.* **49**, 321–337.

Eyal-Giladi, H., Raveh, D., Feinstein, N. & Friedländer, M. (1979). Glycogen metabolism in the prelaid chick embryo. *J. Morph.* **161**, 23–38.

Eyal-Giladi, H. & Spratt, N. T. (1964). Embryo formation in folded chick blastoderms. *Am. Zool.* **4**, 428 (abstract).

Eyal-Giladi, H. & Spratt, N. T. (1965). The embryo-forming potencies of the young chick blastoderm. *J. Embryol. exp. Morph.* **13**, 267–273.

Eyal-Giladi, H. & Wolk, M. (1970). The inducing capacities of the primary hypoblast as revealed by transfilter induction studies. *Wilhelm Roux Arch. EntwMech. Org.* **165**, 226–241.

Fabian, B. & Eyal-Giladi, H. (1981). A SEM study of cell shedding during the formation of the area pellucida in the chick embryo. *J. Embryol. exp. Morph.* **64**, 11–22.

Failly-Crépin, C. & Martin, G. R. (1979). Protein synthesis and differentiation in a clonal line of teratocarcinoma and in preimplantation mouse embryo. *Cell Differ.* **8**, 61–73.

Farfaglio, G. (1963). Experiments on the formation of the ciliated plates in ctenophores. *Acta Embryol. Morph. exp.* **6**, 191–203.

Farinella-Ferruzza, N. (1955). Lo sviluppo embrionale delle Ascidie dopo trattamento con LiCl. *Pubbl. Staz. zool. Napoli* **26**, 42–54.

Farinella-Ferruzza, N. & Reverberi, G. (1969). Gigantic larvae of ascidians from two fused eggs. *Acta Embryol. exp.* 281–290.

Faulhaber, I. (1972). Die Induktionsleistung subzellularer Fraktionen aus der Gastrula von *Xenopus laevis*. *Wilhelm Roux Arch. EntwMech. Org.* **171**, 87–108.

Faulhaber, I. & Lyra, L. (1974). Ein Vergleich der Induktionsfähigkeit von Hüllenmaterial der Dotterplättchen und der Mikrosomenfraktion aus Furchungs- sowie Gastrula- und Neurulastadien des Krallenfrosches *Xenopus laevis*. *Wilhelm Roux Arch. EntwMech. Org.* **176**, 151–157.

Fenderson, B. A., Hahnel, A. C. & Eddy, E. M. (1983). Immunohistochemical localization of two monoclonal antibody-defined carbohydrate antigens during early murine embryogenesis. *Devl Biol.* **100**, 318–327.

Fernández, J. (1980). Embryonic development of the glossiphoniid leech *Thermomyzon rude*: characterization of developmental stages. *Devl Biol.* **76**, 245–262.

Fernández, J. & Stent, G. S. (1980). Embryonic development of the glossiphoniid leech *Theromyzon rude*: structure and development of the germinal bands. *Devl Biol.* **78**, 407–434.

Ferruzza, N. & Farinella, E. (1981). The influence of LiCl on animal halves isolated from ascidian eggs. *Acta Embryol. Morph. exp. n.s.* **2**, 57–67.

Fielding, C. J. (1967). Developmental genetics of the mutant *grandchildless* of *Drosophila subobscura*. *J. Embryol. exp. Morph.* **17**, 375–384.

Fink, R. D. & McClay, D. R. (1985). Three cell recognition changes accompany the ingression of sea urchin primary mesenchyme cells. *Devl Biol.* **107**, 66–74.

Fischel, A. (1903). Entwickelung und Organ-Differenzirung. *Arch. EntwMech. Org.* **15**, 679–750.

Fischer, A. (1974). Activity of a gene in transplanted oocytes in the annelid, *Platynereis*. *Wilhelm Roux Arch. EntwMech. Org.* **174**, 250–251.

Fisher, M. & Solursh, M. (1977). Glycosaminoglycan localization and role in maintenance of tissue spaces in the early chick embryo. *J. Embryol. exp. Morph.* **42**, 195–207.

Fjose, A., McGinnis, W. J. & Gehring, W. J. (1985). Isolation of a homoeo box-containing gene from the *engrailed* region of *Drosophila* and the spatial distribution of its transcripts. *Nature, Lond.* **313**, 284–289.

Flach, G., Johnson, M. H., Braude, P. R., Taylor, R. A. S. & Bolton, V. N.

(1982). The transition from maternal to embryonic control in the 2-cell mouse embryo. *EMBO J.* **1**, 681–686.

Fleming, T. P. (1987). A quantitative analysis of cell allocation to trophectoderm and inner cell mass in the mouse blastocyst. *Devl Biol.* **119**, 520–531.

Fleming, T. P., Warren, P. D., Chisholm, J. C. & Johnson, M. H. (1984). Trophectodermal processes regulate the expression of totipotency within the inner cell mass of the mouse expanding blastocyst. *J. Embryol. exp. Morph.* **84**, 63–90.

Flickinger, R. A. (1969). A regional difference in the time of initial synthesis of ribosomal RNA in frog embryos. *Expl Cell Res.* **55**, 422–423.

Flickinger, R. A. (1970). The role of gene redundancy and number of cell divisions in embryonic determination. In *Changing Syntheses in Development*, ed. M. N. Runner, pp. 12–41. New York: Academic Press.

Flickinger, R. A. (1980). The effect of heparin upon differentiation of ventral halves of frog gastrulae. *Wilhelm Roux Arch. dev. Biol.* **188**, 9–11.

Flickinger, R. A., Daniel, J. C. & Greene, R. F. (1970a). Effect of rate of cell division on RNA synthesis in developing frog embryos. *Nature, Lond.* **228**, 557–559.

Flickinger, R. A., Hatton, E. & Rounds, D. E. (1959). Protein transfer in chimaeric *Taricha-Rana* explants. *Expl Cell Res.* **17**, 30–34.

Flickinger, R. A., Kohl, D. M., Lauth, M. R. & Stambrook, P. J. (1970b). Effect of rate and number of cell divisions on RNA synthesis. *Biochim. biophys. Acta* **209**, 260–262.

Flickinger, R. A., Lauth, M. R. & Stambrook P. J. (1970c). An inverse relation between the rate of cell division and RNA synthesis per cell in developing frog embryos. *J. Embryol. exp. Morph.* **23**, 571–582.

Flickinger, R. A., Miyagi, M., Moser, C. R. & Rollins, E. (1967). The relation of DNA synthesis to RNA synthesis in developing frog embryos. *Devl Biol.* **15**, 414–431.

Flynn, J. M. & Woodland, H. R. (1980). The synthesis of histone H1 during early amphibian development. *Devl Biol.* **75**, 222–230.

Flytzanis, C. N., Brandhorst, B. P., Britten, R. J. & Davidson, E. H. (1982). Developmental patterns of cytoplasmic transcript prevalence in sea urchin embryos. *Devl Biol.* **91**, 27–35.

Foe, V. E. & Alberts, B. M. (1983). Studies of nuclear and cytoplasmic behaviour during the five mitotic cycles that precede gastrulation in *Drosophila* embryogenesis. *J Cell Sci.* **61**, 31–70.

Forbes, D. J., Kornberg, T. B. & Kirschner, M. W. (1983). Small nuclear RNA transcription and ribonucleoprotein assembly in early *Xenopus* development. *J. Cell Biol.* **97**, 62–72.

Ford, C. C., Pestell, R. Q. W. & Benbow, R. M. (1975). Template preferences of DNA polymerase and nuclease activities appearing during early development of *Xenopus laevis*. *Devl Biol.* **43**, 175–188.

Ford, P. J. (1971). Non-coordinated accumulation and synthesis of 5S ribonucleic acid by ovaries of *Xenopus laevis*. *Nature, Lond.* **233**, 561–564.

Ford, P. J., Mathieson, T. & Rosbash, M. (1977). Very long-lived messenger RNA in ovaries of *Xenopus laevis*. *Devl Biol.* **57**, 417–426.

Forman, D. & Slack, J. M. W. (1980). Determination and cellular commitment in the embryonic amphibian mesoderm. *Nature, Lond.* **286**, 492–494.

Forquignon, F. (1981). A maternal effect mutation leading to deficiencies of organs and homeotic transformations in the adults of *Drosophila*. *Wilhelm Roux Arch. dev. Biol.* **190**, 132–138.

Franke, W. W., Grund, C., Kuhn, C., Jackson, B. W. & Illmensee, K. (1982). Formation of cytoskeletal elements during mouse embryogenesis. III. Primary mesenchymal cells and the first appearance of vimentin filaments. *Differentiation* **23**, 43–59.

Franke, W. W., Rathke, P. C., Seib, E., Trendelenburg, M. F., Osborn, M. & Weber, K. (1976). Distribution and mode of arrangement of microfilamentous structures and actin in the cortex of the amphibian oocyte. *Cytobiologie* **14**, 111–130.

Franz, J. K., Gall, L., Williams, M. A., Picheral, B. & Franke, W. W. (1983). Intermediate-size filaments in a germ cell: expression of cytokeratins in oocytes and eggs of the frog *Xenopus*. *Proc. natn. Acad. Sci. U.S.A.* **80**, 6254–6258.

Fraser, R. C. (1954). Studies on the hypoblast of the young chick embryo. *J. exp. Zool.* **126**, 349–399.

Fraser, R. C. (1960). Somite genesis in the chick. III. The role of induction. *J. exp. Zool.* **145**, 151–163.

Frausto da Silva, J. J. R. & Williams, R. J. P. (1976). The uptake of elements by biological systems. *Struct. & Bond.* **29**, 67–121.

Freeman, G. (1976a). The role of cleavage in the localization of developmental potential in the ctenophore *Mnemiopsis leidyi*. *Devl Biol.* **49**, 143–177.

Freeman, G. (1976b). The effects of altering the positions of cleavage planes on the process of localization of developmental potential in ctenophores. *Devl Biol.* **51**, 332–337.

Freeman, G. (1977). The establishment of the oral–aboral axis in the ctenophore embryo. *J. Embryol. exp. Morph.* **42**, 237–260.

Freeman, G. (1978). The role of asters in the localization of the factors that specify the apical tuft and the gut of the nemertine *Cerebratulus lacteus*. *J. exp. Zool.* **206**, 81–107.

Freeman, G. (1979). The multiple roles which cell division can play in the localization of developmental potential. In *Determinants of Spatial Organization*, ed. S. Subtelny & I. R. Konigsberg, pp. 53–76. New York: Academic Press.

Freeman, G. (1981a). The cleavage initiation site establishes the posterior pole of the hydrozoan embryo. *Wilhelm Roux Arch. dev. Biol.* **190**, 123–125.

Freeman, G. (1981b). The role of polarity in the development of the hydrozoan planula larva. *Wilhelm Roux Arch. dev. Biol.* **190**, 168–184.

Freeman, G. (1987). The role of oocyte maturation in the ontogeny of the fertilization site in the hydrozoan *Hydractinia echinata*. *Wilhelm Roux Arch. dev. Biol.* **196**, 83–92.

Freeman, G. & Lundelius, J. W. (1982). The developmental genetics of dextrality and sinistrality in the gastropod *Lymnaea peregra*. *Wilhelm Roux Arch. dev. Biol.* **191**, 69–83.

Freeman, G. & Reynolds, G. T. (1973). The development of bioluminescence in the ctenophore *Mnemiopsis leidyi*. *Devl Biol.* **31**, 61–100.

Frey, A. & Gutzeit, H. (1986). Follicle cells and germ line cells both affect polarity in *dicephalic* chimeric follicles of *Drosophila*. *Wilhelm Roux Arch. dev. Biol.* **195**, 527–531.

Fritz, A., Parisot, R., Newmeyer, D. & de Robertis, E. M. (1984). Small nuclear U-ribonucleoproteins in *Xenopus laevis* development. Uncoupled accumulation of the protein and RNA components. *J. molec. Biol.* **178**, 273–285.

Frohnhöfer, H. G., Lehmann, R. & Nüsslein-Volhard, C. (1986). Manipulating the anteroposterior pattern of the *Drosophila* embryo. *J. Embryol. exp. Morph.* **97**, *Suppl.* 169–179.

Frohnhöfer, H. G. & Nüsslein-Volhard, C. (1986). Organization of anterior pattern in the *Drosophila* embryo by the maternal gene *bicoid*. *Nature, Lond.* **324**, 120–125.

Fujiwara, A. & Yasumasu, I. (1974*a*). Some observations of abnormal embryos induced by short-period treatment with chloramphenicol during early development of sea urchin. *Dev. Growth Differ.* **16**, 83–92.

Fujiwara, A. & Yasumasu, I. (1974*b*). Morphogenetic substances found in the embryos of sea urchin, with special reference to the anti-vegetalizing substance. *Dev. Growth Differ.* **16**, 93–103.

Fullilove, S. L. & Jacobson, A. G. (1971). Nuclear elongation and cytokinesis in *Drosophila montana*. *Devl Biol.* **26**, 560–577.

Fullilove, S. L. & Woodruff, R. C. (1974). Genetic, cytological, and ultrastructural characterization of a temperature-sensitive lethal in *Drosophila melanogaster*. *Devl Biol.* **38**, 291–307.

Gabrielli, F. & Baglioni, C. (1975). Maternal messenger RNA and histone synthesis in embryos of the surf clam *Spisula solidissima*. *Devl Biol.* **43**, 254–263.

Gadenne, M., van Zoelen, E. J. J., Tencer, R. & de Laat, S. W. (1984). Increased rate of capping of concanavalin A receptors during early *Xenopus* development is related to changes in protein and lipid mobility. *Devl Biol.* **104**, 461–468.

Galau, G. A., Klein, W. H., Davis, M. M., Wold, B. J., Britten, R. J. & Davidson, E. H. (1976). Structural gene sets active in embryos and adult tissues of the sea urchin. *Cell* **7**, 487–505.

Galileo, D. S. & Morrill, J. B. (1985). Patterns of cells and extracellular material of the sea urchin *Lytechinus variegatus* (Echinodermata; Echinoidea) embryo, from hatched blastula to late gastrula. *J. Morph.* **185**, 387–402.

Gall, J. G. (1968). Differential synthesis of the genes for ribosomal RNA during amphibian oogenesis. *Proc. natn. Acad. Sci U.S.A.* **60**, 553–560.

Gall, J. G. & Callan, H. G. (1962). H$^3$uridine incorporation in lampbrush chromosomes. *Proc. natn. Acad. Sci. U.S.A.* **48**, 562–570.

Gallera, J. (1959). Le facteur 'temps' dans l'action inductrice du chordo-mésoblaste et l'âge de l'ectoblaste réagissant. *J. Embryol. exp. Morph.* **7**, 487–511.

Gallera, J. (1960). L'action inductrice du chordo-mésoblaste au cours de la gastrulation et de la neurulation et les effets de la culture *in vitro* sur ses manifestations. *J. Embryol. exp. Morph.* **8**, 477–494.

Gallera, J. (1968). Induction neurale chez les Oiseaux. Rapport temporel entre la neurulation du blastoderme-hôte et l'apparition de l'ébauche neurale induite par un fragment de la ligne primitive. *Revue suisse Zool.* **75**, 227–234.

Gallera, J. (1969). Evolution intrinsèque de l'ectoblaste et induction neurale chez les oiseaux. *Acta Embryol. exp.* 5–16.

Gallera, J. (1970*a*). L'action de l'actinomycine D sur le pouvoir inducteur du noeud de Hensen et la compétence neurogène de l'ectoblaste de poulet. *J. Embryol. exp. Morph.* **23**, 473–489.

Gallera, J. (1970*b*). Différence de réactivité à l'inducteur neurogène entre de l'aire opaque et celui de l'aire pellucide chez le Poulet. *Experientia* **26**, 1353–1354.

Gallera, J. & Ivanov, I. (1964). La compétence neurogène du feuillet externe du blastoderme de Poulet en fonction du facteur 'temps'. *J. Embryol. exp. Morph.* **12**, 693–711.

Gallera, J. & Nicolet, G. (1969). Le pouvoir inducteur de l'endoblaste présomptif contenu dans la ligne primitive jeune de Poulet. *J. Embryol. exp. Morph.* **21**, 105–118.

Gallera, J., Nicolet, G. & Baumann, M. (1968). Induction neurale chez les Oiseaux à travers un filtre millipore: étude au microscope optique et électronique. *J. Embryol. exp. Morph.* **19**, 439–450.

Ganion, L. R. & Kessel, R. G. (1972). Intracellular synthesis, transport, and packaging of proteinaceous yolk in oocytes of *Oronectes immunis*. *J. Cell Biol.* **52**, 420–437.

Gans, M., Audit, C. & Masson, M. (1975). Isolation and characterization of sex-linked female-sterile mutants in *Drosophila melanogaster*. *Genetics, Princeton* **81**, 683–704.

Garber, R. L., Kuroiwa, A. & Gehring, W. J. (1983). Genomic and cDNA clones of the homeotic locus *Antennapedia* in *Drosophila*. *EMBO J.* **2**, 2027–2036.

García-Bellido, A. (1975). Genetic control of wing disc development in *Drosophila*. In *Cell Patterning*, Ciba Found. Symp. **29** (new series), 161–182.

García-Bellido, A., Ripoll, P. & Morata, G. (1973). Developmental compartmentalisation of the wing disk of *Drosophila*. *Nature new Biol.* **245**, 251–253.

García-Bellido, A. & Santamaria, P. (1972). Developmental analysis of the wing disc in the mutant *engrailed* of *Drosophila melanogaster*. *Genetics, Princeton* **72**, 87–104.

Gaul, U., Seifert, E., Schuh, R. & Jäckle, H. (1987). Analysis of *Krüppel* protein distribution during early *Drosophila* development reveals posttranscriptional regulation. *Cell* **50**, 639–647.

Gaunt, S. J., Miller, J. R., Powell, D. J. & Duboule, D. (1986). Homoeobox gene expression in mouse embryos varies with position by the primitive streak stage. *Nature, Lond.* **324**, 662–664.

Gautier, J. & Beetschen, J.-C. (1983). Inhibition of protein synthesis elicits early grey crescent formation in the axolotl oocyte. *Wilhelm Roux Arch. dev. Biol.* **192**, 196–199.

Gebhardt, D. O. E. & Nieuwkoop, P. D. (1964). The influence of lithium on the competence of the ectoderm in *Ambystoma mexicanum*. *J. Embryol. exp. Morph.* **12**, 317–331.

Geilenkirchen, W. L. M., Timmermans, L. P. M., van Dongen, C. A. M. & Arnolds, W. J. A. (1971). Symbiosis of bacteria with eggs of *Dentalium* at the vegetal pole. *Expl Cell Res.* **67**, 477–478.

Geilenkirchen, W. L. M, Verdonk, N. H. & Timmermans, L. P. M. (1970). Experimental studies on morphogenetic factors localized in the first and the second polar lobe of *Dentalium* eggs. *J. Embryol. exp. Morph.* **23**, 237–243.

Geithe, H. P., Asashima, M., Asahi, K.I., Born, J., Tiedemann, H. & Tiedemann, H. (1981). A vegetalizing inducing factor. Isolation and chemical properties. *Biochim. biophys. Acta* **676**, 350–356.

Geithe, H. P., Asashima, M., Born, J., Tiedemann, H. & Tiedemann, H. (1975). Isolation of a homogeneous morphogenetic factor, inducing mesoderm and endoderm derived tissues in *Triturus* ectoderm. *Expl Cell Res.* **94**, 447–449.

Geithe, H. P., Tiedemann, H. & Tiedemann, H. (1970). Electrofocusing of the vegetalizing inducing factor. *Biochim. biophys. Acta* **208**, 157–159.

Gerhart, J., Ubbels, G., Black, S., Hara, K. & Kirschner, M. (1981). A reinvestigation of the role of the grey crescent in axis formation in *Xenopus laevis. Nature, Lond.* **292**, 511–516.

Gerhart, J., Wu, M. & Kirschner, M. (1984a). Cell cycle dynamics of an M-phase-specific cytoplasmic factor in *Xenopus laevis* oocytes and eggs. *J. Cell Biol.* **98**, 1247–1255.

Gerhart, J. C., Vincent, J.-P., Scharf, S. R., Black, S. D., Gimlich, R. L. & Danilchik, M. (1984b). Localization and induction in early development of *Xenopus. Phil. Trans. R. Soc. B* **307**, 319–330.

Geyer-Duszyńska, I. (1959). Experimental research on chromosome diminution in Cecidomyiidae (Diptera). *J. exp. Zool.* **141**, 391–441.

Gianni, A. M., Giglioni, B., Ottolenchi, S., Comi, P. & Guidotti, G. G. (1972). Globin α-chain synthesis directed by "supernatant" 10S RNA from rabbit reticulocytes. *Nature new Biol.* **240**, 183–185.

Gilkey, J. C., Jaffe, L. F., Ridgway, E. B. & Reynolds, G. T. (1978). A free calcium wave traverses the activating egg of the medaka, *Oryzias latipes. J. Cell Biol.* **76**, 448–466.

Gill, K. S. (1964). Epigenetics of the promorphology of the egg in *Drosophila melanogaster. J. exp. Zool.* **155**, 91–104.

Gimlich, R. L. (1985). Cytoplasmic localization and chordamesoderm induction in the frog embryo. *J. Embryol. exp. Morph.* **89**, *Suppl.* 89–111.

Gimlich, R. L. (1986). Acquisition of developmental autonomy in the equatorial region of the *Xenopus* embryo. *Devl Biol.* **115**, 340–352.

Gimlich, R. L. & Cooke, J. (1983). Cell lineage and the induction of second nervous systems in amphibian development. *Nature, Lond.* **306**, 471–473.

Gimlich, R. L. & Gerhart, J. C. (1984). Early cellular interactions promote embryonic axis formation in *Xenopus laevis. Devl Biol.* **104**, 117–130.

Gineitis, A. A., Stankevičiute, J. V. & Vorob'ev, V. I. (1976). Chromatin proteins from normal, vegetalized, and animalized sea urchin embryos. *Devl Biol.* **52**, 181–192.

Gingle, A. R. & Robertson, A. (1979). Responses of the early chick embryo to external cAMP sources. *J. Embryol. exp. Morph.* **53**, 353–365.

Ginsburg, M. & Eyal-Giladi, H. (1986). Temporal and spatial aspects of the gradual migration of primordial germ cells from the epiblast into the germinal crescent in the avian embryo. *J. Embryol. exp. Morph.* **95**, 53–71.

Ginzberg, R. D., Morales, E. A., Spray, D. C. & Bennett, M. V. L. (1985). Cell junctions in early embryos of squid (*Loligo pealei*). *Cell Tissue Res.* **239**, 477–484.

Gipouloux, J.-D. & Girard, C. (1986). Effects of the removal of neural crest anlage upon endodermal morphogenesis and primordial germ cells migration in toad embryo. *Wilhelm Roux Arch. dev. Biol.* **195**, 355–358.

Giudice, G. (1973). *Developmental Biology of the Sea Urchin Embryo.* New York: Academic Press.

Giudice, G. & Hörstadius, S. (1965). Effect of actinomycin D on the segregation of animal and vegetal potentialities in the sea urchin egg. *Expl Cell Res.* **39**, 117–120.

Giudice, G., Mutolo, V. & Donatuti, V. (1968). Gene expression in sea urchin development. *Wilhelm Roux Arch. EntwMech. Org.* **161**, 118–128.

Glade, R. W., Burrill, E. M. & Falk, R. J. (1967). The influence of a temperature gradient on bilateral symmetry in *Rana pipiens. Growth* **31**, 231–249.

Glišin, V. R., Glišin, M. V. & Doty, P. (1966). The nature of messenger RNA in the early stages of sea urchin development. *Proc. natn. Acad. Sci. U.S.A.* **56,** 285–289.

Gloor, H. (1947). Phänokopie-Versuche mit Äther an *Drosophila. Revue suisse Zool.* **54,** 637–712.

Gluecksohn-Waelsch, S. (1953). Lethal factors in development. *Q. Rev. Biol.* **28,** 115–135.

Godsave, S. F., Anderton, B. H., Heasman, J. & Wylie, C. C. (1984a). Oocytes and early embryos of *Xenopus laevis* contain intermediate filaments which react with anti-mammalian vimentin antibodies. *J. Embryol. exp. Morph.* **83,** 169–187.

Godsave, S. F., Wylie, C. C., Lane, E. B. & Anderton, B. H. (1984b). Intermediate filaments in the *Xenopus* oocyte: the appearance and distribution of cytokeratin-containing filaments. *J. Embryol. exp. Morph.* **83,** 157–167.

Golbus, M. S., Calarco, P. G. & Epstein, C. J. (1973). The effects of inhibitors of RNA synthesis ($\alpha$-amanitin and actinomycin D) on preimplantation mouse embryogenesis. *J. exp. Zool.* **186,** 207–216.

Golden, L., Schafer, U. & Rosbash, M. (1980). Accumulation of individual $pA^+$ RNAs during oogenesis of *Xenopus laevis. Cell* **22,** 835–844.

Goldstein, E. S. (1978). Translated and sequestered untranslated message sequences in *Drosophila* oocytes and embryos. *Devl Biol.* **63,** 59–66.

Gontcharoff, M. & Mazia, D. (1967). Developmental consequences of introduction of bromouracil into the DNA of sea urchin embryos during early division stages. *Expl Cell Res.* **46,** 315–327.

Goodall, H. & Johnson, M. H. (1984). The nature of intercellular coupling within the preimplantation mouse embryo. *J. Embryol. exp. Morph.* **79,** 53–76.

Gooding, L. R., Hsu, Y.-C. & Edidin, M. (1976). Expression of teratoma-associated antigens on murine ova and early embryos. Identification of two early differentiation markers. *Devl Biol.* **49,** 479–486.

Goodwin, B. C. & Cohen, M. H. (1969). A phase-shift model for the spatial and temporal organization of developing systems. *J. theor. Biol.* **25,** 49–107.

Gossett, L. A., Hecht, R. M. & Epstein, H. F. (1982). Muscle differentiation in normal and cleavage-arrested mutant embryos of *Caenorhabditis elegans. Cell* **30,** 193–204.

Grabowski, C. T. (1956). The effects of the excision of Hensen's node on the early development of the chick embryo. *J. exp. Zool.* **133,** 301–343.

Graham, C. F. & Deussen, Z. A. (1978). Features of cell lineage in preimplantation mouse development. *J. Embryol. exp. Morph.* **48,** 53–72.

Graham, C. F. & Morgan, R. W. (1966). Changes in the cell cycle during early amphibian development. *Devl Biol.* **14,** 439–460.

Grainger, J. L., von Brunn, A. & Winkler, M. M. (1986). Transient synthesis of a specific set of proteins during the rapid cleavage phase of sea urchin development. *Devl Biol.* **114,** 403–415.

Granholm, N. H. & Baker, J. R. (1970). Cytoplasmic microtubules and the mechanism of avian gastrulation. *Devl Biol.* **23,** 563–584.

Grant, P. & Wacaster, J. F. (1972). The amphibian gray crescent region – a site of developmental information? *Devl Biol.* **28,** 454–471.

Graziosi, G. & Micali, F. (1974). Differential responses to ultraviolet irradiation of the polar cytoplasm of *Drosophila* eggs. *Wilhelm Roux Arch. EntwMech. Org.* **175,** 1–11.

Graziosi, G. & Roberts, D. B. (1975). Molecular anisotropy of the early *Drosophila* embryo. *Nature, Lond.* **258,** 157–159.

Green, H., Goldberg, B., Schwartz, M. & Brown, D. D. (1968). The synthesis of collagen during the development of *Xenopus laevis*. *Devl Biol.* **18**, 391–400.

Green, J. (1971). Crustaceans. In *Experimental Embryology of Marine and Fresh-Water Invertebrates*, ed. G. Reverberi, pp. 312–362. Amsterdam: North-Holland.

Gross, K., Ruderman, J., Jacobs-Lorena, M., Baglioni, C. & Gross, P. R. (1973). Cell-free synthesis of histones directed by messenger RNA from sea urchin embryos. *Nature new Biol.* **241**, 272–274.

Gross, P. R. (1964). The immediacy of genomic control during early development. *J. exp. Zool.* **157**, 21–38.

Grunz, H. (1968). Experimentelle Untersuchungen über die Kompetenzverhältnisse früher Entwicklungsstadien des Amphibien-Ektoderms. *Wilhelm Roux Arch. EntwMech. Org.* **160**, 344–374.

Grunz, H. (1970). Abhängigkeit der Kompetenz des Amphibien-Ektoderms von der Proteinsynthese. *Wilhelm Roux Arch. EntwMech. Org.* **165**, 91–102.

Grunz, H. (1973). The ultrastructure of amphibian ectoderm treated with an inductor or actinomycin D. *Wilhelm Roux Arch. EntwMech. Org.* **173**, 283–293.

Grunz, H. (1977). Differentiation of the four animal and the four vegetal blastomeres of the eight-cell-stage of *Triturus alpestris*. *Wilhelm Roux Arch. dev. Biol.* **181**, 267–277.

Grunz, H. (1983). Change in the differentiation pattern of *Xenopus laevis* ectoderm by variation of the incubation time and concentration of vegetalizing factor. *Wilhelm Roux Arch. dev. Biol.* **192**, 130–137.

Grunz, H. (1985). Information transfer during embryonic induction in amphibians. *J. Embryol. exp. Morph.* **89**, *Suppl.* 349–364.

Grunz, H. (1987). The importance of inducing factors for determination, differentiation and pattern formation in early amphibian development. *Zool. Sci.* **4**, 579–591.

Grunz, H. & Staubach, J. (1979*a*). Cell contacts between chordamesoderm and the overlaying neurectoderm (presumptive central nervous system) during the period of primary embryonic induction in amphibians. *Differentiation* **14**, 59–65.

Grunz, H. & Staubach, J. (1979*b*). Changes of the cell surface charge of amphibian ectoderm after induction. *Wilhelm Roux Arch. dev. Biol.* **186**, 77–80.

Grunz, H. & Tacke, L. (1986). The inducing capacity of the presumptive endoderm of *Xenopus laevis* studied by transfilter experiments. *Wilhelm Roux Arch. dev. Biol.* **195**, 467–473.

Gruzova, M. N. (1982). Early stages of oocyte and trophocyte differentiation in the ovaries of insects. *Ontogenez* **13**, 563–571 (in translation 335–342).

Gualandris, L., Duprat, A.-M. & Rougé, P. (1987). Cross-linking of membrane glycoconjugates is not a sufficient condition for neural induction by concanavalin A. *Cell Differ.* **21**, 93–99.

Guerrier, P. (1967). Les facteurs de polarisation dans les premiers stades du développement chez *Parascaris equorum*. *J. Embryol. exp. Morph.* **18**, 121–142.

Guerrier, P. (1970*a*). Les caractères de la segmentation et la détermination de la polarité dorsoventrale dans le développement de quelques Spiralia. II. *Sabellaria alveolata* (Annélide polychète). *J. Embryol. exp. Morph.* **23**, 639–665.

Guerrier, P. (1970*b*). Les caractères de la segmentation et la détermination de la polarité dorsoventrale dans le développement de quelques Spiralia. III.

*Pholas dactylus* et *Spisula subtruncata* (Mollusques Lamellibranches). *J. Embryol. exp. Morph.* **23**, 667–692.

Guerrier, P., van den Biggelaar, J. A. M., van Dongen, C. A. M. & Verdonk, N. H. (1978). Significance of the polar lobe for the determination of dorsoventral polarity in *Dentalium vulgare* (da Costa). *Devl Biol.* **63**, 233–242.

Gulyas, B. J. (1975). A reexamination of cleavage patterns in eutherian mammalian eggs: rotation of blastomere pairs during second cleavage in the rabbit. *J. exp. Zool.* **193**, 235–248.

Guraya, S. S. (1982). Recent progress in the structure, origin, composition, and function of cortical granules in animal egg. *Int. Rev. Cytol.* **78**, 257–360.

Gurdon, J. B. & Fairman, S. (1986). Muscle gene activation by induction and the nonrequirement for cell division. *J. Embryol. exp. Morph.* **97**, *Suppl.* 75–84.

Gurdon, J. B., Fairman, S., Mohun, T. J. & Brennan, S. (1985a). Activation of muscle specific actin genes in *Xenopus* development by an induction between animal and vegetal cells of a blastula. *Cell* **41**, 913–922.

Gurdon, J. B., Mohun, T. J., Brennan, S. & Cascio, S. (1985b). Actin genes in *Xenopus* and their developmental control. *J. Embryol. exp. Morph.* **89**, *Suppl.* 125–136.

Gurdon, J. B., Mohun, T. J., Fairman, S. & Brennan, S. (1985c). All components required for the eventual activation of muscle-specific actin genes are localized in the subequatorial region of an uncleaved amphibian egg. *Proc. natn. Acad. Sci. U.S.A.* **82**, 139–143.

Gurdon, J. B. & Woodland, H. R. (1969). The influence of the cytoplasm on the nucleus during cell differentiation with special reference to RNA synthesis during amphibian cleavage. *Proc. R. Soc. B* **173**, 99–111.

Gustafson, T. (1965). Morphogenetic significance of biochemical patterns in sea urchin embryos. In *The Biochemistry of Animal Development*, vol. 1, ed. R. Weber, pp. 139–202. New York: Academic Press.

Gustafson, T. & Lenicque, P. (1952). Studies on mitochondria in the developing sea urchin egg. *Expl Cell Res.* **3**, 251–274.

Gustafson, T. & Wolpert, L. (1967). Cellular movement and contact in sea urchin morphogenesis. *Biol. Rev.* **42**, 442–498.

Guthrie, S. C. (1984). Patterns of junctional communication in the early amphibian embryo. *Nature, Lond.* **311**, 149–151.

Gutzeit, H. (1986a). The role of microtubules in the differentiation of ovarian follicles during vitellogenesis in *Drosophila*. *Wilhelm Roux Arch. dev. Biol.* **195**, 173–181.

Gutzeit, H. (1986b). The role of microfilaments in cytoplasmic streaming in *Drosophila* follicles. *J. Cell Sci.* **80**, 159–169.

Gutzeit, H. (1986c). Transport of molecules and organelles in meroistic ovarioles of insects. *Differentiation* **31**, 155–165.

Gutzeit, H. O. & Gehring, W. J. (1979). Localized synthesis of specific proteins during oogenesis and early embryogenesis in *Drosophila melanogaster*. *Wilhelm Roux Arch. dev. Biol.* **187**, 151–165.

Hadorn, E. & Müller, G. (1974). Has the yolk an influence on the differentiation of embryonic *Drosophila* blastemas? *Wilhelm Roux Arch. EntwMech. Org.* **174**, 333–335.

Hafen, E., Kuroiwa, A. & Gehring, W. J. (1984). Spatial distribution of transcripts from the segmentation gene *fushi tarazu* during *Drosophila* embryonic development. *Cell* **37**, 833–841.

Hafen, E., Levine, M. & Gehring, W. J. (1984). Regulation of *Antennapedia* transcript distribution by the bithorax complex in *Drosophila*. *Nature, Lond.* **307**, 287–289.

Hagenmaier, H. E. (1969). Der Nucleinsäure-bzw. Ribonucleoproteid-Status während der Frühentwicklung von Fischkeimen (*Salmo irideus* und *Salmo trutta fario*). *Wilhelm Roux Arch. EntwMech. Org.* **162**, 19–40.

Haget, A. (1953). Analyse expérimentale des facteurs de la morphogenèse embryonnaire chez le coléoptère *Leptinotarsa*. *Bull. biol. Fr. Belg.* **87**, 123–217.

Hagström, B. E. (1963). The effect of lithium and *o*-iodosobenzoic acid on the early development of the sea urchin egg. *Biol. Bull. mar. biol. Lab., Woods Hole* **124**, 55–64.

Hagström, B. E. & Lönning, S. (1969). Time-lapse and electron microscopic studies on sea urchin micromeres. *Protoplasma* **68**, 271–288.

Hallberg, R. L. & Smith, D. C. (1975). Ribosomal protein synthesis in *Xenopus laevis* oocytes. *Devl Biol.* **42**, 40–52.

Hama, T. (1949). Explantation of the urodelan organizer and the process of morphological differentiation attendant upon invagination. *Proc. Japan Acad.* **25**, 4–11.

Hama, T., Tsujimura, H., Kanéda, T., Takata, K. & Ohara, A. (1985). Inductive capacities for the dorsal mesoderm of the dorsal marginal zone and pharyngeal endoderm in the very early gastrula of the newt, and presumptive pharyngeal endoderm as an initiator of the organization center. *Dev. Growth Differ.* **27**, 419–433.

Hamasima, N. (1982). Effect of hyaluronic acid on the periimplantational development of mouse embryos *in vitro*. *Dev. Growth Differ.* **24**, 353–357.

Handyside, A. H. (1978). Time of commitment of inside cells isolated from preimplantation mouse embryos. *J. Embryol. exp. Morph.* **45**, 37–53.

Handyside, A. H. (1980). Distribution of antibody- and lectin-binding sites on dissociated blastomeres from mouse morulae: evidence for polarization at compaction. *J. Embryol. exp. Morph.* **60**, 99–116.

Handyside, A. H. (1981). Immunofluorescence techniques for determining the numbers of inner and outer blastomeres in mouse morulae. *J. reprod. Immunol.* **2**, 339–350.

Handyside, A. H. & Johnson, M. H. (1978). Temporal and spatial patterns of the synthesis of tissue-specific polypeptides in the preimplantation mouse embryo. *J. Embryol. exp. Morph.* **44**, 191–199.

Hara, K. (1961). Regional neural differentiation induced by prechordal and presumptive chordal mesoderm in the chick embryo. Ph. D. thesis, Utrecht.

Hara, K. (1971). Cinematographic observation of 'surface contraction waves' (SCW) during the early cleavage of axolotl eggs. *Wilhelm Roux Arch. Entw-Mech. Org.* **167**, 183–186.

Hara, K. (1977). The cleavage pattern of the axolotl egg studied by cinematography and cell counting. *Wilhelm Roux Arch. dev. Biol.* **181**, 73–87.

Hara, K. & Tydeman, P. (1979). Cinematographic observation of an 'activation wave' (AW) on the locally inseminated egg of *Xenopus laevis*. *Wilhelm Roux Arch. dev. Biol.* **186**, 91–94.

Hara, K., Tydeman, P. & Hengst, R. T. M. (1977). Cinematographic observation of 'post-fertilization waves' (PFW) on the zygote of *Xenopus laevis*. *Wilhelm Roux Arch. dev. Biol.* **181**, 189–192.

Hardin, J. D. & Cheng, L. Y. (1986). The mechanisms and mechanics of archenteron elongation during sea urchin gastrulation. *Devl Biol.* **115**, 490–501.

Harding, K., Wedeen, C., McGinnis, W. & Levine, M. (1985). Spatially regulated expression of homeotic genes in *Drosophila*. *Science, N.Y.* **229**, 1236–1242.

Harkey, M. A. & Whiteley, A. H. (1982). Cell-specific regulation of protein synthesis in the sea urchin gastrula: a two-dimensional electrophoretic study. *Devl Biol.* **93**, 453–462.

Harkey, M. A. & Whiteley, A. H. (1983). The program of protein synthesis during the development of the micromere-primary mesenchyme cell line in the sea urchin embryo. *Devl Biol.* **100**, 12–28.

Harris, P. (1979). A spiral cortical fiber system in fertilized sea urchin eggs. *Devl Biol.* **68**, 525–532.

Harris, T. M. (1964). Pregastrular mechanisms in the morphogenesis of the salamander *Ambystoma maculatum*. *Devl Biol.* **10**, 247–268.

Harrison, M. F. & Wilt, F. H. (1982). The program of H1 histone synthesis in *S. purpuratus* embryos and the control of its timing. *J. exp. Zool.* **223**, 245–256.

Harrisson, F., van Hoof, J. & Foidart, J. M. (1984). Demonstration of the interaction between glycosaminoglycans and fibronectin in the basement membrane of the living chick embryo. *Wilhelm Roux Arch. dev. Biol.* **193**, 418–421.

Hartenstein, V., Technau, G. M. & Campos-Ortega, J. A. (1985). Fatemapping in wild-type *Drosophila melanogaster*, III. A fate map of the blastoderm. *Wilhelm Roux Arch. dev. Biol.* **194**, 213–216.

Harvey, E. B. (1933). Development of the parts of sea urchin eggs separated by centrifugal force. *Biol. Bull. mar. biol. Lab., Woods Hole* **64**, 125–148.

Harvey, E. B. (1936). Parthenogenetic merogony or cleavage without nuclei in *Arbacia punctulata*. *Biol. Bull. mar. biol. Lab., Woods Hole* **71**, 101–121.

Harvey, E. B. (1956). *The American Arbacia and other Sea Urchins*. Princeton: The University Press.

Harvey, R. P., Tabin, C. J. & Melton, D. A. (1986). Embryonic expression and nuclear localization of *Xenopus* homoeobox (*Xhox*) gene products. *EMBO J.* **5**, 1237–1244.

Hatt, P. (1932). Essais expérimentaux sur les localisations germinales dans l'oeuf d'un annélide (*Sabellaria alveolata* L.). *Archs Anat. microsc.* **28**, 81–98.

Hatta, K. & Takeichi, M. (1986). Expression of N-cadherin adhesion molecules associated with early morphogenetic events in chick development. *Nature, Lond.* **320**, 447–449.

Hausen, P., Wang, Y. H., Dreyer, C. & Stick, R. (1985). Distribution of nuclear proteins during maturation of the *Xenopus* oocyte. *J. Embryol. exp. Morph.* **89, Suppl.** 17–34.

Hayes, P. H., Sato, T. & Denell, R. E. (1984). Homeosis in *Drosophila*: the *Ultrabithorax* larval syndrome. *Proc. natn. Acad. Sci. U.S.A.* **81**, 545–549.

Haynie, J. L. & Bryant, P. J. (1976). Intercalary regeneration in imaginal wing disk of *Drosophila melanogaster*. *Nature, Lond.* **259**, 659–662.

Heasman, J., Wylie, C. C., Hausen, P. & Smith, J. C. (1984). Fates and states of determination of single vegetal pole blastomeres of *X. laevis*. *Cell* **37**, 185–194.

Hebard, C. N. & Herold, R. C. (1967). The ultrastructure of the cortical cytoplasm in the unfertilized egg and first cleavage zygote of *Xenopus laevis*. *Expl Cell Res.* **46**, 553–570.

Hecht, R. M., Gossett, L. A. & Jeffery, W. R. (1981). Ontogeny of maternal and newly transcribed mRNA analysed by *in situ* hybridization during development of *Caenorhabditis elegans*. *Devl Biol.* **83**, 374–379.

Hecht, R. M., Wall, S. M., Schomer, D. F., Oró, J. A. & Bartel, A. H. (1982).

DNA replication may be uncoupled from nuclear and cellular division in temperature-sensitive embryonic lethal mutants of *Caenorhabditis elegans*. *Devl Biol.* **94**, 183–191.

Hegner, R. W. (1917). The genesis of the organization of the insect egg. *Am. Nat.* **51**, 641–661.

Heider, K. (1900). Das Determinationsproblem. *Verh. dt. zool Ges.* **10**, 45–97.

Heinrich, U.-R., Kaufmann, R. & Gutzeit, H. O. (1983). Cation-distribution in developing follicles of the fruit fly *Drosophila melanogaster*. *Differentiation* **25**, 10–15.

Henley, C. (1946). The effects of lithium chloride on the fertilized eggs of *Nereis limbata*. *Biol. Bull. mar. biol. Lab., Woods Hole* **90**, 188–199.

Henry, J. J. (1986). The role of unequal cleavage and the polar lobe in the segregation of developmental potential during first cleavage in the embryo of *Chaetopterus variopedatus*. *Wilhelm Roux Arch. dev. Biol.* **195**, 103–116.

Herbst, C. (1892). Experimentelle Untersuchungen über den Einfluss der veränderten chemischen Zusammensetzung des umgebenden Mediums auf die Entwickelung der Thiere. I. *Z. wiss. Zool.* **55**, 446–518.

Herbst, C. (1897). Über die zur Entwickelung der Seeigellarven nothwendigen anorganischen Stoffe, ihre Rolle und ihre Vertreibarkeit. I. *Arch. EntwMech. Org.* **5**, 649–793.

Herkovits, J. & Ubbels, G. A. (1979). The ultrastructure of the dorsal yolk-free cytoplasm and the immediately surrounding cytoplasm in the symmetrized egg of *Xenopus laevis*. *J. Embryol. exp. Morph.* **51**, 155–164.

Herlitzka, A. (1897). Sullo sviluppo di embrioni completi da blastomeri isolati di uova di tritone (*Molge cristata*). *Arch. EntwMech. Org.* **4**, 624–658.

Hess, O. (1956a). Die Entwicklung von Halbkeimen bei dem Süsswasser-Prosobranchier *Bithynia tentaculata* L. *Wilhelm Roux Arch. EntwMech. Org.* **148**, 336–361.

Hess, O. (1956b). Die Entwicklung von Exogastrulakeimen bei dem Süsswasser-Prosobranchier *Bithynia tentaculata* L. *Wilhelm Roux Arch. EntwMech. Org.* **148**, 474–488.

Hess, O. (1971). Fresh water gastropoda. In *Experimental Embryology of Marine and Fresh-Water Invertebrates*, ed. G. Reverberi, pp. 215–247. Amsterdam: North-Holland.

Heuser, C. H. & Streeter, G. L. (1941). Development of the macaque embryo. *Contr. Embryol.* **29**, 15–56.

Hille, M. B. (1974). Inhibitor of protein synthesis isolated from ribosomes of unfertilised eggs and embryos of sea urchins. *Nature, Lond.* **249**, 556–558.

Hille, M. B., Hall, D. C., Yablonka-Reuveni, Z., Danilchik, M. V. & Moon, R. T. (1981). Translational control in sea urchin eggs and embryos: initiation is rate limiting in blastula stage embryos. *Devl Biol.* **86**, 241–249.

Hillman, N., Sherman, N. I. & Graham, C. (1972). The effect of spatial arrangement on cell determination during mouse development. *J. Embryol. exp. Morph.* **28**, 263–278.

Hino, A. & Yasumasu, I. (1979). Change in the glycogen content of sea urchin eggs during early development. *Dev. Growth Differ.* **21**, 229–236.

Hirsh, D. (1979). Temperature sensitive maternal effect mutants of early development in *Caenorhabditis elegans*. In *Determinants of Spatial Organization*, ed. S. Subtelny & I. R. Konigsberg, pp. 149–165. New York: Academic Press.

Hishida, T. & Nakano, E. (1954). Respiratory metabolism during fish development. *Embryologia* **2**, 67–79.

Ho, R. K. & Weisblat, D. A. (1987). A provisional epithelium in leech embryo:

cellular origins and influence on a developmental equivalence group. *Devl Biol.* **120**, 520–534.

Hofmeister, F. (1888). Zur Lehre von der Wirkung der Salze. II. *Arch. exp. Path. Pharmak.* **24**, 247–260.

Hogan, J. C. & Trinkaus, J. P. (1977). Intercellular junctions, intramembranous particles and cytoskeletal elements of deep cells of the *Fundulus* gastrula. *J. Embryol. exp. Morph.* **40**, 125–141.

Hogue, M. J. (1910). Über die Wirkung der Centrifugalkraft auf die Eier von *Ascaris megalocephala*. *Arch. EntwMech. Org.* **29**, 109–145.

Holland, P. W. H. & Hogan, B. L. M. (1986). Phylogenetic distribution of *Antennapedia*-like homoeo boxes. *Nature, Lond.* **321**, 251–253.

Holm, Å. (1952). Experimentelle Untersuchungen über die Entwicklung und Entwicklungsphysiologie des Spinnenembryos. *Zool. Bidr. Upps.* **29**, 293–424.

Holoubek, V. & Tiedemann, H. (1978). Electrophoretic spectra of nuclear proteins from embryos of *Xenopus laevis*. *Wilhelm Roux Arch. dev. Biol.* **184**, 171–180.

Holter, H. & Zeuthen, E. (1944). The respiration of the egg and embryo of the ascidian, *Ciona intestinalis* L. *C. r. Trav. Lab. Carlsberg, Sér. chim.* **25**, 33–65.

Holtfreter, J. (1934*a*). Der Einfluss thermischer, mechanischer und chemischer Eingriffe auf die Induzierfähigkeit von *Triton*-Keimteilen. *Wilhelm Roux Arch. EntwMech. Org.* **132**, 225–306.

Holtfreter, J. (1934*b*). Über die Verbreitung induzierender Substanzen und ihre Leistungen im *Triton*keim. *Wilhelm Roux Arch. EntwMech. Org.* **132**, 307–383.

Holtfreter, J. (1936). Regionale Induktionen in xenoplastisch zusammengesetzen Explantaten. *Wilhelm Roux Arch. EntwMech. Org.* **134**, 466–550.

Holtfreter, J. (1938*a*). Differenzierungspotenzen isolierter Teile der Urodelengastrula. *Wilhelm Roux Arch. EntwMech. Org.* **138**, 522–656.

Holtfreter, J. (1938*b*). Differenzierungspotenzen isolierter Teile der Anurengastrula. *Wilhelm Roux Arch. EntwMech. Org.* **138**, 657–738.

Holtfreter, J. (1943). A study of the mechanics of gastrulation. Part I. *J. exp. Zool.* **94**, 261–318.

Holtfreter, J. (1944). A study of the mechanics of gastrulation. Part II. *J. exp. Zool.* **95**, 171–212.

Holtfreter, J. (1945). Neuralization and epidermization of gastrula ectoderm. *J. exp. Zool.* **98**, 161–209.

Holtfreter, J. (1947). Neural induction in explants which have passed through a sublethal cytolysis. *J. exp. Zool.* **106**, 197–222.

Holtfreter, J. & Hamburger, V. (1955). Amphibians. In *Analysis of Development*, ed. B. H. Willier, P. A. Weiss & V. Hamburger, pp. 230–296. Philadelphia: W.B. Saunders Co.

Holtzer, H., Lash, J. & Holtzer, S. (1956). The enhancement of somitic muscle maturation by the embryonic spinal cord. *Biol. Bull. mar. biol. Lab., Woods Hole* **111**, 303–304.

Horder, T. (1976). Pattern formation in animal embryos. In *The Developmental Biology of Plants and Animals*, ed. C. F. Graham & P. F. Wareing, pp. 169–197. Oxford: Blackwell Scientific.

Hornbruch, A., Summerbell, D. & Wolpert, L. (1979). Somite formation in the early chick embryo following grafts of Hensen's node. *J. Embryol. exp. Morph.* **51**, 51–62.

Horowitz, N. H. (1940). Comparison of the oxygen consumption of normal embryos and Dauerblastulae of the sea urchin. *J. cell. comp. Physiol.* **15**, 309–316.

Hörstadius, S. (1927). Studien über die Determination bei *Paracentrotus lividus* Lk. *Wilhelm Roux Arch. EntwMech. Org.* **112**, 239–246.

Hörstadius, S. (1928). Über die Determination des Keimes bei Echinodermen. *Acta zool., Stockh.* **9**, 1–191.

Hörstadius, S. (1935). Über die Determination im Verlaufe der Eiachse bei Seeigeln. *Pubbl. Staz. zool. Napoli* **14**, 251–429.

Hörstadius, S. (1936a). Über die zeitliche Determination im Keim von *Paracentrotus lividus* Lk. *Wilhelm Roux Arch. EntwMech. Org.* **135**, 1–39.

Hörstadius, S. (1936b). Weitere Studien über die Determination im Verlaufe der Eiachse bei Seeigeln. *Wilhelm Roux Arch. EntwMech. Org.* **135**, 40–68.

Hörstadius, S. (1937a). Investigations as to the localization of the micromere-, the skeleton-, and the entoderm-forming material in the unfertilized egg of *Arbacia punctulata. Biol. Bull. mar. biol. Lab., Woods Hole* **73**, 295–316.

Hörstadius, S. (1937b). Experiments on determination in the early development of *Cerebratulus lacteus. Biol. Bull. mar. biol. Lab., Woods Hole* **73**, 317–342.

Hörstadius, S. (1938). Schnürungsversuche an Seeigelkeimen. *Wilhelm Roux Arch. EntwMech. Org.* **138**, 197–258.

Hörstadius, S. (1950). Transplantation experiments to elucidate interactions and regulations within the gradient system of the developing sea urchin egg. *J. exp. Zool.* **113**, 245–276.

Hörstadius, S. (1952). Induction and inhibition of reduction gradients by the micromeres in the sea urchin egg. *J. exp. Zool.* **120**, 421–436.

Hörstadius, S. (1953). Vegetalization of the sea urchin egg by dinitrophenol and animalization by trypsin and ficin. *J. Embryol. exp. Morph.* **1**, 327–348.

Hörstadius, S. (1959). The effect of sugars on differentiation of larvae of *Psammechinus miliaris. J. exp. Zool.* **142**, 141–158.

Hörstadius, S. (1965a). Über fortschreitende Determination in Furchungsstadien von Seeigeleiern. *Z. Naturf.* **20b**, 331–333.

Hörstadius, S. (1965b). Über die animalisierende Wirkung von Trypsin auf Seeigelkeime. *Zool. Jb. Abt. Physiol.* **71**, 241–244.

Hörstadius, S. (1973a). *Experimental Embryology of Echinoderms.* Oxford: Clarendon Press.

Hörstadius, S. (1973b). Two metabolic systems with different reactions to temperature in sea urchin larvae. *Expl Cell Res.* **78**, 251–255.

Hörstadius, S., Lorch, J. & Danielli, J. F. (1953). The effect of enucleation on the development of sea urchin eggs. II. Enucleation of animal or vegetal halves. *Expl Cell Res.* **4**, 263–274.

Hörstadius, S. & Wolsky, A. (1936). Studien über die Determination der Bilateralsymmetrie des jungen Seeigelkeimes. *Wilhelm Roux Arch. Entw-Mech. Org.* **135**, 69–113.

Hough, B. R., Yancey, P. H. & Davidson, E. H. (1973). Persistence of maternal RNA in *Engystomops* embryos. *J. exp. Zool.* **185**, 357–368.

Hough-Evans, B. R., Ernst, S. G., Britten, R. J. & Davidson, E. H. (1979). RNA complexity in developing sea urchin oocytes. *Devl Biol.* **69**, 258–269.

Hough-Evans, B. R., Wold, B. J., Ernst. S. G., Britten, R. J. & Davidson, E. H. (1977). Appearance and persistence of maternal RNA sequences in sea urchin development. *Devl Biol.* **60**, 258–277.

Howlett, S. K. (1986). The effect of inhibiting DNA replication in the one-cell mouse embryo. *Wilhelm Roux Arch. dev. Biol.* **195**, 499–505.

Hubert, J. (1962). Étude histologique des jeunes stades du développement embryonnaire du lézard vivipare (*Lacerta vivipara* Jacquin). *Archs Anat. microsc. Morph. exp.* **51**, 11–26.

Humphrey, R. R. (1966). A recessive factor (o, for ova deficient) determining a complex of abnormalities in the Mexican axolotl (*Ambystoma mexicanum*). *Devl Biol.* **13**, 57–76.

Humphreys, W. J. (1964). Electron microscope studies of the fertilized egg and the two-cell stage of *Mytilus edulis. J. Ultrastruct. Res.* **10**, 244–262.

Hunt, T. E. (1937*a*). The development of gut and its derivatives from the mesectoderm and mesentoderm of early chick blastoderms. *Anat. Rec.* **68**, 349–369.

Hunt, T. E. (1937*b*). The origin of entodermal cells from the primitive streak of the chick embryo. *Anat. Rec.* **68**, 449–459.

Hutchins, R. & Brandhorst, B. P. (1979). Commitment to vegetalized development in sea urchin embryos. *Wilhelm Roux Arch. dev. Biol.* **186**, 95–102.

Hynes, R. O., Greenhouse, G. A., Minkoff, R. & Gross, P. R. (1972). Properties of the three cell types in sixteen-cell sea urchin embryos: RNA synthesis. *Devl Biol.* **27**, 457–478.

Hynes, R. O. & Gross, P. R. (1970). A method for separating cells from early sea urchin embryos. *Devl Biol.* **21**, 383–402.

Hynes, R. O. & Gross, P. R. (1972). Informational RNA sequences in early sea urchin embryos. *Biochim. biophys. Acta* **259**, 104–111.

Ignat'eva, G. M. & Dunaeva, M. E. (1973). Differentiation of teleost blastoderm *in vitro. Ontogenez.* **4**, 304–308 (in translation 279–283).

Ignatieva, G. M. (1960). Regional induction capacity of the chorda-mesoderm in the acipenserid embryos. *Dokl. Akad. Nauk SSSR* **134**, 233–236.

Ignatieva, G. M. (1961). Inductive properties of the chordamesodermal rudiment before onset of invagination and regulation of defects of this rudiment in sturgeon embryos. *Dokl. Akad. Nauk SSSR* **139**, 503–505.

Ikeda, M. (1965). Behaviour of sulfhydryl groups of sea urchin eggs under the blockage of cell division by UV and heat shock. *Expl Cell Res.* **40**, 282–291.

Illmensee, K. & Mahowald. A. P. (1974). Transplantation of posterior polar plasm in *Drosophila*. Induction of germ cells at the anterior pole of the egg. *Proc. natn. Acad. Sci. U.S.A.* **71**, 1016–1020.

Illmensee, K., Mahowald, A. P. & Loomis, M. R. (1976). The ontogeny of germ plasm during oogenesis in *Drosophila. Devl Biol.* **49**, 40–65.

Imoh, H. (1977). Changes in H1 histone during development of newt embryos. *Expl Cell Res.* **108**, 57–62.

Imoh, H. (1984). Appearance and distribution of RNA-rich cytoplasms in the embryo of *Xenopus laevis* during early development. *Dev. Growth Differ.* **26**, 167–176.

Imoh, H. (1985). Formation of the neural plate and the mesoderm in normally developing embryos of *Xenopus laevis. Dev. Growth Differ.* **27**, 1–11.

Imoh, H. (1986). Cell death during normal gastrulation in the newt, *Cynops pyrrhogaster. Cell Differ.* **19**, 35–42.

Inase, M. (1967). Behaviour of the pole plasm in the early development of the aquatic worm, *Tubifex hattai. Sci. Rep. Tôhoku Univ., Ser. IV.* **33**, 223–231.

Infante, A. A. & Heilmann, L. J. (1981). Distribution of messenger ribonucleic acid in polysomes and nonpolysomal particles of sea urchin embryos: translational control of actin synthesis. *Biochemistry, N.Y.* **20**, 1–8.

Ingham, P. W. (1984). A gene that regulates the bithorax complex differentially in larval and adult cells of *Drosophila. Cell* **37**, 815–823.

Ingham, P. W., Howard, K. R. & Ish-Horowicz, D. (1985*a*). Transcription pattern of the *Drosophila* segmentation gene *hairy*. *Nature, Lond.* **318**, 439–445.

Ingham, P. W., Ish-Horowicz, D. & Howard, K. R. (1986). Correlative changes in homeotic and segmentation gene expression in *Krüppel* mutant embryos of *Drosophila*. *EMBO J.* **5**, 1659–1665.

Ingham, P. W. & Martinez-Arias, A. (1986). The correct activation of Antennapedia and bithorax complex genes requires the *fushi tarazu* gene. *Nature, Lond.* **324**, 592–597.

Ingham, P. W., Martinez-Arias, A., Lawrence, P. A. & Howard, K. R. (1985b). Expression of *engrailed* in the parasegment of *Drosophila*. *Nature, Lond.* **317**, 634–636.

Isaeva, V. V. & Presnov, E. V. (1985). Influence of injury to unfertilized sea urchin eggs on the animal–vegetal polarity of the embryos. *Ontogenez* **16**, 597–603 (in translation 361–366).

Isnenghi. E., Cassada, R., Smith, K., Denich, K., Radnia, K. & von Ehrenstein, G. (1983). Maternal effects and temperature-sensitive period of mutations affecting embryogenesis in *Caenorhabditis elegans*. *Devl Biol.* **98**, 465–480.

Ito, S. & Hori, N. (1966). Electrical characteristics of *Triturus* egg cells during cleavage. *J. gen. Physiol.* **49**, 1019–1027.

Ito, S. & Ikematsu, Y. (1980). Inter and intratissue communication during amphibian development. *Dev. Growth Differ.* **22**, 247–256.

Itow, T. & Sekiguchi, K. (1979). Induction of multiple embryos with NaHCO₃ or calcium free sea water in the horseshoe crab. *Wilhelm Roux Arch. dev. Biol.* **187**, 245–254.

Izquierdo, L. (1955). Fixation des oeufs de rat colorés vitalement par le bleu de toluidene. Technique et observations cytologiques. *Archs Biol., Liège.* **66**, 403–438.

Izquierdo, L. & Becker, M. I. (1982). Effect of Li⁺ on preimplantation mouse embryos. *J. Embryol. exp. Morph.* **67**, 51–58.

Jäckle, H. (1980a). *In vitro* translatability of RNP-particles from insect eggs and embryos (*Smittia* spec., Chironomidae, Diptera). *J. exp. Zool.* **212**, 177–182.

Jäckle, H. (1980b). Possible control for tubulin synthesis during early development of an insect. *J. exp. Zool.* **214**, 219–227.

Jäckle, H. & Eagleson, G. W. (1980). Spatial distribution of abundant proteins in oocytes and fertilized eggs of the Mexican axolotl (*Ambystoma mexicanum*). *Devl Biol.* **75**, 492–499.

Jäckle, H. & Kalthoff, K. (1978). Photoreactivation of RNA in UV-irradiated insect eggs (*Smittia* sp., Chironomidae, Diptera). I. Photosensitized production and light-dependent disappearance of pyrimidine dimers. *Photochem. Photobiol.* **27**, 309–315.

Jäckle, H. & Kalthoff, K. (1980). Synthesis of a posterior indicator protein in normal embryos and double abdomens of *Smittia* sp. (Chironomidae, Diptera). *Proc. natn. Acad. Sci U.S.A.* **77**, 6700–6704.

Jäckle, H. & Kalthoff, K. (1981). Proteins foretelling head or abdomen development in the embryo of *Smittia* spec. (Chironomidae, Diptera). *Devl Biol.* **85**, 287–298.

Jäckle, H., Tautz, D., Schuh, R., Seifert, E. & Lehmann, R. (1986). Cross-regulatory interactions among the gap genes of *Drosophila*. *Nature, Lond.* **324**, 668–670.

Jackson, B. W., Grund, C., Schmid, E., Burki, K., Franke, W. W. & Illmensee, K. (1980). Formation of cytoskeletal elements during mouse embryogenesis. Intermediate filaments of the cytokeratin type and desmosomes in preimplantation embryos. *Differentiation* **17**, 161–179.

Jackson, I. J., Schofield, P. & Hogan, B. (1985). A mouse homoeo box gene is expressed during embryogenesis and in adult kidney. *Nature, Lond.* **317**, 745–748.

Jacobson, M. & Rutishauser, U. (1986). Induction of neural cell adhesion molecule (NCAM) in *Xenopus* embryos. *Devl Biol.* **116**, 524–531.

Jacobson, W. (1938). The early development of the avian embryo. II. Mesoderm formation and the distribution of presumptive embryonic material. *J. Morph.* **62**, 445–501.

Jaeger, L. (1945). Glycogen utilization by the amphibian gastrula in relation to invagination and induction. *J. cell. comp. Physiol.* **25**, 97–120.

Jaffe, L. A. & Guerrier, P. (1981). Localization of electrical excitability in the early embryo of *Dentalium*. *Devl Biol.* **83**, 370–373.

Jaffe, L. F. & Stern, C. D. (1979). Strong electrical currents leave the primitive streak of chick embryos. *Science, N.Y.* **206**, 569–570.

Jamrich, M., Sargent, T. D. & Dawid, I. B. (1987). Cell-type-specific expression of epidermal cytokeratin genes during gastrulation of *Xenopus laevis*. *Genes & Dev.* **1**, 124–132.

Janeczek, J., Born, J., John, M., Scharschmidt, M., Tiedemann, H. & Tiedemann, H. (1984a). Ribonucleoprotein particles from *Xenopus* eggs and embryos. Neural-archencephalic inducing activity of the protein moiety. *Eur. J. Biochem.* **140**, 257–264.

Janeczek, J., John, M., Born, J., Tiedemann, H. & Tiedemann, H. (1984b). Inducing activity of subcellular fractions from amphibian embryos. *Wilhelm Roux Arch. dev. Biol.* **193**, 1–12.

Janning, W. (1976). Entwicklungsgenetische Untersuchungen an Gynandern von *Drosophila melanogaster*. IV. Vergleich der morphogenetischen Anlagepläne larvaler und imaginaler Strukturen. *Wilhelm Roux Arch. dev. Biol.* **179**, 349–372.

Jeffery, W. R. (1982). Calcium ionophore polarizes ooplasmic segregation in ascidian eggs. *Science, N.Y.* **216**, 545–547.

Jeffery, W. R. (1984a). Pattern formation by ooplasmic segregation in ascidian eggs. *Biol. Bull. mar. biol. Lab., Woods Hole* **166**, 277–298.

Jeffery, W. R. (1984b). Spatial distribution of messenger RNA in the cytoskeletal framework of ascidian eggs. *Devl Biol.* **103**, 482–492.

Jeffery, W. R. (1985a). The spatial distribution of maternal mRNA is determined by a cortical cytoskeletal domain in *Chaetopterus* eggs. *Devl Biol.* **110**, 217–229.

Jeffery, W. R. (1985b). Identification of proteins and mRNAs in isolated yellow crescents of ascidian eggs. *J. Embryol. exp. Morph.* **89**, 275–287.

Jeffery, W. R., Bates, W. R., Beach, R. L. & Tomlinson, C. R. (1986). Is maternal mRNA a determinant of tissue-specific proteins in ascidian embryos? *J. Embryol. exp. Morph.* **97, Suppl.** 1–14.

Jeffery, W. R. & Capco, D. G. (1978). Differential accumulation and localization of maternal poly(A)-containing RNA during early development of the ascidian, *Styela*. *Devl Biol.* **67**, 152–166.

Jeffery, W. R. & Meier, S. (1983). A yellow crescent cytoskeletal domain in ascidian eggs and its role in early development. *Devl Biol.* **96**, 125–143.

Jeffery, W. R., Tomlinson, C. R. & Brodeur, R. D. (1983). Localization of actin messenger RNA during early ascidian development. *Devl Biol.* **99**, 408–417.

Jeffery, W. R. & Wilson, L. J. (1983). Localization of messenger RNA in the cortex of *Chaetopterus* eggs and early embryos. *J. Embryol. exp. Morph.* **75**, 225–239.

Jenkins, N. A., Kaumeyer, J. F., Young, E. M. & Raff, R. A. (1978). A test for masked message: the template activity of messenger ribonucleoprotein particles isolated from sea urchin eggs. *Devl Biol.* **63**, 279–298.

Jenkinson, J. W. (1911*a*). On the development of isolated pieces of the gastrulae of the Sea-Urchin *Strongylocentrotus lividus*. *Arch. EntwMech. Org.* **32**, 269–297.

Jenkinson, J. W. (1911*b*). On the origin of the polar and bilateral structure of the egg of the sea urchin. *Arch. EntwMech. Org.* **32**, 699–716.

Jiménez, F. & Campos-Ortega, J. A. (1981). A cell arrangement specific to thoracic ganglia in the central nervous system of the *Drosophila* embryo: its behaviour in homoeotic mutants. *Wilhelm Roux Arch. dev. Biol.* **190**, 370–373.

John, M., Janeczek, J., Born, J., Hoppe, P., Tiedemann, H. & Tiedemann, H. (1983). Neural induction in amphibians. Transmission of a neuralizing factor. *Wilhelm Roux Arch. dev. Biol.* **192**, 45–47.

Johnson, K. E. & Smith, E. P. (1976). The binding of concanavalin A to dissociated embryonic amphibian cells. *Expl Cell Res.* **101**, 63–70.

Johnson, L. V. & Calarco, P. G. (1980). Immunological characterization of embryonic cell surface antigens recognized by antiblastocyst serum. *Devl Biol.* **79**, 208–223.

Johnson, M. H. (1979). Molecular differentiation of inside cells and inner cell masses isolated from the preimplantation mouse embryo. *J. Embryol. exp. Morph.* **53**, 335–344.

Johnson, M. H., Chakraborty, J., Handyside, A. H., Willison, K. & Stern P. (1979). The effect of prolonged decompaction on the development of the preimplantation mouse embryo. *J. Embryol. exp. Morph.* **54**, 241–261.

Johnson, M. H., McConnell, J. & van Blerkom, J. (1984). Programmed development in the mouse embryo. *J. Embryol. exp. Morph.* **83**, *Suppl.* 197–231.

Johnson, M. H. & Ziomek, C. A. (1981*a*). The foundation of two distinct cell lineages within the mouse morula. *Cell* **24**, 71–80.

Johnson, M. H. & Ziomek, C. A. (1981*b*). Induction of polarity in mouse 8-cell blastomeres: specificity, geometry and stability. *J. Cell Biol.* **91**, 303–308.

Jones, E. A. (1985). Epidermal development in *Xenopus laevis*: the definition of a monoclonal antibody to an epidermal marker. *J. Embryol. exp. Morph.* **89**, *Suppl.* 155–166.

Jones, K. W. & Elsdale, T. (1963). The culture of small aggregates of amphibian embryonic cells *in vitro*. *J. Embryol. exp. Morph.* **11**, 135–154.

Josefsson, L. & Hörstadius, S. (1969). Morphogenetic substances from sea urchin eggs. Isolation of animalizing and vegetalizing substances from unfertilized eggs of *Paracentrotus lividus*. *Devl Biol.* **20**, 481–500.

Joshi, S. S. (1968). Effect of chloropicrin on the early development of chick embryo. *Wilhelm Roux Arch. EntwMech. Org.* **160**, 237–242.

Jung, E. (1966). Untersuchungen am Ei des Speisebohnenkäfers *Bruchidius obtectus* Say (Coleoptera). II. Entwicklungsphysiologische Ergebnisse der Schnürungsexperimente. *Wilhelm Roux Arch. EntwMech. Org.* **157**, 320–392.

Junquera, P. (1983). Polar plasm and pole cell formation in an insect egg developing with or without follicular epithelium: an ultrastructural study. *J. exp. Zool.* **227**, 441–452.

Jura, C. (1960). Cytological and cytochemical observations on the embryonic development of *Succinea putris* L. (Mollusca) with particular reference to ooplasmic segregation. *Zoologica Pol.* **10**, 95–130.

Jürgens, G. (1985). A group of genes controlling the spatial expression of the bithorax complex in *Drosophila. Nature, Lond.* **316**, 153–155.

Jürgens, G., Lehmann, R., Schardin, M. & Nüsslein-Volhard, C. (1986). Segmental organization of the head in the embryo of *Drosophila melanogaster*. A blastoderm fate map of the cuticle structures of the larval head. *Wilhelm Roux Arch. dev. Biol.* **195**, 359–377.

Kafiani, C. A., Timofeeva, M. J., Neyfakh, A. A., Melnikova, N. L. & Rachkus, J. A. (1969). RNA synthesis in the early embryogenesis of a fish (*Misgurnus fossilis*). *J. Embryol. exp. Morph.* **21**, 295–308.

Kageura, H. & Yamana, K. (1983). Pattern regulation in isolated halves and blastomeres of early *Xenopus laevis. J. Embryol. exp. Morph.* **74**, 221–234.

Kageura, H. & Yamana, K. (1984). Pattern regulation in defect embryos of *Xenopus laevis. Devl Biol.* **101**, 410–415.

Kageura, H. & Yamana, K. (1986). Pattern formation in 8-cell composite embryos of *Xenopus laevis. J. Embryol. exp. Morph.* **91**, 79–100.

Kahle, W. (1908). Der Paedogenesis der Cecidomyiden. *Zoologica Stutt.* **21**, Heft 55, 1–80.

Kajishima, T. (1952). Experimental studies on the embryonic development of the isopod crustacean, *Megaligia exotica* Roux. *Annotnes zool. jap.* **25**, 172–181.

Kalt, M. R. (1971*a*). The relationship between cleavage and blastocoel formation in *Xenopus laevis*. I. Light microscopic observations. *J. Embryol. exp. Morph.* **26**, 37–49.

Kalt, M. R. (1971*b*). The relationship between cleavage and blastocoel formation in *Xenopus laevis*. II. Electron microscopic observations. *J. Embryol. exp. Morph.* **26**, 51–66.

Kalthoff, K. (1971). Photoreversion of UV induction of the malformation 'double abdomen' in the egg of *Smittia* spec. (Diptera, Chironomidae). *Devl Biol.* **25**, 119–132.

Kalthoff, K. (1973). Action spectra for UV induction and photoreversal of a switch in the developmental program of the egg of an insect (*Smittia*). *Photochem. & Photobiol.* **18**, 355–364.

Kalthoff, K. (1979). Analysis of a morphogenetic determinant in an insect embryo (*Smittia* spec., Chironomidae, Diptera). In *Determinants of Spatial Organization*, ed. S. Subtelny & I. R. Konigsberg, pp. 97–126. New York: Academic Press.

Kalthoff, K., Hanel, P. & Zissler, D. (1977). A morphogenetic determinant in the anterior pole of an insect egg (*Smittia* spec., Chironomidae, Diptera). Localization by combined centrifugation and ultraviolet irradiation. *Devl Biol.* **55**, 285–305.

Kalthoff, K., Kandler-Singer, I., Schmidt, O., Zissler, D. & Versen, G. (1975). Mitochondria and polarity in the egg of *Smittia* spec. (Diptera, Chironomidae): UV irradiation, respiration measurements, ATP determinations and application of inhibitors. *Wilhelm Roux Arch. dev. Biol.* **178**, 99–121.

Kalthoff, K., Rau, K.-G. & Edmond, J. C. (1982). Modifying effects of ultraviolet irradiation on the development of abnormal body patterns in centrifuged insect embryos (*Smittia* sp., Chironomidae, Diptera). *Devl Biol.* **91**, 413–422.

Kandler-Singer, I. & Kalthoff, K. (1976). RNase sensitivity of an anterior morphogenetic determinant in an insect egg (*Smittia* spec., Chironomidae, Diptera). *Proc. natn. Acad. Sci. U.S.A.* **73**, 3739–3743.

Kanéda, T. (1980). Studies on the formation and state of determination of the trunk organizer in the newt, *Cynops pyrrhogaster*. II. Inductive effect from the underlying cranial archenteron roof. *Dev. Growth Differ.* **22**, 841–849.

Kanéda, T. (1981). Studies on the formation and state of determination of the trunk organizer in the newt, *Cynops pyrrhogaster*. III. Tangential induction in the dorsal marginal zone. *Dev. Growth Differ.* **23**, 553–564.

Kanéda, T. & Hama, T. (1979). Studies on the formation and state of determination of the trunk organizer in the newt, *Cynops pyrrhogaster*. *Wilhelm Roux Arch. dev. Biol.* **187**, 25–34.

Kanéda, T. & Suzuki, A. S. (1983). Studies on the formation and state of determination of the trunk organizer in the newt, *Cynops pyrrhogaster*. IV. The association of the neural-inducing activity with the mesodermization of the trunk organizer. *Wilhelm Roux Arch. dev. Biol.* **192**, 8–12.

Kao, K. R., Masui, Y. & Elinson, R. P. (1986). Lithium-induced respecification of pattern in *Xenopus laevis* embryos. *Nature, Lond.* **322**, 371–373.

Karasaki, S. (1957). On the mechanism of the dorsalization in the ectoderm of *Triturus* gastrulae caused by precytolytic treatments. I. Cytological and morphogenetic effects of various injurious agents. *Embryologia* **3**, 317–334.

Karasaki, S. & Yamada, T. (1955). Morphogenetic effects of centrifugation on the isolated ectoderm and whole embryo of some anurans. *Experientia* **11**, 191–194.

Karasiewicz, J. & Soltyńska, M. S. (1985). Ultrastructural evidence for the presence of actin filaments in mouse eggs at fertilization. *Wilhelm Roux Arch. dev. Biol.* **194**, 369–372.

Karch, F., Weiffenbach, B., Peifer, M., Bender, W., Duncan, I., Celniker, S., Crosby, M. & Lewis, E. B. (1985). The abdominal region of the bithorax complex. *Cell* **43**, 81–96.

Kaska, D. D. & Triplett, E. L. (1980). Glycoprotein secretion by isolated *Rana pipiens* gastrula chordamesoderm. *Cell Differ.* **9**, 281–290.

Kastern, W. H. & Berry, S. J. (1976). Non-methylated guanosine as the 5′ terminus of capped mRNA from insect oocytes. *Biochem. biophys. Res. Commun.* **71**, 37–44.

Kastern, W. H., Underberg, D. A. & Berry, S. J. (1981). DNA-dependent cytoplasmic RNA polymerases I and II in insect oocytes. *Devl Biol.* **87**, 383–389.

Kato, K.-I. (1958). Studies on the differentiating potencies of the dorsal part of the blastoporal lip in *Triturus*-gastrula. I. Differentiation of mesodermal tissues in relation to the neural tissue. *Mem. Coll. Sci. Kyoto Univ.* B **25**, 1–10.

Kato, K.-I. (1959). Studies on the differentiating potencies of the dorsal part of the blastoporal lip in *Triturus*-gastrula. II. On the differentiation of notochord. *Mem. Coll. Sci. Kyoto Univ.* B **26**, 1–7.

Kato, K.-I. & Okada, T. S. (1956). A mutual relationship between the explanted piece of dorsal blastoporal lip and the enveloping ectoderm in *Triturus*-gastrulae. *Mem. Coll. Sci. Kyoto Univ.* B **23**, 1–9.

Katow, H. (1983). Obstruction of blastodisc formation by cytochalasin B in the zebrafish, *Brachydanio rerio*. *Dev. Growth Differ.* **25**, 477–484.

Katow, H. & Solursh, M. (1980). Ultrastructure of primary mesenchyme cell ingression in the sea urchin *Lytechinus pictus*. *J. exp. Zool.* **213**, 231–246.

Kaufman, T. C., Lewis, R. & Wakimoto, B. (1980). Cytogenetic analysis of chromosome 3 in *Drosophila melanogaster*: the homoeotic gene complex in polytene chromosome interval 84A–B. *Genetics, Princeton* **94**, 115–133.

Kaulenas, M. S., Foor, W. E. & Fairbairn, D. (1969). Ribosomal RNA synthesis during cleavage of *Ascaris lumbricoides* eggs. *Science, N.Y.* **163**, 1201–1203.

Kautzsch, G. (1910). Über die Entwicklung von Spinnenembryonen unter dem Einfluß des Experiments. *Arch. EntwMech. Org.* **30**, 369–388.

Kawakami, I. (1976). Fish swimbladder: an excellent mesodermal inductor in primary embryonic induction. *J. Embryol. exp. Morph.* **36**, 315–320.

Kawakami, I., Iyeiri, S. & Sasaki, N. (1960). Analysis of inductive agents in chick embryo extracts using *Triturus* gastrula ectoderm as reactor. *J. exp. Zool.* **144**, 33–42.

Kawakami, I., Noda, S., Kurihara, K. & Okuma, K. (1977). Vegetalising factor extracted from the fish swimbladder and tested on presumptive ectoderm of *Triturus* embryos. *Wilhelm Roux Arch. dev. Biol.* **182**, 1–7.

Kaye, P. L. & Wales, R. G. (1981). Histone synthesis in preimplantation mouse embryos. *J. exp. Zool.* **216**, 453–459.

Kedes, L. H., Gross, P. R., Cognetti, G. & Hunter, A. L. (1969). Synthesis of nuclear and chromosomal proteins on light polyribosomes during cleavage in the sea urchin embryo. *J. molec. Biol.* **45**, 337–351.

Keller, R. E. (1975). Vital dye mapping of the gastrula and neurula of *Xenopus laevis*. I. Prospective areas and morphogenetic movements of the superficial layer. *Devl Biol.* **42**, 222–241.

Keller, R. E. (1976). Vital dye mapping of the gastrula and neurula of *Xenopus laevis*. II. Prospective areas and morphogenetic movements of the deep layer. *Devl Biol.* **51**, 118–137.

Keller, R. E. (1978). Time-lapse cinemicrographic analysis of superficial cell behavior during and prior to gastrulation in *Xenopus laevis*. *J. Morph.* **157**, 223–247.

Keller, R. E. (1980). The cellular basis of epiboly: an SEM study of deep-cell rearrangement during gastrulation in *Xenopus laevis*. *J. Embryol. exp. Morph.* **60**, 201–234.

Keller, R. E. (1981). An experimental analysis of the role of bottle cells and the deep marginal zone in gastrulation of *Xenopus laevis*. *J. exp. Zool.* **216**, 81–101.

Keller, R. E., Danilchik, M., Gimlich, R. & Shih, J. (1985). The function and mechanism of convergent extension during gastrulation of *Xenopus laevis*. *J. Embryol. exp. Morph.* **89**, Suppl. 185–209.

Kelley, R. O. (1969). An electron microscopic study of chordamesoderm-neurectoderm association in gastrulae of a toad, *Xenopus laevis*. *J. exp. Zool.* **172**, 153–179.

Kelly, S. J. (1975). Studies on the potency of the early cleavage blastomeres of the mouse. In *The Early Development of Mammals*, ed. M. Balls & A. E. Wild, pp. 97–105. Cambridge: The University Press.

Kelly, S. J., Mulnard, J. G. & Graham, C. F. (1978). Cell division and cell allocation in early mouse development. *J. Embryol. exp. Morph.* **48**, 37–51.

Kemler, R., Babinet, C., Eisen, H. & Jacob, F. (1977). Surface antigens in early differentiation. *Proc. natn. Acad. Sci. U.S.A.* **74**, 4449–4452.

Kemphues, K. J., Wolf, N., Wood, W. B. & Hirsh, D. (1986). Two loci required for cytoplasmic organization in early embryos of *Caenorhabditis elegans*. *Devl Biol.* **113**, 449–460.

Kessel, R. G. (1983). The structure and function of annulate lamellae: porous cytoplasmic and intranuclear membranes. *Int. Rev. Cytol.* **82**, 181–303.

Khaner, O. & Eyal-Giladi, H. (1986). The embryo-forming potency of the posterior marginal zone in stages X through XII of the chick. *Devl Biol.* **115**, 275–281.

Khaner, O., Mitrani, E. & Eyal-Giladi, H. (1985). Developmental potencies of area opaca and marginal zone areas of early chick blastoderms. *J. Embryol. exp. Morph.* **89**, 235–241.

Kidder, G. M. (1976). RNA synthesis and the ribosomal cistrons in early molluscan development. *Am. Zool.* **16**, 501–520.

Kidder, G. M. & Pedersen, R. A. (1982). Turnover of embryonic messenger RNA in preimplantation mouse embryos. *J. Embryol. exp. Morph.* **67**, 37–49.

Kiessling, A. A. & Weitlauf, H. M. (1981). Poly(A),oligo(dT)-stimulated DNA polymerase activity in preimplantation mouse embryos. *J. exp. Zool.* **215**, 117–120.

Kimber, S. J., Surani, M. A. H. & Barton, S. C. (1982). Interactions of blastomeres suggest changes in cell surface adhesiveness during the formation of inner cell mass and trophectoderm in the preimplantation mouse embryo. *J. Embryol. exp. Morph.* **70**, 133–152.

Kimelman, D., Kirschner, M. & Scherson, T. (1987). The events of the mid-blastula transition in *Xenopus* are regulated by changes in the cell cycle. *Cell* **48**, 399–407.

Kimmel, C. B. & Law, R. D. (1985a). Cell lineage of zebrafish blastomeres. I. Cleavage pattern and cytoplasmic bridges between cells. *Devl Biol.* **108**, 78–85.

Kimmel, C. B. & Law, R. D. (1985b). Cell lineage of zebrafish blastomeres. II. Formation of the yolk syncytial layer. *Devl Biol.* **108**, 86–93.

Kimmel, C. B. & Law, R. D. (1985c). Cell lineage of zebrafish blastomeres. III. Clonal analysis of the blastula and gastrula stages. *Devl Biol.* **108**, 94–101.

Kimmel, C. B. & Warga, R. M. (1986). Tissue-specific cell lineages originate in the gastrula of the zebrafish. *Science, N.Y.* **231**, 365–368.

King, M. L. & Barklis, E. (1985). Regional distribution of maternal messenger RNA in the amphibian oocyte. *Devl Biol.* **112**, 203–212.

King, R. C. (1972). *Drosophila* oogenesis and its genetic control. In *Oogenesis* ed. J. D. Biggers & A. W. Schuetz, pp. 253–275. Baltimore: University Park Press.

King, R. L. & Beams, H. W. (1938). An experimental study of chromatin diminution in *Ascaris*. *J. exp. Zool.* **77**, 425–443.

Kinoshita, S. & Saiga, H. (1979). The role of proteoglycan in the development of sea urchins. I. Abnormal development of sea urchin embryos caused by the disturbance of proteoglycan synthesis. *Expl Cell Res.* **123**, 229–236.

Kintner, C. R. & Melton, D. A. (1987). Expression of *Xenopus* N-CAM RNA in ectoderm is an early response to neural induction. *Development* **99**, 311–325.

Kirby, D. R. S., Potts, D. M. & Wilson, I. B. (1967). On the orientation of the implanting blastocyst. *J. Embryol. exp. Morph.* **17**, 527–532.

Kirschner, M. W. & Hara, K. (1980). A new method for local vital staining of amphibian embryos using Ficoll and 'crystals' of Nile red. *Mikroskopie* **36**, 12–15.

Klag, J. J. & Ubbels, G. A. (1975). Regional morphological and cytochemical differentiation in the fertilized egg of *Discoglossus pictus* (Anura). *Differentiation* **3**, 15–20.

Klein, S. L. (1987). The first cleavage furrow demarcates the dorsal–ventral axis in *Xenopus* embryos. *Devl Biol.* **120**, 299–304.

Kleinschmidt, J. A., Scheer, U., Dabauvalle, M.-C., Bustin, M. & Franke, W. W. (1983). High mobility group proteins of amphibian oocytes: a large storage pool of a soluble high mobility group-1-like protein and involvement in transcriptional events. *J. Cell Biol.* **97**, 838–848.

Kloc, M. & Matuszewski, B. (1977). Extrachromosomal DNA and the origin of oocytes in the telotrophic-meroistic ovary of *Creophilus maxillosus* (L).

(Staphylinidae, Coleoptera-Polyphaga). *Wilhelm Roux Arch. dev. Biol.* **183**, 351–368.

Klyachko, O. S., Korzh, V. P., Gorgolyuk, S. I., Timofeev, A. V. & Neyfakh, A. A. (1982). Nonuniform distribution of enzymes in fish eggs. *J. exp. Zool.* **222**, 137–148.

Klyachko, O. S. & Neifach, A. A. (1980). Distribution of endogenous and injected lactate dehydrogenase between the blastoderm and the yolk in loach embryos. *Ontogenez* **11**, 476–484 (in translation 284–290).

Klymkowsky, M. W., Maynell, L. A. & Polson, A. G. (1987). Polar asymmetry in the organization of the cortical cytokeratin system of *Xenopus laevis* oocytes and embryos. *Development* **100**, 543–557.

Knight, P. F. & Schechtman, A. M. (1954). The passage of heterologous serum proteins from the circulation into the ovum of the fowl. *J. exp. Zool.* **127**, 271–304.

Knipple, D. C., Seifert, E., Rosenberg, U. B., Preiss, A. & Jäckle, H. (1985). Spatial and temporal patterns of *Krüppel* gene expression in early *Drosophila* embryos. *Nature, Lond.* **317**, 40–44.

Knöchel, W. & Bladauski, D. (1980). A comparison of sequence complexity of nuclear and polysomal poly(A)$^+$RNA from different developmental stages of *Xenopus laevis*. *Wilhelm Roux Arch. dev. Biol.* **188**, 187–193.

Knöchel, W. & Bladauski, D. (1981). Cloning of cDNA sequences derived from poly(A)$^+$ nuclear RNA of *Xenopus laevis* at different developmental stages: evidence for stage specific regulation. *Wilhelm Roux Arch. dev. Biol.* **190**, 97–102.

Knowland, J. & Graham, C. (1972). RNA synthesis at the two-cell stage of mouse development. *J. Embryol. exp. Morph.* **27**, 167–176.

Kobayakawa, Y. (1985). Accumulation of pigment granules around nuclei in early embryos of Anura (Amphibia). *J. Embryol. exp. Morph.* **88**, 293–302.

Kobayakawa, Y. & Kubota, H. Y. (1981). Temporal pattern of cleavage and the onset of gastrulation in amphibian embryos developed from eggs with the reduced cytoplasm. *J. Embryol. exp. Morph.* **62**, 83–94.

Koch, P. & Heinig, S. (1968). Die räumliche Verteilung der Proteine in Feld- und Hausgrilleneiern. *Wilhelm Roux Arch. EntwMech. Org.* **161**, 241–248.

Kochav, S. & Eyal-Giladi, H. (1971). Bilateral symmetry in chick embryo determination by gravity. *Science, N.Y.* **171**, 1027–1028.

Kochav, S., Ginsburg, M. & Eyal-Giladi, H. (1980). From cleavage to primitive streak formation: a complementary normal table and a new look at the first stages of the development of the chick. II. Microscopic anatomy and cell population dynamics. *Devl Biol.* **79**, 296–308.

Kocher-Becker, U. & Tiedemann, H. (1971). Induction of mesodermal and endodermal structures and primordial germ cells in *Triturus* ectoderm by a vegetalizing factor from chick embryos. *Nature, Lond.* **233**, 65–66.

Kocher-Becker, U., Tiedemann, H. & Tiedemann, H. (1965). Exovagination of newt endoderm: cell affinities altered by the mesodermal inducing factor. *Science N.Y.* **147**, 167–169.

Koebke, J. (1977). Über das Differenzierungsverhalten des mesodermalen Keimbezirks verschieden alter Prägastrulationsstadien von *Ambystoma mexicanum*. Aufzucht von unbehandelten und mit Lithium behandelten Isolaten. *Z. mikrosk.-anat. Forsch.* **91**, 215–228.

Komatsu, M. (1976). Wrinkled blastula of the sea-star, *Asterina minor* Hayashi. *Dev. Growth Differ.* **18**, 435–438.

Komazaki, S. (1983). Ultrastructural study of changes in cell morphology and extracellular matrix in prospective endodermal cells of newt embryos (*Cynops pyrrhogaster*) before and during gastrulation. *Dev. Growth Differ.* **25**, 181–192.

Kominami, T. (1983). Establishment of embryonic axes in larvae of the starfish, *Asterina pectinifera. J. Embryol. exp. Morph.* **75**, 87–100.

Kominami, T. (1984). Allocation of mesendodermal cells during early embryogenesis in the starfish, *Asterina pectinifera. J. Embryol. exp. Morph.* **84**, 177–190.

Kominami, T. (1985). The role of cell adhesion in the differentiation of mesendodermal tissues in the starfish, *Asterina pectinifera. Dev. Growth Differ.* **27**, 679–688.

Kornberg, T. (1981). Compartments in the abdomen of *Drosophila* and the role of the *engrailed* locus. *Devl Biol.* **86**, 363–372.

Koser, R. B. & Collier, J. R. (1976). An electrophoretic analysis of RNA synthesis in normal and lobeless *Ilyanassa* embryo. *Differentiation* **6**, 47–52.

Kosher, R. A. & Searls, R. L. (1973). Sulfated mucopolysaccharide synthesis during the development of *Rana pipiens. Devl Biol.* **32**, 50–68.

Kostellow, A. B. & Morrill, G. A. (1968). Intracellular sodium ion concentration changes in the early amphibian embryo and the influence on nuclear metabolism. *Expl Cell Res.* **50**, 639–644.

Kostomarova, A. A. (1969). The differentiation capacity of isolated loach (*Misgurnus fossilis* L.) blastoderm. *J. Embryol. exp. Morph.* **22**, 407–430.

Kostomarova, A. A. & Nechaeva, N. V. (1970). Distribution of ribosomal cytoplasmic RNA in the blastoderm of the loach *Misgurnus fossilis* L. *Ontogenez* **1**, 391–397 (in translation 283–287).

Krause, G. (1939). Die Eitypen der Insekten. *Biol. Zbl.* **59**, 495–536.

Krause, G. (1952). Schnittoperation im Insekten-Ei zum Nachweis komplementärer Induktion bei Zwillingsbildung. *Naturwissenschaften* **39**, 356.

Krigsgaber, M. R., Kostomarova, A. A. & Burakova, T. A. (1975). Protein synthesis in loach embryos isolated from yolk during cultivation *in vitro*. *Ontogenez* **6**, 466–474 (in translation 402–407).

Krigsgaber, M. R. & Neyfakh, A. A. (1972). Investigation of the mode of nuclear control over protein synthesis in early development of loach and sea urchin. *J. Embryol. exp. Morph.* **28**, 491–509.

Krugelis, E. J., Nicholas, J. S. & Vosgian, M. E. (1952). Alkaline phosphatase activity and nucleic acids during embryonic development of *Amblystoma punctatum* at different temperatures. *J. exp. Zool.* **121**, 489–504.

Kubota, H. Y. & Durston, A. J. (1978). Cinematographical study of cell migration in the opened gastrula of *Ambystoma mexicanum. J. Embryol. exp. Morph.* **44**, 71–80.

Kubota, T. (1967). A regional change in the rigidity of the cortex of the egg of *Rana nigromaculata* following extrusion of the second polar body. *J. Embryol. exp. Morph.* **17**, 331–340.

Kühn, A. (1971). *Lectures on Developmental Physiology*, translated by R. Milkman. Berlin: Springer-Verlag.

Kuhtreiber, W. M., van der Bent, J., Dorresteijn, A. W. C., de Graaf, A., van den Biggelaar, J. A. M. & van Dongen, C. A. M. (1986). The presence of extracellular matrix between cells involved in the determination of the mesoderm bands in embryos of *Patella vulgata* (Mollusca, Gastropoda). *Wilhelm Roux Arch. dev. Biol.* **195**, 265–275.

Kunkel, J. G. (1986). Dorsoventral currents are associated with vitellogenesis in cockroach ovaries. In *Ionic Currents in Development*, ed. R. Nuccitelli, pp. 165–172. New York: Alan R. Liss, Inc.

368     *References*

Kunz, W., Trepte, H.-H. & Bier, K. (1970). On the function of the germ-line chromosomes in the oogenesis of *Wachtliella persicariae* (Cecidomyiidae). *Chromosoma* **30**, 180–192.

Kurihara, K. (1981). Transmission of homoiogenetic induction in presumptive ectoderm of newt embryo. *Dev. Growth Differ.* **23**, 361–369.

Küthe, H.-W. (1966). Das Differenzierungszentrum als selbstregulierendes Faktorensystem für den Aufbau der Keimanlage im Ei von *Dermestes frischi* (Coleoptera). *Wilhelm Roux Arch. EntwMech. Org.* **157**, 212–302.

Küthe, H.-W. (1972). Untersuchungen zur Lokalisation und Synthese von Eiproteinen während der frühen Embryonalentwicklung von *Dermestes frischi* (Coleoptera). *Wilhelm Roux Arch. EntwMech. Org.* **170**, 165–174.

Laale, H. W. & Lerner, W. (1981). Teratology and early fish development. *Am. Zool.* **21**, 517–533.

Laasberg, T. & Neuman, T. (1985). Changes in the acetylcholinesterase and choline acetyltransferase activities in the early development of the chick embryo. *Wilhelm Roux Arch. dev. Biol.* **194**, 306–310.

Laasberg, T. A. & Kyarner, Yu. K. (1979). Distribution of acid phosphatase in the chick blastoderm. *Ontogenez* **10**, 302–304 (in translation 272–275).

Labordus, V. & van der Wal, U. P. (1986). The determination of the shell field cells during the first hour in the sixth cleavage cycle of eggs of *Ilyanassa obsoleta. J. exp. Zool.* **239**, 65–75.

Lakshmi, M. S. (1962a). The effect of chloroacetophenone on chick embryos cultured *in vitro. J. Embryol. exp. Morph.* **10**, 373–382.

Lakshmi, M. S. (1962b). The effect of chloroacetophenone on the inducing capacity of Hensen's node. *J. Embryol. exp. Morph.* **10**, 383–388.

Lallier, R. (1954). Chlorure de lithium et biochimie du développement de l'oeuf d'Amphibien. *J. Embryol. exp. Morph.* **2**, 323–339.

Lallier, R. (1955). Effets des ions zinc et cadmium sur le développement de l'oeuf de l'Oursin *Paracentrotus lividus. Archs Biol., Liège* **66**, 75–102.

Lallier, R. (1959). Recherches sur l'animalisation de l'oeuf d'Oursin par les ions zinc. *J. Embryol. exp. Morph.* **7**, 540–548.

Lallier, R. (1961). Les effets du chloramphénicol sur la détermination embryonnaire de l'oeuf de l'oursin *Paracentrotus lividus. C. r. hebd. Séanc. Acad. Sci., Paris* **253**, 3060–3062.

Lallier, R. (1964). Biochemical aspects of animalization and vegetalization in the sea urchin embryo. *Adv. Morphogen.* **3**, 147–196.

Lallier, R. (1972). Effects of concanavalin A on the development of sea urchin egg. *Expl Cell Res.* **72**, 157–163.

Lamb, M. M. & Laird, C. D. (1976). Increase in nuclear poly(A)-containing RNA at syncytial blastoderm in *Drosophila melanogaster* embryos. *Devl Biol.* **52**, 31–42.

Lambert, C. C. (1971). Genetic transcription during the development and metamorphosis of the tunicate *Ascidia callosa. Expl Cell Res.* **66**, 401–409.

Landström, U. & Løvtrup, S. (1974). Oxygen consumption of normal and dwarf embryos of *Xenopus laevis. Wilhelm Roux Arch. EntwMech. Org.* **176**, 1–12.

Landström, U. & Løvtrup, S. (1975). On the determination of the dorso-ventral polarity in *Xenopus laevis* embryos. *J. Embryol. exp. Morph.* **33**, 879–895.

Landström, U. & Løvtrup, S. (1977a). Is heparan sulphate the agent of the amphibian inductor? *Acta Embryol. exp.* 171–178.

Landström, U. & Løvtrup, S. (1977b). Deoxynucleoside inhibition of differentiation in cultured embryonic cells. *Expl Cell Res.* **108**, 201–206.

Landström, U., Løvtrup-Rein, H. & Løvtrup, S. (1975). Control of cell division and cell differentiation by deoxynucleotides in the early embryo of *Xenopus laevis*. *Cell Differ.* **4**, 313–325.

Landström, U., Løvtrup-Rein, H. & Løvtrup, S. (1976). On the determination of the dorso-ventral polarity in the amphibian embryo: suppression by lactate of the formation of Ruffini's flask cells. *J. Embryol. exp. Morph.* **36**, 343–354.

Langelan, R. E. & Whiteley, A. H. (1985). Unequal cleavage and the differentiation of echinoid primary mesenchyme. *Devl Biol.* **109**, 464–475.

Larsen, W. J., Wert, S. E. & Brunner, G. D. (1987). Differential modulation of rat follicle cell gap junction populations at ovulation. *Devl Biol.* **122**, 61–71.

Laskey, R. A., Honda, B. M., Mills, A. D. & Finch, J. T. (1978). Nucleosomes are assembled by an acidic protein which binds histones and transfers them to DNA. *Nature, Lond.* **275**, 416–420.

Laskey, R. A., Mills, A. D., Gurdon, J. B. & Partington, G. A. (1977). Protein synthesis in oocytes of *Xenopus laevis* is not regulated by the supply of messenger RNA. *Cell* **11**, 345–351.

La Spina, R. (1960). Sviluppo di frammenti dell'uovo di *Ciona intestinalis* ottenuti per centrifugazione. *Acta Embryol. Morph. exp.* **3**, 1–11.

La Spina, R. (1963). Development of fragments of the fertilized egg of ctenophores and their ability to form ciliated plates. *Acta Embryol. Morph. exp.* **6**, 204–211.

Laufer, J. S., Bazzicalupo, P. & Wood, W. B. (1980). Segregation of developmental potential in early embryos of *Caenorhabditis elegans*. *Cell* **19**, 569–577.

Laufer, J. S. & von Ehrenstein, G. (1981). Nematode development after removal of egg cytoplasm: absence of localised unbound determinants. *Science, N.Y.* **211**, 402–405.

Laughon, A. & Scott, M. P. (1984). Sequence of a *Drosophila* segmentation gene: protein structure homology with DNA-binding proteins. *Nature, Lond.* **310**, 25–31.

Lawrence, P. A. (1973). A clonal analysis of segment development in *Oncopeltus* (Hemiptera). *J. Embryol. exp. Morph.* **30**, 681–699.

Lawrence, P. A. (1985). Notes on the genetics of pattern formation in the internal organs of *Drosophila*. *Trends Neurosci.* **8**, 267–269.

Lawrence, P. A. & Johnston, P. (1986). Observations on cell lineage of internal organs of *Drosophila*. *J. Embryol. exp. Morph.* **91**, 251–266.

Lawrence, P. A. & Morata, G. (1977). The early development of mesothoracic compartments in *Drosophila*. An analysis of cell lineage and fate mapping and an assessment of methods. *Devl Biol.* **56**, 40–51.

LeBlanc, J. & Brick, I. (1981). Morphologic aspects of adhesion and spreading behavior of amphibian blastula and gastrula cells. *J. Embryol. exp. Morph.* **61**, 145–163.

Lee, H.-Y., Nagele, R. G. & Roisen, F. J. (1985). Nerve growth factor induces neural differentiation in undifferentiated cells of early chick embryos. *J. exp. Zool.* **233**, 83–91.

Legname, A. H. (1968). The effect of arsenite in isolated mitochondria of amphibian embryos. *Acta Embryol. Morph. exp.* **10**, 117–123.

Lehmann, F. E. (1926). Entwicklungsstörungen in der Medullaranlage von *Triton*, erzeugt durch Unterlagerungsdefekte. *Wilhelm Roux Arch. Entw-Mech. Org.* **108**, 243–282.

Lehmann, F. E. (1928). Die Bedeutung der Unterlagerung für die Entwicklung der Medullarplatte von *Triton*. *Wilhelm Roux Arch. EntwMech. Org.* **113**, 123–171.

Lehmann, F. E. (1937). Mesodermisierung des präsumptiven Chordamaterials durch Einwirkung von Lithiumchlorid auf die Gastrula von *Triton alpestris.* *Wilhelm Roux Arch. EntwMech. Org.* **136**, 112–146.

Lehmann, F. E. (1948). Zur Entwicklungsphysiologie der Polplasmen des Eies von *Tubifex. Revue suisse Zool.* **55**, 1–43.

Lehmann, F. E. (1958). Phases of dependent and autonomous morphogenesis in the so-called mosaic-egg of *Tubifex.* In *The Chemical Basis of Development,* ed. W. D. McElroy & B. Glass, pp. 73–93. Baltimore: The Johns Hopkins Press.

Lehmann, R., Jiménez, F., Dietrich, U. & Campos-Ortega, J. A. (1983). On the phenotype and development of mutants of early neurogenesis in *Drosophila melanogaster. Wilhelm Roux Arch. dev. Biol.* **192**, 62–74.

Lehmann, R. & Nüsslein-Volhard, C. (1986). Abdominal segmentation, pole cell formation, and embryonic polarity require the localized activity of *oskar,* a maternal gene in *Drosophila. Cell* **47**, 141–152.

Lehtonen, E. (1980). Changes in cell dimensions and intercellular contacts during cleavage-stage cell cycles in mouse embryonic cells. *J. Embryol. exp. Morph.* **58**, 231–249.

Lehtonen, E. & Badley, R. A. (1980). Localization of cytoskeletal proteins in preimplantation mouse embryos. *J. Embryol. exp. Morph.* **55**, 211–225.

Leikola, A. (1963). The mesodermal and neural competence of isolated gastrula ectoderm studied by heterogenous inductors. *Annls zool. Soc. Vanamo* **25**, 1–50.

Leikola, A. (1965). On the loss of mesodermal competence of the *Triturus* gastrula ectoderm in vivo. *Experientia* **21**, 458.

Leikola, A. & McCallion, D. J. (1968). Inductive capacity of killed Hensen's node and head process in the chick embryo. *Can. J. Zool.* **46**, 205–206.

Leivo, I. I., Vaheri, A., Timpl, R. & Wartiovaara, J. (1980). Appearance and distribution of collagens and laminin in the early mouse embryo. *Devl Biol.* **76**, 100–114.

Lenicque, P. (1959). Studies on homologous inhibition in the chick embryo. *Acta zool., Stockh.* **40**, 141–202.

Leopoldseder, F. (1931). Entwicklung des Eies von *Clepsine* nach Entfernung des vegetativen Polplasmas. *Z. wiss. Zool.* **139**, 210–248.

Lepori, N. G. (1966). Analisi del processo di accorciamento della linea primitiva nel blastodisco di pollo e di anatra. *Acta Embryol. Morph. exp.* **9**, 61–68.

Levak-Švajger, B. & Švajger, A. (1974). Investigation on the origin of the definitive endoderm in the rat embryo. *J. Embryol. exp. Morph.* **32**, 445–459.

Levey, I. L., Troike, D. E. & Brinster, R. L. (1977). Effects of α-amanitin on the development of mouse ova in culture. *J. Reprod. Fert.* **50**, 147–150.

Levy, J. B., Johnson, M. H., Goodall, H. & Maro, B. (1986). The timing of compaction: control of a major developmental transition in mouse early embryogenesis. *J. Embryol. exp. Morph.* **95**, 213–237.

Lewis, C. A., Chia, F.-S. & Schroeder, T. E. (1973). Peristaltic constrictions in fertilized barnacle eggs (*Pollicipes polymerus*). *Experientia* **29**, 1533–1535.

Lewis, E. B. (1963). Genes and developmental pathways. *Am. Zool.* **3**, 33–56.

Lewis, E. B. (1978). A gene complex controlling segmentation in *Drosophila. Nature, Lond.* **276**, 565–570.

Lewis, W. H. & Wright, E. S. (1935). On the early development of the mouse egg. *Contr. Embryol.* **25**, 113–144.

Leyhausen, C. (1987). Early reciprocal interactions between ectoderm and chordamesoderm: statistical evaluation of classical embryological experiments. *Dev. Growth Differ.* **29**, 271–284.

Lillie, F. R. (1902). Differentiation without cleavage in the egg of the annelid *Chaetopterus pergamentaceus*. *Arch. EntwMech. Org.* **14**, 477–499.

Lindahl, P. E. (1932). Zur Kenntnis des Ovarialeies bei dem Seeigel. *Wilhelm Roux Arch. EntwMech. Org.* **126**, 373–390.

Lindahl, P. E. (1933). Über 'animalisierte' und 'vegetativisierte' Seeigellarven. *Wilhelm Roux Arch. EntwMech. Org.* **128**, 661–664.

Lindahl, P. E. (1936). Zur Kenntnis der physiologischen Grundlagen der Determination im Seeigelkeim. *Acta zool., Stockh.* **17**, 179–365.

Lindahl, P. E. (1937). Über eineiige Zwillinge aus Doppeleiern. *Biol. Zbl.* **57**, 389–393.

Lindahl, P. E. & Holter, H. (1940). Beiträge zur enzymatischen Histochemie. XXXIII. Die Atmung animaler und vegetativer Keimhälften von *Paracentrotus lividus*. *C. R. Trav. Lab. Carlsberg, Sér. chim.* **23**, 257–288.

Lipshitz, H. D., Peattie, D. A. & Hogness, D. S. (1987). Novel transcripts from the *Ultrabithorax* domain of the bithorax complex. *Genes & Dev.* **1**, 307–322.

Lockshin, R. A. (1966). Insect embryogenesis: macromolecular synthesis during early development. *Science, N.Y.* **154**, 775–776.

Lohmann, K. (1972). Untersuchungen zur Frage der DNS-Konstanz in der Embryonalentwicklung (Urodela, *Triturus vulgaris*). *Wilhelm Roux Arch. EntwMech. Org.* **169**, 1–40.

Lohs-Schardin, M. (1982). *Dicephalic* – a *Drosophila* mutant affecting polarity in follicle organization and embryonic patterning. *Wilhelm Roux Arch. dev. Biol.* **191**, 28–36.

Lohs-Schardin, M., Sander, K., Cremer, C., Cremer, T. & Zorn, C. (1979). Localized ultraviolet laser microbeam irradiation of early *Drosophila* embryos: fate maps based on location and frequency of adult defects. *Devl Biol.* **68**, 533–545.

Lois, P. & Izquierdo, L. (1984). Cell membrane regionalization and cytoplasm polarization in the rat early embryo. *Wilhelm Roux Arch. dev. Biol.* **193**, 205–210.

Long, W. L. (1980*a*). Analysis of yolk syncytium behavior in *Salmo* and *Catostomus*. *J. exp. Zool.* **214**, 323–331.

Long, W. L. (1980*b*). Proliferation, growth, and migration of nuclei in the yolk syncytium of *Salmo* and *Catostomus*. *J. exp. Zool.* **214**, 333–343.

Long, W. L. (1983). The role of the yolk syncytial layer in determination of the plane of bilateral symmetry in the rainbow trout, *Salmo gairdneri* Richardson. *J. exp. Zool.* **228**, 91–97.

Long, W. L. (1984). Cell movements in teleost fish development. *BioScience* **34**, 84–88.

Long, W. L. & Speck, N. A. (1984). Determination of the plane of bilateral symmetry in the teleost fish, *Oryzias latipes*. *J. exp. Zool.* **229**, 241–245.

Longo, F. J. & Chen, D.-Y. (1985). Development of cortical polarity in mouse eggs: involvement of the meiotic apparatus. *Devl Biol.* **107**, 382–394.

Lovell-Badge, R. H., Evans, M. J. & Bellairs, R. (1985). Protein synthetic patterns of tissues in the early chick embryo. *J. Embryol. exp. Morph.* **85**, 65–80.

Lovett, J. A. & Goldstein, E. S. (1977). The cytoplasmic distribution and characterization of poly(A)$^+$RNA in oocytes and embryos of *Drosophila*. *Devl Biol.* **61**, 70–78.

Løvtrup, S. (1955). Chemical differentiation during amphibian embryogenesis. *C. r. Trav. Lab. Carlsberg, Sér. chim.* **29**, 262–314.

Løvtrup, S. (1958). A physiological interpretation of the mechanism involved in the determination of bilateral symmetry in amphibian embryos. *J. Embryol. exp. Morph.* **6**, 15–27.

Løvtrup, S. (1965). Morphogenesis in the amphibian embryo: Determination of the site of invagination. *Acta zool., Stockh.* **46**, 119–165.

Løvtrup, S., Landström, U. & Løvtrup-Rein, H. (1978). Polarities, cell differentiation and primary induction in the amphibian embryo. *Biol. Rev.* **53**, 1–42.

Løvtrup, S. & Perris, R. (1983). Instructive induction or permissive activation? Differentiation of ectodermal cells isolated from the axolotl blastula. *Cell Differ.* **12**, 171–176.

Løvtrup, S. & Pigon, A. (1958). Inversion of the dorso-ventral axis in amphibian embryos by unilateral restriction of oxygen supply. *J. Embryol. exp. Morph.* **6**, 486–490.

Løvtrup, S., Rehnholm, A. & Perris, R. (1984). Induction of the synthesis of melanin and pteridine in cells isolated from the axolotl embryo. *Dev. Growth Differ.* **26**, 445–450.

Løvtrup-Rein, H. & Løvtrup, S. (1980). Energy sources in the sea urchin embryo. A critique of current views. *Acta Embryol. Morph. exp. n.s.* **1**, 93–101.

Løvtrup-Rein, H. & Nelson, L. (1982). Changes in mitochondrial respiration during the development of *Xenopus laevis*. *Cell Differ.* **11**, 125–133.

Löwkvist, B., Emanuelsson, H., Egiházi, E., Sjöberg, J., Långström, E. & Heby, O. (1986). Transcriptional inhibition in early chick embryos as a result of polyamine depletion. *Devl Biol.* **116**, 291–301.

Löwkvist, B., Emanuelsson, H. & Heby, O. (1983). Effects of polyamine limitation on nucleolar development and morphology in early chick embryos. *Cell Differ.* **12**, 19–26.

Lubin, M. (1963). A priming reaction in protein synthesis. *Biochim. biophys. Acta* **72**, 345–348.

Luchtel, D. L. (1976). An ultrastructural study of the egg and early cleavage stages of *Lymnaea stagnalis*, a pulmonate mollusc. *Am. Zool.* **16**, 405–419.

Lundmark, C. (1986). Role of bilateral zones of ingressing superficial cells during gastrulation of *Ambystoma mexicanum*. *J. Embryol. exp. Morph.* **97**, 47–62.

Lundquist, A. & Löwkvist, B. (1983). Nuclear kinetics during cleavage in a dipteran egg. *J. exp. Zool.* **228**, 151–155.

Luther, W. (1935). Entwicklungsphysiologische Untersuchungen am Forellenkeim: die Rolle des Organisationszentrums bei der Entstehung der Embryonalanlage. *Biol. Zbl.* **55**, 114–137.

Luther, W. (1936a). Potenzprüfungen an isolierten Teilstücken der Forellenkeimscheibe. *Wilhelm Roux Arch. EntwMech. Org.* **135**, 359–383.

Luther, W. (1936b). Austausch von präsumptiver Epidermis und Medullarplatte beim Forellenkeim. *Wilhelm Roux Arch. EntwMech. Org.* **135**, 384–388.

Luther, W. (1937). Transplantations- und Defektversuche am Organisationszentrum der Forellenkeimscheibe. *Wilhelm Roux Arch. EntwMech. Org.* **137**, 404–424.

Lutz, H. (1949). Sur la production expérimentale de la polyembryonie et de la monstruosité double chez les oiseaux. *Archs Anat. microsc. Morph. exp.* **38**, 79–144.

Lutz, H. (1953). L'orientation des axes embryonnaires dans la gémellité expérimentale chez les oiseaux. *Bull. biol. Fr. Belg.* **87**, 34–67.

Lutz, H. & Lutz-Ostertag, Y. (1957). Gémellité expérimentale et *situs inversus viscerum* chez l'embryon d'Oiseau. *C. r. hebd. Séanc. Acad. Sci., Paris* **244**, 1543–1545.

Lützeler, I. E. & Malacinski, G. M. (1974). Modulations in the electrophoretic spectrum of newly synthesized protein in early axolotl (*Ambystoma mexicanum*) development. *Differentiation* **2**, 287–297.

Lynn, D. A., Angerer, L. M., Bruskin, A. M., Klein, W. H. & Angerer, R. C. (1983). Localization of a family of mRNAs in a single cell type and its precursors in sea urchin embryos. *Proc. natn. Acad. Sci. U.S.A.* **80**, 2656–2660.

Lyon, E. P. (1907). Results of centrifugalizing eggs. *Arch. EntwMech. Org.* **23**, 151–173.

Lyon, M. F. (1972). X-chromosome inactivation and developmental patterns in mammals. *Biol. Rev.* **47**, 1–35.

Maas, A.-H. (1949). Über die Auslösbarkeit von Temperaturmodifikationen während der Embryonalentwicklung von *Drosophila melanogaster* Meigen. *Wilhelm Roux Arch. EntwMech. Org.* **143**, 515–572.

Macdonald, P. M. & Struhl, G. (1986). A molecular gradient in early *Drosophila* embryos and its role in specifying the body pattern. *Nature, Lond.* **324**, 537–545.

Macgregor, H. C. & Andrews, C. (1977). The arrangement and transcription of 'middle repetitive' DNA sequences on lampbrush chromosomes of *Triturus*. *Chromosoma* **63**, 109–126.

Macgregor, H. C. & Stebbings, H. (1970). A massive system of microtubules associated with cytoplasmic movement in telotrophic ovarioles. *J. Cell Sci.* **6**, 431–449.

Mackay, S. & Newrock, K. M. (1982). Histone subtypes and switches in synthesis of histone subtypes during *Ilyanassa* development. *Devl Biol.* **93**, 430–437.

Macmurdo-Harris, H. & Zalik, S. E. (1971). Microelectrophoresis of early amphibian embryonic cells. *Devl Biol.* **24**, 335–347.

Magnuson, T., Demsey, A. & Stackpole, C. W. (1977). Characterization of intercellular junctions in the preimplantation mouse embryo by freeze-fracture and thin-section electron microscopy. *Devl Biol.* **61**, 252–261.

Magnuson, T. & Epstein, C. J. (1981). Genetic control of very early mammalian development. *Biol. Rev.* **56**, 369–408.

Magnuson, T., Jacobson, J. B. & Stackpole, C. W. (1978). Relationship between intercellular permeability and junction organization in the preimplantation mouse embryo. *Devl Biol.* **67**, 214–224.

Mahowald, A. P. (1968). Polar granules of *Drosophila*. II. Ultrastructural changes during early embryogenesis. *J. exp. Zool.* **167**, 237–262.

Mahowald, A. P. (1971*a*). Polar granules of *Drosophila*. III. The continuity of polar granules during the life cycle of *Drosophila*. *J. exp. Zool.* **176**, 329–343.

Mahowald, A. P. (1971*b*). Polar granules of *Drosophila*. IV. Cytochemical studies showing loss of RNA from polar granules during early stages of embryogenesis. *J. exp. Zool.* **176**, 345–352.

Mahowald, A. P., Allis, C. D., Karrer, K. M., Underwood, E. M. & Waring, G. L. (1979*a*). Germ plasm and pole cells of *Drosophila*. In *Determinants of Spatial Organization*, ed. S. Subtelny & I. R. Konigsberg, pp. 127–146. New York: Academic Press.

Mahowald, A. P., Caulton, J. H. & Gehring, W. J. (1979*b*). Ultrastructural studies of oocytes and embryos derived from female flies carrying the *grandchildless* mutation in *Drosophila subobscura*. *Devl Biol.* **69**, 118–132.

Mahowald, A. P. & Hennen, S. (1971). Ultrastructure of the 'germ plasm' in eggs and embryos of *Rana pipiens*. *Devl Biol.* **24**, 37–53.

Maisonhaute, C. (1977). Comparaison des effets de deux inhibiteurs de la synthèse d'ARN (actinomycine D et α-amanitine) sur le développement embryonnaire d'un insecte: déterminisme génique du début de l'embryo-

374    References

génèse de *Leptinotarsa decemlineata* (Coleoptera). *Wilhelm Roux Arch. dev. Biol.* **183**, 61–77.

Malacinski, G. M. (1971). Genetic control of qualitative changes in protein synthesis during early amphibian (Mexican axolotl) embryogenesis. *Devl Biol.* **26**, 442–451.

Malacinski, G. M. (1972). Deployment of maternal template during early amphibian embryogenesis. *J. exp. Zool.* **181**, 409–420.

Malacinski, G. M. (1984). Axis specification in amphibian eggs. In *Pattern Formation: A Primer in Developmental Biology*, ed. G. M. Malacinski & S. V. Bryant, pp. 435–456. New York: Macmillan.

Malacinski, G. M., Allis, C. D. & Chung, H.-M. (1974). Correction of developmental abnormalities resulting from localized ultra-violet irradiation of an amphibian egg. *J. exp. Zool.* **189**, 249–254.

Malacinski, G. M., Benford, H. & Chung, H.-M. (1975). Association of an ultraviolet irradiation sensitive cytoplasmic localization with the future dorsal side of the amphibian egg. *J. exp. Zool.* **191**, 97–110.

Malacinski, G. M., Chung, H.-M. & Asashima, M. (1980). The association of primary embryonic organizer activity with the future dorsal side of amphibian eggs and early embryos. *Devl Biol.* **77**, 449–462.

Malacinski, G. M., Ryan, B. & Chung, H.-M. (1978). Surface coat movements in unfertilized amphibian eggs. *Differentiation* **10**, 101–107.

Malacinski, G. M. & Spieth, J. (1979). Maternal effect genes in the Mexican axolotl (*Ambystoma mexicanum*). In *Maternal Effects in Development*, ed. D. R. Newth & M. Balls, pp. 241–267. Cambridge: The University Press.

Maleyvar, R. P. & Lowery, R. (1973). The patterns of mitosis and DNA synthesis in the presumptive neurectoderm of *Xenopus laevis* (Daudin). In *The Cell Cycle in Development and Differentiation*, ed. M. Balls & F. S. Billett, pp. 249–255. Cambridge: The University Press.

Maller, J. L. (1985). Regulation of amphibian oocyte maturation. *Cell Differ.* **16**, 211–221.

Mancuso, V. (1964). The distribution of the ooplasmic components in the unfertilised, fertilised and 16-cell stage egg of *Ciona intestinalis*. *Acta Embryol. Morph. exp.* **7**, 71–82.

Mancuso, V. (1974). Formation of the ultrastructural components of *Ciona intestinalis* tadpole test by the animal embryo. *Experientia* **30**, 1078.

Manen, C.-A. & Russell, D. H. (1973). Spermine is major polyamine in sea urchins: studies of polyamines and their synthesis in developing sea urchins. *J. Embryol. exp. Morph.* **29**, 331–345.

Manes, C. & Menzel, P. (1982). Spontaneous release of nucleosome cores from embryoblast chromatin. *Devl Biol.* **92**, 529–538.

Manes, M. E. & Barbieri, F. D. (1976). Symmetrisation in the amphibian egg by disrupted sperm cells. *Devl Biol.* **53**, 138–141.

Manes, M. E. & Barbieri, F. D. (1977). On the possibility of sperm aster involvement in dorso-ventral polarization and pronuclear migration in the amphibian egg. *J. Embryol. exp. Morph.* **40**, 187–197.

Manes, M. E. & Elinson, R. P. (1980). Ultraviolet light inhibits grey crescent formation on the frog egg. *Wilhelm Roux Arch. dev. Biol.* **189**, 73–76.

Manes, M. E., Elinson, R. P. & Barbieri, F. D. (1978). Formation of the amphibian grey crescent: effects of colchicine and cytochalasin B. *Wilhelm Roux Arch. dev. Biol.* **185**, 99–104.

Mangold, O. (1923). Transplantationsversuche zur Frage der Spezifität und Bildung der Keimblätter bei *Triton*. *Arch. mikrosk. Anat. EntwMech.* **100**, 198–301.

Mangold, O. (1932). Autonome und komplementäre Induktionen bei Amphibien. *Naturwissenschaften* **20**, 371–375.

Mangold, O. & Spemann, H. (1927). Über Induktion von Medullarplatte durch Medullarplatte im jüngeren Keim, ein Beispiel homöogenetischer oder assimilatorischer Induktion. *Wilhelm Roux Arch. EntwMech. Org.* **111**, 341–422.

Marinos, E. & Billett, F. S. (1981). Mitochondrial number, cytochrome oxidase and succinic dehydrogenase activity in *Xenopus laevis* oocytes. *J. Embryol. exp. Morph.* **62**, 395–409.

Mark, E. L. (1890). Studies on *Lepidosteus*. Part I. *Bull. Mus. comp. Zool. Harvard* **19**, 1–127.

Markman, B. (1961*a*). Regional differences in isotopic labelling of nucleic acid and protein in early sea urchin development. An autoradiographic study. *Expl Cell Res.* **23**, 118–124.

Markman, B. (1961*b*). Differences in isotopoic labelling of nucleic acid and protein in sea urchin embryos developing from animal and vegetal egg halves. *Expl Cell Res.* **25**, 224–227.

Maro, B., Johnson, M. H., Pickering, S. J. & Flach, G. (1984). Changes in actin distribution during fertilization of the mouse egg. *J. Embryol. exp. Morph.* **81**, 211–237.

Maro, B., Johnson, M. H., Webb, M. & Flach, G. (1986). Mechanism of polar body formation in the mouse oocyte: an interaction between the chromosomes, the cytoskeleton and the plasma membrane. *J. Embryol. exp. Morph.* **92**, 11–32.

Marsh, J. L. & Wieschaus, E. (1977). Germ-line dependence of the *maroon-like* maternal effect in *Drosophila*. *Devl Biol.* **60**, 396–403.

Marthy, H.-J. (1970). Beobachtungen beim Transplantieren von Organanlagen am Embryo von *Loligo vulgaris* (Cephalopoda, Decapoda). *Experientia* **26**, 160–161.

Marthy, H.-J. (1972). Sur la localisation et la stabilité du plan d'ebauches d'organes chez l'embryon de *Loligo vulgaris*. *C. r. hebd. Séanc. Acad. Sci., Paris D* **275**, 1291–1293.

Marthy, H.-J. (1973). An experimental study of eye development in the cephalopod *Loligo vulgaris*: determination and regulation during formation of the primary optic vesicle. *J. Embryol. exp. Morph.* **29**, 347–361.

Marthy, H.-J. (1975). Organogenesis in Cephalopoda: further evidence of blastodisc-bound developmental information. *J. Embryol. exp. Morph.* **33**, 75–83.

Marthy, H.-J. (1978). Recherches sur le rôle morphogénétique du més-endo-derme dans l'embryogenèse de *Loligo vulgaris* (Céphalopode). *C. r. hebd. Séanc. Acad. Sci., Paris D* **287**, 1345–1348.

Martin, G. R., Smith, S. & Epstein, C. J. (1978). Protein synthetic patterns in teratocarcinoma stem cells and mouse embryos at early stages of development. *Devl Biol.* **66**, 8–16.

Martindale, M. Q. (1986). The ontogeny and maintenance of adult symmetry properties in the ctenophore, *Mnemiopsis mccradyi*. *Devl Biol.* **118**, 556–576.

Martindale, M. Q., Doe, C. Q. & Morrill, J. B. (1985). The role of animal-vegetal interaction with respect to the determination of dorsoventral polarity in the equal-cleaving spiralian, *Lymnaea palustris*. *Wilhelm Roux Arch. dev. Biol.* **194**, 281–295.

Martinez-Arias, A., Ingham, P. W., Scott, M. P. & Akam, M. E. (1987). The spatial and temporal deployment of *Dfd* and *Scr* transcripts throughout development of *Drosophila*. *Development* **100**, 673–683.

Martinez-Arias, A. & Lawrence, P. A. (1985). Parasegments and compartments in the *Drosophila* embryo. *Nature, Lond.* **313**, 639–642.

Maruyama, Y. K., Nakaseko, Y. & Yagi, S. (1985). Localization of cytoplasmic determinants responsible for primary mesenchyme formation and gastrulation in the unfertilized egg of the sea urchin *Hemicentrotus pulcherrimus*. *J. exp. Zool.* **236**, 155–163.

Masuda, M. (1979). Species specific patterns of ciliogenesis in developing sea urchin embryos. *Dev. Growth Differ.* **21**, 545–552.

Masui, Y. (1960a). Differentiation of the prechordal tissue under influence of lithium chloride. *Mem. Konan Univ. Sci. Ser.* **4**, 65–78.

Masui, Y. (1960b). Alteration of the differentiation of gastrula ectoderm under influence of lithium chloride. *Mem. Konan Univ. Sci. Ser.* **4**, 79–102.

Masui, Y. (1961). Mesodermal and endodermal differentiation of the presumptive ectoderm of *Triturus* gastrula through influence of lithium ion. *Experientia* **17**, 458–459.

Masui, Y. & Clarke, H. J. (1979). Oocyte maturation. *Int. Rev. Cytol.* **57**, 185–282.

Mazabraud, A., Wegnez, M. & Denis, H. (1975). Biochemical research on oogenesis. RNA accumulation in the oocytes of teleosts. *Devl Biol.* **44**, 326–332.

Mazur, G. D., Regier, J. C. & Kafatos, F. C. (1980). The silkmoth chorion: morphogenesis of surface structures and its relation to synthesis of specific proteins. *Devl Biol.* **76**, 305–321.

Mazur, G. D., Regier, J. C. & Kafatos, F. C. (1982). Order and defect in the silkmoth chorion, a biological analogue of a cholesteric liquid crystal. In *Insect Ultrastructure*, vol. 1, ed. R. C. King & H. Akai, pp. 150–185. New York: Plenum Press.

McCallion, D. J. & Shinde, V. A. (1973). Induction in the chick by quail Hensen's node. *Experientia* **29**, 321–322.

McClay, D. R. & Chambers, A. F. (1978). Identification of four classes of cell surface antigens appearing at gastrulation in sea urchin embryos. *Devl Biol.* **63**, 179–186.

McClendon, J. F. (1910). The development of isolated blastomeres of the frog's egg. *Am. J. Anat.* **10**, 425–430.

McClendon, J. F. (1912). An attempt towards the physical chemistry of the production of one-eyed monstrosities. *Am. J. Physiol.* **29**, 289–297.

McCormick, P. J. & Babiarz, B. (1984). Expression of a glucose-regulated cell surface protein in early mouse embryos. *Devl Biol.* **105**, 530–534.

McCue, P. A. & Sherman, M. I. (1982). Effect of antimetabolites on programming of inner cells of the mouse blastocyst. *J. exp. Zool.* **224**, 445–450.

McDonald, R. I. (1973). Chemical modification of cell type determination in amphibian embryos. *J. exp. Zool.* **186**, 175–186.

McGinnis, W., Levine, M. S., Hafen, E., Kuroiwa, A. & Gehring, W. J. (1984). A conserved DNA sequence in homoeotic genes of *Drosophila* Antennapedia and bithorax complexes. *Nature, Lond.* **308**, 428–433.

McGrath, J. & Solter, D. (1984). Completion of mouse embryogenesis requires both the maternal and paternal genomes. *Cell* **37**, 179–183.

McLachlin, J. R., Caveney, S. & Kidder, G. M. (1983). Control of gap junction formation in early mouse embryos. *Devl Biol.* **98**, 155–164.

Mee, J. E. (1986). Pattern formation in fragmented eggs of the short germ insect *Schistocerca gregaria*. *Wilhelm Roux Arch. dev. Biol.* **195**, 506–512.

Meedel, T. H. (1983). Myosin expression in the developing ascidian embryo. *J. exp. Zool.* **227**, 203–211.

Meedel, T. H., Crowther, R. J. & Whittaker, J. R. (1987). Determinative properties of muscle lineages in ascidian embryos. *Development* **100**, 245–260.

Meedel, T. H. & Whittaker, J. R. (1978). Messenger RNA synthesis during early ascidian development. *Devl Biol.* **66**, 410–421.

Meedel, T. H. & Whittaker, J. R. (1979). Development of acetylcholinesterase during embryogenesis of the ascidian *Ciona intestinalis*. *J. exp. Zool.* **210**, 1–10.

Meedel, T. H. & Whittaker, J. R. (1984). Lineage segregation and developmental autonomy in expression of functional muscle acetylcholinesterase mRNA in the ascidian embryo. *Devl Biol.* **105**, 479–487.

Meier, S. (1979). Development of the chick embryo mesoblast. Formation of the embryonic axis and establishment of the metameric pattern. *Devl Biol.* **73**, 25–45.

Meijer, L. & Guerrier, P. (1984). Maturation and fertilization in starfish oocytes. *Int. Rev. Cytol.* **86**, 129–196.

Meinhardt, H. (1977). A model of pattern formation in insect embryogenesis. *J. Cell Sci.* **23**, 117–139.

Meinhardt, H. (1982). *Models of Biological Pattern Formation.* London: Academic Press.

Meinhardt, H. (1986). Hierarchical inductions of cell states: a model for segmentation in *Drosophila*. *J. Cell Sci. Suppl.* **4**, 357–381.

Melekhova, O. P. (1976). Free radical processes in the embryogenesis of Anura. *Ontogenez* **7**, 131–140 (in translation 109–116).

Mel'nikova, N. L., Timofeeva, M. Ya., Rott, N. N. & Ignat'eva, G. M. (1972). Synthesis of ribosomal RNA in the early embryogenesis of the trout. *Ontogenez* **3**, 85–94 (in translation 67–74).

Melton, C. G. & Smorul, R. P. (1974). Functional volume of frog eggs: equivalence of metabolite diffusion space in chemically demembranated embryos and aqueous phase (non-yolk) volume. *J. exp. Zool.* **187**, 239–247.

Melton, D. A. (1987). Translocation of a localized maternal mRNA to the vegetal pole of *Xenopus* oocytes. *Nature, Lond.* **328**, 80–82.

Mergner, H. (1971). Cnidaria. In *Experimental Embryology of Marine and Fresh-Water Invertebrates*, ed. G. Reverberi, pp. 1–84. Amsterdam: North-Holland.

Mermod, J. J., Schatz, G. & Crippa, M. (1980). Specific control of messenger translation in *Drosophila* oocytes and embryos. *Devl Biol.* **75**, 177–186.

Merriam, R. W. & Sauterer, R. A. (1983). Localization of a pigment-containing structure near the surface of *Xenopus* eggs which contracts in response to calcium. *J. Embryol. exp. Morph.* **76**, 51–65.

Merriam, R. W., Sauterer, R. A. & Christensen, K. (1983). A subcortical pigment-containing structure in *Xenopus* eggs with contractile properties. *Devl Biol.* **95**, 439–446.

Meshcheryakov, V. N. (1978a). Orientation of cleavage spindles in pulmonate mollusks. I. Role of blastomere shape in orientation of second division spindles. *Ontogenez* **9**, 558–566 (in translation 487–494).

Meshcheryakov, V. N. (1978b). Orientation of cleavage spindles in pulmonate mollusks. II. Role of structure of intercellular contacts in orientation of the third and fourth cleavage spindles. *Ontogenez* **9**, 567–575 (in translation 494–501).

Meshcheryakov, V. N. & Belousov, L. V. (1973). Effect of trypsin on spatial organization of early cleavage of the mollusks *Limnaea stagnalis* L. and *Physa fontinalis* L. *Ontogenez* **4**, 359–372 (in translation 330–340).

Meshcheryakov, V. N. & Beloussov, L. V. (1975). Asymmetrical rotations of

378    References

blastomeres in early cleavage of Gastropoda. *Wilhelm Roux Arch. dev. Biol.* **177**, 193–203.

Meuler, D. C. & Malacinski, G. M. (1985). An analysis of protein synthesis patterns during early embryogenesis of the urodele – *Ambystoma mexicanum. J. Embryol. exp. Morph.* **89**, 71–92.

Miller, J. H. & Epel, D. (1973). Studies of oogenesis in *Urechis caupo*. II. Accumulation during oogenesis of carbohydrate, RNA, microtubule protein, and soluble mitochondrial, and lysosomal enzymes. *Devl Biol.* **32**, 331–344.

Miller, O. L. (1973). The visualization of genes in action. *Scient. Am.* **228**, no. 3, 34–42.

Miller, O. L. & Bakken, A. H. (1972). Morphological studies of transcription. *Acta Endocr., Copenh. Suppl.* **168**, 155–177.

Mills, R. M. & Brinster, R. L. (1967). Oxygen consumption of preimplantation mouse embryos. *Expl Cell Res.* **47**, 337–344.

Minganti, A. (1950). Acidi nucleici e fosfatasi nello sviluppo della *Limnaea. Riv. Biol.* **42**, 295–319.

Minganti, A. (1951). Ricerche istochimiche sulla localizzione del territorio presuntivo degli organi sensoriali nelle larve di Ascidie. *Pubbl. Staz. zool. Napoli* **23**, 52–57.

Minganti, A. (1954a). Fosfatasi alcaline nello sviluppo delle Ascidie. *Pubbl. Staz. zool. Napoli* **25**, 9–17.

Minganti, A. (1954b). Fosfatasi alcaline nei semiembrioni animali e vegetativi di Ascidie. *Pubbl. Staz. zool. Napoli* **25**, 438–443.

Minganti, A. (1959a). Androgenetic hybrids in ascidians. I. *Ascidia malaca* (♀)×*Phallusia mamillata* ♂. *Acta Embryol. Morph. exp.* **2**, 244–256.

Minganti, A. (1959b). Lo sviluppo embryonale e il comportamento dei cromosomi in ibridi tra 5 specie di Ascidie. *Acta Embryol. Morph. exp.* **2**, 269–301.

Mintz, B. (1964). Synthetic processes and early development in the mammalian egg. *J. exp. Zool.* **157**, 85–100.

Minuth, M. & Grunz, H. (1980). The formation of mesodermal derivatives after induction with vegetalizing factor depends on secondary cell interactions. *Cell Differ.* **9**, 229–238.

Minuth, W. W. (1978). Mesodermalization of amphibian gastrula ectoderm in transfilter experiments. *Med. Biol.* **56**, 349–354.

Misumi, Y., Kurata, S. & Yamana, K. (1980). Initiation of ribosomal RNA synthesis and cell division in *Xenopus laevis* embryos. *Dev. Growth Differ.* **22**, 773–780.

Mita-Miyazawa, I. & Satoh, N. (1986). Mass isolation of muscle lineage blastomeres from ascidian embryos. *Dev. Growth Differ.* **28**, 483–488.

Mitrani, E. (1984). Mitosis in the formation and function of the primary hypoblast of the chick. *Wilhelm Roux Arch. dev. Biol.* **193**, 402–405.

Mitrani, E. & Eyal-Giladi, H. (1981). Hypoblastic cells can form a disk inducing an embryonic axis in chick epiblast. *Nature, Lond.* **289**, 800–802.

Mitrani, E., Shimoni, Y. & Eyal-Giladi, H. (1983). Nature of the hypoblastic influence on the chick embryo epiblast. *J. Embryol. exp. Morph.* **75**, 21–30.

Mitsunaga, K., Fujiwara, A., Yoshimi, T. & Yasumasu, I. (1983). Stage specific effects on sea urchin embryogenesis of $Zn^{2+}$, $Li^+$, several inhibitors of cAMP-phosphodiesterase and inhibitors of protein synthesis. *Dev. Growth Differ.* **25**, 249–260.

Miura, Y. & Wilt, F. H. (1969). Tissue interaction and the formation of the first erythroblasts of the chick embryo. *Devl Biol.* **19**, 201–211.

Miura, Y. & Wilt, F. H. (1971). The effects of 5-bromodeoxyuridine on yolk sac erythropoiesis in the chick embryo. *J. Cell Biol.* **48**, 523–532.

Miwa, J., Schierenberg, E., Miwa, S. & von Ehrenstein, G. (1980). Genetics and mode of expression of temperature-sensitive mutations arresting embryonic development in *Caenorhabditis elegans*. *Devl Biol.* **76**, 160–174.

Miyagawa, N. & Suzuki, A. (1969). Changes of inducing potency and self-differentiation of dorsal mesoderm of late *Triturus* gastrula with time. *Kumamoto J. Sci. B* **9**, 109–116.

Miyamoto, D. M. & van der Meer, J. M. (1982). Early egg contractions and patterned parasynchronous cleavage in a living insect egg. *Wilhelm Roux Arch. dev. Biol.* **191**, 95–102.

Mizoguchi, H. & Yasumasu, I. (1982). Exogut formation by the treatment of sea urchin embryos with ascorbate and $\alpha$-ketoglutarate. *Dev. Growth Differ.* **24**, 359–368.

Mizoguchi, H. & Yasumasu, I. (1983a). Effect of $\alpha$, $\alpha'$-dipyridyl on exogut formation in vegetalized embryos of the sea urchin. *Dev. Growth Differ.* **25**, 57–64.

Mizoguchi, H. & Yasumasu, I. (1983b). Inhibition of archenteron formation by the inhibitors of prolyl hydroxylase in sea urchin embryos. *Cell Differ.* **12**, 225–231.

Mizuno, S., Lee, Y. R., Whiteley, A. H. & Whiteley, H. R. (1974). Cellular distribution of RNA populations in 16-cell stage embryos of the sand dollar, *Dendraster excentricus*. *Devl Biol.* **37**, 18–27.

Mlodzik, M., Fjose, A. & Gehring, W. J. (1985). Isolation of *caudal*, a *Drosophila* homeo box-containing gene with maternal expression, whose transcripts form a concentration gradient at the pre-blastoderm stage. *EMBO J.* **4**, 2961–2969.

Moen, T. L. & Namenwirth, M. (1977). The distribution of soluble proteins along the animal–vegetal axis of frog eggs. *Devl Biol.* **58**, 1–10.

Mohler, J. & Wieschaus, E. F. (1986). Dominant maternal-effect mutations of *Drosophila melanogaster* causing the production of double-abdomen embryos. *Genetics, Princeton* **112**, 803–822.

Mohler, J. D. (1977). Developmental genetics of the *Drosophila* egg. I. Identification of 59 sex-linked cistrons with maternal effects on embryonic development. *Genetics, Princeton* **85**, 259–272.

Mohun, T. J., Brennan, S., Dathan, N., Fairman, S. & Gurdon, J. B. (1984a). Cell type-specific activation of actin genes in the early amphibian embryo. *Nature, Lond.* **311**, 716–721.

Mohun, T. J., Brennan, S. & Gurdon, J. B. (1984b). Region-specific regulation of the actin multi-gene family in early amphibian embryos. *Phil. Trans. R. Soc. B* **307**, 337–342.

Mohun, T. J., Tilly, R., Mohun, R. & Slack, J. M. W. (1980). Cell commitment and gene expression in the axolotl embryo. *Cell* **22**, 9–15.

Monesi, V. & Salfi, V. (1967). Macromolecular syntheses during early development in the mouse embryo. *Expl Cell Res.* **46**, 632–635.

Moody, S. A. (1987a). Fates of the blastomeres of the 16-cell stage *Xenopus* embryo. *Devl Biol.* **119**, 560–578.

Moody, S. A. (1987b). Fates of the blastomeres of the 32-cell-stage *Xenopus* embryo. *Devl Biol.* **122**, 300–319.

Moody, W. J. (1985). The development of calcium and potassium currents during oogenesis in the starfish, *Leptasterias hexactis*. *Devl Biol.* **112**, 405–413.

Moore, N. W., Adams, C. E. & Rowson, L. E. A. (1968). Developmental potential of single blastomeres of the rabbit egg. *J. Reprod. Fert.* **17**, 527–531.

Moran, D. (1985). Rapid induction of morphogenetic movement in amphibian gastrulae with $Ca^{2+}$ ionophores. *Wilhelm Roux Arch. dev. Biol.* **194**, 271–274.

Moran, D. & Mouradian, W. E. (1975). A scanning electron microscopic study of the appearance and localization of cell surface material during amphibian gastrulation. *Devl Biol.* **46**, 422–429.

Morata, G. & Kerridge, S. (1981). Sequential functions of the bithorax complex of *Drosophila*. *Nature, Lond.* **290**, 778–781.

Morata, G. & Lawrence, P. A. (1975). Control of compartment development by the *engrailed* gene in *Drosophila*. *Nature, Lond.* **255**, 614–617.

Morata, G., Sánchez-Herrero, E. & Casanova, J. (1986). The bithorax complex of *Drosophila*: an overview. *Cell Differ.* **18**, 67–78.

Morgan, T. H. (1893). Experimental studies on echinoderm eggs. *Anat. Anz.* **9**, 141–152.

Morgan, T. H. (1927). *Experimental Embryology*. New York: Columbia University Press.

Morgan, T. H. (1934). *Embryology and Genetics*. New York: Columbia University Press.

Morgan, T. H. & Spooner, G. B. (1909). The polarity of the centrifuged egg. *Arch. EntwMech. Org.* **28**, 104–117.

Morgan, T. H. & Tyler, A. (1930). The point of entrance of the spermatozoön in relation to the orientation of the embryo in eggs with spiral cleavage. *Biol. Bull. mar. biol. Lab., Woods Hole* **58**, 59–73.

Morgan, T. H. & Tyler, A. (1938). The relation between entrance point of the spermatozoön and bilaterality of the egg of *Chaetopterus*. *Biol. Bull. mar. biol. Lab., Woods Hole* **74**, 401–402.

Mori, Y. (1932). Entwicklung isolierter Blastomeren und teilweise abgetöteter älterer Keime von *Clepsine sexoculata*. *Z. wiss. Zool.* **141**, 399–431.

Moritz, K. B. (1967). Die Blastomerendifferenzierung für Soma und Keimbahn bei *Parascaris equorum*. *Wilhelm Roux Arch. EntwMech. Org.* **159**, 203–266.

Morrill, J. B. (1963) Development of centrifuged *Limnaea stagnalis* eggs with giant polar bodies. *Expl Cell Res.* **31**, 490–498.

Morrill, J. B., Blair, C. A. & Larsen, W. J. (1973). Regulative development in the pulmonate gastropod, *Lymnaea palustris*, as determined by blastomere deletion experiments. *J. exp. Zool.* **183**, 47–55.

Morrill, J. B. & Perkins, F. O. (1973). Microtubules in the cortical region of the egg of *Lymnaea* during cortical segregation. *Devl Biol.* **33**, 206–212.

Morrill, J. B., Rubin, R. W. & Grandi, M. (1976). Protein synthesis and differentiation during pulmonate development. *Am. Zool.* **16**, 547–561.

Motomura, I. (1949). Artificial alteration of the embryonic axis in the centrifuged eggs of sea urchins. *Sci. Rep. Tôhoku Univ. Biol.* **18**, 117–125.

Muchmore, W. B. (1957). Differentiation of the trunk mesoderm in *Ambystoma maculatum*. II. Relation of the size of presumptive somite explants to subsequent differentiation. *J. exp. Zool.* **134**, 293–313.

Mulherkar, L. (1958). Induction by regions lateral to the streak in the chick embryo. *J. Embryol. exp. Morph.* **6**, 1–14.

Müller, K. J. (1932). Über normale Entwicklung, inverse Asymmetrie und Doppelbildungen bei *Clepsine sexoculata*. *Z. wiss. Zool.* **142**, 425–490.

Müller, M. M., Carrasco, A. E. & de Robertis, E. M. (1984). A homeo-box-containing gene expressed during oogenesis in *Xenopus*. *Cell* **39**, 157–162.

Müller-Holtkamp, F., Knipple, D. C., Seifert, E. & Jäckle, H. (1985). An early role of maternal mRNA in establishing the dorsoventral pattern in *pelle* mutant *Drosophila* embryos. *Devl Biol.* **110**, 238–246.

Mulnard, J. & Huygens, R. (1978). Ultrastructural localization of non-specific alkaline phosphatase during cleavage and blastocyst formation in the mouse. *J. Embryol. exp. Morph.* **44**, 121–131.

Naidet, C., Sémériva, M., Yamada, K. M. & Thiery, J. P. (1987). Peptides containing the cell-attachment recognition signal Arg-Gly-Asp prevent gastrulation in *Drosophila* embryos. *Nature, Lond.* **325**, 348–350.

Nakamura, O. (1942). Die Entwicklung der hinteren Körperhälfte bei Urodelen. *Annotnes zool. jap.* **21**, 169–236.

Nakamura, O., Aochi, M. & Shiomi, H. (1970). Association of blastomeres as a basic factor in differentiation of cell species in amphibian morulae. *Proc. Japan Acad.* **46**, 965–970.

Nakamura, O. & Matsuzawa, T. (1967). Differentiation capacity of the marginal zone in the morula and blastula of *Triturus pyrrhogaster*. *Embryologia* **9**, 223–237.

Nakamura, O. & Takasaki, H. (1971a). Effects of actinomycin on development of amphibian morulae and blastulae, with special reference to formation of the organizer. I. *Proc. Japan Acad.* **47**, 92–97.

Nakamura, O. & Takasaki, H. (1971b). Analysis of causal factors giving rise to the organizer. I. Removal of polar blastomeres from 32 cell embryos of *Xenopus laevis*. *Proc. Japan Acad.* **47**, 499–504.

Nakamura, O., Takasaki, H. & Ishihara, M. (1971a). Formation of the organizer from combinations of presumptive ectoderm and endoderm. I. *Proc. Japan Acad.* **47**, 313–318.

Nakamura, O., Takasaki, H. & Nagata, A. (1978). Further studies of the prospective fates of blastomeres at the 32-cell stage of *Xenopus laevis* embryos. *Med. Biol.* **56**, 355–360.

Nakamura, O., Takasaki, H., Okumoto, T. & Iida, H. (1971b). Differentiation during cleavage in *Xenopus laevis*. II. Development of inductive activity of the organizer. *Proc. Japan Acad.* **47**, 203–208.

Nakatsuji, N. (1974). Studies on the gastrulation of amphibian embryos: pseudopodia in the gastrula of *Bufo bufo japonicus* and their significance to gastrulation. *J. Embryol. exp. Morph.* **32**, 795–804.

Nakatsuji, N. (1975). Studies on the gastrulation of amphibian embryos: light and electron microscopic observation of a urodele, *Cynops pyrrhogaster*. *J. Embryol. exp. Morph.* **34**, 669–685.

Nakatsuji, N. (1976). Studies on the gastrulation of amphibian embryos: ultrastructure of the migrating cells of anurans. *Wilhelm Roux Arch. dev. Biol.* **180**, 229–240.

Nakatsuji, N. (1979). Effects of injected inhibitors of microfilament and microtubule function on the gastrulation movement in *Xenopus laevis*. *Devl Biol.* **68**, 140–150.

Nakatsuji, N. (1984). Cell locomotion and contact guidance in amphibian gastrulation. *Am. Zool.* **24**, 615–627.

Nakatsuji, N., Gould, A. C. & Johnson, K. E. (1982). Movement and guidance of migrating mesodermal cells in *Ambystoma maculatum* gastrulae. *J. Cell Sci.* **56**, 207–222.

Nakatsuji, N., Hashimoto, K. & Hayashi, M. (1985). Laminin fibrils in newt gastrulae visualized by the immunofluorescent staining. *Dev. Growth Differ.* **27**, 639–643.

Nakatsuji, N., Snow, M. H. L. & Wylie, C. C. (1986). Cinemicrographic study of the cell movement in the primitive-streak-stage mouse embryo. *J. Embryol. exp. Morph.* **96**, 99–109.

Nakauchi, M. & Takeshita, T. (1983). Ascidian one-half embryos can develop into functional adult ascidians. *J. exp. Zool.* **227**, 155–158.

Nash, M. A, Kozak, S. E., Angerer, L. M., Angerer, R. C., Schatten, H., Schatten, G. & Marzluff, W. F. (1987). Sea urchin maternal and embryonic U1 RNAs are spatially segregated in early embryos. *J. Cell Biol.* **104**, 1133–1142.

Nath, J. & Rebhun, L. I. (1973). Studies on the uptake and metabolism of adenosine 3':5'-cyclic monophosphate and $N^6$, $O^2$-dibutyryl 3':5' cyclic adenosine monophosphate in sea urchin eggs. *Expl Cell Res.* **82**, 73–78.

Needham, J. (1942). *Biochemistry and Morphogenesis.* Cambridge: The University Press.

Needham, J., Waddington, C. H. & Needham, D. M. (1934). Physico-chemical experiments on the amphibian organizer. *Proc. R. Soc. B.* **114**, 393–422.

Neff, A. W., Malacinski, G. M., Wakahara, M. & Jurand, A. (1983). Pattern formation in amphibian embryos prevented from undergoing the classical 'rotation response' to egg activation. *Devl Biol.* **97**, 103–112.

Neff, A. W., Wakahara, M., Jurand, A. & Malacinski, G. M. (1984). Experimental analyses of cytoplasmic rearrangements which follow fertilization and accompany symmetrization of inverted *Xenopus* eggs. *J. Embryol. exp. Morph.* **80**, 197–224.

Nemer, M. (1986). An altered series of ectodermal gene expressions accompanying the reversible suspension of differentiation in the zinc-animalized sea urchin embryo. *Devl Biol.* **114**, 214–224.

Nemer, M. & Surrey, S. (1976). mRNAs containing and lacking poly(A) function as separate and distinct classes during embryonic development. *Prog. nucl. Acid Res. mol. Biol.* **19**, 119–122.

Nemer, M., Travaglini, E. C., Rondinelli, E. & d'Alonzo, J. (1984). Developmental regulation, induction and embryonic tissue specificity of sea urchin metallothionein gene expression. *Devl Biol.* **102**, 471–482.

Nemer, M., Wilkinson, D. G. & Travaglini, E. C. (1985). Primary differentiation and ectoderm-specific gene expression in the animalized sea urchin embryo. *Devl Biol.* **109**, 418–427.

Neufang, O., Born, J., Tiedemann, H. & Tiedemann, H. (1978). A proteoglycan with affinity for the vegetalizing factor: characterization by density gradient centrifugation. *Med. Biol.* **56**, 361–365.

Neuman, T., Laasberg, T. & Kärner, J. (1983). Ultrastructural localization of adenylate cyclase and cAMP phosphodiesterase in the gastrulating chick embryo. *Wilhelm Roux Arch. dev. Biol.* **192**, 42–44.

New, D. A. T. (1959). The adhesive properties and expansion of the chick blastoderm. *J. Embryol. exp. Morph.* **7**, 146–164.

Newport, G. (1854). Researches on the impregnation of the ovum in the Amphibia; and on the early stages of development of the embryo. *Phil. Trans. R. Soc.* **144**, 229–244.

Newport, J. & Kirschner, M. (1982). A major developmental transition in early *Xenopus* embryos: I. Characterization and timing of cellular changes at the midblastula stage. *Cell* **30**, 675–686.

Newport, J. W. & Kirschner, M. W. (1984). Regulation of the cell cycle during early *Xenopus* development. *Cell* **37**, 731–742.

Newrock, K. M., Alfageme, C. R., Nardi, R. V. & Cohen, L. H. (1978).

Histone changes during chromatin remodelling in embryogenesis. *Cold Spring Harb. Symp. quant. Biol.* **42**, 421–431.

Newrock, K. M. & Raff, R. A. (1975). Polar lobe specific regulation of translation in embryos of *Ilyanassa obsoleta. Devl Biol.* **42**, 242–261.

Neyfakh, A. A. (1959). X-ray inactivation of nuclei as a method for studying their function in the early development of fishes. *J. Embryol. exp. Morph.* **7**, 173–192.

Neyfakh, A. A. (1964). Radiation investigation of nucleo-cytoplasmic interrelations in morphogenesis and biochemical differentiation. *Nature, Lond.* **201**, 880–884.

Neyfakh, A. A. (1971). Steps of realization of genetic information in early development. *Curr. Top. dev. Biol.* **6**, 45–77.

Neyfakh, A. A., Kostomarova, A. A. & Burakova, T. A. (1972). Transfer of RNA from nucleus to cytoplasm in the early development of fish. An autoradiographical study. *Expl Cell Res.* **72**, 223–232.

Nicander, L., Afzelius, B. A. & Sjödén. I. (1968). Fine structure and early fertilization changes of the animal pole in eggs of the river lamprey, *Lampetra fluviatilis. J. Embryol. exp. Morph.* **19**, 319–326.

Nicholas, J. S. & Hall, B. V. (1942). Experiments on developing rats. II. The development of isolated blastomeres and fused eggs. *J. exp. Zool.* **90**, 441–459.

Nicholas, J. S. & Oppenheimer, J. M. (1942). Regulation and reconstitution in *Fundulus. J. exp. Zool.* **90**, 127–157.

Nicolet, G. (1970). Is the presumptive notochord responsible for somite genesis in the chick? *J. Embryol. exp. Morph.* **24**, 467–478.

Nicolet, G. (1971). Avian gastrulation. *Adv. Morphogen.* **9**, 231–262.

Nicosia, S. V., Wolf, D. P. & Inoue, M. (1977). Cortical granule distribution and cell surface characteristics in mouse eggs. *Devl Biol.* **57**, 56–74.

Nieuwkoop, P. D. (1953). The influence of the Li ion on the development of the egg of *Ascidia malaca. Pubbl. Staz. zool. Napoli* **24**, 101–141.

Nieuwkoop, P. D. (1958). Neural competence of the gastrula ectoderm in *Amblystoma mexicanum.* An attempt at quantitative analysis of morphogenesis. *Acta Embryol. Morph. exp.* **2**, 13–53.

Nieuwkoop, P. D. (1969a). The formation of the mesoderm in urodelean amphibians. I. Induction by the endoderm. *Wilhelm Roux Arch. EntwMech. Org.* **162**, 341–373.

Nieuwkoop, P. D. (1969b). The formation of the mesoderm in urodelean amphibians. II. The origin of the dorso-ventral polarity of the mesoderm. *Wilhelm Roux Arch. EntwMech. Org.* **163**, 298–315.

Nieuwkoop, P. D. (1970). The formation of the mesoderm in urodelean amphibians. III. The vegetalizing action of the Li ion. *Wilhelm Roux Arch. EntwMech. Org.* **166**, 105–123.

Nieuwkoop, P. D. (1973). The 'organization center' of the amphibian embryo: its origin, spatial organization, and morphogenetic action. *Adv. Morphogen.* **10**, 1–39.

Nieuwkoop, P. D. (1977). Origin and establishment of embryonic polar axes in amphibian development. *Curr. Top. dev. Biol.* **11**, 115–132.

Nieuwkoop, P. D. & Florschütz, P. A. (1950). Quelques caractères spéciaux de la gastrulation et de la neurulation de l'oeuf de *Xenopus laevis* Daud. et de quelques autres anoures. 1ère partie. Étude descriptive. *Archs Biol., Liège* **61**, 113–150.

Nieuwkoop, P. D. & Sutasurya, L. A. (1979). *Primordial Germ Cells in the Chordates.* Cambridge: The University Press.

Nieuwkoop, P. D. & Ubbels, G. A. (1972). The formation of the mesoderm in urodelean amphibians. IV. Qualitative evidence for the purely 'ectodermal' origin of the entire mesoderm and of the pharyngeal endoderm. *Wilhelm Roux Arch. EntwMech. Org.* **169**, 185–199.

Nieuwkoop, P. D. & van der Grinten, S. J. (1961). The relationship between the 'activating' and the 'transforming' principle in the neural induction process in amphibians. *Embryologia* **6**, 51–66.

Nieuwkoop, P. D. & Weijer, C. J. (1978). Neural induction, a two-way process. *Med. Biol.* **56**, 366–371.

Nigon, V., Guerrier, P. & Monin, H. (1960). L'architecture polaire de l'oeuf et les mouvements des constituants cellulaires au cours des premieres étapes du développement chez quelques nématodes. *Bull. biol. Fr. Belg.* **94**, 131–202.

Nijhawan, P. & Marzluff, W. F. (1979). Metabolism of low molecular weight ribonucleic acids in early sea urchin embryos. *Biochemistry, N.Y.* **18**, 1353–1360.

Niki, Y. & Okada, M. (1981). Isolation and characterization of *grandchildless*-like mutants in *Drosophila melanogaster. Wilhelm Roux Arch. dev. Biol.* **190**, 1–10.

Nishida, H. (1987). Cell lineage analysis in ascidian embryos by intracellular injection of a tracer enzyme. III. Up to the tissue restricted stage. *Devl Biol.* **121**, 526–541.

Nishida, H. & Satoh, N. (1983). Cell lineage analysis in ascidian embryos by intracellular injection of a tracer enzyme. I. Up to the eight-cell stage. *Devl Biol.* **99**, 382–394.

Nishida, H. & Satoh, N. (1985). Cell lineage analysis in ascidian embryos by intracellular injection of a tracer enzyme. II. The 16- and 32-cell stages. *Devl Biol.* **110**, 440–454.

Nishikata, T., Mita-Miyazawa, I., Deno, T. & Satoh, N. (1987*a*). Muscle cell differentiation in ascidian embryos analysed with a tissue-specific monoclonal antibody. *Development* **99**, 163–171.

Nishikata, T., Mita-Miyazawa, I., Deno, T. & Satoh, N. (1987*b*). Monoclonal antibodies against components of the myoplasm of eggs of the ascidian *Ciona intestinalis* partially block the development of muscle-specific acetylcholinesterase. *Development* **100**, 577–586.

Niu, M. C. (1956). New approaches to the problem of embryonic induction. In *Cellular Mechanisms in Differentiation and Growth*, ed. D. Rudnick, pp. 155–171. Princeton: The University Press.

Niu, M. C. & Twitty, V. C. (1953). The differentiation of gastrula ectoderm in medium conditioned by axial mesoderm. *Proc. natn. Acad. Sci. U.S.A.* **39**, 985–989.

Noronha, J. M., Sheys, G. H. & Buchanan, J. M. (1972). Induction of a reductive pathway for deoxyribonucleotide synthesis during early embryogenesis of the sea urchin. *Proc. natn. Acad. Sci. U.S.A.* **69**, 2006–2010.

Novikoff, A. B. (1938). Embryonic determination in the annelid, *Sabellaria vulgaris.* II. Transplantation of polar lobes and blastomeres as a test of their inducing capacities. *Biol. Bull. mar. biol. Lab., Woods Hole* **74**, 211–234.

Novikoff, A. B. (1940). Morphogenetic substances or organizers in annelid development. *J. exp. Zool.* **85**, 127–155.

Nüsslein-Volhard, C. (1977). Genetic analysis of pattern-formation in the embryo of *Drosophila melanogaster.* Characterization of the maternal-effect mutant *bicaudal. Wilhelm Roux Arch. dev. Biol.* **183**, 249–268.

Nüsslein-Volhard, C. (1979). Maternal effect mutations that alter the spatial coordinates of the embryo of *Drosophila melanogaster*. In *Determinants of Spatial Organization*, ed. S. Subtelny & I. R. Konigsberg, pp. 185–211. New York: Academic Press.

Nüsslein-Volhard, C., Lohs-Schardin, M., Sander, K. & Cremer, C. (1980). A dorso-ventral shift of embryonic primordia in a new maternal-effect mutant of *Drosophila*. *Nature, Lond.* **283**, 474–476.

Nüsslein-Volhard, C. & Wieschaus, E. (1980). Mutations affecting segment number and polarity in *Drosophila*. *Nature, Lond.* **287**, 795–801.

Nyholm, M., Saxén, L., Toivonen, S. & Vainio, T. (1962). Electron microscopy of transfilter neural induction. *Expl Cell Res.* **28**, 209–212.

O'Dell, D. S., Tencer, R., Monroy, A. & Brachet, J. (1974). The pattern of concanavalin A-binding sites during the early development of *Xenopus laevis*. *Cell Differ.* **3**, 193–198.

O'Dor, R. K. & Wells, M. J. (1973). Yolk protein synthesis in the ovary of *Octopus vulgaris* and its control by the optic gland gonadotropin. *J. exp. Biol.* **59**, 665–674.

Ogi, K. I. (1958*a*). The effect of sodium thiocyanate on isolates of the presumptive ectoderm and medio-ventral marginal zone of *Triturus* gastrulae. *J. Embryol. exp. Morph.* **6**, 412–417.

Ogi, K. I. (1958*b*). Inductive ability of the isolated ventral mesoderm of *Triturus* gastrulae dorsalized by various agents. *Embryologia* **4**, 161–173.

Ogi, K. I. (1961). Vegetalization of the presumptive ectoderm of the *Triturus*-gastrula by exposure to lithium chloride solution. *Embryologia* **5**, 384–396.

Ogi, K. I. (1967). Determination in the development of the amphibian embryo. *Sci. Rep. Tôhoku Univ., Ser. IV* **33**, 239–247.

Ohara, A. (1980). Brain induction in variously aged presumptive ectoderms by head organizer in *Cynops pyrrhogaster*. *Dev. Growth Differ.* **22**, 805–812.

Ohara, A. (1981). Inducing time for neural tissues in gastrula ectoderm of *Cynops pyrrhogaster*. *Dev. Growth Differ.* **23**, 51–58.

Ohara, A. & Hama, T. (1979*a*). The effect of aging on the neural competence of the presumptive ectoderm and the effect of aged ectoderm on the differentiation of the trunk organizer in *Cynops pyrrhogaster*. *Wilhelm Roux Arch. dev. Biol.* **187**, 13–23.

Ohara, A. & Hama, T. (1979*b*). Effect of aging on neural competence of presumptive ectoderm and effect of aged ectoderm on differentiation of the trunk organizer in *Cynops pyrrhogaster*. II. Role of extreme posterior of the archenteron roof in the slit-blastopore stage in normal development. *Dev. Growth Differ.* **21**, 509–517.

Okada, M., Kleinman, I. A. & Schneiderman, H. A. (1974*a*). Restoration of fertility in sterilized *Drosophila* eggs by transplantation of polar cytoplasm. *Devl Biol.* **37**, 43–54.

Okada, M., Kleinman, I. A. & Schneiderman, H. A. (1974*b*). Chimeric *Drosophila* adults produced by transplantation of nuclei into specific regions of fertilized eggs. *Devl Biol.* **39**, 286–294.

Okada, M. & Togashi, S. (1985). Isolation of a factor inducing pole cell formation from *Drosophila* embryos. *Int. J. Invertebr. Reprod. Dev.* **8**, 207–217.

Okada, T. S. (1953). Role of the mesoderm in the differentiation of endodermal organs. *Mem. Coll. Sci. Kyoto Univ. B* **20**, 157–162.

Okada, T. S. (1954). Experimental studies on the differentiation of the endodermal organs in Amphibia. I. Significance of the mesenchymatous tissue to the differentiation of the presumptive endoderm. *Mem. Coll. Sci. Kyoto Univ. B* **21**, 1–6.

Okada, T. S. (1955a). Experimental studies on the differentiation of the endodermal organs in Amphibia. III. The relation between the differentiation of pharynx and head-mesenchyme. *Mem. Coll. Sci. Kyoto Univ. B* **22**, 17–22.

Okada, T. S. (1955b). Experimental studies on the differentiation of the endodermal organs in Amphibia. IV. The differentiation of the intestine from the fore-gut. *Annotnes zool. jap.* **28**, 210–214.

Okada, T. S. (1957). The pluripotency of the pharyngeal primordium in urodelan neurulae. *J. Embryol. exp. Morph.* **5**, 438–448.

Okada, T. S. (1960). Epithelio-mesenchymal relationships in the regional differentiation of the digestive tract in the amphibian embryo. *Wilhelm Roux Arch. EntwMech. Org.* **152**, 1–21.

Okada, Y. K. (1927). Versuche über die Wirkung der Dotterwegnahme am meroblastischen Ei (Ei von *Loligo bleekeri* Keperstein). *Zool. Anz.* **73**, 280–284.

Okada, Y. K. & Hama, T. (1943). Examination of regional differences in the inductive activity of the organizer by means of transplantation into ectodermal vesicles. *Proc. imp. Acad. Japan* **19**, 48–53.

Okada, Y. K. & Hama, T. (1945a). Regional differences in the inductive capacity of the dorsal roof of the archenteron of the urodele, *Triturus pyrrhogaster*. *Proc. Japan Acad.* **21**, 240–247.

Okada, Y. K. & Hama, T. (1945b). Prospective fate and inductive capacity of the dorsal lip of the blastopore of the *Triturus* gastrula. *Proc. Japan Acad.* **21**, 342–348.

Okada, Y. K. & Takaya, H. (1942). Experimental investigation of regional differences in the inductive capacity of the organizer. *Proc. imp. Acad. Japan* **18**, 505–513.

Okazaki, K. (1975). Spicule formation by isolated micromeres of the sea urchin embryo. *Am. Zool.* **15**, 567–581.

Okazaki, K., Fukushi, T. & Dan, K. (1962). Cyto-embryological studies of sea urchins. IV. Correlation between the shape of the ectodermal cells and the arrangement of the primary mesenchyme cells in sea urchin larvae. *Acta Embryol. Morph. exp.* **5**, 17–31.

Okazaki, R. (1955). A critical study on the inducing capacity of the devitalized organiser. *Expl Cell Res.* **9**, 579–582.

Oliver, B. C. & Shen, S. S. (1986). Cytoplasmic control of chromosome diminution in *Ascaris suum*. *J. exp. Zool.* **239**, 41–55.

Olszańska, B. & Kludkiewicz, B. (1983). The effect of transcription inhibitors on early development of the avian embryo. *Cell Differ.* **12**, 115–120.

Olszańska, B., Szolajska, E. & Lassota, Z. (1984). Effect of spatial position of uterine quail blastoderms cultured *in vitro* on bilateral symmetry formation. *Wilhelm Roux Arch. dev. Biol.* **193**, 108–110.

O'Melia, A. F. (1972). Changes in esterase and cholinesterase isozymes in normally developing, animalized and radialized embryos of *Arbacia punctulata*. *Expl Cell Res.* **73**, 469–474.

O'Melia, A. F. (1979). The synthesis of 5S RNA and its regulation during early sea urchin development. *Dev. Growth Differ.* **21**, 99–108.

O'Melia, A. F. (1984). The effects of chemical animalizing and vegetalizing agents and of cell dissociation on the synthesis of 5S RNA and transfer RNA in cleaving sea urchin embryos. *Dev. Growth Differ.* **26**, 73–80.

Ooi, V. E. C., Sanders, E. J. & Bellairs, R. (1986). The contribution of the primitive streak to the somites in the avian embryo. *J. Embryol. exp. Morph.* **92,** 193–206.

Oppenheimer, J. M. (1934). Experimental studies on the developing perch (*Perca flavescens* Mitchill). *Proc. Soc. exp. Biol. Med.* **31,** 1123–1124.

Oppenheimer, J. M. (1936a). The development of isolated blastoderms of *Fundulus heteroclitus*. *J. exp. Zool.* **72,** 247–269.

Oppenheimer, J. M. (1936b). Transplantation experiments on developing teleosts (*Fundulus* and *Perca*). *J. exp. Zool.* **72,** 409–437.

Oppenheimer, J. M. (1947). Organization of the teleost blastoderm. *Q. Rev. Biol.* **22,** 105–118.

Ortolani, G. (1955). I movimenti corticali dell'uovo di ascidie alla fecondazione. *Riv. Biol.* **47,** 169–180.

Ortolani, G. (1958). Cleavage and development of egg fragments in ascidians. *Acta Embryol. Morph. exp.* **1,** 247–272.

Ortolani, G. (1959). Ricerche sulla induzione del sistema nervoso nelle larve delle Ascidie. *Boll. Zool.* **26,** 341–348.

Ortolani, G. (1964). Origin dell'organo apicale e di derivati mesodermici nello sviluppo embrionale di ctenofori. *Acta Embryol. Morph. exp.* **7,** 191–200.

Ortolani, G. (1969). The action of sodium thiocyanate (NaSCN) on the embryonic development of the ascidians. *Acta Embryol. exp.* 27–34.

Ortolani, G., O'Dell, D. S. & Monroy, A. (1977). Localized binding of *Dolichos* lectin to the early *Ascidia* embryo. *Expl Cell Res.* **106,** 402–404.

Ortolani, G., Patricolo, E. & Mansueto, C. (1979). Trypsin-induced cell surface changes in ascidian embryonic cells. *Expl Cell Res.* **122,** 137–147.

Orts-Llorca, F. (1963). Influence of the endoderm on heart differentiation during the early stages of development of the chick embryo. *Wilhelm Roux Arch. EntwMech. Org.* **154,** 533–551.

Osborn, J. C. (1977). The influence of lithium on the morphology, ultrastructure and biochemistry of the early development of the South African clawed toad, *Xenopus laevis*. Ph.D. thesis, University of Liverpool.

Osborn, J. C., Wall, R. & Stanisstreet, M. (1979). Quantitative aspects of RNA synthesis in normal and lithium-treated embryos of *Xenopus laevis*. *Wilhelm Roux Arch. dev. Biol.* **187,** 269–282.

Oshima, R. G., Howe, W. E., Klier, F. G., Adamson, E. D. & Shevinsky, L. H. (1983). Intermediate filament protein synthesis in preimplantation murine embryos. *Devl Biol.* **99,** 447–455.

Ostroumova, T. V. & Belousov, L. V. (1971). Determination of morphological polarity in embryogenesis of hydroids. *Zh. obshch. Biol.* **32,** 323–331.

Ostroumova, T. V., Belousov, L. V. & Mikhailova, E. G. (1977). Dynamics of regional metabolic differences in the early development of the sea urchin. *Ontogenez* **8,** 323–334 (in translation 277–287).

Overall, R. & Jaffe, L. F. (1985). Patterns of ionic current through *Drosophila* follicles and eggs. *Devl Biol.* **108,** 102–119.

Paglia, L. M., Berry, S. J. & Kastern, W. H. (1976). Messenger RNA synthesis, transport, and storage in silkmoth ovarian follicles. *Devl Biol.* **51,** 173–181.

Pai, S. (1928). Die Phasen des Lebenscyclus der *Anguillula aceti* Ehrbg. und ihre experimentell-morphologische Beeinflussung. *Z. wiss. Zool.* **131,** 293–344.

Painter, T. S. & Taylor, A. N. (1942). Nucleic acid storage in the toad's egg. *Proc. natn. Acad. Sci. U.S.A.* **28,** 311–317.

Paleček, J., Habrová, V., Nedvídek, J. & Romanovsky, A. (1985). Dynamics of

388     References

tubulin structures in *Xenopus laevis* oogenesis. *J. Embryol. exp. Morph.* **87**, 75–86.

Paleček, J., Ubbels, G. A. & Rzehak, K. (1978). Changes of the external and internal pigment pattern upon fertilization in the egg of *Xenopus laevis*. *J. Embryol. exp. Morph.* **45**, 203–214.

Parisi, E., Filosa, S., de Petrocellis, B. & Monroy, A. (1978). The pattern of cell division in the early development of the sea urchin, *Paracentrotus lividus*. *Devl Biol.* **65**, 38–49.

Pasteels, J. (1932). Expériences de piqures localisées sur l'oeuf non-segmenté des Anoures. *C. r. Ass. Anat.* **27**, 440–449.

Pasteels, J. (1938). Recherches sur les facteurs initiaux de la morphogenèse chez les amphibiens anoures. I. Résultats de l'expérience de Schultze et leur interprétation. *Archs Biol., Liège* **49**, 629–667.

Pasteels, J. (1953). Les effets de la centrifugation sur la blastula et la jeune gastrula des amphibiens. III et IV. *J. Embryol. exp. Morph.* **2**, 122–148. aires aux dépens de l'ectoblaste. *J. Embryol. exp. Morph.* **1**, 5–24.

Pasteels, J. (1954). Les effets de la centrifugation sur la blastula et la jeune gastrula des amphibiens. III et-IV. *J. Embryol. exp. Morph.* **2**, 122–148.

Pasteels, J. J. (1937). Etudes sur la gastrulation des Vertebrés méroblastiques. II. Reptiles. *Archs Biol., Liège* **48**, 105–184.

Pasteels, J. J. (1957*a*). La formation de l'endophylle et de l'endoblaste vitellin chez les reptiles, cheloniens et lacertiliens. *Acta anat.* **30**, 601–612.

Pasteels, J. J. (1957*b*). Une table analytique du developpement des reptiles. I. Stades de gastrulation chez les Chéloniens et les Lacertiliens. *Annls Soc. r. zool. Belg.* **87**, 217–241.

Pasteels, J. J. (1964). The morphogenetic role of the cortex of the amphibian egg. *Adv. Morphogen.* **3**, 363–388.

Pasteels, J. J. & de Harven, E. (1963). Etude au microscope électronique du cytoplasme de l'oeuf vierge et fécondé de *Barnea candida* (Mollusque bivalve). *Archs Biol., Liège* **74**, 415–437.

Pasternak, L. & McCallion, D. J. (1962). Heterogeneous inductions in the chick embryo. *Can. J. Zool.* **40**, 585–591.

Paterson, M. C. (1957). Animal–vegetal balance in amphibian development. *J. exp. Zool.* **134**, 183–205.

Paul, M., Goldsmith, M. R., Hunsley, J. R. & Kafatos, F. C. (1972). Specific protein synthesis in cellular differentiation. Production of eggshell proteins by silkmoth follicular cells. *J. Cell Biol.* **55**, 653–680.

Peaucellier, G., Guerrier, P. & Bergerard, J. (1974). Effects of cytochalasin B on meiosis and development of fertilized and activated eggs of *Sabellaria alveolata* L. (Polychaete Annelid). *J. Embryol. exp. Morph.* **31**, 61–74.

Pedersen, R. A. & Spindle, A. I. (1980). Role of the blastocoele microenvironment in early mouse embryo differentiation. *Nature, Lond.* **284**, 550–552.

Pedersen, R. A., Wu, K. & Bałakier, H. (1986). Origin of the inner cell mass in mouse embryos: cell lineage analysis by microinjection. *Devl Biol.* **117**, 581–595.

Pedrazzi, G. (1957). Sull'azione dell'urea nello sviluppo embrionale dei ricci di mare. *Rc. Ist. lomb. Sci. Lett.* **91**, 672–679.

Pehrson, J. R. & Cohen, L. H. (1986). The fate of the small micromeres in sea urchin development. *Devl Biol.* **113**, 522–526.

Penner, P. L. & Brick, I. (1984). Acetylcholinesterase and polyingression in the epiblast of the primitive streak chick embryo. *Wilhelm Roux Arch. dev. Biol.* **193**, 234–241.

Penners, A. (1922*a*). Die Furchung von *Tubifex rivulorum* Lam. *Zool. Jb. Abt. Anat. Ont.* **43**, 323–368.

Penners, A. (1922b). Über Doppelbildungen bei *Tubifex rivulorum* Lam. mit Demonstration von Präparaten. *Verh. dt. zool. Ges.* **27**, 46–49.

Penners, A. (1926). Experimentelle Untersuchungen zum Determinationsproblem am Keim von *Tubifex rivulorum* Lam. II. Die Entwicklung teilweise abgetöteter Keime. *Z. wiss. Zool.* **127**, 1–140.

Penners, A. & Schleip, W. (1928a). Die Entwicklung der Schultze'schen Doppelbildungen aus dem Ei von *Rana fusca*. Teil I–VI. *Z. wiss. Zool.* **130**, 307–454.

Penners, A. & Schleip, W. (1928b). Die Entwicklung der Schultze'schen Doppelbildungen aus dem Ei von *Rana fusca*. Teil V und VI. *Z. wiss. Zool.* **131**, 1–156.

Percy, J., Kuhn, K. L. & Kalthoff, K. (1986). Scanning electron microscopic analysis of spontaneous and UV-induced abnormal segment patterns in *Chironomus samoensis* (Diptera, Chironomidae). *Wilhelm Roux Arch. dev. Biol.* **195**, 92–102.

Perkowska, E., Macgregor, H. C. & Birnstiel, M. L. (1968). Gene amplification in the oocyte nucleus of mutant and wild-type *Xenopus laevis. Nature, Lond.* **217**, 649–650.

Perlman, S. & Rosbash, M. (1978). Analysis of *Xenopus laevis* ovary and somatic cell polyadenylated RNA by molecular hybridization. *Devl Biol.* **63**, 197–212.

Perrimon, N., Engstrom, L. & Mahowald, A. P. (1984). The effects of zygotic lethal mutations on female germ-line functions in *Drosophila. Devl Biol.* **105**, 404–414.

Perrimon, N., Engstrom, L. & Mahowald, A. P. (1985). A pupal lethal mutation with a paternally influenced maternal effect on embryonic development in *Drosophila melanogaster. Devl Biol.* **110**, 480–491.

Perry, H. E. & Melton, D. A. (1983). A rapid increase in acetylcholinesterase mRNA during ascidian embryogenesis as demonstrated by microinjection into *Xenopus laevis* oocytes. *Cell Differ.* **13**, 233–238.

Pestell, R. Q. W. (1975). Microtubule protein synthesis during oogenesis and early embryogenesis in *Xenopus laevis. Biochem. J.* **145**, 527–534.

Peters, N. K. & Kleinsmith, L. J. (1984). Kinetic model for the study of gene expression in the developing sea urchin. *Devl Biol.* **102**, 433–437.

Peyriéras, N., Hyafil, F., Louvard, D., Ploegh, H. L. & Jacob, F. (1983). Uvomorulin: a nonintegral membrane protein of early mouse embryo. *Proc. natn. Acad. Sci. U.S.A.* **80**, 6274–6277.

Pflüger, E. (1883). Ueber den Einfluss der Schwerkraft auf die Theilung der Zellen und auf die Entwicklung des Embryo. *Arch. ges. Physiol.* **32**, 1–79.

Phillips, C. R. (1982). The regional distribution of poly(A) and total RNA concentrations during early *Xenopus* development. *J. exp. Zool.* **223**, 265–275.

Phillips, C. R. (1985). Spatial changes in poly(A) concentrations during early embryogenesis in *Xenopus laevis*: analysis by *in situ* hybridization. *Devl Biol.* **109**, 299–310.

Picard, J. J. (1975). *Xenopus laevis* cement gland as an experimental model for embryonic differentiation. I. *In vitro* stimulation of differentiation by ammonium chloride. *J. Embryol. exp. Morph.* **33**, 957–967.

Pierandrei-Amaldi, P., Campioni, N., Beccari, E., Bozzoni, I. & Amaldi, F. (1982). Expression of ribosomal-protein genes in *Xenopus laevis* development. *Cell* **30**, 163–171.

Pittman, D. & Ernst, S. G. (1984). Developmental time, cell lineage, and environment regulate the newly synthesized proteins in sea urchin embryos. *Devl Biol.* **106**, 236–242.

Poccia, D., Salik, J. & Krystal, G. (1981). Transitions in histone variants of the male pronucleus following fertilization and evidence for a maternal store of cleavage-stage histones in the sea urchin egg. *Devl Biol.* **82**, 287–296.

Poelmann, R. E. (1980). Differential mitosis and degeneration patterns in relation to the alterations in the shape of the embryonic ectoderm of early postimplantation mouse embryos. *J. Embryol. exp. Morph.* **55**, 33–51.

Pohl, V. & Brachet, J. (1962). Etude du rôle des groupes sulfhydriles dans la morphogénèse de l'embryon de poulet. *Devl Biol.* **4**, 549–568.

Poole, S. J., Kauvar, L. M., Drees, B. & Kornberg, T. (1985). The *engrailed* locus of *Drosophila*: structural analysis of an embryonic transcript. *Cell* **40**, 37–43.

Posakony, J. W., Scheller, R. H., Anderson, D. M., Britten, R. J. & Davidson, E. H. (1981). Repetitive sequences of the sea urchin genome. III. Nucleotide sequences of cloned repeat elements. *J. molec. Biol.* **149**, 41–67.

Potter, D. D., Furshpan, E. J. & Lennox, E. S. (1966). Connections between cells of the developing squid as revealed by electrophysiological methods. *Proc. natn. Acad. Sci. U.S.A.* **55**, 328–336.

Poulson, D. F. (1950). Histogenesis, organogenesis, and differentiation in the embryo of *Drosophila melanogaster* Meigen. In *Biology of Drosophila*, ed. M. Demerec, pp. 168–274. New York: John Wiley & Sons.

Poupko, J. M., Kostellow, A. B. & Morrill, G. A. (1977). Histone acetylation associated with gastrulation in *Rana pipiens*. *Differentiation* **8**, 167–174.

Pratt, H. P. M. (1982). Preimplantation mouse embryos synthesize membrane sterols. *Devl Biol.* **89**, 101–110.

Pratt, H. P. M., Chakraborty, J. & Surani, M. A. H. (1981). Molecular and morphological differentiation of the mouse blastocyst after manipulations of compaction with cytochalasin D. *Cell* **26**, 279–292.

Pratt, H. P. M., Keith, J. & Chakraborty, J. (1980). Membrane sterols and the development of the preimplantation mouse embryo. *J. Embryol. exp. Morph.* **60**, 303–319.

Preiss, A., Rosenberg, U. B., Kienlin, A., Seifert, E. & Jäckle, H. (1985). Molecular genetics of *Krüppel*, a gene required for segmentation of the *Drosophila* embryo. *Nature, Lond.* **313**, 27–32.

Priess, J. R. & Thomson, J. N. (1987). Cellular interactions in early *C. elegans* embryos. *Cell* **48**, 241–250.

Prothero, J. W. & Tamarin, A. (1977). The blastomere pattern in echinoderms: cleavages one to four. *J. Embryol. exp. Morph.* **40**, 23–34.

Pucci-Minafra, I., Minafra, S. & Collier, J. R. (1969). Distribution of ribosomes in the egg of *Ilyanassa obsoleta*. *Expl Cell Res.* **57**, 167–178.

Quatrano, R. S., Brawley, S. H. & Hogsett, W. E. (1979). The control of the polar deposition of a sulfated polysaccharide in *Fucus* zygotes. In *Determinants of Spatial Organization*, ed. S. Subtelny & I. R. Konigsberg, p. 77–96. New York: Academic Press.

Raff, R. A. (1972). Polar lobe formation by embryos of *Ilyanassa obsoleta*. Effects of inhibitors of microtubule and microfilament function. *Expl Cell Res.* **71**, 455–459.

Raff, R. A., Colot, H. V., Selvig, S. E. & Gross, P. R. (1972). Oogenetic origin of messenger RNA for embryonic synthesis of microtubule proteins. *Nature, Lond.* **235**, 211–214.

Raff, R. A., Greenhouse, G., Gross, K. W. & Gross, P. R. (1971). Synthesis and storage of microtubule proteins by sea urchin embryos. *J. Cell Biol.* **50**, 516–527.

Ranzi, S. (1928). Suscettibilita differenziale nello sviluppo dei Cefalopodi. *Pubbl. Staz. zool. Napoli* **9**, 81–159.

Ranzi, S. (1944). Effetto di KSCN sullo sviluppo di embrioni di cefalopodi. *Boll. Soc. ital. Biol. sper.* **19**, 68–70.

Ranzi, S. (1957). Early determination in development under normal and experimental conditions. In *The Beginnings of Embryonic Development*, ed. A. Tyler, R. C. von Borstel & C. B. Metz, pp. 291–318. Washington: American Association for the Advancement of Science.

Ranzi, S. (1962). The proteins in embryonic and larval development. *Adv. Morphogen.* **2**, 211–257.

Ranzi, S. & Ferreri, G. (1944). Effetto di LiCl sullo sviluppo embrionale delle ascidie. *Boll. Soc. ital. Biol. sper.* **19**, 287–288.

Ranzi, S. & Ferreri, G. (1945). Evocazione nello sviluppo embrionale delle ascidie. *Boll. Soc. ital. Biol. sper.* **20**, 153–156.

Ranzi, S. & Janeselli, L. (1941). Effetto di LiCl sullo sviluppo dei ciclostomi. *Rc. Ist. lomb. Sci. Lett., Ser. 3,* **74**, 403–436.

Rao, B. R. (1968). The appearance and extension of neural differentiation tendencies in the neurectoderm of the early chick embryo. *Wilhelm Roux Arch. EntwMech. Org.* **160**, 187–236.

Rappaport, R. (1971). Cytokinesis in animal cells. *Int. Rev. Cytol.* **31**, 169–213.

Rasilo, M.-L. & Leikola, A. (1976). Neural induction by previously induced epiblast in avian embryo *in vitro. Differentiation* **5**, 1–7.

Rattenbury, J. C. & Berg, W. E. (1954). Embryonic segregation during early development of *Mytilus edulis. J. Morph.* **95**, 393–414.

Rau, K.-G. & Kalthoff, K. (1980). Complete reversal of antero-posterior polarity in a centrifuged insect embryo. *Nature, Lond.* **287**, 635–637.

Raveh, D., Friedländer, M. & Eyal-Giladi, H. (1976). Nucleolar ontogenesis in the uterine chick germ correlated with morphogenetic events. *Expl Cell Res.* **100**, 195–203.

Raven, C. P. (1938). Experimentelle Untersuchungen über die 'bipolare Differenzierung' des Polychaeten- und Molluskeneies. *Acta neerl. Morph.* **1**, 337–357.

Raven, C. P. (1948). The chemical and experimental embryology of *Limnaea. Biol. Rev.* **23**, 333–369.

Raven, C. P. (1952). Morphogenesis in *Limnaea stagnalis* and its disturbance by lithium. *J. exp. Zool.* **121**, 1–78.

Raven, C. P. (1956). Effects of monovalent cations on the eggs of *Limnaea. Pubbl. Staz. zool. Napoli* **28**, 136–168.

Raven, C. P. (1961). *Oogenesis: The Storage of Developmental Information.* Oxford: Pergamon Press.

Raven, C. P. (1963*a*). The nature and origin of the cortical morphogenetic field in *Limnaea. Devl Biol.* **7**, 130–143.

Raven, C. P. (1963*b*). Differentiation in mollusc eggs. *Symp. Soc. exp. Biol.* **17**, 274–284.

Raven, C. P. (1966). *Morphogenesis: The Analysis of Molluscan Development*, 2nd edn. Oxford: Pergamon Press.

Raven, C. P. (1967). The distribution of special cytoplasmic differentiations of the egg during early cleavage in *Limnaea stagnalis. Devl Biol.* **16**, 407–437.

Raven, C. P. (1970). The cortical and subcortical cytoplasm of the *Lymnaea* egg. *Int. Rev. Cytol.* **28**, 1–44.

Raven, C. P. (1974). Further observations on the distribution of cytoplasmic substances among the cleavage cells in *Lymnaea stagnalis. J. Embryol. exp. Morph.* **31**, 37–59.

Raven, C. P. (1976). Morphogenetic analysis of spiralian development. *Am. Zool.* **16**, 395–403.

Raven, C. P. & van der Wal, U. P. (1964). Analysis of the formation of the animal pole plasm in the eggs of *Limnaea stagnalis*. *J. Embryol. exp. Morph.* **12**, 123–139.

Raynaud, A. (1961). Quelques phases du développement des oeufs chez l'orvet (*Anguis fragilis* L.). *Bull. biol. Fr. Belg.* **95**, 365–387.

Rebagliati, M. R., Weeks, D. L., Harvey, R. P. & Melton, D. A. (1985). Identification and cloning of localized maternal RNAs from *Xenopus* eggs. *Cell* **42**, 769–777.

Recanzone, G. & Harris, W. A. (1985). Demonstration of neural induction using nuclear markers in *Xenopus*. *Wilhelm Roux Arch. dev. Biol.* **194**, 344–354.

Reeve, W. J. D. (1981). Cytoplasmic polarity develops at compaction in rat and mouse embryos. *J. Embryol. exp. Morph.* **62**, 351–367.

Reeve, W. J. D. & Kelly, F. P. (1983). Nuclear position in the cells of the mouse early embryo. *J. Embryol. exp. Morph.* **75**, 117–139.

Reeve, W. J. D. & Ziomek, C. A. (1981). Distribution of microvilli on dissociated blastomeres from mouse embryos: evidence for surface polarization at compaction. *J. Embryol. exp. Morph.* **62**, 339–350.

Regier, J. C., Mazur, G. D. & Kafatos, F. C. (1980). The silkmoth chorion: morphological and biochemical characterization of four surface regions. *Devl Biol.* **76**, 286–304.

Regulski, M., Harding, K., Kostriken, R., Karch, F., Levine, M. & McGinnis, W. (1985). Homeo box genes of the Antennapedia and bithorax complexes of *Drosophila*. *Cell* **43**, 71–80.

Render, J. A. (1983). The second polar lobe of the *Sabellaria cementarium* embryo plays an inhibitory role in apical tuft formation. *Wilhelm Roux Arch. dev. Biol.* **192**, 120–129.

Render, J. A. & Elinson, R. P. (1986). Axis determination in polyspermic *Xenopus laevis* eggs. *Devl Biol.* **115**, 425–433.

Render, J. A. & Guerrier, P. (1984). Size regulation and morphogenetic localization in the *Dentalium* polar lobe. *J. exp. Zool.* **232**, 79–86.

Reporter, M. & Rosenquist, G. C. (1972). Adenosine 3′,5′-monophosphate: regional differences in chick embryos at the head process stage. *Science, N.Y.* **178**, 628–630.

Reverberi, G. (1936). La segmentazione dei frammenti dell'uovo non fecondato di Ascidie. *Pubbl. Staz. zool. Napoli* **15**, 198–216.

Reverberi, G. (1937). Ricerche sperimentali sulla struttura dell'uovo fecondato delle ascidie. *Commentat. pontif. Acad. Scient.* **1**, 135–172.

Reverberi, G. (1956). The mitochondrial pattern in the development of the ascidian egg. *Experientia* **12**, 55–56.

Reverberi, G. (1957). Mitochondrial and enzymatic segregation through the embryonic development in ctenophores. *Acta Embryol. Morph. exp.* **1**, 134–142.

Reverberi, G. (1958). Selective distribution of mitochondria during the development of the egg of *Dentalium*. *Acta Embryol. Morph. exp.* **2**, 79–87.

Reverberi, G. (1970). The ultrastructure of *Dentalium* egg at the trefoil stage. *Acta Embryol. exp.* 31–43.

Reverberi, G. (1971). Ctenophores. In *Experimental Embryology of Marine and Fresh-Water Invertebrates*, ed. G. Reverberi, pp. 85–103. Amsterdam: North–Holland.

Reverberi, G. & Gorgone, I. (1962). Gigantic tadpoles from Ascidian eggs fused at the 8-cell stage. *Acta Embryol. Morph. exp.* **5**, 104–112.

Reverberi, G. & La Spina, R. (1959). Normal larvae obtained from dark fragments of centrifuged *Ciona* eggs. *Experientia* **15**, 122.

Reverberi, G. & Minganti, A. (1947*a*). Le potenze dei quartetti animale e vegetativo isolati di *Ascidiella aspersa*. *Pubbl. Staz. zool. Napoli* **20**, 135–151.

Reverberi, G. & Minganti, A. (1947*b*). Fenomeni di evocazione nello sviluppo dell'uovo di Ascidie. Risultati dell'indagine sperimentale sull'uovo di *Ascidiella aspersa* e di *Ascidia malaca* allo stadio di otto blastomeri. *Pubbl. Staz. zool. Napoli* **20**, 199–252.

Reverberi, G. & Minganti, A. (1947*c*). La distribuzione delle potenze nel germe di Ascidie allo stadio di otto blastomeri, analizzata mediante le combinazione e i trapianti di blastomeri. *Pubbl. Staz. zool. Napol* **21**, 1–35.

Reverberi, G. & Ortolani, G. (1962). Twin larvae from halves of the same egg in ascidians. *Devl Biol.* **5**, 84–100.

Reverberi, G. & Ortolani, G. (1963). On the origin of the ciliated plates and the mesoderm in the ctenophores. *Acta Embryol. Morph. exp.* **6**, 175–190.

Reverberi, G., Ortolani, G. & Farinella-Ferruzza, N. (1960). The causal formation of the brain in the ascidian larva. *Acta Embryol. Morph. exp.* **3**, 296–336.

Reyss-Brion, M. (1964). L'effet des rayons X sur les potentialités respectives de l'ectoderme compétent et de son inducteur naturel chez la jeune gastrula d'amphibien. *Archs Anat. microsc. Morph. exp.* **53**, 397–465.

Rhumbler, L. (1902). Zur Mechanik des Gastrulationsvorganges insbesondere der Invagination. Eine entwicklungsmechanische Studie. *Arch. EntwMech. Org.* **14**, 401–476.

Rice, T. B. & Garen, A. (1975). Localized defects of blastoderm formation in maternal effect mutants of *Drosophila*. *Devl Biol.* **43**, 277–286.

Richa, J., Damsky, C. H., Buck, C. A., Knowles, B. B. & Solter, D. (1985). Cell surface glycoproteins mediate compaction, trophoblast attachment and endoderm formation during early mouse development. *Devl Biol.* **108**, 513–521.

Richard, L., Devillers, C. & Colas, J. (1956). Comportement en culture *in vitro* du blastoderme de Truite complet ou fragmenté. *C. r. hebd. Séanc. Acad. Sci., Paris* **243**, 1918–1920.

Richter, J. D. & Smith, L. D. (1984). Reversible inhibition of translation by *Xenopus* oocyte-specific proteins. *Nature, Lond.* **309**, 378–380.

Rickoll, W. L. (1976). Cytoplasmic continuity between embryonic cells and the primitive yolk sac during early gastrulation in *Drosophila melanogaster*. *Devl Biol.* **49**, 304–310.

Rickoll, W. L. & Counce, S. J. (1980). Morphogenesis in the embryo of *Drosophila melanogaster* – germ band extension. *Wilhelm Roux Arch. dev. Biol.* **188**, 163–177.

Roberson, M., Neri, A. & Oppenheimer, S. B. (1975). Distribution of concanavalin A receptor sites on specific populations of embryonic cells. *Science, N.Y.* **189**, 639–640.

Roberson, M. & Oppenheimer, S. B. (1975). Quantitative agglutination of specific populations of sea urchin embryo cells with concanavalin A. *Expl Cell Res.* **91**, 263–268.

Robertson, A. (1979). Waves propagated during vertebrate development: observations and comments. *J. Embryol. exp. Morph.* **50**, 155–167.

Robertson, N. (1978). Labilization of the superficial layer and reduction in size of yolk platelets during early development of *Xenopus laevis*. *Cell Differ.* **7**, 185–192.

Robinson, K. R. (1979). Electrical currents through full-grown and maturing *Xenopus* oocytes. *Proc. natn. Acad. Sci. U.S.A.* **76**, 837–841.

394    *References*

Rodgers, W. H. & Gross, P. R. (1978). Inhomogeneous distribution of egg RNA sequences in the early embryo. *Cell* **14**, 279–288.

Roeder, R. G. (1974). Multiple forms of deoxyribonucleic acid-dependent ribonucleic acid polymerase in *Xenopus laevis*. Levels of activity during oocyte and embryonic development. *J. biol. Chem.* **249**, 249–256.

Rogers, K. T. (1963). Experimental production of perfect cyclopia in the chick by means of LiCl, with a survey of the literature on cyclopia produced experimentally by various means. *Devl Biol.* **8**, 129–150.

Rollins, J. W. & Flickinger, R. A. (1973). Stimulation of RNA synthesis in developing frog embryos by microinjection of ribosomes. *J. exp. Zool.* **183**, 193–199.

Romani, S., Campuzano, S. & Modolell, J. (1987). The *achaete-scute* complex is expressed in neurogenic regions of *Drosophila* embryos. *EMBO J.* **6**, 2085–2092.

Roosen-Runge, E. C. (1938). On the early development – bipolar differentiation and cleavage – of the zebra fish, *Brachydanio rerio*. *Biol. Bull. mar. biol. Lab., Woods Hole* **75**, 119–133.

Rosbash, M. & Ford, P. J. (1974). Polyadenylic acid-containing RNA in *Xenopus laevis* oocytes. *J. molec. Biol.* **85**, 87–101.

Rose, S. M. (1939). Embryonic induction in the Ascidia. *Biol. Bull. mar. biol. Lab., Woods Hole* **77**, 216–232.

Rosenberg, U. B., Preiss, A., Seifert, E., Jäckle, H. & Knipple, D. C. (1985). Production of phenocopies by *Krüppel* antisense RNA injection into *Drosophila* embryos. *Nature, Lond.* **313**, 703–706.

Rosenberg, U. B., Schröder, C., Preiss, A., Kienlin, A., Côté, S., Riede, I. & Jäckle, H.. (1986). Structural homology of the product of the *Drosophila Krüppel* gene with *Xenopus* transcription factor IIIa. *Nature, Lond.* **319**, 336–339.

Rosenquist, G. C. (1966). A radioautographic study of labeled grafts in the chick blastoderm. Development from primitive-streak stages to stage 12. *Contr. Embryol.* **38**, 71–110.

Rossant, J. (1975). Investigation of the determinative state of the mouse inner cell mass. II. The fate of isolated inner cell masses transferred to the oviduct. *J. Embryol. exp. Morph.* **33**, 991–1001.

Rossant, J. & Vijh, K. M. (1980). Ability of outside cells from preimplantation mouse embryos to form inner cell mass derivatives. *Devl Biol.* **76**, 475–482.

Rostedt, I. (1971). Responses of ectoderm to heterogeneous inductors in chick as compared with *Triturus*. *Annls Med. exp. Biol. Fenn.* **49**, 186–203.

Roth, G. E. & Moritz, K. B. (1981). Restriction enzyme analysis of the germ line limited DNA of *Ascaris suum*. *Chromosoma* **83**, 169–190.

Roth, T. F. & Porter, K. R. (1964). Yolk protein uptake in the oocyte of the mosquito *Aedes aegypti* L. *J. Cell Biol.* **20**, 313–332.

Rott, N. N., Bozhkova, B. P., Kvavilashvili, I. Sh., Khariton, V. Yu. & Sharova, L. V. (1978). Development of isolated blastoderms of loach during cultivation in different saline media. *Ontogenez* **9**, 457–469 (in translation 394–404).

Rounds, D. E. & Flickinger, R. A. (1958). Distribution of ribonucleoprotein during neural induction of the frog embryo. *J. exp. Zool.* **137**, 479–499.

Roux, W. (1888). Beitrage zur Entwickelungsmechanik des Embryo. *Arch. path. Anat. Physiol.* **114**, 113–153 & 246–291.

Roux, W. (1903). Ueber die Ursachen der Bestimmung der Hauptrichtungen des Embryo im Froschei. *Anat. Anz.* **23**, 65–91.

Ruddell, A. & Jacobs-Lorena, M. (1984). Preferential expression of actin genes during oogenesis of *Drosophila*. *Devl Biol.* **105**, 115–120.

Rudensey, L. M. & Infante, A. A. (1979). Translational efficiency of cytoplasmic nonpolysomal messenger ribonucleic acid from sea urchin embryos. *Biochemistry, N.Y.* **18**, 3056–3063.

Ruderman, J. V. & Gross, P. R. (1974). Histones and histone synthesis in sea urchin development. *Devl Biol.* **36**, 286–298.

Runnström, J. (1929). Über Selbstdifferenzierung und Induktion bei dem Seeigelkeim. *Wilhelm Roux Arch. EntwMech. Org.* **117**, 123–145.

Runnström, J. (1931). Zur Entwicklungsmechanik des Skelletmusters bei dem Seeigelkeim. *Wilhelm Roux Arch. EntwMech. Org.* **124**, 273–297.

Runnström, J. (1964). Genetic and epigenetic factors involved in the early differentiation of the sea urchin egg (*Paracentrotus lividus, Psammechinus miliaris*). In *Acidi Nucleici e loro Funzione Biologica*, pp. 342–351. Milan: Istituto Lombardo.

Runnström, J., Hörstadius, S., Immers, J. & Fudge-Mastrangelo, M. (1964). An analysis of the role of sulfate in the embryonic differentiation of the sea urchin *Paracentrotus lividus*. *Revue suisse Zool.* **71**, 21–54.

Runnström, J. & Kriszat, G. (1952). Animalizing action of iodosobenzoic acid in the sea urchin development. *Expl Cell Res.* **3**, 497–499.

Runnström, J., Nuzzolo, C. & Citro, G. (1972). Ribosomes and polyribosomes in sea urchin development, particularly in embryos undergoing animalization or vegetalization. *Expl Cell Res.* **72**, 252–256.

Russell, D. H. (1971). Putrescine and spermidine biosynthesis in the development of normal and anucleolate mutants of *Xenopus laevis*. *Proc. natn. Acad. Sci. U.S.A.* **68**, 523–527.

Rustad, R. C. (1960). Dissociation of the mitotic time-schedule from the micromere 'clock' with X-rays. *Acta Embryol. Morph. exp.* **3**, 155–158.

Ruud, G. (1925). Die Entwicklung isolierter Keimfragmente frühester Stadien von *Triturus taeniatus*. *Wilhelm Roux Arch. EntwMech. Org.* **105**, 209–293.

Sabour, M. (1972). RNA synthesis and heterochromatization in early development of a mealybug. *Genetics, Princeton* **70**, 291–298.

Sagata, N., Okuyama, K. & Yamana, K. (1981). Localization and segregation of maternal RNA's during early cleavage of *Xenopus laevis* embryos. *Dev. Growth Differ.* **23**, 23–32.

Sakai, H. (1968). Contractile properties of protein threads from sea urchin eggs in relation to cell division. *Int. Rev. Cytol.* **23**, 89–112.

Sakai, Y. T. (1964). Studies on the ooplasmic segregation in the egg of the fish, *Oryzias latipes*. I. Ooplasmic segregation in egg fragments. *Embryologia* **8**, 129–134.

Sakata, S. & Kotani, M. (1985). Decrease in the number of primordial germ cells following injection of the animal pole cytoplasm into the vegetal pole region of *Xenopus* eggs. *J. exp. Zool.* **233**, 327–330.

Sakoyama, Y. & Okubo, S. (1981). Two-dimensional gel patterns of protein species during development of *Drosophila* embryos. *Devl Biol.* **81**, 361–365.

Sánchez, S. S., Cabada, M. O. & Barbieri, F. D. (1983). Newly synthesised extracellular ribonucleic acids in the amphibian gastrula. *Cell Differ.* **13**, 149–157.

Sanchez-Herrero, E., Vernós, I., Marco, R. & Morata, G. (1985). Genetic organization of *Drosophila* bithorax complex. *Nature, Lond.* **313**, 108–113.

Sander, K. (1959). Analyse des ooplasmatischen Reaktionssystems von *Euscelis plebejus* Fall. (Cicadina) durch Isolieren und Kombinieren von Keimteilen. I.

Die Differenzierungsleistungen vorderer und hinterer Eiteile. *Wilhelm Roux Arch. EntwMech. Org.* **151**, 430–497.

Sander, K. (1960). Analyse des ooplasmatischen Reaktionssystems von *Euscelis plebejus* Fall. (Cicadina) durch Isolieren und Kombinieren von Keimteilen. II. Die Differenzierungsleistungen nach Verlagern von Hinterpolmaterial. *Wilhelm Roux Arch. EntwMech. Org.* **151**, 660–707.

Sander, K. (1961). New experiments concerning the ooplasmic reaction system of *Euscelis plebejus* F. (Cicadina). In *Symposium on the Germ Cells and Earliest Stages of Development*, pp. 338–353. Pallanza: Institut International d'Embryologie.

Sander, K. (1971). Pattern formation in longitudinal halves of leafhopper eggs (Homoptera) and some remarks on the definition of 'embryonic regulation'. *Wilhelm Roux Arch. EntwMech. Org.* **167**, 336–352.

Sander, K. (1976). Specification of the basic body pattern in insect embryogenesis. *Adv. Insect Physiol.* **12**, 125–238.

Sander, K., Herth, W. & Vollmar, H. (1969). Abwandlungen des metameren Organisationsmusters in fragmentieren und in abnormen Insekteneiern. *Zool. Anz. Suppl.* **33**, 46–52.

Sanders, E. J. & Prasad, S. (1986). Epithelial and basement membrane responses to chick embryo primitive streak grafts. *Cell Differ.* **18**, 233–242.

Santamaria, P. & Nüsslein-Volhard, C. (1983). Partial rescue of *dorsal*, a maternal effect mutation affecting the dorso-ventral pattern of the *Drosophila* embryo, by the injection of wild-type cytoplasm. *EMBO J.* **2**, 1695–1699.

Sardet, C. & Chang, P. (1987). The egg cortex: from maturation through fertilization. *Cell Differ.* **21**, 1–19.

Sargent, T. D. & Dawid, I. B. (1983). Differential gene expression in the gastrula of *Xenopus laevis*. *Science, N.Y.* **222**, 135–139.

Sargent, T. D., Jamrich, M. & Dawid, I. B. (1986). Cell interactions and the control of gene activity during early development of *Xenopus laevis*. *Devl Biol.* **114**, 238–246.

Sargent, T. D. & Raff, R. A. (1976). Protein synthesis and messenger RNA stability in activated, enucleate sea urchin eggs are not affected by actinomycin D. *Devl Biol.* **48**, 327–335.

Sasaki, N., Iwamoto, K., Noda, S. & Kawakami, I. (1976). Staging of newt blastula embryos and first appearance of primary mesodermal competence. *Dev. Growth Differ.* **18**, 457–465.

Sathananthan, A. H. (1970). Studies on mitochondria in early development of the slug *Arion ater rufus* L. *J. Embryol. exp. Morph.* **24**, 555–582.

Satoh, N. (1978). Cellular morphology and architecture during early morphogenesis of the ascidian egg: an SEM study. *Biol. Bull. mar. biol. Lab., Woods Hole* **155**, 608–614.

Satoh, N. (1982a). DNA replication is required for tissue-specific enzyme development in ascidian embryos. *Differentiation* **21**, 37–40.

Satoh, N. (1982b). Timing mechanisms in early embryonic development. *Differentiation* **22**, 156–163.

Satoh, N. & Ikegami, S. (1981a). A definite number of aphidicolin-sensitive cell-cyclic events are required for acetylcholinesterase development in the presumptive muscle cells of the ascidian embryos. *J. Embryol. exp. Morph.* **61**, 1–13.

Satoh, N. & Ikegami, S. (1981b). On the 'clock' mechanism determining the time of tissue-specific enzyme development during ascidian embryogenesis. II. Evidence for association of the clock with the cycle of DNA replication. *J. Embryol. exp. Morph.* **64**, 61–71.

Sauer, G. (1961). Ordnungsvorgänge und ihre Regulationsmöglichkeit bei der

Bildung der Keimanlage auf der Eioberfläche von *Gryllus domesticus*. *Verh. dt. zool. Ges.* **54**, 113–121.

Savoini, A., Micali, F., Marzari, R., de Cristini, F. & Graziosi, G. (1981). Low variability of the protein species synthesized by *Drosophila melanogaster* embryos. *Wilhelm Roux Arch. dev. Biol.* **190**, 161–167.

Sawada, T. (1983). How ooplasm segregates bipolarly in ascidian eggs. *Bull. biol. Stn. Asamushi* **17**, 123–140.

Sawada, T. & Osanai, K. (1981). The cortical contraction related to the ooplasmic segregation in *Ciona intestinalis* eggs. *Wilhelm Roux Arch. dev. Biol.* **190**, 208–214.

Sawada, T. & Osanai, K. (1984). Cortical contraction and ooplasmic movement in centrifuged or artificially constricted eggs of *Ciona intestinalis*. *Wilhelm Roux Arch. dev. Biol.* **193**, 127–132.

Sawada, T. & Osanai, K. (1985). Distribution of actin filaments in fertilized eggs of the ascidian *Ciona intestinalis*. *Devl Biol.* **111**, 260–265.

Sawai, T. (1982). Wavelike propagation of stretching and shrinkage in the surface of the newt's egg before the first cleavage. *J. exp. Zool.* **222**, 59–68.

Saxén, L. (1961). Transfilter neural induction of amphibian ectoderm. *Devl Biol* **3**, 140–152.

Saxén, L. & Toivonen, S. (1962). *Primary Embryonic Induction*, London: Academic Press.

Saxén, L., Toivonen, S. & Vainio, T. (1964). Initial stimulus and subsequent interactions in embryonic induction. *J. Embryol. exp. Morph.* **12**, 333–338.

Schaeffer, B. E., Schaeffer, H. E. & Brick, I. (1973). Cell electrophoresis of amphibian blastula and gastrula cells; the relationship of surface charge and morphogenetic movement. *Devl Biol.* **34**, 66–76.

Scharf, S. R. & Gerhart, J. C. (1980). Determination of the dorsal–ventral axis in eggs of *Xenopus laevis*: complete rescue of uv-impaired eggs by oblique orientation before first cleavage. *Devl Biol.* **79**, 181–198.

Scharf, S. R. & Gerhart, J. C. (1983). Axis-determination in eggs of *Xenopus laevis*: a critical period before first cleavage, identified by the common effects of cold, pressure and ultraviolet irradiation. *Devl Biol.* **99**, 75–87.

Schatten, G. (1982). Motility during fertilization. *Int. Rev. Cytol.* **79**, 59–163.

Schechtman, A. M. (1935). Mechanism of ingression in the egg of *Triturus torosus*. *Proc. Soc. exp. Biol. Med.* **32**, 1072–1073.

Schierenberg, E. (1984). Altered cell division rates after laser-induced cell fusion in nematode embryos. *Devl Biol.* **101**, 240–245.

Schierenberg, E. (1986). Developmental strategies during early embryogenesis of *Caenorhabditis elegans*. *J. Embryol. exp. Morph.* **97**, Suppl. 31–44.

Schierenberg, E. (1987). Reversal of cellular polarity and early cell–cell interaction in the embryo of *Caenorhabditis elegans*. *Devl Biol.* **122**, 452–463.

Schierenberg, E., Carlson, C. & Sidio, W. (1984). Cellular development of a nematode: 3-D computer reconstruction of living embryos. *Wilhelm Roux Arch. dev. Biol.* **194**, 61–68.

Schierenberg, E., Miwa, J. & von Ehrenstein, G. (1980). Cell lineages and developmental defects of temperature-sensitive embryonic arrest mutants in *Caenorhabditis elegans*. *Devl Biol.* **76**, 141–159.

Schierenberg, E. & Wood, W. B. (1985). Control of cell-cycle timing in early embryos of *Caenorhabditis elegans*. *Devl Biol.* **107**, 337–354.

Schindler, J. & Sherman, M. I. (1981). Effects of α-amanitin on programming of mouse blastocyst development. *Devl Biol.* **84**, 332–340.

Schleip, W. (1914). Die Furchung des Eies der Rüsselegel. *Zool. Jb. Abt. Anat. Ont.* **37**, 313–368.

Schleip, W. (1929). *Die Determination der Primitiventwicklung.* Leipzig: Akademische Verlagsgesellschaft M.B.H.

Schmekel, L. (1970). Elektronmikroskopie der Makromeren-Mikromeren-grenze des Seeigelkeimes. *Zool. Anz. Suppl.* **33**, 141–144.

Schmidt, B. A., Kelly, P. T., May, M. C., Davis, S. E. & Conrad, G. W. (1980). Characterization of actin from fertilized eggs of *Ilyanassa obsoleta* during polar lobe formation and cytokinesis. *Devl Biol.* **76**, 126–140.

Schmidt, G. A. (1933). Schnürungs- und Durchschneidungsversuche am Anurenkeim. *Wilhelm Roux Arch. EntwMech. Org.* **129**, 1–44.

Schmidt, O. & Jäckle, H. (1978). RNA synthesised during oogenesis and early embryogenesis in an insect egg (*Euscelis plebejus*). *Wilhelm Roux Arch. dev. Biol.* **184**, 143–153.

Schmidt, O., Zissler, D., Sander, K. & Kalthoff, K. (1975). Switch in pattern formation after puncturing the anterior pole of *Smittia* eggs (Chironomidae, Diptera). *Devl Biol.* **46**, 216–221.

Schneider, E. D., Nguyen, H. T. & Lennarz, W. J. (1978). The effect of tunicamycin, an inhibitor of protein glycosylation, on embryonic development in the sea urchin. *J. biol. Chem.* **253**, 2348–2355.

Schnetter, M. (1934). Physiologische Untersuchungen über das Differenzierungszentrum in der Embryonalentwicklung der Honigbiene. *Wilhelm Roux Arch. EntwMech. Org.* **131**, 285–323.

Schnetter, W. (1965). Experimente zur Analyse der morphogenetischen Funktion der Ooplasmabestandteile in der Embryonalentwicklung des Kartoffelkäfers (*Leptinotarsa decemlineata* Say). *Wilhelm Roux Arch. EntwMech. Org.* **155**, 637–692.

Schneuwly, S., Kuroiwa, A., Baumgartner, P. & Gehring, W. J. (1986). Structural organization and sequence of the homeotic gene *Antennapedia* of *Drosophila melanogaster*. *EMBO J.* **5**, 733–739.

Schreuer, M. & Czihak, G. (1978). Effect of 5-bromodeoxyuridine on differentiation. I. Probability distribution of BUdR-containing DNA-strands in subsequent divisions. *Differentiation* **11**, 89–101.

Schroeder, T. E. (1980a). Expression of the prefertilization polar axis in sea urchin eggs. *Devl Biol.* **79**, 428–443.

Schroeder, T. E. (1980b). The jelly canal marker of polarity for sea urchin oocytes, eggs, and embryos. *Expl Cell Res.* **128**, 490–494.

Schroeder, T. E. (1985). Cortical expressions of polarity in the starfish oocyte. *Dev. Growth Differ.* **27**, 311–321.

Schubiger, G. (1976). Adult differentiation from partial *Drosophila* embryos after egg ligation during stages of nuclear multiplication and cellular blastoderm. *Devl Biol.* **50**, 476–488.

Schubiger, G. & Newman, S. M. (1982). Determination in *Drosophila* embryos. *Am. Zool.* **22**, 47–55.

Schuh, R., Aicher, W., Gaul, U., Côté, S., Preiss, A., Maier, D., Seifert, E., Nauber, U., Schröder, C., Kemler, R. & Jäckle, H. (1986). A conserved family of nuclear proteins containing structural elements of the finger protein encoded by *Krüppel*, a *Drosophila* segmentation gene. *Cell* **47**, 1025–1032.

Schultz, G. A. & Tucker, E. B. (1977). Protein synthesis and gene expression in preimplantation rabbit embryos. In *Development in Mammals*, vol. 1, ed. M. H. Johnson, pp. 69–97. Amsterdam: North-Holland.

Schultz, R. M., Letourneau, G. E. & Wassarman, P. M. (1979). Program of early development in the mammal: changes in patterns and absolute rates of tubulin and total protein synthesis during oogenesis and early embryogenesis in the mouse. *Devl Biol.* **68**, 341–359.

Schultze, O. (1894). Die künstliche Erzeugung von Doppelbildungen bei Froschlarven mit Hilfe abnormer Gravitationswirkung. *Arch. EntwMech. Org.* **1**, 269–305.

Schüpbach, T. (1987). Germ line and soma cooperate during oogenesis to establish the dorsoventral pattern of egg shell and embryo in *Drosophila melanogaster. Cell* **49**, 699–707.

Schüpbach, T. & Wieschaus, E. (1986). Maternal-effect mutations altering the anterior–posterior pattern of the *Drosophila* embryo. *Wilhelm Roux Arch. dev. Biol.* **195**, 302–317.

Scott, M. P. & Weiner, A. J. (1984). Structural relationships among genes that control development: Sequence homology between the *Antennapedia, Ultrabithorax*, and *fushi tarazu* loci of *Drosophila. Proc. natn. Acad. Sci. U.S.A.* **81**, 4115–4119.

Scott, M. P., Weiner, A. J., Hazelrigg, T. I., Polisky, B. A., Pirrotta, V., Scalenghe, F. & Kaufman, T. C. (1983). The molecular organization of the *Antennapedia* locus of *Drosophila. Cell* **35**, 763–776.

Searle, R. F. & Jenkinson, E. J. (1978). Localization of trophoblast-defined surface antigens during early mouse embryogenesis. *J. Embryol. exp. Morph.* **43**, 147–156.

Seidel, F. (1926). Die Determinierung der Keimanlage bei Insekten. I. *Biol. Zbl.* **46**, 321–343.

Seidel, F. (1934). Das Differenzierungszentrum im Libellenkeim. I. Die dynamischen Voraussetzungen der Determination und Regulation. *Wilhelm Roux Arch. EntwMech. Org.* **131**, 135–187.

Seidel, F. (1936). Entwicklungsphysiologie des Insekten-Keims. *Verh. dt. zool. Ges.* **38**, 291–336.

Seidel, F. (1952). Die Entwicklungspotenzen eine isolierten Blastomere des Zweizellenstadiums im Säugetierei. *Naturwissenschaften* **39**, 355–356.

Seidel, F. (1961). Entwicklungsphysiologische Zentren im Eisystem der Insekten. *Verh. dt. zool. Ges.* **54**, 121–142.

Seidel, F., Bock, E. & Krause, G. (1940). Die Organisation des Insekteneies. *Naturwissenschaften* **28**, 433–446.

Sekiguchi, K. (1957). Reduplication in spider eggs produced by centrifugation. *Sci. Rep. Tokyo Kyoiku Daig.* **8**, 227–280.

Senger, D. R., Arceci, R. J. & Gross, P. R. (1978). Histones of sea urchin embryos. Transients in transcription, translation, and the composition of chromatin. *Devl Biol.* **65**, 416–425.

Senger, D. R. & Gross, P. R. (1978). Macromolecule synthesis and determination in sea urchin blastomeres at the sixteen-cell stage. *Devl Biol.* **65**, 404–415.

Shankland, M. & Weisblat, D. A. (1984). Stepwise commitment of blast cell fates during the positional specification of the O and P cell lines in the leech embryo. *Devl Biol.* **106**, 326–342.

Sharpe, C. R., Fritz, A., de Robertis, E. M. & Gurdon, J. B. (1987). A homeobox-containing marker of posterior neural differentiation shows the importance of predetermination in neural induction. *Cell* **50**, 749–758.

Shaver, J. R. (1957). Some observations on cytoplasmic particles in early echinoderm development. In *The Beginnings of Embryonic Development*, ed. A. Tyler, R. C. von Borstel & C. B. Metz, pp. 263–290. Washington: American Association for the Advancement of Science.

Shepherd, G. W. & Flickinger, R. A. (1979). Post-transcriptional control of messenger RNA diversity in frog embryos. *Biochim. biophys. Acta* **563**, 413–421.

Shepherd, G. W. & Nemer, M. (1980). Developmental shifts in frequency distribution of polysomal mRNA and their posttranscriptional regulation in the sea urchin embryo. *Proc. natn. Acad. Sci. U.S.A.* **77**, 4653–4656.

Sherbet, G. V. & Mulherkar, L. (1963). The morphogenetic action of follicle-stimulating hormone on post-nodal fragments of early chick blastoderms. *Wilhelm Roux Arch. EntwMech. Org.* **154**, 506–512.

Sherman, M. I., Gay, R., Gay, S. & Miller, E. J. (1980). Association of collagen with preimplantation and peri-implantation mouse embryos. *Devl Biol.* **74**, 470–478.

Shimizu, T. (1978*a*). Deformation movement induced by divalent ionophore A23187 in the *Tubifex* egg. *Dev. Growth Differ.* **20**, 27–33.

Shimizu, T. (1978*b*). Mode of microfilament-arrangement in normal and cytochalasin-treated eggs of *Tubifex* (Annelida, Oligochaeta). *Acta Embryol. exp.* 59–74.

Shimizu, T. (1982*a*). Ooplasmic segregation in the *Tubifex* egg: mode of pole plasm accumulation and possible involvement of microfilaments. *Wilhelm Roux Arch. dev. Biol.* **191**, 246–256.

Shimizu, T. (1982*b*). Development in the freshwater oligochaete *Tubifex*. In *Developmental Biology of Freshwater Invertebrates*, ed. F. W. Harrison & R. R. Cowden, pp. 283–316. New York: Alan R. Liss Inc.

Shimizu, T. (1984). Dynamics of the actin microfilament system in the *Tubifex* egg during ooplasmic segregation. *Devl Biol.* **106**, 414–426.

Shimizu, T. (1986). Bipolar segregation of mitochondria, actin network, and surface in the *Tubifex* egg: role of cortical polarity. *Devl Biol.* **116**, 241–251.

Shiokawa, K. (1983). Mobilization of maternal mRNA in amphibian eggs with special reference to the possible role of membraneous supramolecular structures. *FEBS Lett.* **151**, 179–184.

Shiokawa, K., Misumi, Y. & Yamana, K. (1981*a*). Mobilization of newly synthesized RNAs into polysomes in *Xenopus laevis* embryos. *Wilhelm Roux Arch. dev. Biol.* **190**, 103–110.

Shiokawa, K., Misumi, Y. & Yamana, K. (1981*b*). Demonstration of rRNA synthesis in pre-gastrular embryos of *Xenopus laevis*. *Dev. Growth Differ.* **23**, 579–587.

Shiokawa, K. & Yamana, K. (1979). Differential initiation of rRNA gene activity in progenies of different blastomeres of early *Xenopus* embryos: evidence for regulated synthesis of rRNA. *Dev. Growth Differ.* **21**, 501–507.

Shirai, H. & Kanatani, H. (1980). Effect of local application of 1-methyladenine on the site of polar body formation in starfish oocyte. *Dev. Growth Differ.* **22**, 555–560.

Shirayoshi, Y., Okada, T. S. & Takeichi, M. (1983). The calcium-dependent cell–cell adhesion system regulates inner cell mass formation and cell surface polarization in early mouse development. *Cell* **35**, 631–638.

Shmukler, Yu. B., Chailakhyan, L. M., Smolianinov, V. V., Bliokh, Zh. L., Karpovich, A. L., Gusareva, É. V., Naidenko, T. Kh., Khashaev, Kh. M. & Medvedeva, T. D. (1981). Intercellular interactions in early sea urchin embryos. II. Dated mechanical separation of blastomeres. *Ontogenez* **12**, 398–403 (in translation 258–262).

Shott, R. J., Lee, J. J., Britten, R. J. & Davidson, E. H. (1984). Differential expression of the actin gene family of *Strongylocentrotus purpuratus*. *Devl Biol.* **101**, 295–306.

Shur, B. D. (1977). Cell surface glycosyltransferases in gastrulating chick embryos. I. Temporally and spatially specific patterns of four endogenous glycosyltransferase activities. *Devl Biol.* **58**, 23–39.

Siegel, G., Grunz, H., Grundmann, U., Tiedemann, H. & Tiedemann, H. (1985). Embryonic induction and cation concentrations in amphibian embryos. *Cell Differ.* **17**, 209–219.

Signoret, J. & Lefresne, J. (1971). Contribution a l'étude de la segmentation de l'oeuf d'Axolotl. I. Définition de la transition blastuléenne. *Ann. Embryol. Morphogen.* **4**, 113–123.

Simcox, A. A. & Sang, J. H. (1983). When does determination occur in *Drosophila* embryos? *Devl Biol.* **97**, 212–221.

Simpson, S. (1983). Maternal–zygotic gene interactions during formation of the dorsoventral pattern in *Drosophila* embryos. *Genetics, Princeton* **105**, 615–632.

Slack, J. M. W. (1983). *From Egg to Embryo: Determinative Events in Early Development.* Cambridge: The University Press.

Slack, J. M.W. (1984*a*). Regional biosynthetic markers in the early amphibian embryo. *J. Embryol. exp. Morph.* **80**, 289–319.

Slack, J. M. W. (1984*b*). *In vitro* development of isolated ectoderm from axolotl gastrulae. *J. Embryol. exp. Morph.* **80**, 321–330.

Slack, J. M. W. (1985). Peanut lectin receptors in the early amphibian embryo: regional markers for the study of embryonic induction. *Cell* **41**, 237–247.

Slack, J. M. W., Darlington, B. G., Heath, J. K. & Godsave, S. F. (1987). Mesoderm induction in early *Xenopus* embryos by heparin-binding growth factors. *Nature, Lond.* **326**, 197–200.

Smith, J. C. (1987). A mesoderm-inducing factor is produced by a *Xenopus* cell line. *Development* **99**, 3–14.

Smith, J. C. & Malacinski, G. M. (1983). The origin of the mesoderm in an anuran, *Xenopus laevis*, and a urodele, *Ambystoma mexicanum*. *Devl Biol.* **98**, 250–254.

Smith, J. C. & Slack, J. M. W. (1983). Dorsalization and neural induction: properties of the organizer in *Xenopus laevis*. *J. Embryol. exp. Morph.* **78**, 299–317.

Smith, J. C. & Watt, F. M. (1985). Biochemical specificity of *Xenopus* notochord. *Differentiation* **29**, 109–115.

Smith, J. L., Osborn, J. C. & Stanisstreet, M. (1976). Scanning electron microscopy of lithium-induced exogastrulae of *Xenopus laevis*. *J. Embryol. exp. Morph.* **36**, 513–522.

Smith, K. D. (1967). Genetic control of macromolecular synthesis during development of an ascidian: *Ascidia nigra*. *J. exp. Zool.* **164**, 393–405.

Smith, L. D. (1966). The role of a 'germinal plasm' in the formation of primordial germ cells in *Rana pipiens*. *Devl Biol.* **14**, 330–347.

Smith, L. D. & Ecker, R. E. (1970). Uterine suppression of biochemical and morphogenetic events in *Rana pipiens*. *Devl Biol.* **22**, 622–637.

Smith, L. D. & Williams, M. (1979). Germinal plasm and germ cell determinants in anuran amphibians. In *Maternal Effects in Development*, ed. D. R. Newth & M. Balls, pp. 167–197. Cambridge: The University Press.

Smith, L. J. (1980). Embryonic axis orientation in the mouse and its correlation with blastocyst relationships to the uterus. I. Relationships between 82 hours and $4\frac{1}{4}$ days. *J. Embryol. exp. Morph.* **55**, 257–277.

Smith, L. J. (1985). Embryonic axis orientation in the mouse and its correlation with blastocyst relationships to the uterus. II. Relationships from $4\frac{1}{4}$ to $9\frac{1}{2}$ days. *J. Embryol. exp. Morph.* **89**, 15–35.

Smith, R. & McLaren, A. (1977). Factors affecting the time of formation of the mouse blastocoele. *J. Embryol. exp. Morph.* **41**, 79–92.

Smith, R. C. (1986). Protein synthesis and messenger RNA levels along the animal–vegetal axis during early *Xenopus* development. *J. Embryol. exp. Morph.* **95**, 15–35.

Smith, R. C. & Knowland, J. (1984). Protein synthesis in dorsal and ventral regions of *Xenopus laevis* embryos in relation to dorsal and ventral differentiation. *Devl Biol.* **103**, 355–368.

Smith, R. C., Neff, A. W. & Malacinski, G. M. (1986). Accumulation, organization and deployment of oogenetically derived *Xenopus* yolk/nonyolk proteins. *J. Embryol. exp. Morph.* **97**, *Suppl.* 45–64.

Smith, R. K. W. & Johnson, M. H. (1985). DNA replication and compaction in the cleaving embryo of the mouse. *J. Embryol. exp. Morph.* **89**, 133–148.

Snow, M. H. L. (1977). Gastrulation in the mouse: growth and regionalization of the epiblast. *J. Embryol. exp. Morph.* **42**, 293–303.

Snow, M. H. L. (1981). Autonomous development of parts isolated from primitive-streak-stage mouse embryos. Is development clonal? *J. Embryol. exp. Morph.* **65**, *Suppl.* 269–287.

Snow, M. H. L. & Bennett, D. (1978). Gastrulation in the mouse: assessment of cell populations in the epiblast of $t^{w18}/t^{w18}$ embryos. *J. Embryol. exp. Morph.* **47**, 39–52.

Sobel, J. S. (1983). Localization of myosin in the preimplantation mouse embryo. *Devl Biol.* **95**, 227–231.

Soltyńska, M. S. (1982). The possible mechanism of cell positioning in mouse morulae: an ultrastructural study. *J. Embryol. exp. Morph.* **68**, 137–147.

Solursh, M. & Morriss, G. M. (1977). Glycosaminoglycan synthesis in rat embryos during the formation of the primary mesenchyme and neural folds. *Devl Biol.* **57**, 75–86.

Solursh, M. & Revel, J. P. (1978). A scanning electron microscope study of cell shape and cell appendages in the primitive streak region of the rat and chick embryo. *Differentiation* **11**, 185–190.

Spek, J. (1926). Über gesetzmässige Substanzverteilungen bei der Furchung des Ctenophoreneies und ihre Beziehungen zu den Determinationsproblemen. *Wilhelm Roux Arch. EntwMech. Org.* **107**, 54–73.

Spek, J. (1930). Zustandsänderungen der Plasmakolloide bei Befruchtung und Entwicklung des *Nereis*-Eies. *Protoplasma* **9**, 370–427.

Spek, J. (1933). Die bipolare Differenzierung des Protoplasmas des Teleosteer-Eies und ihre Entstehung. *Protoplasma* **18**, 497–545.

Spek, J. (1934). Die bipolare Differenzierung des Cephalopoden- und des Prosobranchiereies (Vitalfärbungsversuche mit Indikatoren an den Eiern von *Loligo vulgaris* und *Columbella avara*). *Wilhelm Roux Arch. EntwMech. Org.* **131**, 362–372.

Speksnijder, J. E. & Dohmen, M. R. (1983). Local surface modulation correlated with ooplasmic segregation in eggs of *Sabellaria alveolata* (Annelida, Polychaeta). *Wilhelm Roux Arch. dev. Biol.* **192**, 248–255.

Speksnijder, J. E., Mulder, M. M., Dohmen, M. R., Hage, W. J. & Bluemink, J. G. (1985). Animal–vegetal polarity in the plasma membrane of a molluscan egg: a quantitative freeze fracture study. *Devl Biol.* **108**, 38–48.

Spemann, H. (1901). Entwicklungsphysiologische Studien am *Triton*-Ei. *Arch. EntwMech. Org.* **12**, 224–264.

Spemann, H. (1918). Über die Determination der ersten Organanlagen des Amphibienembryo. I–VI. *Arch. EntwMech. Org.* **43**, 448–555.

Spemann, H. (1925). Some factors of animal development. *Br. J. exp. Biol.* **2**, 493–504.

Spemann, H. (1931). Über den Anteil von Implantat und Wirtskeim an der Orientierung und Beschaffenheit der induzierten Embryonalanlage. *Wilhelm Roux Arch. EntwMech. Org.* **123**, 389–517.

Spemann, H. (1938). *Embryonic Development and Induction.* New Haven: Yale University Press.

Spemann, H. & Mangold, H. (1924). Über Induktion von Embryonalanlagen durch Implantation artfremder Organisatoren. *Arch. mikrosk. Anat. Entw-Mech.* **100**, 599–638.

Spiegel, E., Burger, M. & Spiegel, M. (1980). Fibronectin in the developing sea urchin embryo. *J. Cell Biol.* **87**, 309–313.

Spiegel, E., Burger, M. M. & Spiegel, M. (1983). Fibronectin and laminin in the extracellular matrix and basement membrane of sea urchin embryos. *Expl Cell Res.* **144**, 47–55.

Spiegel, M. & Rubinstein, N. A. (1972). Synthesis of RNA by dissociated cells of the sea urchin embryo. *Expl Cell Res.* **70**, 423–430.

Spiegel, M. & Spiegel, E. (1978). Sorting out of sea urchin embryonic cells according to cell type. *Expl Cell Res.* **117**, 269–271.

Spiegelman, M. & Bennett, D. (1974). Fine structural study of cell migration in the early mesoderm of normal and mutant mouse embryos (*T*-locus: $t^9/t^9$). *J. Embryol. exp. Morph.* **32**, 723–738.

Spieth, J. & Whiteley, A. H. (1980). Effect of 3'-deoxyadenosine (cordycepin) on the early development of the sand dollar, *Dendraster excentricus. Devl Biol.* **79**, 95–106.

Spirin, A. S., Belitsina, N. V. & Aitkhozhin, M. A. (1964). Messenger RNA in early embryogenesis. *Zh. obshch. Biol.* **25**, 321–338. (In translation in *Fed. Proc.* **24**, Pt. 2, T907–915.)

Spirin, A. S. & Nemer, M. (1965). Messenger RNA in early sea-urchin embryos: cytoplasmic particles. *Science, N.Y.* **150**, 214–217.

Spofford, W. R. (1948). Observations on the posterior part of the neural plate in *Amblystoma.* II. The inductive effect of the intact posterior part of the chorda-mesodermal axis on competent prospective ectoderm. *J. exp. Zool.* **107**, 123–163.

Spratt, N. T. (1955). Studies on the organizer center of the early chick embryo. In *Aspects of Synthesis and Order in Growth*, ed. D. Rudnick, pp. 209–231. Princeton: The University Press.

Spratt, N. T. (1957). Analysis of the organization center in the early chick embryo. III. Regulative properties of the chorda and somite centers. *J. exp. Zool.* **135**, 319–353.

Spratt, N. T. (1958). Analysis of the organizer center in the early chick embryo. IV. Some differential enzyme activities of node center cells. *J. exp. Zool.* **138**, 51–79.

Spratt, N. T. (1966). Some problems and principles of development. *Am. Zool.* **6**, 9–19.

Spratt, N. T. & Haas, H. (1960*a*). Morphogenetic movements in the lower surface of the unincubated and early chick blastoderm. *J. exp. Zool.* **144**, 139–158.

Spratt, N. T. & Haas, H. (1960*b*). Integrative mechanisms in development of the early chick blastoderm. I. Regulative potentiality of separated parts. *J. exp. Zool.* **145**, 97–138.

Spratt, N. T. & Haas, H. (1961*a*). Integrative mechanisms in development of the early chick blastoderm. II. Role of morphogenetic movements and regenerative growth in synthetic and topographically disarranged blastoderms. *J. exp. Zool.* **147**, 57–93.

Spratt, N. T. & Haas, H. (1961*b*). Integrative mechanisms in development of the early chick blastoderm. III. Role of cell population size and growth potentiality in synthetic systems larger than normal. *J. exp. Zool.* **147**, 271–293.

404    References

Spratt, N. T. & Haas, H. (1962a). Integrative mechanisms in development of
the early chick blastoderm. IV. Synthetic systems composed of parts of different developmental age. Synchronization of development rates. *J. exp.
Zool.* **149**, 75–102.

Spratt, N. T. & Haas, H. (1962b). Primitive streak and germ layer formation in
the chick. A reappraisal. *Anat. Rec.* **142**, 327.

Srivastava, M. D. L. & Srivastava, S. (1969). Medio-lateral distribution of
inductive capacity in the organizer of the frog, *Rana cyanophlyctis. Experientia* **25**, 888–889.

Stableford, L. T. (1948). The potency of the vegetal hemisphere of the
*Amblystoma punctatum* embryo. *J. exp. Zool.* **109**, 385–426.

Standart, N. M., Bray, S. J., George, E. L., Hunt, T. & Ruderman, J. V.
(1985). The small subunit of ribonucleotide reductase is encoded by one of the
most abundant translationally regulated maternal RNAs in clam and sea
urchin eggs. *J. Cell Biol.* **100**, 1968–1976.

Stanisstreet, M. & Deuchar, E. M. (1972). Appearance of antigenic material in
gastrula ectoderm after neural induction. *Cell Differ.* **1**, 15–18.

Stanisstreet, M., Jumah, H. & Kurais, A. R. (1980). Properties of cells from
inverted embryos of *Xenopus laevis* investigated by scanning electron
microscopy. *Wilhelm Roux Arch. dev. Biol.* **189**, 181–186.

Steiner, E. (1976). Establishment of compartments in the developing leg
imaginal discs of *Drosophila melanogaster. Wilhelm Roux Arch. dev. Biol.*
**180**, 9–30.

Stephens, L., Hardin, J., Keller, R. & Wilt, F. (1986). The effects of aphidicolin
on morphogenesis and differentiation in the sea urchin embryo. *Devl Biol.*
**118**, 64–69.

Stern, C. D. (1981). Behaviour and motility of cultured chick mesoderm cells in
steady electrical fields. *Expl Cell Res.* **136**, 343–350.

Stern, C. D. & Goodwin, B. C. (1977). Waves and periodic events during
primitive streak formation in the chick. *J. Embryol. exp. Morph.* **41**,
15–22.

Stern, C. D., Manning, S. & Gillespie, J. I. (1985). Fluid transport across the
epiblast of the early chick embryo. *J. Embryol. exp. Morph.* **88**, 365–384.

Stern, S., Biggers, J. D. & Anderson, E. (1971). Mitochondria and early
development of the mouse. *J. exp. Zool.* **176**, 179–191.

Stevens, N. M. (1909). The effects of ultra-violet light upon the developing eggs
of *Ascaris megalocephala. Arch. EntwMech. Org.* **27**, 622–639.

Steward, R., McNally, F. J. & Schedl, P. (1984). Isolation of the *dorsal* locus of
*Drosophila. Nature, Lond.* **311**, 262–265.

Stockard, C. R. (1907). The artificial production of a single median cyclopean
eye in the fish embryo by means of sea water solutions of magnesium chloride.
*Arch. EntwMech. Org.* **23**, 249–258.

Strelkov, L. A. & Ignat'eva, G. M. (1975). Early activation of ribosomal RNA
synthesis in axolotl embryos. *Ontogenez* **6**, 519–523 (in translation 443–447).

Strickland, S., Reich, E. & Sherman, M. I. (1976). Plasminogen activator in
early embryogenesis: enzyme production by trophoblast and parietal endoderm. *Cell* **9**, 231–240.

Strome, S. (1986a). Fluorescence visualization of the distribution of microfilaments in gonads and early embryos of the nematode *Caenorhabditis elegans.*
*J. Cell Biol.* **103**, 2241–2252.

Strome, S. (1986b). Asymmetric movements of cytoplasmic components in
*Caenorhabditis elegans* zygotes. *J. Embryol. exp. Morph.* **97**, Suppl. 15–29.

Strome, S. & Wood, W. B. (1982). Immunofluorescence visualization of germ-line-specific cytoplasmic granules in embryos, larvae, and adults of *Caenorhabditis elegans*. *Proc. natn. Acad. Sci. U.S.A.* **79**, 1558–1562.

Strome, S. & Wood, W. B. (1983). Generation of asymmetry and segregation of germ-line granules in early *C. elegans* embryos. *Cell* **35**, 15–25.

Struhl, G. (1981*a*). A gene product required for correct initiation of segmental determination in *Drosophila*. *Nature, Lond.* **293**, 36–41.

Struhl, G. (1981*b*). A blastoderm fate map of compartments and segments of the *Drosophila* head. *Devl Biol.* **84**, 386–396.

Struhl, G. (1983). Role of the $esc^+$ gene product in ensuring the selective expression of segment-specific homeotic genes in *Drosophila*. *J. Embryol. exp. Morph.* **76**, 297–331.

Struhl, G. & Akam, M. (1985). Altered distributions of *Ultrabithorax* transcripts in *extra sex combs* mutant embryos of *Drosophila*. *EMBO J.* **4**, 3259–3264.

Struhl, G. & Brower, D. (1982). Early role of the $esc^+$ gene product in the determination of segments in *Drosophila*. *Cell* **31**, 285–292.

Struhl, G. & White, R. A. H. (1985). Regulation of the *Ultrabithorax* gene of *Drosophila* by other bithorax complex genes. *Cell* **43**, 507–519.

Sturtevant, A. H. (1923). Inheritance of direction of coiling in *Limnaea*. *Science, N.Y.* **58**, 269–270.

Styron, C. E. (1967). Effects on development of inhibiting polar lobe formation by compression in *Ilyanassa obsoleta* Stimpson. *Acta Embryol. Morph. exp.* **9**, 246–254.

Sudarwati, S. & Nieuwkoop, P. D. (1971). Mesoderm formation in the anuran *Xenopus laevis* (Daudin). *Wilhelm Roux Arch. EntwMech. Org.* **166**, 189–204.

Sugiyama, K. (1972). Occurrence of mucopolysaccharides in the early development of the sea urchin embryo and its role in gastrulation. *Dev. Growth Differ.* **14**, 63–73.

Sulston, J. E., Schierenberg, E., White, J. G. & Thomson, J. N. (1983). The embryonic cell lineage of the nematode *Caenorhabditis elegans*. *Devl Biol.* **100**, 64–119.

Sulston, J. E. & White, J. G. (1980). Regulation and cell autonomy during postembryonic development of *Caenorhabditis elegans*. *Devl Biol.* **78**, 577–597.

Sumiya, M. (1976). Differentiation of the digestive tract epithelium of the chick embryo cultured *in vitro* enveloped in a fragment of the vitelline membrane, in the absence of mesenchyme. *Wilhelm Roux Arch. dev. Biol.* **179**, 1–17.

Summers, M. C., Bedian, V. & Kauffman, S. A. (1986). An analysis of stage-specific protein synthesis in the early *Drosophila* embryo using high-resolution, two-dimensional gel electrophoresis. *Devl Biol.* **113**, 49–63.

Summers, R. G., Hylander, B. L., Colwin, L. H. & Colwin, A. L. (1975). The functional anatomy of the echinoderm spermatozoon and its interaction with the egg at fertilization. *Am. Zool.* **15**, 523–551.

Surani, M. A. H. (1979). Glycoprotein synthesis and inhibition of glycosylation by tunicamycin in preimplantation mouse embryos: compaction and trophoblast adhesion. *Cell* **18**, 217–227.

Surani, M. A. H. & Barton, S. C. (1984). Spatial distribution of blastomeres is dependent on cell division order and interactions in mouse morulae. *Devl Biol.* **102**, 335–343.

Surani, M. A. H., Barton, S. C. & Burling, A. (1980). Differentiation of 2-cell and 8-cell mouse embryos arrested by cytoskeletal inhibitors. *Expl Cell Res.* **125**, 275–286.

Surani, M. A. H., Barton, S. C. & Norris, M. L. (1984). Development of reconstituted mouse eggs suggests imprinting of the genome during gameto-genesis. *Nature, Lond.* **308**, 548–550.

Surrey, S., Ginzburg, I. & Nemer, M. (1979). Ribosomal RNA synthesis in pre- and post-gastrula-stage sea urchin embryos. *Devl Biol.* **71**, 83–99.

Sutasurja, L. A. & Nieuwkoop, P. D. (1974). The induction of the primordial germ cells in the urodeles. *Wilhelm Roux Arch. EntwMech. Org.* **175**, 199–220.

Sutherland, A. E. & Calarco-Gillam, P. G. (1983). Analysis of compaction in the preimplantation mouse embryo. *Devl Biol.* **100**, 328–338.

Suzuki, A. (1968). Studies on primary induction of *Triturus* embryo. I. Changes of inducing potency and metabolism of dorsal mesoderm of early *Triturus* gastrula. *Kumamoto J. Sci B* **9**, 1–8.

Suzuki, A. & Ikeda, K. (1979). Neural competence and cell lineage of gastrula ectoderm of newt embryo. *Dev. Growth Differ.* **21**, 175–188.

Suzuki, A. & Kuwabara, K. (1974). Mitotic activity and cell proliferation in primary induction of newt embryo. *Dev. Growth Differ.* **16**, 29–40.

Suzuki, A. S., Mifune, Y. & Kanéda, T. (1984). Germ layer interactions in pattern formation of amphibian mesoderm during primary embryonic induction. *Dev. Growth Differ.* **26**, 81–94.

Suzuki, A. S., Ueno, T. & Matsusaka, T. (1986). Alteration of cell adhesion system in amphibian ectoderm cells during primary embryonic induction: changes in reaggregation pattern of induced neurectoderm cells and ultrastructural features of the reaggregate. *Wilhelm Roux Arch. dev. Biol.* **195**, 85–91.

Švajger, A. & Levak-Švajger, B. (1975). Technique of separation of germ layers in rat embryonic shields. *Wilhelm Roux Arch. dev. Biol.* **178**, 303–308.

Švajger, A. & Levak-Švajger, B. (1976). Differentiation in renal homografts of isolated parts of rat embryonic ectoderm. *Experientia* **32**, 378–380.

Švajger, A., Levak-Švajger, B., Kostović-Knežević, L. & Bradamante, Ž. (1981). Morphogenetic behaviour of the rat embryonic ectoderm as a renal homograft. *J. Embryol. exp. Morph.* **65**, *Suppl.* 243–267.

Švajger, A., Levak-Švajger, B. & Škreb, N. (1986). Rat embryonic ectoderm as renal isograft. *J. Embryol. exp. Morph.* **94**, 1–27.

Svetlov, P. G., Bystrov, V. D. & Korsakova, G. F. (1962). Some data on morphology and physiology of early developmental stages in embryogenesis of teleosts. *Arkh. Anat. Gistol. Embriol.* **42**, 22–37.

Swalla, B. J., Moon, R. T. & Jeffery, W. R. (1985). Developmental significance of a cortical cytoskeletal domain in *Chaetopterus* eggs. *Devl Biol.* **111**, 434–450.

Swanson, M. M. & Poodry, C. A. (1981). The *shibire*[ts] mutant of *Drosophila*: a probe for the study of embryonic development. *Devl Biol.* **84**, 465–470.

Tadano, M. (1962). Artificial inversion of the primary polarity in *Ascaris* eggs. *Jap. J. Zool.* **13**, 329–355.

Tadano, Y. & Tadano, M. (1974). On the mechanism of determination of embryonic polarity in *Parascaris* and *Rhabditis*. *J. Embryol. exp. Morph.* **32**, 603–617.

Taddei, C., Gambino, R., Metafora, S. & Monroy, A. (1973). Possible role of ribosomal bodies in the control of protein synthesis in pre-vitellogenic oocytes of the lizard *Lacerta sicula* Raf. *Expl Cell Res.* **78**, 159–167.

Takata, C. (1960). The differentiation *in vitro* of the isolated endoderm under the influence of the mesoderm in *Triturus pyrrhogaster*. *Embryologia* **5**, 38–70.

Takata, C. & Yamada, T. (1960). Endodermal tissues developed from the isolated newt ectoderm under the influence of guinea pig bone marrow. *Embryologia* **5**, 8–20.

Takata, K., Yamamoto, K. Y., Ishii, I. & Takahashi, N. (1984). Glycoproteins responsive to the neural-inducing effect of Concanavalin A in *Cynops* presumptive ectoderm. *Cell Differ.* **14**, 25–31.

Takata, K., Yamamato, K. Y. & Ozawa, R. (1981). Use of lectins as probes for analyzing embryonic induction. *Wilhelm Roux Arch. dev. Biol.* **190**, 92–96.

Takaya, H. (1953a). On the notochord-forming potency of the prechordal plate in *Triturus* gastrulae. *Proc. Japan Acad.* **29**, 374–380.

Takaya, H. (1953b). On the loss of notochord-forming potency in the prechordal plate of the *Triturus* gastrula. *Annotnes zool. jap.* **26**, 202–207.

Takaya, H. (1955). Formation of the brain from the prospective spinal cord of amphibian embryos. *Proc. Japan Acad.* **31**, 360–365.

Takaya, H. (1956a). Two types of neural differentiation produced in connection with mesenchymal tissue. *Proc. Japan Acad.* **32**, 282–286.

Takaya, H. (1956b). On the types of neural tissue developed in connection with mesodermal tissues. *Proc. Japan Acad.* **32**, 287–292.

Takaya, H. (1957). Regionalization of the central nervous system in amphibian embryo. *Jap. J. exp. Morph.* **11**, 1–24 (in Japanese).

Takaya, H. (1959). Regional specificities in the differentiating neural-plate ectoderm of amphibian embryos. *Jap. J. Zool.* **12**, 345–359.

Takaya, H. (1978). Dynamics of the organizer. A. Morphogenetic movements and specificities in induction and differentiation of the organizer. In *Organizer: A Milestone of a Half-Century from Spemann*, ed. O. Nakamura & S. Toivonen, pp. 49–70. Amsterdam: Elsevier/North-Holland Biomedical.

Takaya, H. & Watanabe, T. (1961). Differential proliferation of the ependyma in the developing neural tube of amphibian embryo. *Embryologia* **6**, 169–176.

Takeichi, T. & Kubota, H. Y. (1984). Structural basis of the activation wave in the egg of *Xenopus laevis*. *J. Embryol. exp. Morph.* **81**, 1–16.

Tam, P. P. L. & Beddington, R. S. P. (1987). The formation of mesodermal tissues in the mouse embryo during gastrulation and early organogenesis. *Development* **99**, 109–126.

Tanaka, Y. (1976). Effects of the surfactants on the cleavage and further development of the sea urchin embryos. I. The inhibition of micromere formation at the fourth cleavage. *Dev. Growth Differ.* **18**, 113–122.

Tanaka, Y. (1979). Effects of the surfactants on the cleavage and further development of the sea urchin embryos. II. Disturbance in the arrangement of cortical vesicles and change in cortical appearance. *Dev. Growth Differ.* **21**, 331–342.

Tanaka, Y. (1981). Distribution and redistribution of pigment granules in the development of sea urchin embryos. *Wilhelm Roux Arch. dev. Biol.* **190**, 267–273.

Tarin, D. (1973). Histochemical and enzyme digestion studies on neural induction in *Xenopus laevis*. *Differentiation* **1**, 109–126.

Tarkowski, A. K. (1959). Experiments on the development of isolated blastomeres of mouse eggs. *Nature, Lond.* **184**, 1286–1287.

Tarkowski, A. K. (1961). Mouse chimaeras developed from fused eggs. *Nature, Lond.* **190**, 857–860.

Tarkowski, A. K. (1977). *In vitro* development of haploid mouse embryos produced by bisection of one-cell fertilized eggs. *J. Embryol. exp. Morph.* **38**, 187–202.

Tarkowski, A. K. & Rossant, J. (1976). Haploid mouse blastocysts developed from bisected zygotes. *Nature, Lond.* **259**, 663–665.

Tarkowski, A. K. & Wróblewska, J. (1967). Development of blastomeres of mouse eggs isolated at the 4- and 8-cell stage. *J. Embryol. exp. Morph.* **18**, 155–180.

Tato, F., Gandini, D. A. & Tocchini-Valentini, G. P. (1974). Major DNA polymerases common to different *Xenopus laevis* cell types. *Proc. natn. Acad. Sci. U.S.A.* **71**, 3706–3710.

Tautz, D., Lehmann, R., Schnürch, H., Schuh, R., Seifert, E., Kienlin, A., Jones, K. & Jäckle, H. (1987). Finger protein of novel structure encoded by *hunchback*, a second member of the gap class of *Drosophila* segmentation genes. *Nature, Lond.* **327**, 383–389.

Taylor, G. T. & Anderson, E. (1969). Cytochemical and fine structural analysis of oogenesis in the gastropod, *Ilyanassa obsoleta. J. Morph.* **129**, 211–247.

Taylor, M. A., Johnson, A. D. & Smith, L. D. (1985). Growing *Xenopus* oocytes have spare translational capacity. *Proc. natn. Acad. Sci. U.S.A.* **82**, 6586–6589.

Tazima, Y. (1964). *The Genetics of the Silkworm.* London: Logos Press.

Technau, G. M. (1987). A single cell approach to problems of cell lineage and commitment during embryogenesis of *Drosophila melanogaster. Development* **100**, 1–12.

Technau, G. M. & Campos-Ortega, J. A. (1986). Lineage analysis of transplanted individual cells in embryos of *Drosophila melanogaster.* II. Commitment and proliferative capabilities of neural and epidermal cell progenitors. *Wilhelm Roux Arch. dev. Biol.* **195**, 445–454.

Teissier, G. (1931). Étude expérimentale du développement de quelques hydraires. *Annls Sci. nat. Sér. X Zool.* **14**, 5–60.

Telfer, W. H. (1954). Immunological studies of insect metamorphosis. II. The role of a sex-limited blood protein in egg formation by the cecropia silkworm. *J. gen. Physiol.* **37**, 539–558.

Telfer, W. H., Woodruff, R. I. & Huebner, E. (1981). Electrical polarity and cellular differentiation in meroistic ovaries. *Am. Zool.* **21**, 675–686.

Terman, S. A. (1970). Relative effect of transcription-level and translation-level control of protein synthesis during early development of the sea urchin. *Proc. natn. Acad. Sci. U.S.A.* **65**, 985–992.

Thiery, J.-P., Délouvée, A., Gallin, W. J., Cunningham, B. A. & Edelman, G. M. (1984). Ontogenetic expression of cell adhesion molecules: L-CAM is found in epithelia derived from the three primary germ layers. *Devl Biol.* **102**, 61–78.

Thoman, M. & Gerhart, J. C. (1979). Absence of dorso-ventral differences in energy metabolism in early embryos of *Xenopus laevis. Devl Biol.* **68**, 191–202.

Thomas, R. J. (1968). Yolk distribution and utilization during early development of a teleost embryo (*Brachydanio rerio*). *J. Embryol. exp. Morph.* **19**, 203–215.

Thomas, T. L., Posakony, J. W., Anderson, D. M., Britten, R. J. & Davidson, E. H. (1981). Molecular structure of maternal RNA. *Chromosoma* **84**, 319–335.

Thomas, V., Heasman, J., Ford, C., Nagajski, D. & Wylie, C. C. (1983). Further analysis of the effect of ultra-violet irradiation on the formation of the germ line in *Xenopus laevis. J. Embryol. exp. Morph.* **76**, 67–81.

Tiedemann, H. (1968). Factors determining embryonic differentiation. *J. cell. Physiol.* **72**, *Suppl.* **1**, 129–144.

Tiedemann, H. (1978). Chemical approach to the inducing agents. In *Organizer: A Milestone of a Half-Century from Spemann*, ed. O. Nakamura & S. Toivonen, pp. 91–117. Amsterdam: Elsevier/North-Holland Biomedical.

Tiedemann, H. (1982). Signals of cell determination in embryogenesis. In *Biochemistry of Differentiation and Morphogenesis*, ed. L. Jaenicke, pp. 275–287. Berlin: Springer-Verlag.

Tiedemann, H., Becker, U. & Tiedemann, H. (1963). Chromatographic separation of a hind-brain inducing substance into mesodermal and neural-inducing subfractions. *Biochim. biophys. Acta* **74**, 557–560.

Tiedemann, H. & Born, J. (1978). Biological activity of vegetalizing and neuralizing inducing factors after binding to BAC-cellulose and CNBr-Sepharose. *Wilhelm Roux Arch. dev. Biol.* **184**, 285–299.

Tiedemann, H., Born, J. & Tiedemann, H. (1967). Embryonale Induktion und Hemmung der Ribonucleinsäure-Synthese durch Actinomycin D. *Z. Naturf.* **22b**, 649–659.

Tiedemann, H., Born, J. & Tiedemann, H. (1972). Mechanisms of cell differentiation: affinity of a morphogenetic factor to DNA. *Wilhelm Roux Arch. EntwMech. Org.* **171**, 160–169.

Tiedemann, H. & Tiedemann, H. (1964). Das Induktionsvermögen gereinigter Induktionsfaktoren im Kombinationsversuch. *Revue suisse Zool.* **71**, 117–137.

Tiedemann, H., Tiedemann, H., Born, J. & Kocher-Becker, U. (1969). Wirkung von Sulfhydrylverbindungen auf embryonale Induktionsfaktoren. *Wilhelm Roux Arch. EntwMech. Org.* **163**, 316–324.

Timofeeva, M. Ya. & Solovjeva, I. A. (1973). Transfer RNA synthesis in early embryogenesis. *FEBS Lett.* **33**, 327–330.

Timourian, H. & Watchmaker, G. (1975). The sea urchin blastula: extent of cellular determination. *Am. Zool.* **15**, 607–627.

Titlebaum, A. (1928). Artificial production of Janus embryos of *Chaetopterus*. *Proc. natn. Acad. Sci. U.S.A.* **14**, 245–247.

Tluczek, L., Lau, Y. T. & Horowitz, S. B. (1984). Water, potassium, and sodium during amphibian oocyte development. *Devl Biol.* **104**, 97–105.

Tobler, H., Müller, F., Back, E. & Aeby, P. (1985). Germ line–soma differentiation in *Ascaris*: a molecular approach. *Experientia* **41**, 1311–1319.

Togashi, S., Kobayashi, S. & Okada, M. (1986). Functions of maternal mRNA as a cytoplasmic factor responsible for pole cell formation in *Drosophila* embryos. *Devl Biol.* **118**, 352–360.

Toivonen, S. (1978). Regionalization of the embryo. In *Organizer: A Milestone of a Half-Century from Spemann*, ed. O. Nakamura & S. Toivonen, pp. 119–156. Amsterdam: Elsevier/North-Holland Biomedical.

Toivonen, S. (1979). Transmission problem in primary induction. *Differentiation* **15**, 177–181.

Toivonen, S., Kohonen, J., Saukkonen, J., Saxén, L. & Vainio, T. (1961). Preliminary observations of the inhibition of neural induction by 5-fluorouracil. *Embryologia* **6**, 177–184.

Toivonen, S. & Saxén, L. (1955a). Ueber die Induktion des Neuralrohrs bei *Triturus*keimen als simultane Leistung des Leber-und Knochenmarkgewebes vom Meerschweinchen. *Annls Acad. Sci. Fenn.* A **30**, 1–29.

Toivonen, S. & Saxén, L. (1955b). The simultaneous inducing action of liver and bone-marrow of the guinea-pig in implantation and explantation experiments with embryos of *Triturus*. *Expl Cell Res. Suppl.* **3**, 346–357.

Toivonen, S. & Saxén, L. (1966). Late tissue interactions in the segregation of the central nervous system. *Annls Med. exp. Biol. Fenn.* **44**, 128–130.

Toivonen, S. & Saxén, L. (1968). Morphogenetic interaction of presumptive neural and mesodermal cells mixed in different ratios. *Science, N.Y.* **159**, 539–540.

Toivonen, S., Tarin, D., Saxén, L., Tarin, P. J. & Wartiovaara, J. (1975). Transfilter studies on neural induction in the newt. *Differentiation* **4**, 1–7.

Toivonen, S. & Wartiovaara, J. (1976). Mechanisms of cell interaction during primary embryonic induction studied in transfilter experiments. *Differentiation* **5**, 61–66.

Tomkins, R. & Rodman, W. P. (1971). The cortex of *Xenopus laevis* embryos: regional differences in composition and biological activity. *Proc. natn. Acad. Sci. U.S.A.* **68**, 2921–2923.

Trelstad, R. L., Hay, E. D. & Revel, J.-P. (1967). Cell contact during early morphogenesis in the chick embryo. *Devl Biol.* **16**, 78–106.

Trinkaus, J. P. (1984). *Cells into Organs: The Forces that Shape the Embryo*, 2nd edn. Englewood Cliffs, New Jersey: Prentice-Hall.

Trinkaus, J. P. & Lentz, T. L. (1967). Surface specializations of *Fundulus* cells and their relation to cell movements during gastrulation. *J. Cell Biol.* **32**, 139–153.

Triplett, R. L. & Meier, S. (1982). Morphological analysis of the development of the primary organizer in avian embryos. *J. exp. Zool.* **220**, 191–206.

Trumbly, R. J. & Jarry, B. (1983). Stage-specific protein synthesis during early embryogenesis in *Drosophila melanogaster*. *EMBO J.* **2**, 1281–1290.

Tucker, C. H. (1942). The histology of the gonads and development of the egg envelopes of an ascidian (*Styela plicata* Lesueur). *J. Morph.* **70**, 81–113.

Tucker, J. B. & Meats, M. (1976). Microtubules and control of insect egg shape. *J. Cell Biol.* **71**, 207–217.

Tufaro, F. & Brandhorst, B. P. (1979). Similarities of proteins synthesized by isolated blastomeres of early sea urchin embryos. *Devl Biol.* **72**, 390–397.

Tung, T.-C. (1934). Recherches sur les potentialités des blastomères chez *Ascidiella scabra*. Expériences de translocation, de combinaison et d'isolement de blastomères. *Archs Anat. microsc.* **30**, 381–410.

Tung, T.-C. (1955). Experiments on the developmental potencies of egg-fragments and isolated blastomeres of *Fundulus heteroclitus*. *Acta Biol. exp. sin.* **4**, 129–150.

Tung, T.-C., Chang, C.-Y. & Tung, Y. F. Y. (1945). Experiments on the developmental potencies of blastoderms and fragments of teleostean eggs separated latitudinally. *Proc. zool. Soc. Lond.* **115**, 175–188.

Tung, T.-C., Lee, C. Y. & Tung, Y. F. Y. (1955). Further studies on the developmental potencies of *Carassius* eggs. *Acta Biol. exp. sin.* **4**, 107–128.

Tung, T.-C. & Tung, Y.F.Y. (1944). The development of egg-fragments, isolated blastomeres and fused eggs in the goldfish. *Proc. zool. Soc. Lond.* **114**, 46–64.

Tung, T. C., Wu, S. C. & Tung, Y. F. Y. (1955). The development of the isolated fragments of *Carassius* eggs centrifuged after fertilization. *Acta Biol. exp. sin.* **4**, 365–383.

Tung, T. C., Wu, S. C. & Tung, Y. F. Y. (1958). The development of isolated blastomeres of amphioxus. *Scientia sin.* **7**, 1280–1320.

Tung, T. C., Wu, S. C. & Tung, Y. Y. F. (1960a). The developmental potencies of the blastomere layers in amphioxus egg at the 32-cell stage. *Scientia sin.* **9**, 119–141.

Tung, T. C., Wu, S. C. & Tung, Y. Y. F. (1960b). Rotation of the animal blastomeres in amphioxus egg at the 8-cell stage. *Sci. Rec. Acad. sin.* **4**, 389–394.

Tung, T. C., Wu, S. C. & Tung, Y. Y. F. (1961). Differentiation of the prospective ectodermal and entodermal cells after transplantation to new surroundings in amphioxus. *Acta Biol. exp. sin.* **7**, 259–261.

Tung, T. C., Wu, S. C. & Tung, Y. Y. F. (1962*a*). The presumptive areas of the egg of amphioxus. *Scientia sin.* **11**, 629–644.

Tung, T. C., Wu, S. C. & Tung, Y. Y. F. (1962*b*). Experimental studies on the neural induction in amphioxus. *Scientia sin.* **11**, 805–820.

Tung, T.-C., Wu, S.-C., Yeh, Y.-F., Li, K.-S. & Hsu, M.-C. (1977). Cell differentiation in ascidian studied by nuclear transplantation. *Scientia sin.* **20**, 222–233.

Tupper, J., Saunders, J. W. & Edwards, C. (1970). The onset of electrical communication between cells in the developing starfish embryo. *J. Cell Biol.* **46**, 187–191.

Tupper, J. T. & Saunders, J. W. (1972). Intercellular permeability in the early *Asterias* embryo. *Devl Biol.* **27**, 546–554.

Turner, F. R. & Mahowald, A. P. (1977). Scanning electron microscopy of *Drosophila melanogaster* embryogenesis. II. Gastrulation and segmentation. *Devl Biol.* **57**, 403–416.

Tyler, A. (1930). Experimental production of double embryos in annelids and mollusks. *J. exp. Zool.* **57**, 347–407.

Tyler, A. (1967). Masked messenger RNA and cytoplasmic DNA in relation to protein synthesis and processes of fertilization and determination in embryonic development. In *Control Mechanisms in Developmental Processes*, ed. M. Locke, pp. 170–226. New York: Academic Press.

Ubbels, G. A. (1977). Symmetrisation of the fertilized egg of *Xenopus laevis* (studied by cytological, cytochemical and ultrastructural methods). *Mém. Soc. zool. Fr.* **41**, 103–115.

Ubbels, G. A., Bezem, J. J. & Raven, C. P. (1969). Analysis of follicle cell patterns in dextral and sinistral *Limnaea peregra*. *J. Embryol. exp. Morph.* **21**, 445–466.

Ubbels, G. A., Hara, K., Koster, C. H. & Kirschner, M. W. (1983). Evidence for a functional role of the cytoskeleton in determination of the dorsoventral axis in *Xenopus laevis* eggs. *J. Embryol. exp. Morph.* **77**, 15–37.

Ubbels, G. A. & Hengst, R. T. M. (1978). A cytochemical study of the distribution of glycogen and mucosubstances in the early embryo of *Ambystoma mexicanum*. *Differentiation* **10**, 109–121.

Ueda, R. & Okada, M. (1982). Induction of pole cells in sterilized *Drosophila* embryos by injection of subcellular fractions from eggs. *Proc. natn. Acad. Sci. U.S.A.* **79**, 6946–6950.

Uemura, I. & Endo, Y. (1976). Electron microscopic observations on the extragranular zone of the embryo of the sea urchin, *Hemicentrotus pulcherrimus*. *Dev. Growth Differ.* **18**, 399–406.

Underwood, E. M., Caulton, J. H., Allis, C. D. & Mahowald, A. P. (1980). Developmental fate of pole cells in *Drosophila melanogaster*. *Devl Biol.* **77**, 303–314.

Ursprung, H., Leone, J. & Stein, L. (1968). Blastular arrest and chromosome abnormalities produced by X-rays in two amphibians: *Rana pipiens* and *Xenopus laevis*. *J. exp. Zool.* **168**, 379–386.

Vacquier, V. D. (1981). Dynamic changes of the egg cortex. *Devl Biol.* **84**, 1–26.

412     *References*

Vainio, T., Saxén, L., Toivonen, S. & Rapola, J. (1962). The transmission problem in primary embryonic induction. *Expl Cell Res.* **27**, 527–538.

Vakaet, L. (1955). Recherches cytologiques et cytochimiques sur l'organisation de l'oocyte I de *Lebistes reticulatus*. *Archs Biol.*, *Liège* **66**, 1–73.

Vakaet, L. (1962). Some new data concerning the formation of the definitive endoblast in the chick embryo. *J. Embryol. exp. Morph.* **10**, 38–57.

Vakaet, L. (1964). Diversité fonctionelle de la ligne primitive du blastoderme de Poulet. *C. r. Séanc. Soc. Biol.* **158**, 1964–1966.

Vakaet, L. (1965). Resultats de la greffe de noeuds de Hensen d'âge différent sur le blastoderme de Poulet. *C. r. Séac. Soc. Biol.* **159**, 232–233.

Vakaet, L. (1973). Démonstration expérimentale de l'invagination blastoporale à travers la ligne primitive du poulet. *C. r. Séanc. Soc. Biol.* **167**, 781–783.

van Blerkom, J., Barton, S. C. & Johnson, M. H. (1976). Molecular differentiation in the preimplantation mouse embryo. *Nature, Lond.* **259**, 319–321.

van Blerkom, J. & Bell, H. (1986). Regulation of development in the fully grown mouse oocyte: chromosome-mediated temporal and spatial differentiation of the cytoplasm and plasma membrane. *J. Embryol. exp. Morph.* **93**, 213–238.

van Blerkom, J. & Brockway, G. O. (1975). Qualitative patterns of protein synthesis in the preimplantation mouse embryo. I. Normal pregnancy. *Devl Biol.* **44**, 148–157.

van Blerkom, J. & Manes, S. C. (1974). Development of preimplantation rabbit embryos *in vivo* and *in vitro*. II. A comparison of qualitative aspects of protein synthesis. *Devl Biol.* **40**, 40–51.

van Dam, W. I. & Verdonk, N. H. (1982). The morphogenetic significance of the first quartet micromeres for the development of the snail *Bithynia tentaculata*. *Wilhelm Roux Arch. dev. Biol.* **191**, 112–118.

Vandebroek, G. (1936). Les mouvements morphogénétiques au cours de la gastrulation chez *Scyllium canicula* Cuv. *Archs Biol.*, *Liège* **47**, 499–584.

van den Biggelaar, J. A. M. (1971). Timing of the phases of the cell cycle during the period of asynchronous division up to the 49-cell stage in *Lymnaea*. *J. Embryol. exp. Morph.* **26**, 367–391.

van den Biggelaar, J. A. M. (1976a). Development of dorsoventral polarity preceding the formation of the mesentoblast in *Lymnaea stagnalis*. *Proc. K. ned. Akad. Wet.* C **79**, 112–126.

van den Biggelaar, J. A. M. (1976b). The fate of maternal RNA containing ectosomes in relation to the appearance of dorsoventrality in the pond snail, *Lymnaea stagnalis*. *Proc. K. ned. Akad. Wet.* C **79**, 421–426.

van den Biggelaar, J. A. M. (1977a). Development of dorsoventral polarity and mesentoblast determination in *Patella vulgata*. *J. Morph.* **154**, 157–186.

van den Biggelaar, J. A. M. (1977b). Significance of cellular interactions for the differentiation of the macromeres prior to the formation of the mesentoblast in *Lymnaea stagnalis*. *Proc. K. ned. Akad. Wet.* C **80**, 1–12.

van den Biggelaar, J. A. M. (1978). The determinative significance of the geometry of the cell contacts in early molluscan development. *Biol. cell.* **32**, 155–161.

van den Biggelaar, J. A. M. & Guerrier, P. (1979). Dorsoventral polarity and mesentoblast determination as concomitant results of cellular interactions in the mollusk *Patella vulgata*. *Devl Biol.* **68**, 462–471.

van den Biggelaar, J. A. M. & Guerrier, P. (1983). Origin of spatial organization. In *The Mollusca*, vol. 3 *Development*, ed. N. H. Verdonk, J. A. M. van den Biggelaar & A. S. Tompa, pp. 179–213. New York: Academic Press.

van der Meer, J. M. (1979). The specification of metameric order in the insect

*Callosobruchus maculatus* Fabr. (Coleoptera). I. Incomplete segment patterns can result from constriction-induced cytological damage to the egg. *J. Embryol. exp. Morph.* **51**, 1–26.

van der Meer, J. M. (1984). Parameters influencing reversal of segment sequence in posterior egg fragments of *Callosobruchus* (Coleoptera). *Wilhelm Roux Arch. dev. Biol.* **193**, 339–356.

van der Meer, J. M. & Miyamoto, D. M. (1984). The specification of metameric order in the insect *Callosobruchus maculatus* Fabr. (Coleoptera). II The effects of temporary constriction on segment number. Wilhelm Roux Arch. dev. Biol. **193**, 326–338.

van Deusen, E. (1973). Experimental studies on a mutant gene (*e*) preventing the differentiation of eye and normal hypothalamus primordia in the axolotl. *Devl Biol.* **34**, 135–158.

van Dongen, C. A. M. (1976). The development of *Dentalium* with special reference to the significance of the polar lobe. VII. Organogenesis and histogenesis in lobeless embryos of *Dentalium vulgare* (da Costa) as compared to normal development. *Proc. K. ned. Akad. Wet. C* **79**, 454–465.

van Dongen, C. A. M. & Geilenkirchen, W. L. M. (1974). The development of *Dentalium* with special reference to the significance of the polar lobe. I–III. Division chronology and development of the cell pattern in *Dentalium dentale* (Scaphopoda). *Proc. K. ned. Akad. Wet. C* **77**, 57–100.

van Dongen, W. M. A. M., Moorman, A. F. M. & Destrée, O. H. J. (1983). Histone gene expression in early development of *Xenopus laevis*. Analysis of histone mRNA in oocytes and embryos by blot-hybridization and cell-free translation. *Differentiation* **24**, 226–233.

van Hoof, J. & Harrisson, F. (1986). Interaction between epithelial basement membrane and migrating mesoblast cells in the avian blastoderm. *Differentiation* **32**, 120–124.

Vanroelen, C. & Vakaet, L. C. A. (1981). Incorporation of $^{35}$S-sulphate in chick blastoderms during elongation and during shortening of the primitive streak. *Wilhelm Roux Arch. dev. Biol.* **190**, 233–236.

Verdonk, N. H. (1968a). The effect of removing the polar lobe in centrifuged eggs of *Dentalium*. *J. Embryol. exp. Morph.* **19**, 33–42.

Verdonk, N. H. (1968b). The determination of bilateral symmetry in the head region of *Limnaea stagnalis*. *Acta Embryol. Morph. exp.* **10**, 211–227.

Verdonk, N. H. & Cather, J. N. (1973). The development of isolated blastomeres in *Bithynia tentaculata* (Prosobranchia, Gastropoda). *J. exp. Zool.* **186**, 47–61.

Verdonk, N. H. & Cather, J. N. (1983). Morphogenetic determination and differentiation. In *The Mollusca*, vol. 3 *Development*, ed. N. H. Verdonk, J. A. M. van den Biggelaar & A. S. Tompa, pp. 215–252. New York: Academic Press.

Verdonk, N. H., Geilenkirchen, W. L. M. & Timmermans, L. P. M. (1971). The localization of morphogenetic factors in uncleaved eggs of *Dentalium*. *J. Embryol. exp. Morph.* **25**, 57–63.

Verdonk, N. H. & van den Biggelaar, J. A. M. (1983). Early development and the formation of the germ layers. In *The Mollusca*, vol. 3 *Development*, ed. N. H. Verdonk, J. A. M. van den Biggelaar & A. S. Tompa, pp. 91–122. New York: Academic Press.

Vestweber, D. & Kemler, R. (1984). Rabbit antiserum against a purified surface glycoprotein decompacts mouse preimplantation embryos and reacts with specific adult tissues. *Expl Cell Res.* **152**, 169–178.

Vialleton, M. L. (1888). Recherches sur les premières phases du développement de la seiche (*Sepia officinalis*). *Annls Sci. nat. Zool. Sér.* 10, **6**, 165–280.

Vincent, J.-P., Oster, G. F. & Gerhart, J. C. (1986). Kinematics of grey crescent formation in *Xenopus* eggs: the displacement of subcortical cytoplasm relative to the egg surface. *Devl Biol.* 113, 484–500.

Vintemberger, P. (1936). Sur le développement comparé des micromères de l'oeuf de *Rana fusca* divisé en huit a) après isolement, b) après transplantation sur un socle de cellules vitellines. *C. r. Séanc. Soc. Biol.* 122, 927–930.

Visschedijk, A. H. J. (1953). The effect of a heat shock on morphogenesis in *Limnaea stagnalis*. *Proc. K. ned. Akad. Wet.* C **56**, 590–597.

Viswanath, J. R., Leikola, A. & Rostedt, I. (1968). Induction by killed Hensen's node in chick embryo ectoderm. *Annls Zool. Fenn.* **5**, 384–388.

Viswanath, J. R., Leikola, A. & Toivonen, S. (1969). The inductive action of alcohol-killed chick Hensen's node on amphibian ectoderm. *Experientia* **25**, 38.

Vitorelli, M. L., Matranga, V., Feo, S., Giudice, G. & Noll, H. (1980). Diverse effects on thymidine incorporation in dissociated blastula cells of the sea urchin *Paracentrotus lividus* induced by butanol treatment and Fab addition. *Cell Differ.* **9**, 63–70.

Vivien, J. & Hay, D. (1954). Monstruosités doubles et polyembryologie obtenues expérimentalement chez un Sélacien, *Scylliorhinus canicula*. *C. r. hebd. Séanc. Acad. Sci., Paris* **238**, 1914–1916.

Voet, A. (1937). Quantitative lyotropy. *Chem. Rev.* **20**, 169–179.

Vogel, O. (1977). Regionalisation of segment-forming capacities during early embryogenesis in *Drosophila melanogaster*. *Wilhelm Roux Arch. dev. Biol.* **182**, 9–32.

Vogel, O. (1978). Pattern formation in the egg of the leafhopper *Euscelis plebejus* Fall. (Homoptera): developmental capacities of fragments isolated from the polar egg regions. *Devl Biol.* **67**, 357–370.

Vogel, O. (1982*a*). Experimental test fails to confirm gradient interpretation of embryonic patterning in leafhopper eggs. *Devl Biol.* **90**, 160–164.

Vogel, O. (1982*b*). Development of complete embryos in drastically deformed leafhopper eggs. *Wilhelm Roux Arch. dev. Biol.* **191**, 134–136.

Vogel, O. (1983). Pattern formation by interaction of three cytoplasmic factors in the egg of the leafhopper *Euscelis plebejus*. *Devl Biol.* **99**, 166–171.

Vogt, W. (1929). Gestaltungsanalyse am Amphibienkeim mit örtlicher Vitalfärbung. II. Gastrulation und Mesodermbildung bei Urodelen und Anuren. *Wilhelm Roux Arch. EntwMech. Org.* **120**, 385–706.

von Ubisch, L. (1925). Entwicklungsphysiologische Studien an Seeigelkeimen. III. Die normale und durch Lithium beeinflusste Anlage der Primitivorgane bei animalen und vegetativen Halbkeimen von *Echinocyamus pusillus*. *Z. wiss. Zool.* **124**, 469–486.

von Ubisch, L. (1932). Untersuchungen über Formbildung mit Hilfe experimentell erzeugter Keimblattchimären von Echinodermenlarven. II. *Wilhelm Roux Arch. EntwMech. Org.* **129**, 19–68.

von Ubisch, L. (1938). Über Keimverschmelzungen an *Ascidiella aspersa*. *Wilhelm Roux Arch. EntwMech. Org.* **138**, 18–36.

von Ubisch, L. (1963). Über Induktion bei Amphioxus und Ascidien. *Wilhelm Roux Arch. EntwMech. Org.* **154**, 466–494.

Vorobyev, V. I., Gineitis, A. A. & Vinogradova, I. A. (1969). Histones in early embryogenesis. *Expl Cell Res.* **57**, 1–7.

Waddington, C. H. (1932). Experiments on the development of chick and duck embryos, cultured *in vitro*. *Phil. Trans. R. Soc.* B **221**, 179–230.

Waddington, C. H. (1933a). Induction by the endoderm in birds. *Wilhelm Roux Arch. EntwMech. Org.* **128**, 502–521.

Waddington, C. H. (1933b). Induction by coagulated organisers in the chick embryo. *Nature, Lond.* **131**, 275–276.

Waddington, C. H. (1934). Experiments on embryonic induction. III. A note on inductions by chick primitive streak transplanted to the rabbit embryo. *J. exp. Biol.* **11**, 224–227.

Waddington, C. H. (1937). Experiments on determination in the rabbit embryo. *Archs Biol.*, Liège **48**, 273–290.

Waddington, C. H. (1956). *Principles of Embryology*. London: Allen & Unwin.

Waddington, C. H., Needham, J. & Brachet, J. (1936). Studies on the nature of the amphibian organization centre. III. The activation of the evocator. *Proc. R. Soc.* B **120**, 173–198.

Waheed, M. A. & McCallion, D. J. (1969). Alteration of the inductive capacity of post-nodal fragments of the chick embryo by lithium chloride. *Annls Zool. Fenn.* **6**, 448–451.

Waheed, M. A. & Mulherkar, L. (1967). Studies on induction by substances containing sulphydryl groups in post-nodal pieces of chick blastoderm. *J. Embryol. exp. Morph.* **17**, 161–169.

Wahn, H. L., Lightbody, L. E., Tchen, T. T. & Taylor, J. D. (1975). Induction of neural differentiation in cultures of amphibian undetermined presumptive epidermis by cyclic AMP derivatives. *Science, N.Y.* **188**, 366–369.

Wahn, H. L., Taylor, J. D. & Tchen, T. T. (1976). Acceleration of amphibian embryonic melanophore development by melanophore-stimulating hormone, $N^6,O^2$-dibutyryl adenosine $3',5'$-monophosphate and theophylline. *Devl Biol.* **49**, 470–478.

Wakahara, M. (1986). Modification of dorsal–ventral polarity in *Xenopus laevis* embryos following withdrawal of egg contents before first cleavage. *Dev. Growth Differ.* **28**, 543–554.

Wakahara, M., Neff, A. W. & Malacinski, G. M. (1984). Topology of the germ plasm and development of primordial germ cells in inverted amphibian eggs. *Differentiation* **26**, 203–210.

Wall, R. (1973). Physiological gradients in development – a possible role for messenger ribonucleoprotein. *Adv. Morphogen.* **10**, 41–114.

Wall, R. (1979). RNA metabolism in early embryos of *Xenopus laevis*: studies of translation level controls in development. Ph.D. thesis, University of Liverpool.

Wall, R. & Faulhaber, I. (1976). Inducing activity of fractionated microsomal material from the *Xenopus laevis* gastrula stage. *Wilhelm Roux Arch. dev. Biol.* **180**, 207–212.

Wallace, R. A. (1961). Enzymatic patterns in the developing frog embryo. *Devl Biol.* **3**, 486–515.

Wallace, R. A. (1972). The role of protein uptake in vertebrate oocyte growth and yolk formation. In *Oogenesis*, ed. J. D. Biggers & A. W. Schuetz, pp. 339–359. Baltimore: University Park Press.

Wang, Y.-H., Mo, H.-Y. & Shen, J.-Y. (1963). Studies on the mesoderm-inducing agent from mammalian liver. *Acta Biol. exp. sin.* **8**, 356–367.

Waring, G. L., Allis, C. D. & Mahowald, A. P. (1978). Isolation of polar granules and the identification of polar-granule-specific protein. *Devl Biol.* **66**, 197–206.

Warn, R. M., Bullard, B. & Magrath, R. (1980). Changes in the distribution of cortical myosin during the cellularization of the *Drosophila* embryo. *J. Embryol. exp. Morph.* **57**, 167–176.

Warn, R. M., Smith, L. & Warn, A. (1985). Three distinct distributions of F-actin occur during the divisions of polar surface caps to produce pole cells in *Drosophila* embryos. *J. Cell Biol.* **100**, 1010–1015.

Warner, A. & Gurdon, J. B. (1987). Functional gap junctions are not required for muscle gene activation by induction in *Xenopus* embryos. *J. Cell Biol.* **104**, 557–564.

Warner, A. E. (1985). The role of gap junctions in amphibian development. *J. Embryol. exp. Morph.* **89**, *Suppl.* 365–380.

Warner, A. E., Guthrie, S. C. & Gilula, N. B. (1984). Antibodies to gap-junctional protein selectively disrupt junctional communication in the early amphibian embryo. *Nature, Lond.* **311**, 127–131.

Warner, C. M. & Spannaus, D. J. (1984). Demonstration of *H-2* antigens on preimplantation mouse embryos using conventional antisera and monoclonal antibody. *J. exp. Zool.* **230**, 37–52.

Wartiovaara, J., Leivo, I. & Vaheri, A. (1979). Expression of the cell surface-associated glycoprotein, fibronectin, in the early mouse embryo. *Devl Biol.* **69**, 247–257.

Wartiovaara, J., Nordling, S., Lehtonen, E. & Saxén, L. (1974). Transfilter induction of kidney tubules: correlation with cytoplasmic penetration into Nucleopore filters. *J. Embryol. exp. Morph.* **31**, 667–682.

Wassarman, P. M., Hollinger, T. G. & Smith, L. D. (1972). RNA polymerases in the germinal vesicle contents of *Rana pipiens* oocytes. *Nature new Biol.* **240**, 208–210.

Wassarman, P. M. & Mrozak, S. C. (1981). Program of early development in the mammal: synthesis and intracellular migration of histone H4 during oogenesis in the mouse. *Devl Biol.* **84**, 364–371.

Watanabe, M., Bertolini, D. R., Kew, D. & Turner, R. S. (1982). Changes in the nature of the cell adhesions of the sea urchin embryo. *Devl Biol.* **91**, 278–285.

Watasé, S. (1891). Studies on cephalopods. I. Cleavage of the ovum. *J. Morph.* **4**, 247–303.

Watterson, R. L., Goodheart, C. R. & Lindberg, G. (1955). The influence of adjacent structures upon the shape of the neural tube and neural plate of chick embryos. *Anat. Rec.* **122**, 539–560.

Weeks, D. L. & Melton, D. A. (1987a). An mRNA localized to the animal pole of *Xenopus* eggs encodes a subunit of a mitochondrial ATPase. *Proc. natn. Acad. Sci. U.S.A.* **84**, 2798–2802.

Weeks, D. L. & Melton, D. A. (1987b). A maternal mRNA localized to the vegetal hemisphere in *Xenopus* eggs codes for a growth factor related to TGF-β. *Cell* **51**, 861–867.

Wegnez, M., Monier, R. & Denis, H. (1972). Sequence heterogeneity of 5S RNA in *Xenopus laevis*. *FEBS Lett.* **25**, 13–20.

Weinberger, C. & Brick, I. (1982). Primary hypoblast development in the chick. I. Scanning electron microscopy of normal development. *Wilhelm Roux Arch. dev. Biol.* **191**, 119–126.

Weir, M. P. & Kornberg, T. (1985). Patterns of *engrailed* and *fushi tarazu* transcripts reveal novel intermediate stages in *Drosophila* segmentation. *Nature, Lond.* **318**, 433–439.

Weisblat, D. A. & Blair, S. S. (1984). Developmental indeterminacy in embryos of the leech *Helobdella triserialis*. *Devl Biol.* **101**, 326–335.

Weisblat, D. A., Kim, S. Y. & Stent, G. S. (1984). Embryonic origins of cells in the leech *Helobdella triserialis*. *Devl Biol.* **104**, 65–85.

Weiss, P. (1958). Cell contact. *Int. Rev. Cytol.* **7**, 391–423.

Weissenberg, R. (1934). Untersuchungen über den Anlageplan beim Neunau-genkeim: Mesoderm, Rumpfdarmbildung und Übersicht der centralen Anla-gezonen. *Anat. Anz.* **79**, 177–199.

Wessel, G. M., Marchase, R. B. & McClay, D. R. (1984). Ontogeny of the basal lamina in the sea urchin embryo. *Devl Biol.* **103**, 235–245.

Wessel, G. M. & McClay, D. R. (1985). Sequential expression of germ-layer specific molecules in the sea urchin embryo. *Devl Biol.* **111**, 451–463.

Weyer, C. J., Nieuwkoop, P. D. & Lindenmayer, A. (1977). A diffusion model for mesoderm induction in amphibian embryos. *Acta biotheor.* **26**, 164–180.

White, R. A. H., Perrimon, N. & Gehring, W. J. (1984). Differentiation markers in the *Drosophila* ovary. *J. Embryol. exp. Morph.* **84**, 275–286.

White, R. A. H. & Wilcox, M. (1984). Protein products of the bithorax complex in *Drosophila*. *Cell* **39**, 163–171.

Whittaker, J. R. (1973*a*). Segregation during ascidian embryogenesis of egg cytoplasmic information for tissue-specific enzyme development. *Proc. natn. Acad. Sci. U.S.A.* **70**, 2096–2100.

Whittaker, J. R. (1973*b*). Tyrosinase in the presumptive pigment cells of ascidian embryos: tyrosine accessibility may initiate melanin synthesis. *Devl Biol.* **30**, 441–454.

Whittaker, J. R. (1977). Segregation during cleavage of a factor determining endodermal alkaline phosphatase development in ascidian embryos. *J. exp. Zool.* **202**, 139–153.

Whittaker, J. R. (1979*a*). Quantitative control of end products in the melano-cyte lineage of the ascidian embryo. *Devl Biol.* **73**, 76–83.

Whittaker, J. R. (1979*b*). Development of vestigial tail muscle acetyl-cholinesterase in embryos of an anural ascidian species. *Biol. Bull. mar. biol. Lab., Woods Hole* **156**, 393–407.

Whittaker, J. R. (1979*c*). Development of tail muscle acetylcholinesterase in ascidian embryos lacking mitochondrial localization and segregation. *Biol. Bull. mar. biol. Lab., Woods Hole* **157**, 344–355.

Whittaker, J. R. (1979*d*). Cytoplasmic determinants of tissue differentiation in the ascidian egg. In *Determinants of Spatial Organization*, ed. S. Subtelny & I. R. Konigsberg, pp. 29–51. New York: Academic Press.

Whittaker, J. R. (1982). Muscle lineage cytoplasm can change the develop-mental expression in epidermal lineage cells of ascidian embryos. *Devl Biol.* **93**, 463–470.

Whittaker, J. R. (1983). Quantitative regulation of acetylcholinesterase development in the muscle lineage cells of cleavage-arrested ascidian embryos. *J. Embryol. exp. Morph.* **76**, 235–250.

Whittaker, J. R., Ortolani, G. & Farinella-Ferruzza, N. (1977). Autonomy of acetylcholinesterase differentiation in muscle lineage cells of ascidian embryos. *Devl Biol.* **55**, 196–200.

Whitten, W. K. (1957). Culture of tubal ova. *Nature, Lond.* **179**, 1081–1082.

Whitten, W. K. & Dagg, C. P. (1972). Influence of spermatozoa on the cleavage rate of mouse eggs. *J. exp. Zool.* **148**, 173–183.

Wierzejski, A. (1905). Embryologie von *Physa fontinalis* L. *Z. wiss. Zool.* **83**, 502–706.

Wieschaus, E. & Gehring, W. (1976). Clonal analysis of primordial disc cells in the early embryo of *Drosophila melanogaster*. *Devl Biol.* **50**, 249–263.

Wieschaus, E., Nüsslein-Volhard, C. & Kluding, H. (1984). *Krüppel*, a gene whose activity is required early in the zygotic genome for normal embryonic segmentation. *Devl Biol.* **104**, 172–186.

Wilcox, M. & Leptin, M. (1985). Tissue-specific modulation of a set of related cell surface antigens in *Drosophila*. *Nature, Lond.* **316**, 351–354.

Wilde, C. E. (1955). The urodele neuroepithelium. II. The relationship between phenyl alanine metabolism and the differentiation of neural crest cells. *J. Morph.* **97**, 313–344.

Wiley, L. M. (1978). Apparent trophoblast giant cell production *in vitro* by core cells isolated from cultured mouse inner cell masses. *J. exp. Zool.* **206**, 13–16.

Willadsen, S. M. (1981). The developmental capacity of blastomeres from 4- and 8-cell sheep embryos. *J. Embryol. exp. Morph.* **65**, 165–172.

Wilson, C. B. (1900). The habits and early development of *Cerebratulus lacteus* (Verrill). A contribution to physiological morphology. *Q. Jl microsc. Sci.* **43**, 97–198.

Wilson, E. B. (1892). The cell-lineage of *Nereis*. A contribution to the cytogeny of the annelid body. *J. Morph.* **6**, 361–480.

Wilson, E. B. (1893). Amphioxus, and the mosaic theory of development. *J. Morph.* **8**, 579–638.

Wilson, E. B. (1903). Experiments on cleavage and localization in the nemertine egg. *Arch. EntwMech. Org.* **16**, 411–460.

Wilson, E. B. (1904). Experimental studies on germinal localization. I. The germ regions in the egg of *Dentalium*. *J. exp. Zool.* **1**, 1–72.

Wilson, E. B. (1925). *The Cell in Development and Heredity*, 3rd edn. New York: Macmillan.

Wilson, E. B. (1929). The development of egg-fragments in annelids. *Wilhelm Roux Arch. EntwMech. Org.* **117**, 179–210.

Wilson, I. B., Bolton, E. & Cuttler, R. H. (1972). Preimplantation differentiation in the mouse egg as revealed by microinjection of vital markers. *J. Embryol. exp. Morph.* **27**, 467–479.

Wilt, F. H. (1965a). Erythropoiesis in the chick embryo: the role of endoderm. *Science, N.Y.* **147**, 1588–1590.

Wilt, F. H. (1965b). Regulation of the initiation of chick embryo haemoglobin synthesis. *J. molec. Biol.* **12**, 331–341.

Wilt, F. H. (1970). The acceleration of ribonucleic acid synthesis in cleaving sea urchin embryos. *Devl Biol.* **23**, 444–455.

Winter, H. (1974). Ribonucleoprotein-Partikel aus dem telotroph-meroistischen Ovar von *Dysdercus intermedius* Dist. (Heteroptera, Pyrrhoc.) und ihr Verhalten im zellfreien Proteinsynthesesystem. *Wilhelm Roux Arch. Entw-Mech. Org.* **175**, 103–127.

Wintrebert, P. (1922). La polarité mécanique du germe des Sélaciens (*Scylliorhinus canicula*, L. Gill) au temps de la gastrulation. *C. r. hebd. Séanc. Acad. Sci., Paris* **175**, 411–413.

Wittenberg, C., Kohl, D. M. & Triplett, E. L. (1978). Amphibian embryo protease inhibitor. V. Effect of calcium on the distribution of amphibian trypsin inhibitor during fertilization and subsequent development of *Rana pipiens*. *Cell Differ.* **7**, 11–20.

Woerdeman, M. W. (1933). Über den Glykogenstoffwechsel des Organisationszentrums in der Amphibiengastrula. *Proc. Sect. Sci. K. ned. Akad. Wet.* **36**, 189–194.

Wolf, N., Priess, J. & Hirsh, D. (1983). Segregation of germline granules in early embryos of *Caenorhabditis elegans*: an electron microscopic analysis. *J. Embryol. exp. Morph.* **73**, 297–306.

Wolf, R. (1980). Migration and division of cleavage nuclei in the gall midge, *Wachtliella persicariae*. II. Origin and ultrastructure of the migration cytaster. *Wilhelm Roux Arch. dev. Biol.* **188**, 65–73.

Wolf, R. (1985). Migration and division of cleavage nuclei in the gall midge, *Wachtliella persicariae*. III. Patterns of anaphase-triggering waves altered by temperature gradients and local gas exchange. *Wilhelm Roux Arch. dev. Biol.* **194**, 257–270.

Wolgemuth, D. J., Jagiello, G. M. & Henderson, A. S. (1980). Baboon late diplotene oocytes contain micronucleoli and a low level of extra rDNA templates. *Devl Biol.* **78**, 598–604.

Wolk, M. & Eyal-Giladi, H. (1977). The dynamics of antigenic changes in the epiblast and hypoblast of the chick during the processes of hypoblast, primitive streak and head process formation, as revealed by immunofluorescence. *Devl Biol.* **55**, 33–45.

Wolpert, L. (1969). Positional information and the spatial pattern of cellular differentiation. *J. theor. Biol.* **25**, 1–47.

Wolpert, L. & Gustafson, T. (1961). Studies on the cellular basis of morphogenesis of the sea urchin embryo. The formation of the blastula. *Expl Cell Res.* **25**, 374–382.

Wolpert, L. & Mercer, E. H. (1963). An electron microscope study of the development of the blastula of the sea urchin embryo and its radial polarity. *Expl Cell Res.* **30**, 280–300.

Wood, W. B., Hecht, R., Carr, S., Vanderslice, R., Wolf, N. & Hirsh, D. (1980). Parental effects and phenotypic characterization of mutations that affect early development in *Caenorhabditis elegans*. *Devl Biol.* **74**, 446–469.

Woodland, H. R. (1974). Changes in the polysome content of developing *Xenopus laevis* embryos. *Devl Biol.* **40**, 90–101.

Woodland, H. R. & Adamson, E. D. (1977). The synthesis and storage of histones during the oogenesis of *Xenopus laevis*. *Devl Biol.* **57**, 118–135.

Woodland, H. R. & Ballantine, J. E. M. (1980). Paternal gene expression in developing hybrid embryos of *Xenopus laevis* and *Xenopus borealis*. *J. Embryol. exp. Morph.* **60**, 359–372.

Woodland, H. R. & Gurdon, J. B. (1968). The relative rates of synthesis of DNA, sRNA and rRNA in the endodermal region and other parts of *Xenopus laevis* embryos. *J. Embryol. exp. Morph.* **19**, 363–385.

Woodland, H. R. & Gurdon, J. B. (1969). RNA synthesis in an amphibian nuclear-transplant hybrid. *Devl Biol.* **20**, 89–104.

Woodruff, R. I. & Telfer, W. H. (1973). Polarized intercellular bridges in ovarian follicles of the cecropia moth. *J. Cell Biol.* **58**, 172–188.

Woodruff, R. I. & Telfer, W. H. (1980). Electrophoresis of proteins in intercellular bridges. *Nature, Lond.* **286**, 84–86.

Wourms, J. P. (1972). The developmental biology of annual fishes. II. Naturally occurring dispersion and reaggregation of blastomeres during the development of annual fish eggs. *J. exp. Zool.* **182**, 169–200.

Wright, T. R. F. (1970). The genetics of embryogenesis in *Drosophila*. *Adv. Genet.* **15**, 261–395.

Wu, T.-C., Wan, Y.-T., Chung, A. E. & Damjanov, I. (1983). Immunohistochemical localization of entactin and laminin in mouse embryos and fetuses. *Devl Biol.* **100**, 496–505.

Wudl, L. & Chapman, V. (1976). The expression of β–glucuronidase during preimplantation development of mouse embryos. *Devl Biol.* **48**, 104–109.

Wylie, C. C. (1972). The appearance and quantitation of cytoplasmic ribonucleic acid in the early chick embryo. *J. Embryol. exp. Morph.* **28**, 367–384.

Wylie, C. C., Brown, D., Godsave, S. F., Quarmby, J. & Heasman, J. (1985a).

The cytoskeleton of *Xenopus* oocytes and its role in development. *J. Embryol. exp. Morph.* **89**, *Suppl.* 1–15.

Wylie, C. C., Heasman, J., Snape, A., O'Driscoll, M. & Holwill, S. (1985*b*). Primordial germ cells of *Xenopus laevis* are not irreversibly determined early in development. *Devl Biol.* **112**, 66–72.

Yajima, H. (1960). Studies on embryonic determination of the harlequin-fly, *Chironomus dorsalis.* I. Effects of centrifugation and its combination with constriction and puncturing. *J. Embryol. exp. Morph.* **8**, 198–215.

Yajima, H. (1964). Studies on embryonic determination of the harlequin-fly, *Chironomus dorsalis.* II. Effects of partial irradiation of the egg by ultra-violet light. *J. Embryol. exp. Morph.* **12**, 89–100.

Yajima, H. (1970). Study of the development of the internal organs of the double malformations of *Chironomus dorsalis* by fixed and sectioned materials. *J. Embryol. exp. Morph.* **24**, 287–303.

Yajima, H. (1983). Induction of longitudinal double malformations by centrifugation or by partial UV-irradiation of eggs in the chironomid species, *Chironomus samoensis* (Diptera: Chironomidae). *Entomol. gen.* **8**, 171–191.

Yamada, T. (1938). Induktion der sekundären Embryonalanlage im Neunaugenkeim. *Okajimas Folia anat. jap.* **17**, 369–388.

Yamada, T. (1940). Beeinflussung der Differenzierungsleistung des isolierten Mesoderms von Molchkeimen durch zugefügtes Chorda- und Neuralmaterial. *Okajimas Folia anat. jap.* **19**, 131–197.

Yamada, T. (1958). Induction of specific differentiation by samples of proteins and nucleoproteins in the isolated ectoderm of *Triturus* gastrulae. *Experientia* **14**, 81–87.

Yamada, T. (1961). A chemical approach to the problem of the organizer. *Adv. Morphogen.* **1**, 1–53.

Yamada, T. (1962). The inductive phenomenon as a tool for understanding the basic mechanism of differentiation. *J. cell. comp. Physiol.* **60**, *Suppl.* 1, 49–64.

Yamagishi, H., Kunisada, T., Iwakura, Y., Nishimune, Y., Ogiso, Y. & Matsushiro, A. (1983). Emergence of the extrachromosomal circular DNA complexes as one of the earliest signals of cellular differentiation in the early development of mouse embryo. *Dev. Growth Differ.* **25**, 563–569.

Yamaguchi, Y., Murakami, K., Furusawa, M. & Miwa, J. (1983). Germline-specific antigens identified by monoclonal antibodies in the nematode *Caenorhabditis elegans*. *Dev. Growth Differ.* **25**, 121–131.

Yao, T. (1950). Cytochemical studies on the embryonic development of *Drosophila melanogaster.* II. Alkaline and acid phosphatases. *Q. Jl microsc. Sci.* **91**, 79–88.

Yasumasu, I., Hino, A., Suzuki, A. & Mita, M. (1984). Changes in the triglyceride level in sea urchin eggs and embryos during early development. *Dev. Growth Differ.* **26**, 525–532.

Yatsu, N. (1904). Experiments on the development of egg fragments in *Cerebratulus. Biol. Bull. mar. biol. Lab., Woods Hole* **6**, 123–136.

Yatsu, N. (1910). Experiments on germinal localization in the egg of *Cerebratulus. J. Coll. Sci. imp. Univ. Tokyo* **27**, Art. 17.

Yatsu, N. (1912). Observations and experiments on the ctenophore egg. The structure of the egg and experiments on cell-division. *J. Coll. Sci. imp. Univ. Tokyo* **32**, Art. 3.

Yoneda, M. & Schroeder, T. E. (1984). Cell cycle timing in colchicine-treated sea urchin eggs: persistent coordination between the nuclear cycles and the rhythm of cortical stiffness. *J. exp. Zool.* **231**, 367–378.

Yoshimi, T. & Yasumasu, I. (1978). Vegetalization of sea urchin larvae induced with cAMP phosphodiesterase inhibitors. *Dev. Growth Differ.* **20**, 213–218.
Yoshimi, T. & Yasumasu, I. (1979). Prevention by hydroxyurea of vegetalization of sea urchin larvae induced by cAMP phosophodiesterase inhibitors. *Dev. Growth Differ.* **21**, 271–280.
Yoshimi, T. & Yasumasu, I. (1981). Vegetalization of sea urchin larvae induced with cycloheximide. *Dev. Growth Differ.* **23**, 137–147.
Yoshizaki, N. (1976). Effect of actinomycin D on the differentiation of hatching gland cell and cilia cell in the frog embryo. *Dev. Growth Differ.* **18**, 133–143.
Yoshizaki, N. (1981). Ionic induction of the frog cement-gland cell from presumptive ectodermal tissues. *J. Embryol. exp. Morph.* **61**, 249–258.
Youn, B. W. & Malacinski, G. M. (1980). Action spectrum for ultraviolet irradiation inactivation of a cytoplasmic component(s) required for neural induction in the amphibian egg. *J. exp. Zool.* **211**, 369–377.
Zagris, N. & Eyal-Giladi, H. (1982). 5-bromodeoxyuridine inhibition of the epiblast competence for primitive streak formation in the young chick blastoderm. *Devl Biol.* **91**, 208–214.
Zagris, N. & Matthopoulos, D. (1985). Patterns of protein synthesis in the chick blastula: a comparison of the component areas of the epiblast and the primary hypoblast. *Devl Genet.* **5**, 209–217.
Zalik, S. E., Milos, N. & Ledsham, I. (1983). Distribution of two $\beta$-D-galactoside-binding lectins in the gastrulating chick embryo. *Cell Differ.* **12**, 121–127.
Zalokar, M. (1974). Effect of colchicine and cytochalasin B on ooplasmic segregation of ascidian eggs. *Wilhelm Roux Arch. EntwMech. Org.* **175**, 243–248.
Zalokar, M. (1976). Autoradiographic study of protein and RNA formation during early development of *Drosophila* eggs. *Dev. Biol.* **49**, 425–437.
Zalokar, M., Audit, C. & Erk, I. (1975). Developmental defects of female-sterile mutants of *Drosophila melanogaster. Devl Biol.* **47**, 419–432.
Zalokar, M. & Erk, I. (1976). Division and migration of nuclei during early embryogenesis of *Drosophila melanogaster. J. Microsc. Biol. cell.* **25**, 97–106.
Zalokar, M. & Sardet, C. (1984). Tracing of cell lineage in embryonic development of *Phallusia mammillata* (Ascidia) by vital staining of mitochondria. *Devl Biol.* **102**, 195–205.
Zeleny, C. (1904). Experiments on the localization of developmental factors in the nemertine egg. *J. exp. Zool.* **1**, 293–329.
Zetter, B. R. & Martin, G. R. (1978). Expression of a high molecular weight cell surface glycoprotein (LETS protein) by preimplantation mouse embryos and teratocarcinoma stem cells. *Proc. natn. Acad. Sci. U.S.A.* **75**, 2324–2328.
Ziegler, H. E. (1895). Untersuchungen über die ersten Entwicklungsvorgänge der Nematoden. *Z. wiss. Zool.* **60**, 351–410.
Ziomek, C. A. & Johnson, M. H. (1981). Properties of polar and apolar cells from the 16-cell mouse morula. *Wilhelm Roux Arch dev. Biol.* **190**, 287–296.
Ziomek, C. A., Johnson, M. H. & Handyside, A. H. (1982). The developmental potential of mouse 16-cell blastomeres. *J. exp. Zool.* **221**, 345–355.
Zissler, D. & Sander, K. (1973). The cytoplasmic architecture of the egg cell of *Smittia* spec. (Diptera, Chironomidae). I. Anterior and posterior pole regions. *Wilhelm Roux Arch. EntwMech. Org.* **172**, 175–186.
Zissler, D. & Sander, K. (1977). The cytoplasmic architecture of the egg cell of *Smittia* spec. (Diptera, Chironomidae). II. Periplasm and yolk endoplasm. *Wilhelm Roux Arch. dev. Biol.* **183**, 233–248.
Zoja, R. (1895). Sullo sviluppo dei blastomeri isolati dalle uova di alcune meduse (e di altri organismi). *Arch. EntwMech. Org.* **2**, 1–37.

zur Strassen, O. (1896). Embryonalentwickelung der *Ascaris megalocephala*. *Arch. EntwMech. Org.* **3**, 27–105 & 133–190.

zur Strassen, O. (1951). Der Erbgang der Nematoden-Asymmetrie. *Zool. Anz. Suppl.* **16**, 77–81.

zur Strassen, O. L. (1898). Über die Riesenbildung bei *Ascaris*-Eiern. *Arch. EntwMech. Org.* **7**, 642–676.

Zusman, S. B. & Wieschaus, E. F. (1985). Requirements for zygotic gene activity during gastrulation in *Drosophila melanogaster*. *Devl Biol.* **111**, 359–371.

Züst, B. & Dixon, K. E. (1975). The effect of u.v. irradiation of the vegetal pole of *Xenopus laevis* eggs on the presumptive primordial germ cells. *J. Embryol. exp. Morph.* **34**, 209–220.

# Index

*abd-A* and *Abd-B* genes 179–81
abdomen, in insects 25–6, 121, 169–70, 187
  experimental effects on 122–5, 127, 173–4, 176
  mutational effects on 21–2, 24, 180–1
ablation 76, 97, 117; *see also* deletion of cells
acetylcholinesterase 100–6 *passim*, 298
*Acheta* 121, 127, 171–2
*Acipenser* 264
actin 6, 7, 44, 48, 126
  expression of genes for, in amphibians 55, 194, 241, 242, 244; in ascidians 11–12, 42–3, 104; in sea urchins 161, 256–8
activation of a meridian 23, 63, 316; *see also* egg activation, gene activation
activator RNA 317–18
adhesiveness 166, 187, 188, 206, 244, 254
alkaline phosphatase 100–1, 104–5, 261
*Ambystoma* 226, 227; *see also* axolotl
*Amia* 264
ammonium chloride 228
amnion and amnioserosa 170, 260
amoeboid movements 39, 64
amphibians 45–9, 57–63, 133–8, 188–97, 211–52
  oocytes of 5, 6, 7, 10, 11
amphioxus 16, 107–9, 198–9, 262–3
*Amphisbetia* 16, 112–13
*Anguis* 284
animal pole and area 10, 27, 38, 272, 315
animalisation 306, 316
  in amphibians 191–2, 224, 227, 228
  in sea urchins 150–1, 154–61 *passim*, 165, 166, 258
animal–vegetal organisation 10, 13–16, 32–8, 315–16
  in amphibians, 46, 133–4, 188–9, 211–12, 215; biochemical, of egg and oocyte 11, 12, 17, 49; of embryo

137, 194, 196–7, 241, 244, 245; experimental 57–8, 65, 136–7, 189–92, 220
  in amphioxus 108–9, 198–9
  in ascidians 41, 43–4, 52–3, 97–102 *passim*, 105, 197–8
  in meroblastic embryos 267, 269–70, 271, 273–5, 277
  in nematodes 56, 113–14, 116, 117–18
  in sea urchins 127–9, 146, 253–5; experimental 19, 54, 56, 130–1, 147–58 *passim*; physiological and biochemical 159–66 *passim*, 258
  in spiralians, lobe-bearing 51–2, 64, 82, 88; others 50, 57, 77–8, 92
  in other embryos 109, 112–13, 133, 188
annelids 19, 69; *see also specific groups and genera*
annulate lamellae 7, 31
ANT-C complex 181–6 *passim*
*Antennapedia* gene 182
anteriorisation 223–6, 228, 243, 246
antero-posterior organisation
  in amphibians 221–6, 229–31, 233–4, 243, 248–50, 251
  in ascidians 41, 53, 97–100, 102, 107, 263
  in birds 285–95 *passim*
  in fish 264, 275–80, 283
  in hydrozoans 112–13, 204, 265–6
  in insects 10–15 *passim*, 121, 127, 185–7; experimental 21–2, 24–6, 121–5, 170–84, 259
  in other embryos 108, 283–4, 299, 300
*Antheraea* 9
anurans 28, 45–6, 65, 213–15, 218, 234
apical tuft
  in ctenophores 109–10, 111
  in sea urchins 54, 130–1, 154
  in spiralians 50, 51–2, 64, 73, 77–81, 203
*Apis* 120, 171
arachnids 187–8; *see also* spiders

*Hydractinia* 204, 266
hydration of ions 160
hydrozoans 34, 112–13, 142, 204, 265–6
hyperblastula 136, 276, 278, 305
hypoblast 285–6, 288, 295–6, 299;
  inductive role of 292–3, 294, 295–6
hypodermis 115, 119
hypothalamus 247

ICM 205–8, 299
*Ilyanassa* 13, 35, 37, 93–4, 262
  experimental studies of 55, 81–2, 86–9
implantation studies 61, 219–20, 229,
  230, 264
inductions 317–18; *see also* interactions,
  neuralisation, vegetalisation
  in amphibians 190–4, 216–46 *passim*
  in amphioxus 198, 263
  in ascidians 100, 101–2, 197, 263
  in birds 292–5, 295–7, 298
  in cephalopods 270, 272, 273
  in sea urchins 148, 162
  in spiralians 88, 201, 261–2
  in teleosts 278, 279–80
  in other embryos 117, 259–61, 264,
  300
ingression 189, 211, 253, 265, 283, 285
  at primitive streak 287–8, 298, 299,
  300–1
inhibition in development
  of amphibians 232, 239, 248
  of spiralians 80–2, 88, 92
  of other embryos 100, 118, 197
initiation and initiation factors 6, 164,
  316
insects 119–27, 168–87, 258–61
  oocyte and egg 3–4, 9, 10, 12, 13
inside-outside position 139–40, 141; *see
  also* concentric organisation
interactions 7, 144, 317; *see also*
  inductions, inhibition
  in amphibians 190, 224–6
  in ascidians and amphioxus 105, 199
  in insects 172, 174
  in sea urchins 148–52, 158
  in spiralians 89, 200–4
  in other embryos 116–17, 187–8, 207
intermediate filaments 7, 43, 48
intermediate-germ insects 120, 122, 169
intestine 71, 86–8, 115, 151, 219
intravitelline cleavage 120, 121–3
invagination 21, 29, 211–12, 241, 254,
  258; *see also* involution
inversion of eggs and embryos 57–9,
  278; *see also* rotations
involution 211–15, 268, 283–4
  effects on cell properties 223–6, 246,
  264
iodosobenzoic acid 154, 156, 188

ionophores 38, 44–5
ions 11, 17, 154, 160, 203, 227; *see also*
  specific ions
isolation studies; *see also* explantation,
  fragmentation and ligation
  in amphibians 134–7, 189–90, 223
  in ascidians and amphioxus 99–101,
  108
  in fish 276–8, 282, 283
  in hydrozoans 113, 204
  in mammals 139–40, 299–300
  in sea urchins 129–31, 147, 152–6
  *passim*
  in spiralians 50, 75–82, 84, 86, 94
  in other embryos 111, 133, 290, 291
isozymes 254

jelly canal 13, 55
junctions 8, 138, 140, 194, 280; *see also*
  desmosomes, gap junctions, tight
  junctions

*knirps* gene 183
*Krüppel* gene 183, 184, 186, 314

lamellibranchs 16, 84; *see also specific
  genera*
laminin 166, 213
lampbrush chromosomes 2–5
*Lampetra* 34
lampreys 33, 263–4
larvae; *see also* pluteus, tadpole
  from egg fragments/fusions 18, 50–6
  from isolated blastomeres 77, 108,
  113, 129–35 *passim*
  from later part embryos 87, 204, 265
  genetic effects on 19, 22, 24, 178,
  180, 183
  other citations 64, 107, 109, 112–13,
  119
late gene set 317
lateral plate mesoderm 134, 214–15,
  218–19, 251, 264, 289
lateral structures 23, 129, 156, 169, 184
*Lebistes* 278
lectins 43, 48, 238, 298
leeches 34, 72–4, 76, 84–5, 89–90, 203;
  *see also Clepsine*
left/right organisation 77, 108, 130, 133–
  4, 143, 153, 278
  asymmetry of 114–15, 116–17, 289–95
  *passim*
legs 177–8
lens induction 217
*Leptinotarsa* 177, 184, 185
ligations; *see also* isolation studies
  of cephalopods 270, 271
  of insects 121–3, 170–6, 259